PERSPECTIVE AS PRACTICE

TECHNE

KNOWLEDGE, TECHNIQUE, AND MATERIAL CULTURE

1

Perspective as Practice

Renaissance Cultures of Optics

Edited by

SVEN DUPRÉ

BREPOLS

This project has received funding from the European Research Council (ERC)
under the European Union's Horizon 2020 research and innovation programme
(grant agreement No. 648718).

European Research Council
Established by the European Commission

Cover illustration: Astrolabe quadrant by Ieremias Arscenius, 1573.
Gilt brass, 170 mm. Photo by J.N. Lamas, MUHNAC, inv. nr.
MUHNAC-UL-DEP262. Courtesy MUHNAC.

Cover design: Johan Van Looveren
Typesetting: Crius Group, Hulshout

D/2019/0095/104
ISBN 978-2-503-58107-1
eISBN 978-2-503-58145-3
DOI 10.1484/M.TECHNE-EB.5.116014

Printed in the EU on acid-free paper.

Contents

II
Writing on Perspective

III
Drawing, Constructing, Painting

Plates

Acknowledgements

This book is the outcome of a working group conceived by me and Jeanne Peiffer in 2012 as a collaboration between the Centre Alexandre Koyré (CAK) in Paris and my Research Group 'Art and Knowledge in Premodern Europe' at the Max Planck Institute for the History of Science (MPIWG) in Berlin. It follows from two workshops held in October 2012 at the MPIWG and in September 2013 at the CAK. It has subsequently been developed with the support of fellowships held by several working group members at the MPIWG and a senior visiting fellowship awarded to Georges Farhat at the Descartes Centre for the History and Philosophy of Science and the Humanities at Utrecht University. The working group has also benefited from [Perspectiva+], a collaborative digital project hosted at the Biblioteca Hertziana in Rome, conceived in collaboration with the late Andreas Thielemann, and developed by Klaus E. Werner of the Max Planck Research Group.

I wish to acknowledge the support of these institutes and, in particular, thank Jeanne Peiffer for all the intellectual and organizational work undertaken during the initial planning and development of this project. Unfortunately, due to personal circumstances, she could not assist in the final stages of editing the book for publication, but the intellectual set-up of the working group is as much her work as it is mine, and the final result would have been much less without her efforts. The last stages of the working group project were made possible at Utrecht University and the University of Amsterdam by a five-year European Research Council Consolidator Grant awarded to me. My ARTECHNE project has received funding from the European Research Council (ERC) under the European Union's Horizon 2020 research and innovation programme (grant agreement No. 648718). I also wish to thank all participants who were involved in the workshops: at the risk of forgetting someone, Hans Belting, Vincenzo De Risi, Ruth Ezra, Robert Felfe, Francesca Fiorani, Alexa Greist, Wolfgang Lefevre, Peter Scholz, J.B. Shank, and David Summers.

I am grateful to Christophe Lebbe and Robert Halleux for the initial interest they expressed in publishing this book with Brepols, and to Koen Vermeir and Dániel Margócsy for accepting this book for publication in their new series 'Techne: Knowledge, Technique, and Material Culture'. I also wish to thank the two anonymous referees whose comments and insights have allowed us to improve the book and saved us from several significant oversights. The editor at Brepols, Alexander Sterkens, has guided us through the publication process with tact and efficiency. This book would not have been possible without the invaluable editorial support of Jill Briggeman, Nicholas Forshaw, and Gina Partridge Grzimek.

Sven Dupré
December 2018

SVEN DUPRÉ

Introduction*

This book is about the development of optics and perspective between the fifteenth and seventeenth centuries. It intervenes in two distinct historiographies: firstly, the history of perspective, an interdisciplinary field of study, to which primarily art historians and historians of science have contributed, and which developed in the wake of Erwin Panofsky's foundational study *Perspective as Symbolic Form* (1927); and secondly, the history of optics, a sub-field within the history of science, of which the contours have been outlined in David Lindberg's classic study of the history of optics *Theories of Vision. From al-Kindi to Kepler* (1976).[1] Both fields of study come with their defining experiments and canonical texts. For Panofsky, Filippo Brunelleschi's peephole and panel experiments in front of the Baptistery and the Palazzo Vecchio in early fifteenth-century Florence marked the invention of linear perspective. He cast Brunelleschi's invention as pointing forward to the first codification of perspectival procedures in Leon Battista Alberti's *On Painting*, a method of construction which, based on nineteenth-century German scholarship, Panofsky elevated to the status of 'costruzione legittima'. Lindberg, in his turn, considered the medieval texts of Roger Bacon, John Pecham, and Witelo as the canon of optics (or *perspectiva*, foremost to be defined as a theory of vision) connecting the reception of Ibn al-Haytham (known as Alhacen) with the work of Johannes Kepler at the beginning of the seventeenth century.

Panofsky wrote *Perspective as Symbolic Form* at a time when scholarship on the history of optics, and medieval optics and the reception of Alhacen in particular, was still largely non-existent. However, more is at hand in the separation of the histories of optics and perspective than simply ignorance, which can be remedied by the progress of scholarship. Looked at from the other side of the divide, Lindberg discussed perspective as an impoverished application of optical theory with no development of its own and very little consequence for the route taken by the discipline of optics. More recent scholarship in the history of optics by Gérard Simon and A. Mark Smith is more attentive to perspective and

* This project has received funding from the European Research Council (ERC) under the European Union's Horizon 2020 research and innovation programme (grant agreement No 648718).

1 Erwin Panofsky, *Perspective as Symbolic Form*, trans. by Christopher S. Wood (New York: Zone Books, 1997); David C. Lindberg, *Theories of Vision. From al-Kindi to Kepler* (Chicago: University of Chicago Press, 1976). I have discussed the historiography in Sven Dupré, 'The Historiography of Perspective and "Reflexy-Const" in Netherlandish Art', *Nederlands Kunsthistorisch Jaarboek*, 61 (2011), 35–60.

Sven Dupré Utrecht University and University of Amsterdam, S.G.M.Dupre@uu.nl

Perspective as Practice. Renaissance Cultures of Optics, ed. by Sven Dupré, Turnhout, 2019 (*Techne. Knowledge, Technique, and Material Culture*, 1), p. 9-15

DOI 10.1484/M.*Techne*-EB.5.117720

art and disagrees with Lindberg's embracement of continuity of optics from Antiquity to Kepler, while leaving Lindberg's basic assumption of a well-defined separation between optics and perspective intact.[2] Panofsky's own position in *Perspective as Symbolic Form* is rather ambivalent: while he recognizes the multiplicity of perspectives, especially ancient and modern ones, as well as different ways of constructing perspective, the one often more natural than the other, there are equally numerous passages in the book in which he privileges the *costruzione legittima* and a teleological view separating the histories of optics and perspective. Remarkably, despite his own ambivalence, Panofsky's most whiggish definition of perspective still haunts present-day scholarship on the history of perspective. Most recently, Hans Belting even revived Panofsky's notion of perspective as 'symbolic form', implying that it was 'expressive' of Renaissance culture.[3] While Ibn al-Haytham plays a crucial role for Belting, the argument hinges on a problematically essentialist definition of two cultures, as well as an equally problematic separation of optics and perspective.

Recent scholarship in the history of perspective has more strongly embedded the rise of perspective in the history of optics. In *L'Hypothèse d'Oxford* (1998) Dominique Raynaud attributed the highest importance to optics, a well-developed discipline in the Middle Ages in the hands of the Franciscans Robert Grosseteste, Roger Bacon and John Pecham, as a body of knowledge available to the artisans and craftsmen inventing linear perspective. And in a more recent book, Raynaud showed why, given the importance of the diffusion of Franciscan optics to the rise of perspective, it emerged in central Italy (rather than Oxford, the cradle of Franciscan optics, or the medieval Islamic world).[4] Most recently, Pietro Roccasecca proposed a new reading of Alberti's *On Painting*, replacing Panofsky's interpretation of Alberti's perspective as *costruzione legittima* with an emphasis on the importance of Alhacen's optics to Alberti's perspective.[5]

Most importantly, these studies have downplayed the significance of the invention of perspective as a singular moment in the hands of one individual (Brunelleschi). Contrary to Panofsky's elevation of Alberti's perspective as the *costruzione legittima*, it has been shown, on the basis of the study of the material practices of painters in imitating and representing the effects of light and space, that Renaissance artists used several, sometimes incompatible techniques to create the illusion of three dimensions on a two-dimensional surface.[6] Moreover, Renaissance authors on perspective used several concepts and aspects of perspective in the

2 A. Mark Smith, *From Sight to Light: The Passage from Ancient to Modern Optics* (Chicago: University of Chicago Press, 2014); Gérard Simon, *Archéologie de la Vision: L' optique, le corps, la peinture* (Paris: Editions du Seuil, 2003).

3 Hans Belting, *Florence and Baghdad: Renaissance Art and Arab Science*, trans. by Deborah Lucas Schneider (Cambridge, Mass.: Belknap Press of Harvard University Press, 2011).

4 Dominique Raynaud, *L'hypothèse d'Oxford. Essai sur les origins de la perspective* (Paris: Presses universitaires de France, 1998); Dominique Raynaud, *Optics and the Rise of Perspective: A Study in Network Knowledge Diffusion* (Oxford: The Bardwell Press, 2014).

5 Pietro Roccasecca, *Filosofi, oratori e pittori. Una nuova lettura del De Pictura di Leon Battista Alberti* (Rome, Campisano Editore, 2016).

6 Pietro Roccasecca, 'Gentile da Fabriano, A Miracle of Saint Nicholas: A Rigorous Nonperspective Spatial Representation', *Center: Record of Activities and Research Reports, National Gallery of Art, Washington*, 21 (2001), 126–30; Pietro Roccasecca, 'Not Albertian', *Center: Record of Activities and Research Reports, National Gallery of Art, Washington*, 22 (2002), 167–69.

most diverse ways rather than working towards a Cartesian conceptualisation of space.[7] Several central concepts and theorems, most notably that of the vanishing point, were only acquired in the late sixteenth and early seventeenth centuries; they should not be projected back into the early fifteenth century and the work of Brunelleschi and Alberti. In short, essentialist and teleological tendencies have coloured the historiography of perspective in the last century following Panofsky's *Perspective as Symbolic Form*. Instead, the point of departure of this book is the recognition of the polysemy of perspective, that is, the plurality of meanings of perspective, building on the ground-breaking work of Raynaud and Roccasecca already mentioned, of Filippo Camerota's on 'prospettiva aedificandi', and that of Jeanne Peiffer on 'Messung'.[8]

To say that this book is about perspective might be as confusing as it is to state that it is about the history of optics. Both optics and perspective come with present-day associations as well as connotations emerging from the historiographies in which optics and perspective have been clearly separated. If we want to avoid these confusing associations and connotations, we could write that it is about *perspectiva*, which is a period term used interchangeably for texts, things and thoughts which today we would classify, without hesitation, as either optics or perspective. It is *perspectiva* which we have, for ease, translated as 'perspective' in the title of this book. There is, unfortunately, no less ambiguous term in English.

To bring forward the polysemy of perspective, this book treats the history of *perspectiva* in terms of practices, a conglomerate of material, social, literary and reproductive practices, through which knowledge claims in perspective were produced, promoted, legitimated and circulated in and through a variety of sites and institutions. The ways optical knowledge was used by different groups in different places (such as the university classroom, the anatomist's dissection table, the goldsmith's workshop, and the astronomer's observatory) defined the meanings of Renaissance perspective. As this period was characterized by widespread 'optical literacy', perspective was defined in different ways in different places and sites by various groups of practitioners.

This book aims to reveal the polysemy of perspective by focusing on three different aspects of perspective as practice. Section 1 focuses on different sites in which perspective is practiced. It aims to elucidate not only the widespread optical literacy of the period, but especially the site-dependent meanings of *perspectiva*. Most interestingly, sites such as the theatre, the instrument maker's workshop and the courtly garden were home to practices of perspective which have remained on the margin, or even completely invisible, in the historiographies of optics and perspective. Other sites have been privileged in scholarship, and among those the astronomical observatory in particular has received ample attention. It has been shown that astrology, and its connection to understandings of the physics of rays in optics, was important to Renaissance image theories.[9] Even more

7 James Elkins, *The Poetics of Perspective* (Ithaca: Cornell University Press, 1994).

8 Filippo Camerota, *La prospettiva del Rinascimento: Arte, architettura, scienza* (Verona: Electa, 2006); Jeanne Peiffer, 'Projections Embodied in Technical Drawings: Dürer and his Followers', in *Picturing Machines 1400–1700*, ed. by Wolfgang Lefèvre, (Cambridge, Mass.: The MIT Press, 2004), pp. 245–75.

9 Mary Quinlan-McGrath, *Influences: Art, Optics, and Astrology in the Italian Renaissance* (Chicago: University of Chicago Press, 2013).

to the point, most recently, Raz Chen-Morris argued that Kepler's optics is fundamentally providing the epistemological foundations of his astronomical research program as well as a response to widely shared anxieties about vision and knowledge in Renaissance culture.[10] True knowledge is gained not by direct access to the world via the sense of sight but by artificially construed observation — that is, by measuring shadows in the camera obscura — Kepler maintains, according to Chen-Morris. As an important site of observation, the astronomical observatory is the prime locus in which the epistemological implications of perspective were considered. However, precisely because it has received so much attention, this book does not include a chapter specifically devoted to the astronomical observatory and shifts attention to the sites which have remained more marginal in scholarship. This is not to say that the anxieties about the instability of vision and knowledge, which Stuart Clark has argued preoccupied the period between 1430 and 1670, are not present in any of the other sites than astronomical observatories.[11] This is most certainly not the case given the ubiquity of these anxieties, but the question which this book raises is a different one: in what ways do the plurality of sites make a difference to our understanding of the polysemy of perspective?

The first site to be discussed is that of the *trecento* urban piazza in Marvin Trachtenberg's contribution to this book. Architectural site planning was shaped by a set of Euclidean-Vitruvian optical principles, and the urban piazza was laid out, according to Trachtenberg, to create particular points of view following these principles.

The two following chapters examine the artisanal workshop. On the basis of an analysis of a variety of practices of working emeralds and of knowing their optical properties, Marjolijn Bol shows that the artisanal workshop was a site of knowledge of light and colour, thereby focusing on domains of *perspectiva* different from the geometry of vision adopted by Brunelleschi and Alberti. Bol also shows how this knowledge reached scholars and natural philosophers, who applied it in the material refurbishing of their studies. More important to the central argument of this book, Bol's chapter defies the historiographical demarcation between optics and perspective and brings to the fore the aspects of perspective which a focus on scholarship on geometry, sight and projective space had obscured. In the next chapter, Samuel Gessner focuses on a different type of workshop, not that of the jeweller, goldsmith or painter discussed by Bol, but of the mathematical instrument maker, namely, Ieremias Arsenius. Well-connected to highly placed patrons as well as circles of mathematicians, such as Gemma Frisius and Gerard Mercator, Gessner shows how the Arsenius workshop and the circles connected to it, employed an understanding of perspective which included 'planisphaeric' or 'stereographic' projection used for the design of instruments.

The following chapters in this section consider three different sites: the anatomy theatre, the courtly theatre, and the courtly garden. In opposition to the historiography of optics following Lindberg, as discussed above, in which the role of anatomists in the development of visual theory is downplayed, and geometry is privileged, Tawrin Baker

10 Raz Chen-Morris, *Measuring Shadows: Kepler's Optics of Invisibility* (University Park, Pennsylvania: Penn State University Press, 2016).

11 Stuart Clark, *Vanities of the Eye: Vision in Early Modern European Culture* (Oxford: Oxford University Press, 2007).

shows how, over the course of the sixteenth century, the anatomy theatre at the University of Padua became the site of dissemination, disputation and teaching of an approach to visual theory integrating ocular anatomy, mathematical optics and natural philosophy. In Jaime Cuenca's chapter we move to another type of theatre, the aristocratic theatre in which perspective was applied to scenery on stage. Cuenca shows how perspective in the courtly theatre had a political function. He traces how the privileged point of view in the perspectival theatre became identified with the prince's seat, and how perspective again lost its political significance in the eighteenth century. Finally, the site of perspective at the centre of Juliet Odgers' attention is the courtly garden. Odgers discusses the design of the Sayes Court garden near London by the seventeenth-century English polymath and member of the Royal Society, John Evelyn. In line with Gessner's discussion of the meaning of perspective employed in the mathematical instrument maker's workshop, Odgers shows that Evelyn's garden is home to a practice in which perspective is embedded in the broader field of the projective mathematical arts.

Section 2 deals with writing as one of the most important practices of *perspectiva*. The chapters in this section concentrate on textual carriers and vehicles of the transmission of *perspectiva* and on how textual transmission entails appropriation resulting in changing meanings of *perspectiva*. Challenging essentialist definitions of Western linear perspective as compared with the image cultures of the Islamic East, Elaheh Kheirandish looks at the transmission of key concepts and aspects of perspective in a variety of textual sources in Arabic and Persian to bring out various practices of perspective in the Islamic Middle Ages.

In his chapter, A. Mark Smith points to the importance of extra-textual conduits of transmission of *perspectiva*. By the mid-thirteenth century the association of *perspectiva* with optics was firmly established, that is, two centuries before it also became connected with linear perspective in the canonical texts of medieval optics already mentioned. They were disseminated in academic milieus via university teaching in the European Middle Ages. However, as A. Mark Smith argues, oral transmission of optical knowledge via literary texts, such as most famously Dante's 'Divine Comedy', often read out loud to an audience of listeners, and especially in church sermons, resulted in a widespread optical literacy.

The next three chapters in section 2 discuss more specific texts, instantiating kinds of writing or genres, and how these textual vehicles of transmission shaped ideas of optics. Dominique Raynaud argues for the existence of a textual source for Leonardo's theory of the penumbra, thereby focusing on a field within perspective — the science of shadows — which has traditionally remained outside the scope of studies of the transmission of optics. Since the source was originally in Arabic, though known to Leonardo through a fourteenth-century Latin translation, Raynaud's chapter highlights the role of translation in the transmission of optical knowledge. Sven Dupré discusses different types of text, recipes and secrets, in his contribution to this book. Through books of secrets, flooding the print market in the sixteenth century, optical knowledge travelled more easily and widely than ever before. These secrets also re-packaged the experiential basis of optical knowledge and changed the meaning of optics. Breaking up optical texts, secrets led to new conceptual possibilities as well as the idea that optics was primarily about the manipulation of instruments to create visual effects. Finally, a notebook of the Spanish mid-sixteenth-century architect Hernan Ruiz II is the object of analysis in Jose Calvo Lopez's chapter. Offering a space for experimentation, the notebook shows various practices of perspective at work in visualising architecture.

Section 3 focuses on the practices of drawing, painting and constructing. Chapters in this section take the visual problems painters, draughtsmen and gardeners face as their point of departure and bring out the differences between codifications of *perspectiva* and practice. These constructional problems, and how they connect to bodies of optical knowledge, rather than the epistemological implications of perspectival art, are the centre of attention in this book. Firstly, there were a variety of non-Albertian constructions to create the illusion of space, as exemplified in the first three chapters of this section.

Filippo Camerota revisits a locus classicus in the historiography, Massaccio's *Trinity* fresco. In contrast to its traditional place in the historiography, Camerota shows that Massaccio did not apply the 'costruzione legittima' but instead applied the constructive tools available in the mathematical culture of the *abaco* tradition. Nevertheless, as Pietro Roccasecca points out in his chapter re-interpreting the perspective codified in Alberti's 'On Painting', perspective entailed a broader engagement with Alhacen's visual theory. Finally, even a painter of the mathematical accomplishment of Piero della Francesca deviated from the rigorous application of geometry if needed, as J.V. Field argues.

Secondly, other types of optical knowledge and experience were as important to artists as the geometry of perspective, as we will already have seen in Bol's chapter. In this section, Field shows how important knowledge of shadows and the reflection and refraction of light was to Piero's painting. In a similar vein, Paul Hills argues that Venetian painting around 1500 was the result of a practice of perspective paying special attention to light. Hill's chapter in this book scrutinizes the singular importance of Luca Pacioli for this practice of perspective. According to Hills, Pacioli's understanding of proportion is in agreement with the geometry and modulation of colour found in Venetian paintings around 1500, perhaps most apparent in works by Giovanni Bellini. Thus, taken together, these chapters show that other domains of *perspectiva* were important to painters. Nor was perspective constrained to the two-dimensional plane, which is obvious from the meaning of perspective and its uses in the context of mathematical instrument design and garden construction in the chapters by Gessner and Odgers, respectively.

In the final chapter of this book Georges Farhat continues this line of inquiry. Based on an analysis of André Le Nôtre's design of the Grand Canal of Versailles, Farhat argues that a specific appropriation of optical knowledge was at work there, which he calls 'topographic perspective', a practice which included the construction of optical devices, visual alignment and the application of anamorphic schemes. The garden is perhaps the best site to show that in the early modern period perspective was not tied to two-dimensional graphic representation. Thus, garden design underscores the polysemy of perspective central to this book.

Bibliography

Secondary Works

Belting, Hans, *Florence and Baghdad: Renaissance Art and Arab Science*, trans. by Deborah Lucas Schneider (Cambridge, Mass.: Belknap Press of Harvard University Press, 2011).

Camerota, Filippo, *La prospettiva del Rinascimento : Arte, architettura, scienza* (Verona : Electa, 2006).

Chen-Morris, Raz, *Measuring Shadows: Kepler's Optics of Invisibility* (University Park, Pennsylvania: Penn State University Press, 2016).

Clark, Stuart, *Vanities of the Eye: Vision in Early Modern European Culture* (Oxford: Oxford University Press, 2007).

Dupré, Sven, 'The Historiography of Perspective and "Reflexy-Const" in Netherlandish Art', *Nederlands Kunsthistorisch Jaarboek*, 61 (2011), 35–60.

Elkins, James, *The Poetics of Perspective* (Ithaca: Cornell University Press, 1994).

Lindberg, David C., *Theories of Vision. From al-Kindi to Kepler* (Chicago: University of Chicago Press, 1976).

Panofsky, Erwin, *Perspective as Symbolic Form*, trans. by Christopher S. Wood (New York: Zone Books, 1997).

Peiffer, Jeanne, 'Projections Embodied in Technical Drawings: Dürer and his Followers', in *Picturing Machines 1400-1700*, ed. by Wolfgang Lefèvre, (Cambridge, Mass.: The MIT Press, 2004), pp. 245–75.

Quinlan-McGrath, Mary, *Influences: Art, Optics, and Astrology in the Italian Renaissance* (Chicago: University of Chicago Press, 2013).

Raynaud, Dominique, *L'hypothèse d'Oxford. Essai sur les origins de la perspective* (Paris: Presses universitaires de France, 1998)

Raynaud, Dominique, *Optics and the Rise of Perspective: A Study in Network Knowledge Diffusion* (Oxford: The Bardwell Press, 2014).

Roccasecca, Pietro, 'Gentile da Fabriano, A Miracle of Saint Nicholas: A Rigorous Nonperspective Spatial Representation', *Center: Record of Activities and Research Reports, National Gallery of Art, Washington*, 21 (2001), 126–30;

Roccasecca, Pietro, 'Not Albertian', *Center: Record of Activities and Research Reports, National Gallery of Art, Washington*, 22 (2002), 167–69.

Roccasecca, Pietro, *Filosofi, oratori e pittori. Una nuova lettura del De Pictura di Leon Battista Alberti* (Rome, Campisano Editore, 2016).

Simon, Gérard, *Archéologie de la Vision: L' optique, le corps, la peinture* (Paris: Editions du Seuil, 2003).

Smith, A. Mark, *From Sight to Light: The Passage from Ancient to Modern Optics* (Chicago: University of Chicago Press, 2014).

I

Sites of Perspective

MARVIN TRACHTENBERG

Perspective as Artistic Form

Optical Theory and Visual Culture from Giotto to Alberti

Part I – Theory

This essay challenges the conventional view that the influence of optical science on the visual arts began in the quattrocento. Optical theories play an important role a century earlier, as this paper seeks to argue, but beyond pictorial perspective. In Florence, all the visual arts of the trecento — guided by theories of vision derived from ancient *cum* medieval learning — inflected form to situated perception.

Standard historiography holds that during the Renaissance, optical science informs 'linear' (or 'rational') perspective in painting.[1] By this one means that the pictorial surface is constructed as a 'window' into a 'visual pyramid' of orthogonal lines theoretically receding to a vanishing point (Figures 1 and 2). The image is formed on a virtual two-dimensional plane (the 'vela') intersecting this pyramid vertically. Relative size of objects is governed by a pavement grid. The spacing of horizontal lines on this grid is determined by diagonals run from a 'distance point' which is determined by the ideal distance between the viewer and the picture surface, thereby incorporating the position of the former into the pictorial structure.

This new system of pictorial representation is believed to have been adumbrated by the architect Filippo Brunelleschi in two demonstration panels of the 1420s (now lost but described by historical sources) depicting the Florentine Baptistery and Piazza della Signoria. Although not theoretically demonstrated until 1480 (by Piero della Francesca, *De prospectiva pingendi*, Proposition I, 13) — indeed not fully codified until 1563 (by the Florentine Academia del Disegno), the method was converted into a quasi-theoretical

1 For the standard origin story (in several variations), see Miriam Schild Bunim, *Space in Medieval Painting and the Forerunners of Perspective* (New York: Columbia University Press, 1940); John White, *The Birth and Rebirth of Pictorial Space* (New York: Thomas Yoseloff, 1958); Alessandro Parronchi, *Studi sulla dolce prospettiva* (Milan: Aldo Martello, 1964); Samuel Y. Edgerton, Jr., *The Renaissance Rediscovery of Linear Perspective* (New York: Harper and Row, 1975); Martin Kemp, *The Science of Art* (London: Yale University Press, 1990), Part I and J. V. Field, *The Invention of Infinity: Mathematics and Art in the Renaissance* (Oxford: Oxford University Press, 1997). Compare the critical assessments of B. B. Johannsen and M. Marcussen, 'A Critical Survey of the Theoretical and Practical Origins of Renaissance Linear Perspective', *Acta ad archaeologiam et artium historiam pertinentia* (1981), 191–227 and Jehane R. Kuhn, 'Measured Appearances: Documentation and Design in Early Perspective Drawing', *Journal of the Warburg and Courtauld Institutes*, 53 (1990), 114–32.

Marvin Trachtenberg Institute of Fine Arts, NYU, mltberg@gmail.com

Perspective as Practice. Renaissance Cultures of Optics, ed. by Sven Dupré, Turnhout, 2019 (*Techne. Knowledge, Technique, and Material Culture*, 1), p. 19-70

DOI 10.1484/M.*Techne*-EB.5.117721

Fig. 1 Neroccio de' Landi, *Annunciation* (c. 1480), Yale University Art Gallery, New Haven.

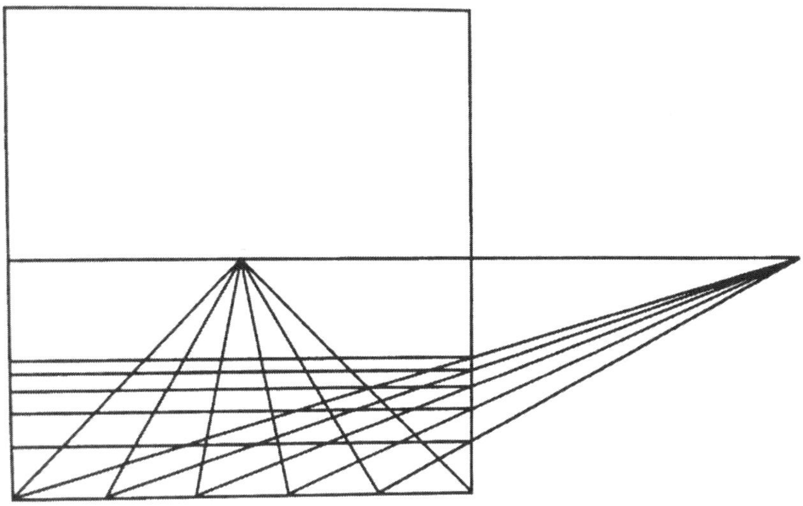

Fig. 2 Albertian perspective construction.

playbook by Leon Battista Alberti in *Della pittura* and dedicated to Brunelleschi in 1436.[2] The spatial construction of many paintings and relief sculptures reflect the new procedure, beginning with pre-Albertian works such as Masaccio's *Trinity* and Donatello's *Salome* panel, although few works conform perfectly to the rule. Architecture and urbanism are subsequently transformed as well: witness Bramante's illusionistic choir in San Satiro in Milan and the Baltimore, Berlin, and Urbino panels of ideal city views.

2 J. V. Field, 'Alberti, the Abacus and Piero della Francesca's Proof of Perspective', *Renaissance Studies*, 2 (1997), 61–88.

According to standard historiography, this precise conjunction of image, picture plane, and viewer replaced a trecento practice of unfocused, 'empirical' perspective. In such perspective *manqué*, lines never recede to a single point, nor are apparent sizes of objects rationally determined. There is thus no coherent space, and the subject-position of the viewer is left indeterminate.

Scholars did identify significant exceptions to this origin story. For example, Dominique Raynaud has recently found double-vanishing point construction in several works of the long trecento, noting instances of 'binocular perspective' which reflect contemporary 'binocular' optical theory.[3] He has also found near-perfect cases of single point perspective in several works attributed to Giotto, lacking only the coincidence of the horizon and eye-level: three out of the four hallmarks of Albertian perspective are thus present.[4] These observations are reinforced by claims of theoretical knowledge by (or for) notable artists, including Giovanni Pisano and Giotto. Such cases indicate a transition to Albertianism in the trecento. This line of analysis reinforces a teleology regarding the so-called 'vanishing axis' of the trecento, a significant step toward the 'vanishing point' construction of 'true' perspective.[5]

The Trecento Refusal of Rational Perspective

But if so, why did the final step require another century to come about? Such a scenario would depend on trecento painters having been incapable of attaining a desired goal in sight. Given the accomplishments of the period any such incapacity would require demonstration. In this paper, I proceed instead on the premise that Florentine artists of the long trecento were no less intelligent than their descendants, and thus their anomalous practice requires a different explanation than the teleological one.

Following the argument I have advanced elsewhere, this 'delay' in the transition from 'empirical' to 'rational' perspective resulted not from weakness but rather from a reasoned *unwillingness* to take the final step.[6] In contrast to modern thinking, which regards the transition to focused perspective as inevitable and salutary, it would likely not have been regarded as an advance in the trecento period eye.

Instead of providing theoretical precision, trecento 'empirical' perspective was flexible, and its inconsistencies may often have been intended.[7] Empirical perspective enabled all orthogonal forms — including ground, ceiling, left and right planes — to be set at any angle to each other as well as to the picture plane (not unlike the fluid geometry of computer spatial modeling). A scene could thereby be presented at any 'stage' angle, including full

3 Dominique Raynaud, 'Geometrical and Arithmetical Methods in Early Medieval Practice', *Physis. Rivista internazionale di storia della scienza*, 45 (2008), 29–55 (p. 51) and idem, 'Optique et perspective avant Alberti', in *Le printemps de la Renaissance. La sculpture et les arts à Florence*, ed. by B. Paolozzi Strozzi and M. Bormand (Paris: Musée du Louvre éditions, 2013), pp. 165–71.

4 Raynaud, 'Optique', p. 170 and Dominique Raynaud, *Optics and the Rise of Perspective: A Study in Network Knowledge Diffusion* (Oxford: The Bardwell Press, 2013), passim.

5 Cf. White, *Birth and Rebirth*.

6 Marvin Trachtenberg, *Dominion of the Eye: Urbanism, Art and Power in Early Modern Florence* (Cambridge: Cambridge University Press, 1997), Chapter 4.

7 White, *Birth and Rebirth*, Chapter 1.

Fig. 3 Giotto, *Feast of Herod*, Peruzzi Chapel, Santa Croce.

Fig. 4 Piero del Pollaiuolo, *Annunciation*, Gemäldegalerie, Berlin.

frontality, oblique and intermediate angles, or, alternatively, in a subtle pivoting of viewpoint (described by John White as 'softened oblique' perspective). Taddeo Gaddi's *Presentation of the Virgin* in the Baroncelli Chapel as well as Giotto's *Herod's Feast* and other scenes in the Peruzzi Chapel of Santa Croce exploit this technical possibility brilliantly (Figures 3 and 24). Trecento painters did not always pursue this variation, but they worked in the knowledge that they could. Their practice was thus a realm of freedom with respect to the construction of space and the settings of narrative scenes.[8]

Albertian perspective, to the contrary, constructs a rigid-rectilinear cage that demands a strict frontality of presentation and an undeviating recession of orthogonals, whose axial member is always perpendicular to the picture plane, just as the transverse lines always run parallel to it (Figure 4). This rule is at the heart of its brilliant conceptual effect in the form of an alluring, transparent rationality. But it requires that the scene never be turned as in Giotto; no softened oblique perspective or other morphing of spatial construction is allowed. Otherwise, among other problems, the pavement grid, actual or virtual, would be sliced along the picture plane on a ragged bias. In Albertian practice, at best the pavement axis may be shifted off center, which famously occurs in Tintoretto's *Finding the Body of Saint Mark* and *Christ Washing His Disciples' Feet*, a device seen earlier in, for example, Paolo Uccello's *Miracle of the Desecrated Host*. All Albertian space remains oriented to the frontal picture plane, with which the architectural frame always is aligned.

I propose that trecento artists understood this dilemma and thus recognised that a fully rational perspectival system posed serious drawbacks.[9] It was not merely that it demanded a more consistent spatial construction — that would have been the easy part. More objectionable was its rigidity of framing, focus, and orientation — altogether a new inflexibility in composing the spatial structure of their pictorial dramas. Thus, even though at times the painters of this century worked close to a fully integrated linear perspective, they refused the final step, and the situational history suggests that this occurred because in the final analysis they were unwilling to sacrifice compositional freedom to theoretical consistency. Brunelleschi, Alberti, and their adherents were motivated by a new set of values that elevated systematic integrity over compositional pragmatism. It would have been this new outlook, not a sudden upgrade in pictorial intelligence, that enabled the conceptual breakthrough, or rather, *reduction* to linear perspective.

Other Modes of Perspectivism

In searching for traces of optical theory in pre-Renaissance artistic practice, historians make a strategic error in limiting investigation to the genealogy of linear perspective. Thereby not only is the reading compromised by the teleology discussed above, wide fields of artistic practice are also excluded from consideration. The present investigation will show that it is precisely in such areas — architecture, urbanism, sculpture, and aspects of painting itself — that new entanglements of theory and practice can be discerned. That scholars may have been looking for the wrong things in the wrong places is all the more regrettable

8 Trachtenberg, *Dominion*, Chapter 4.

9 Martin Kemp suggests that trecento painters understood that they did not 'need' a 'rational' system. See Kemp, *Science of Art*, 12.

since it is well established that optical theory and related knowledges were widely diffused by 1300.[10] When we widen our historical gaze (following the lead, for example, of Paul Hills regarding colour and light), we find abundant signs of the intersecting of theory and practice.[11]

To expand the field of investigation, it is productive to articulate the meaning of 'perspective', a word which comes from the optical sciences (the Latin *perspectiva* derives from *perspicere* ['to see through or clearly'] that Boethius chose for the Greek, *optikē*).[12] In my usage, 'perspectivism' will be a more appropriate term for what was more than just a mode of rationalised spatial illusionism that originated in Italian Renaissance painting. A broader practice historically, 'perspectivism' concerns the theoretical positioning — the *perspective* — of the spectator in the facture of images, and conversely the inflection of the work to the spectator's eye, as well as the shaping of the spatial environment of this perceptual interaction. My focus is thus not on the picture 'itself' or on other works of art considered as absolute objects. Rather it concerns image production in/by the eye of the observer as conditioned by the shaping of both the thing-seen and the ideal viewing position of the person who sees it. It is with respect to this process that I shall consider the role of optical science.

In the present study, I expand the phenomenon of perspectivism in chronological span and intermediality, while revising the role of optics in the visual arts in Italy. I make a diachronic shift in surveying optical theories during the long century before 1400–1600, the default time frame in standard historiography. I do so by widening the interface with optical theory to include architecture, where many of the most inventive works are conceived. Most of my observations concern Florentine art, not only for economy of space, but because the city served as the hinge between the two epochs. Florence was the home of Cimabue, Giotto, and Gaddi as well as Brunelleschi, Ghiberti, and Masaccio. Even Alberti, born and educated in north Italy, was of a noble Florentine family, and the city was the site of his encounter with the Renaissance movement and where he wrote *De pictura* in the 1430s.[13]

The trecento appears to have drawn on much the same set of ancient/medieval textual-theoretical sources that informed practice after 1400, although not necessarily in the same manner. This continuity is not surprising, given the existence of all the relevant textual sources by the end of the thirteenth century. By then all the great medieval optical

10 By the fourteenth century, optics becomes incorporated in the university curriculum, for which Pecham's *Perspectiva communis* was 'intended as an elementary textbook and probably served that purpose from the very beginning' throughout Europe. See David C. Lindberg, *Theories of Vision from Al-Kindi to Kepler* (Chicago: The University of Chicago Press, 1976), pp. 120–21.

11 Paul Hills studies the theorists' understanding of relationships of viewing distance and object clarity in connection with pictorial practices, as well as the possible role of scholastic categories of *lux, lumen*, etc. He notes the presence of optical texts in the Papal court and in Padua in Giotto's time there, and he sees a 'new confidence' among artists in 'tackling the projection of three-dimensional form on a flat surface' as having been abetted by 'an understanding of the relationship between vision and geometry' found in optical theory. See Paul Hills, *The Light of Early Italian Painting* (New Haven: Yale University Press, 1987), Chapter 4.

12 B. A. R. Carter, 'Perspective', in *The Oxford Companion to Art*, ed. by Harald Osborn (Oxford: Oxford University Press, 1970), p. 840.

13 On Alberti's experience in Florence and encounter with (and 'discovery of') the new art movement, see Marvin Trachtenberg, *Building-in-Time: From Giotto to Alberti and Modern Oblivion* (London: Yale University Press, 2010), Chapter 8.

theorists (including Grosseteste, Bacon, Witelo and Pecham) had passed away. Although controversies over sticking points continued to simmer (such as the extramission/intromission question), thereafter no significant perspectivist theory appeared until Kepler in the late sixteenth century.[14] As is apparent in the table of contents of every survey, in optical science the age of Brunelleschi and Alberti is considered to have been a dead period.

As an alternative to this dour assessment, I propose to consider optical science and the visual arts as a single discursive field of imbricated theory and practice in this interval, one in which optical thought migrates in Italy around 1300 from pure textual discourse to the realms of artistic thinking and material practice.[15] This dynamic aesthetic stream might be considered a branching or breakout from the mainstream textual-theoretical tradition. But artistic developments were more than an appendage to optical thought, especially from the standpoint of the recent turn in the history of science led by Pamela Smith and others, which incorporates artisanal practices.[16] In this manner, optical discourse would have remained alive in its transmutation of certain aspects of artistic production in fourteenth and fifteenth century Italy — just as, conversely, artistic practice was reinvigorated by optical theory. This imbrication of practices would have paralleled a shift observed *within* textual theory away from resolving technical issues such as light-ray geometry to Aristotelian, philosophical concerns regarding the phenomenon of vision.[17]

Theoretical Texts of Optical Theory Expanded

Theoretical-textual sources on optical and related themes relevant to perspectival practice in the visual arts of the trecento were not limited to high optical theory cited in the standard reference. They include the following three:

1. Vitruvius, the only ancient architectural theorist whose work survived. Widely known in the middle ages, copies of *De architectura* were owned, for example, by Petrarch and Boccaccio. It was cited in relation to the present set of issues by the Florentine chronicler, Filippo Villani.[18] Vitruvius provided a *general theory of visual analysis* of an architectural design and its inflection to conditions of visibility. The author's goal was the perception of correct proportions. When architectural representation becomes a standard component of painting in the trecento, Vitruvianism may be present.

---

14 Controversy continues regarding specific issues, including the perennial intromission/extramission disputation hardwired into the discourse. David C. Lindberg shows that Kepler's theory of the retinal image (*c.* 1600) was a 'new solution (but not a new kind of solution) to a medieval problem, defined some 600 years earlier by Alhacen'. See Lindberg, *Theories of Vision*, 208.

15 Lindberg notes much crossover among the several branches of visual theory (perspectivist, Aristotelian, and theological) in the fourteenth and fifteenth centuries. See Lindberg, *Theories of Vision*, 143.

16 Pamela Smith, *The Body of the Artisan: Art and Experience in the Scientific Revolution* (Chicago: University of Chicago Press, 2004).

17 Lindberg, *Theories of Vision*, 143–46.

18 Vitruvius, *Ten Books on Architecture*, ed. and trans. by Morris Hicky Morgan (New York: Dover Publications, 1960). Cf. Kenneth J. Conant, 'The Afterlife of Vitruvius in the Middle Ages', *Journal of the Society of Architectural Historians*, 27 (1) (1968), 33–38; Carol Herselle Krinsky, 'Seventy-Eight Vitruvius Manuscripts', *Journal of the Warburg and Courtauld Institutes*, 30 (1967), 36–70 and Hanno-Walter Kruft, *History of Architectural Theory* (Princeton: Princeton Architectural Press, 1996), Chapter 2. For the Villani citation, see Filippo Villani, *Liber de origine civitatis Florentiae et euisdem famosis civibus*, ed. by G. C. Galleti (Florence, 1847), p. 36.

2. Medieval optics, stemming from antiquity, comprising Euclid, Aristotle, and Ptolemy as elaborated by Islamic intermediaries, Ibn al-Haytham (Alhacen) in particular. John Pecham's widely circulated late-thirteenth century *Perspectiva communis* incorporated Roger Bacon, who drew on Alhacen.[19] Pecham and the perspectivist current of medieval optical theory posited *specific parameters of vision* that establish the conditions of visibility of art and architectural form, which constituted a geometricised interpretation of image-production by the eye originating in Euclid and other ancient theorists. Distinct from the philosophical, theological, and medical study of vision and the eye, this branch of optical thought tended to be preoccupied with 'ray-tracing' of light from the object through the multilayered ocular apparatus and other geometrically construed aspects of vision.[20]

3. Surveying textbooks for middle school education in Italian cities, part of the mathematical-geometrical skills useful for commerce, trade, and other occupations.[21] These pedagogical exercises *promoted a mental habit of seeing* buildings in precise terms of their distance/height relationship and the viewing angle, in alignment with the parameters of the visual field according to optical science.

These three categories of ancient and medieval writings appear to have diversely informed the visual arts. Some of the texts present dense passages of technical Latin, unlikely to have been accessible to the painters and other artisans of the period: yet key ideas would have penetrated the episteme in the way that advanced theories often filter down to non-specialists. Given the continuity of educational practices there is no reason to exclude the possibility that certain trecento artisans would have access to the optical texts in the way that Brunelleschi, Ghiberti, Piero della Francesca, and Leonardo, none university educated, evidently did in the next century.

The Sources: Relevant Doctrines, Passages and Illustrations

Specific passages and doctrines relevant to artistic production included the following items, categorised by author and/or area of relevance, and amplified by citations of theoretical texts:

Vitruvius, *De architectura* (A general theory of optical adjustments):

Book III, Chapter 5, Section 9 (on the proportions of the Ionic order, including its entablature):

> For the higher that the eye has to climb, the less easily can it make its way through the thicker and thicker mass of air [...] when the height is great [...] it conveys to

19 Cf. the insightful overview of medieval optics by David C. Lindberg, 'The Science of Optics', in *Science in the Middle Ages*, ed. by David C. Lindberg (Chicago: University of Chicago Press, 1978), pp. 338–68. For an informative, authoritative history, see Lindberg, *Theories of Vision*, and A. Mark Smith, *From Sight to Light: The Passage from Ancient to Modern Optics* (Chicago: University of Chicago Press, 2014).

20 Cf. Lindberg, *Theories of Vision*, 122–46.

21 Cf. Trachtenberg, *Dominion*, 225–32 and Michael Baxandall, *Painting and Experience in Fifteenth-Century Italy* (Oxford: Clarendon Press, 1972), whose general thesis regarding the diffusion of these texts applies to the identical pedagogical culture of the pre-1400 period.

> the mind only a confused estimate of the dimensions. Hence there must always be a corresponding increase in the symmetrical proportions of the members […].

Book III, Chapter 5, Section 13 (diminution of apparent size with distance above eye level):

> In the case of flat surfaces located above the level of the eye, the parts farther away appear lower [paraphrase].

Book V, Chapter 6, Section 7 (on the plan of the theater):

> It is not possible […] that in all theaters these rules of symmetry [= proportions] should answer all conditions and purposes, but the architect ought to consider […] to what extent it may be modified to suit the nature of the site or the size of the work.

Book VI, Chapter 2, Section 1 (modifications in proportions to suit the site):

> There is nothing to which an architect should devote more thought than to the exact proportions of his building […] [he must] modify the plan by diminutions or additions in such a manner that these diminutions or additions in the symmetrical relations may be seen to be made on correct principles […].

Book VI, Chapter 2, Section 2:

> The look of a building when seen close at hand is one thing, on a height it is another […] and in all these cases it takes much judgment to decide what is to be done.

Book VI, Chapter 2, Section 4:

> Since things are sometimes represented by the eyes as other than they are, I think it certain that diminutions or additions should be made to suit the nature or needs of the site, but in such fashion that the buildings lose nothing thereby. These results, however, are also attainable by flashes of genius, and not only by mere science.[22]

Vitruvius thus posits that 'correct' proportions are imperative in architectural design, yet he asserts repeatedly that what is correct not only varies according to 'the nature of the site or the size of the work' but also how it is seen. Although proportional correctness is an absolute value for Vitruvius, such correctness is not vested in the work 'itself' but in the perceived image of the work in the eye of the beholder. In his anti-Platonic view, correct appearance is all. 'Diminutions or additions' of building components serve this function, yet there are no precise rules regarding their deployment. 'Much judgement' is required, even 'flashes of genius', and 'not only mere science'. This indeterminacy of methodology left the door open for a more precise protocol regarding the production of perceived proportional correctness.[23]

---

22 All Vitruvius translation is by Morris Hicky Morgan.

23 The Vitruvian rule regarding the apparent visual compression of high-placed elements on buildings derives from Euclid, whose Propositions 5, 10, and 11 describe such diminution albeit along a horizontal plane (Figure 5). Hard to miss in Euclid, this observation would have reinforced the reception of Vitruvius. Intriguingly, these theorems are associated with the 'distance point' technique of Albertian perspective construction. Cf. Lindberg, *Theories of Vision*, 150.

MEDIEVAL OPTICAL THEORY (Perspectivist doctrines on the structure and behavior of light, vision and image formation, including Euclid's foundational geometricised theory, with emphasis on ray-tracing):[24]

1. Linearity of light rays:

Euclid, *Optics* (geometricised theory of vision):

Axiom 1: 'Visual lines [rays] can be drawn in straight lines to the object' [paraphrased]

John Pecham, *Perspectiva communis*:[25]

Proposition I. 14: 'Rays of primary light and color are always propagated in straight lines unless bent by variations in the medium [refraction]'
Proposition I. 27: 'In fact, a ray is nothing but the species of a visible object fashioned into a straight line by extension'
Proposition I. 32: '[...] vision occurs by straight lines incident perpendicularly on the eye'

2. The ninety degree pyramid/cone of vision:

Euclid, *Optics*:

Axiom 2: 'The form of vision is a cone, with its apex in the eye and the base at the limits of vision [paraphrased]'

John Pecham, *Perspectiva communis*:

Proposition I. 32: 'the species of the visible object is incident on the eye in the form of a pyramid'
Propostition I. 38: 'Visible objects are perceived by means of the pyramid of radiation [...] Every visible object is seen under an angle or triangular figure'
Proposition I. 39: 'It is not under every angle that objects are seen' [goes on to explain the ninety degree limit of clear vision][26]

3. Parameters of relative size within the pyramid/cone:

---

24 Absent from this list are important precepts, including those addressing the central technical issue of medieval optics, 'the problem of a multiplicity of rays emanating from every point in the visual field influencing all parts of the eye', which had been tenuously resolved by Alhacen but was not finalised until Kepler. See Lindberg, *Theories of Vision*, 168. It was this problem, not the question of directionality itself, that kept the intromission-extramission debate alive.

25 Pecham's propositions are cited from John Pecham, *John Pecham and the Science of Optics: Perspectiva communis*, ed. and trans. by David C. Lindberg (Madison: University of Wisconsin Press, 1970).

26 'Therefore if lines are drawn from the extremities of this opening to the center [of the sphere of the uvea], they form a right angle there [...]'. This assertion is modified several lines down in the text: 'Therefore the largest angle under which vision by [direct] radiation occurs is smaller than a right angle [...]'. See Pecham, *Perspectiva communis*, ed. by Lindberg, Proposition 39, p. 123. Lindberg records that Roger Bacon agrees with Pecham that 'the field of vision is slightly less than 90 degrees', see Pecham, *Perspectiva communis*, ed. and trans. by Lindberg, p. 251 n. 106. Under certain conditions, peripheral vision, beyond ninety degrees, occurs (Pecham, *Perspectiva communis*, ed. and trans. by Lindberg, Proposition 42, p. 125f) through refracted (rather than direct) rays, which register objects 'weakly'.

John Pecham, *Perspectiva communis*:

> Proposition I. 50: 'Nothing is seen unless it is of appropriate size... [To be seen, an object] must have [not only] size, but sufficient size. It must not be minute because such a thing would not suffice to impress the eye painfully and powerfully, as Proposition 43 maintains: and [conversely] a body of great size cannot be seen at a single glance, as Proposition 39 makes clear'

4. The (clear) visual field limit of ninety degrees, peripheral vision, and anamorphosis:

John Pecham, *Perspectiva communis*:

> Proposition III. 15: '[...] things at the side, not contained within the aforesaid pyramid, are seen imperfectly [...] things seen in this manner appear weakly'.

Levi ben Gershon, ('Gersonides', early fourteenth century) reasoned that 'the angle of vision cannot exceed 90° degrees without border distortions arising. The geometric representation of objects lying beyond an angle of vision of 90° will seem incorrect'.[27]

5. The importance of the centric ray or axis of vision (due to its unrefracted, hence uncompromised and unweakened clarity):

John Pecham, *Perspectiva communis*:

> Proposition 38: 'Perception is certified by the axis' [of the visual pyramid][28]

Dante, *Convivio* (*c.* 1304–1308), Book II, p. 10:

> [the centric ray] alone is stamped upon the imagination

Alberti, *Della pittura* (1436):

> [the centric ray is] [...] the prince of rays[29]

These precepts, which were among the easiest for a layman to grasp, would have provided a measure of the specificity regarding perception missing in Vitruvius. In his theory, as noted, all is a matter of individual judgment and even 'genius', which permit the realization of correct proportions as experienced, hence the production of authentic beauty of a building in the eye of the beholder. Optical theory, to the contrary, presupposes no

27 Johannsen and Marcussen, 'A Critical Survey', p. 201f; Cf. Erwin Panofsky, *Perspective as Symbolic Form*, trans. by C. S. Wood (New York: MIT Press, 1991), p. 201f; James Elkins, *The Poetics of Perspective* (Ithaca: Cornell University Press, 1994), p. 69 ff.

28 In Pecham, *Perspectiva communis*, ed. and trans. by Lindberg, Proposition I.38, scanning the object by 'turning the eye about' so that the centric ray passes over it is necessary for its full 'certification' ('perception is certified by the axis [of the pyramid] being conveyed over the visible object'), although the writer notes that 'in common parlance it is said that every visible object is seen under an angle or triangular figure'. When pyramid of vision and centric ray theory is transposed to artistic and architectural practice (in Alberti as well as the trecento), however, the eye is treated *as if* immobilised, thereby providing a conceptual model, not for sight but for representation and artistic production, in a manner typical of the period. Cf. Trachtenberg, *Dominion*, 238 f.

29 Leon Battista Alberti, *Della pittura*, trans. by J. R. Spencer (New Haven: Yale University Press, 1954), p. 48.

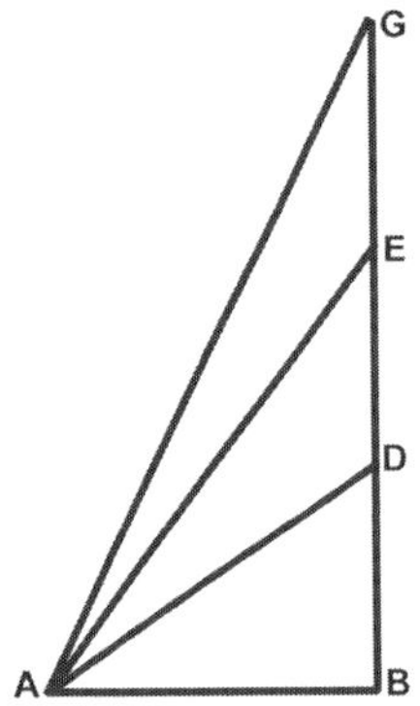

Fig. 5 Euclid, *Optics*, Proposition 11.

such aesthetic goals or standards. Indeed, it does not recognise buildings or works of art as distinct from other physical entities in view. Instead it provides a conceptual framework for understanding how all things are seen, and in a reliable way. In trecento visual media the parameters of this perceptual framework appear sometimes to be adopted in inventive ways as norms of design, as shown later in this paper.

The foundational axiom of this theoretical program is that the visible qualities of things are relayed to the eye through straight lines, or rays (Protocol 1 above). That light is posited by premodern science as traveling in straight lines might seem to be iterating a self-evident point, but it is possible to imagine a world-condition in which light does not always travel in such linear paths (as in relativity theory). In such a regime most capacities of vision, especially the ones that enable the eye to produce trustworthy knowledge of the world, would be crippled. In medieval optics linear light-rays enable the potential *accuracy* of vision, the capacity of the eye to see shapes for what they 'are' and to gauge the relative size of things seen according to geometric reckoning. Linearity was thus essential to the contested assertion of the reliability of vision.[30] Far from being a trivial consideration, it was foundational, as was recognised by Euclid in the primary axiom of his *Optics*, whose geometric construction of sight (mainly using triangulated figures) would have been impossible on any other basis. Certainly, ray-linearity was the most accessible optical principle for the outsider, although the others cited here would not have been far behind.

Such light rays are not received (or projected, as in extramission theory) by the eye in a package of indeterminate shape, but instead via a conical field converging to focus within the eye (cf. Protocol 2 above, and Euclid's second axiom). This 'pyramid of vision', as it was often termed, is vested with several traits. Its visual field does not vary according to circumstances (as in modern science of vision), but forms a cone/pyramid of sight that subtends a ninety degree angle (or slightly less, see Figure 6).[31] When we open our eyes, regardless of what we are observing, this is our normal visual field (although part of it may be 'empty' of object-rays, and scanning with the centric ray is necessary for an accurate image of an object within it).[32]

30 According to Paul Hills, in medieval thought '[...] there was one area of science where the unaided human intellect could hope to achieve certainty [...] If it could be demonstrated that one of the senses operated according to mathematic or geometric principles its trustworthiness was enhanced [...] In optics this means distinguishing how far vision could take place according to the most elementary geometrical principles. Since the most elementary geometric relationship between two points is a straight line, if vision is to be reliable it must follow a straight line from a point in the eye to a point in the object [...] all investigators [thus] believe that light propagates itself along straight lines. See Paul Hills, *The Light in Early Italian Painting* (New Haven: Yale University Press, 1987), p. 13.

31 The slight deviation from the ideal geometry owed to a technicality of the medieval structuring of the eye. See Pecham, *Perspectiva communis*, ed. and trans. by Lindberg, Proposition I. 39).

32 See note 30.

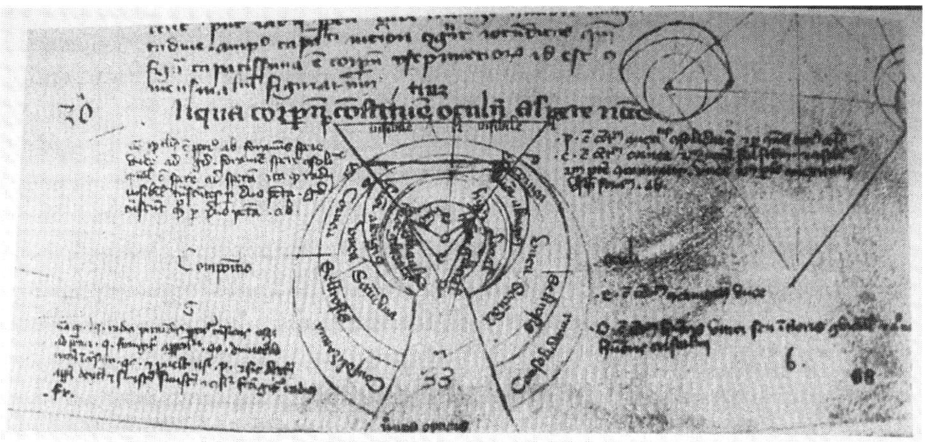

Fig. 6 The eye, with ninety degree field of vision, John Pecham, *Perspectiva communis*, fourteenth century. Vatican Library, MS Vat lat 3102, fol. 66.

Although it is stipulated that vision might spread beyond the ninety degree field, such supernumerary information as received would be 'weak' and tainted by what today is termed anamorphosis (Protocol 4). Such distortions corrupt the flow of information, introducing error to the otherwise reliable optical apparatus. The threshold of distortion is construed as rather abrupt, which makes the power of vision solid and reliable within its legitimate angular parameters, but unreliable at any point beyond it. This would mean that the standard ninety degree visual field is not only real but optimal.

The ninety degree visual pyramid, however, was not an undifferentiated apparatus. The power of vision varied within the ninety degree visual field. Of the innumerable visual rays that entered the cone (or were 'projected' from it in extramission), only the centric, axial ray possessed veritable clarity and provided accurate visual information. In Pecham's words, 'Perception is certified' by this axial information, or as Dante puts it, only this ray is 'stamped upon the imagination'. To direct one's gaze at an object meant, in effect, to direct the centric ray toward it.

At the same time, however, the rest of the ninety degree visual field remained in play. Although its rays were not as accurate a conveyor of information as the centric gaze, their contribution remained essential to the total image. The centric ray was critical to perceptual judgement, but to see at all depended on the object attaining a certain size in the visual field. As communicated in Protocol 3, the eye needs to be 'impressed painfully and powerfully' for an object to be seen, and that power is achieved by sufficient object size. Conversely, for an object 'of great size' to occupy more than a ninety degree field would surpass the limit of what can be seen in a 'single glance'. The optimal visible object would need, therefore, to be large, but not too large. This parameter is indirectly defined. By implication, the optimum size of an object with respect to the dictum that an object have 'sufficient size' would be for it to fill or to span the ninety degree visual field, but no more; and for its principal feature to be aligned with and thereby to produce the centric ray of vision.

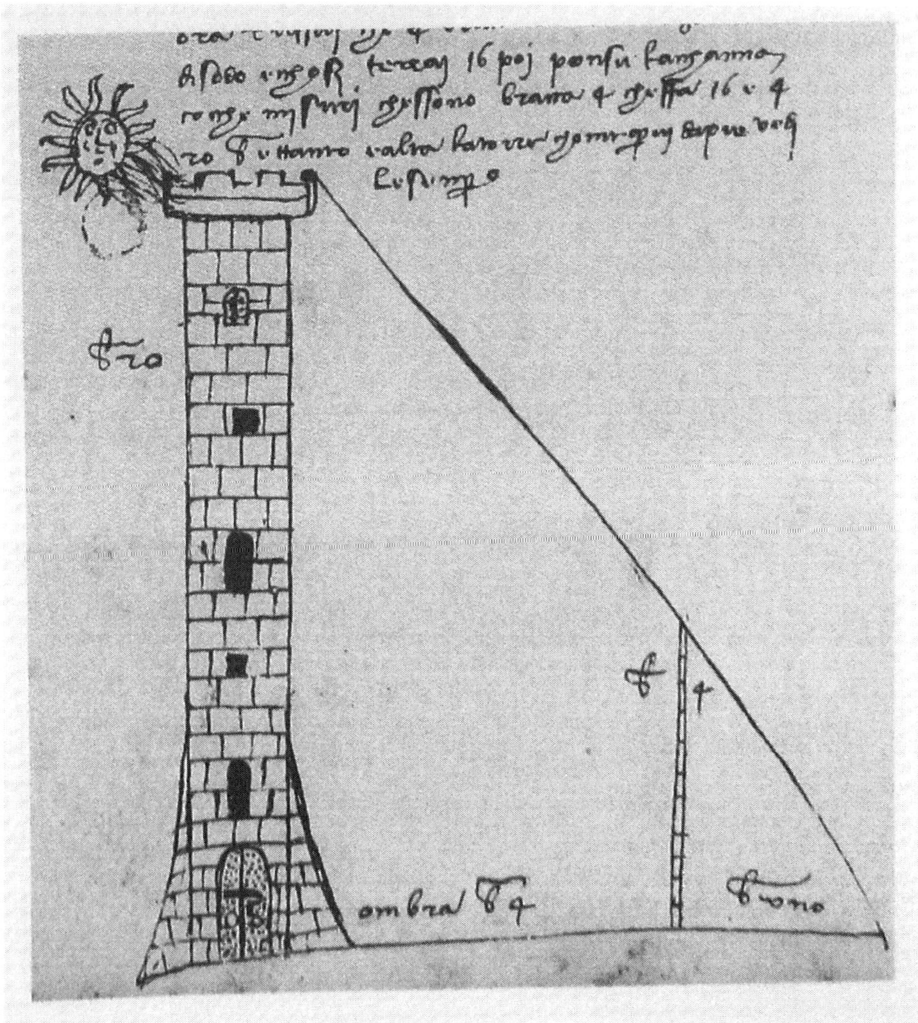

Fig. 7 Measuring a tower using its shadow. *Trattato d'aritmetica* (1434/36). Biblioteca Nazionale, Rome, Conv. Sopp., A.8.1837, 73V.

To integrate Vitruvius and medieval optical theory, the ancient writer was concerned with visual aesthetics as engaged with the undefined power of the senses, whereas medieval optics closely studied the senses themselves but was oblivious to aesthetics, Vitruvian or otherwise. In effect, the two doctrines were complimentary. Vitruvius provided an aesthetic ideal — perfect, harmonious proportions — but no defined means to achieve it, whereas medieval optics entertained no aesthetic program, yet it embodied an explanation of vision in which the power of the image — *any* image — is explained and theoretically enabled.[33]

33 Cf. Katherine H. Tachau, *Vision and Certitude in the Age of Ockham: Optics, Epistemology, and the Foundation of Semantics, 1250–1345* (Leiden: Brill, 1988), passim.

Fig. 8 Giotto and Andrea Pisano, *The Astrologer*, Campanile, 1334-7.

Late Medieval/Trecento Surveying Textbooks

These *promoted a mental habit of seeing* buildings in precise terms of the distance/height relationship and the viewing angle. This message was communicated vividly in the illustrations of numerous manuscripts, which depict the calculation of the height of a tower (Figure 7). In all of these illustrations the reader is presented with the image of a monumental building observed in the angular relationship formed by its height and a specific distance from it. One technique of such 'tower surveying' involved use of the quadrant, a simplified form of the astrolabe. Made principally for astronomical observation, these instruments were so well known to the public that a quadrant could be used to identify one of the main figures in the Campanile reliefs (1334–40) as an astrologer (Figure 8). This angular technique of surveying appears in a number of *abbaco* texts. In an elaborate example, both astrological and terrestrial surveying functions appear in a single illustration, associating surveying with the 'higher' intellectual realm (Figure 9). All such illustrations present the image of a monumental building, not unlike those erected during the period, being observed at a fixed distance in the precise angle of view occupied in vision as the product of the structure's absolute size and distance from the eye.

This intellectually grounded, widely disseminated schema prefigures the built scenographic perspective practice of trecento Florence. A forty-five degree vertical viewing angle is common to both surveying theory and the built viewing spaces. The

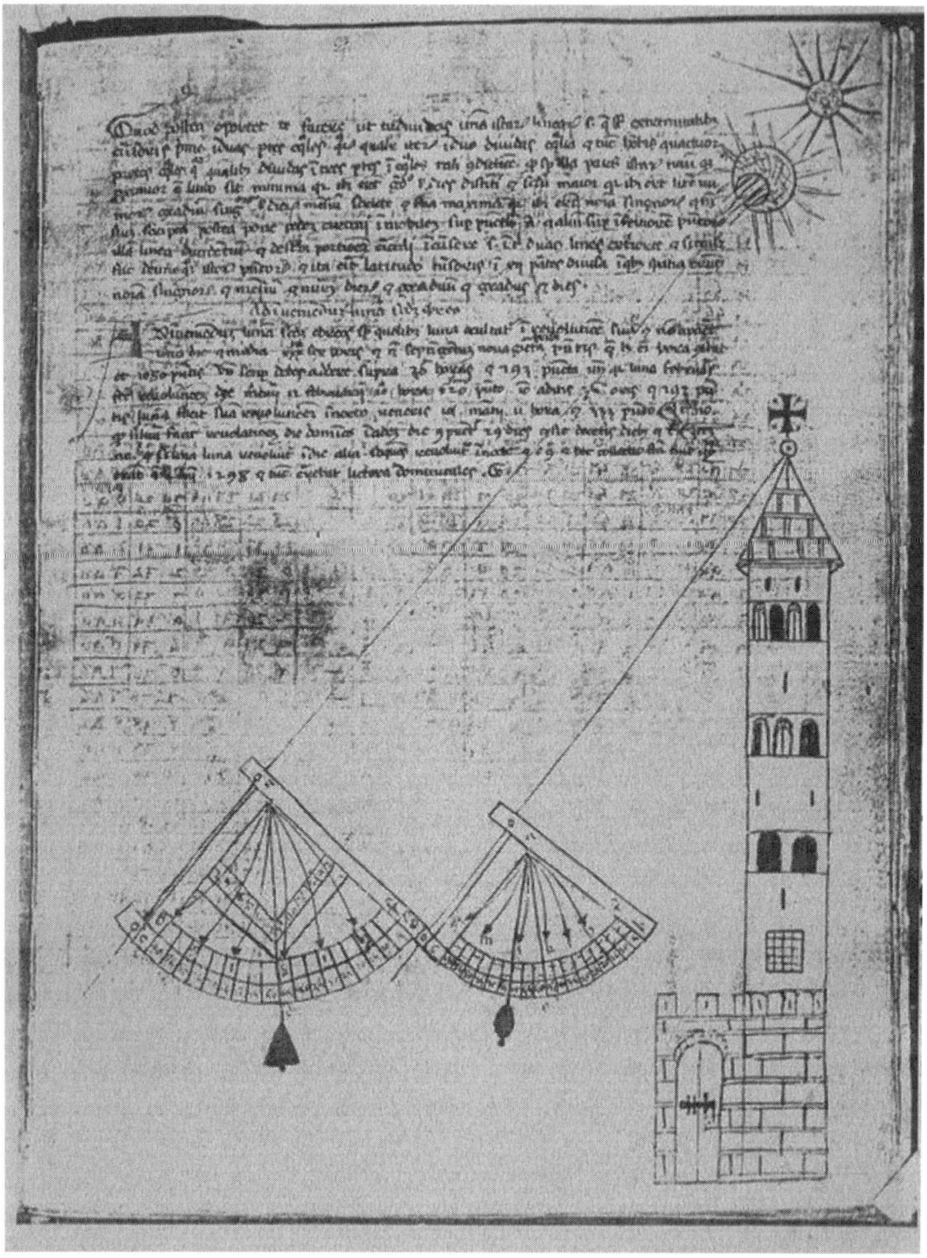

Fig. 9 Measuring a tower-building and sighting the heavens, using quadrants (one with a shadow square), Fibonacci manuscript, Verona, 1389. Biblioteca Medicea Laurenziana, Florence, Ashburnham 356, 25.

scope of the imagined eye of the surveyor/spectator thus aligns with the geometry of vision in contemporary optical theory as outlined above — i.e., the ninety degree visual pyramid — allowing that the eye is imagined 'as if' looking straight ahead (the ideal eye

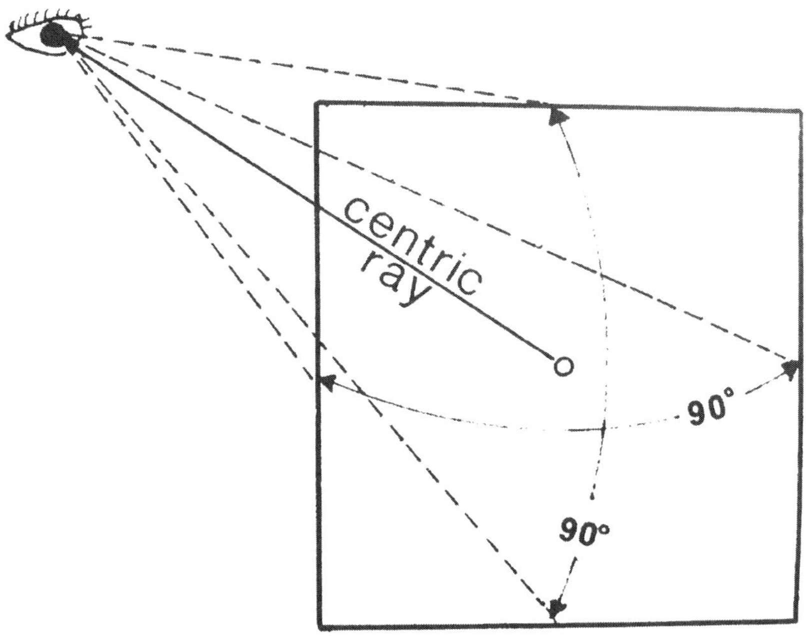

Fig. 10 'The Pyramid of Vision of Medieval Optical Theory'.

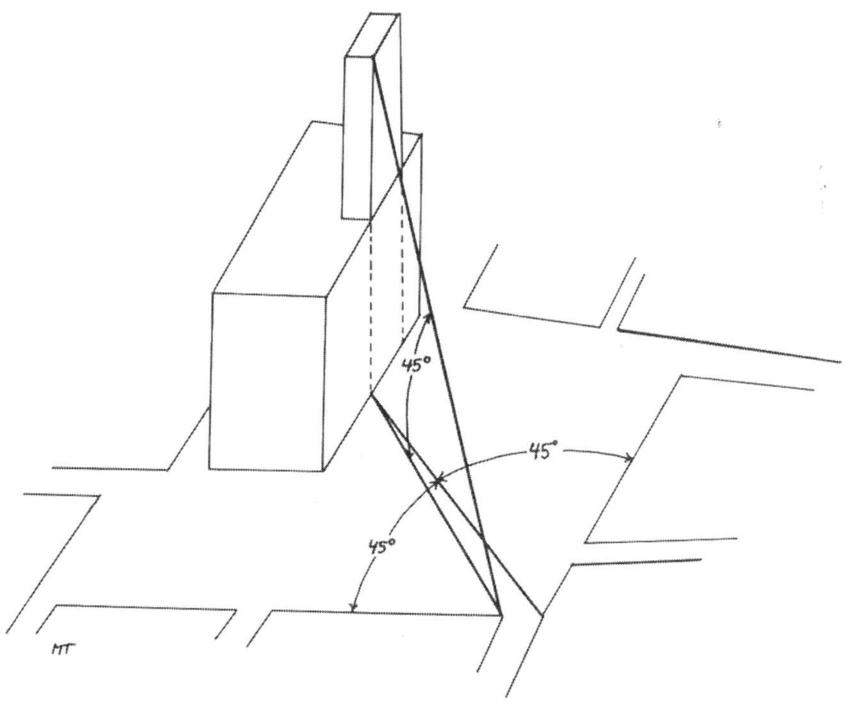

Fig. 11 Piazza della Signoria, Scenographic structure, angle of view 90 degree horizontally, 45 degrees vertically.

of theory) with the building rising forty-five degrees above the horizon, the foreground falling into the angular measure below (Figures 10 and 11).[34] The 'pyramid' here did not function as model for actual seeing but as a *conceptual model* for the production of space and architectural scenography.[35]

Thus, although surveying theory was in itself neither an overt model of visuality (like optical theory) nor an aesthetic doctrine (like Vitruvius), it provided a simple, clear model for visual experience — and for the construction of physical conditions that enabled such experience. It conjured an architectural perspective which contributed to a habit of coordinating eye position, viewing distance, and viewing angle of a spectator on an architectural object in precise numerical and proportional association.

Part II – Case Studies of Trecento Perspectivism as Practice Across Visual Media

The Linearity of Light Rays, Occlusion, and Giotto

Because of the emphasis in scholarship on Albertian techniques of perspective, it is seldom noted that the illusion of spatial depth in trecento painting was produced not only by architectural means, i.e., by the 'recession' of orthogonal lines in pavements, walls, and arcades. Among other devices was *occlusion*, or the apparent overlap or obscuring of one form by another in a painting, which the eye interprets not as one object eating away at another at a single plane (the 'naive' impression) but as the completely (or more completely) seen object partly masking an object situated 'behind' it in space. Occlusion is a function of 'perspective' in its dependence not only on the 'layering' of bodies but a *specific point of view on them*, illusory though that position may be. The technique is so common in European painting that it is rarely commented on, and its historical implications remain unexplored.[36]

In the context of the present study, it is critical to recognise that at a theoretical level this technique is *founded on the linearity of light rays* (i.e., on Euclid's first axiom). It may be regarded as a *function of that linearity*. Conversely, linearity is axiomatic to the practice. As is self-evident, bending rays would make it impossible to correctly sense the size, shape and/or distance of the (partly) occluded object (as noted above regarding non-linearity), much as they would with respect to non-occluded bodies. Here I will study a few examples of pictorial occlusion and related techniques in order to gauge the degree to which this aspect of optical theory and its implications may have affected pictorial practice in the trecento.

34 Cf. Elkins, *Poetics of Perspective*, 72.

35 Cf. Kuhn, 'Measured Appearances', p. 116.

36 The perceptual scientist, Barbara Gillam, observes that 'Occlusion is rarely discussed as a major issue in art, yet it could be regarded as the major issue in depicting a three-dimensional scene on a picture plane' and discusses a few points regarding its deployment in the trecento. See Barbara Gillam, 'Occlusion Issues in Early Renaissance Art', *i-Perception*, 2 (9) (2011), 1076–97 (p. 1076). See also Jules Lubbock, *Storytelling in Christian Art from Giotto to Donatello* (London: Yale University Press, 2006), p. 37 and Appendix.

Occlusion was a pictorial technique long in use before the development of architecturally constructed illusion of depth by Giotto, the Roman school, and the Sienese trecento. For example, in a standard Byzantine Virgin and Child, depth is indicated by the way the child held by the Virgin partly occludes her, thereby producing a taut spatial relationship 'behind' the painting surface. The same spatial interweave holds for more complex groupings of bodies, including their spatial relationship to their illusory environment. Far from being discontinued during the advent of architecturally 'constructed' Giottesque space, occlusion not only continues in the trecento as a standard space-producing device, it becomes so prominent that it attracts attention to itself as a passage of technical *excess*. Spatial artifice escapes its normal boundaries, and is dramatised as a virtual subject parallel to the ostensible theme of the painting.

Giotto himself, although identified with the rise of 'constructed' proto-Albertian pictorial space, was a master of this device. Occlusion is almost always present in his work, where it serves to orchestrate action by gathering secondary figures in layered groups around the principal actors in narrative scenes. In a number of Giotto's paintings, however, occlusion seems developed well beyond the parameters of dramatic stage management or iconic emphasis. Perhaps the extreme case in terms of sheer density of effect occurs in the Berlin *Death of the Virgin* in the sprawling group of saints and angels on the right, where the discerning viewer can trace no less than nine layers of halos and faces, interlayered with dazzling virtuosity (Figure 12). Some of these figures are occluded to the limit of visibility, which draws the puzzled eye deeper into the maze-like pattern of overlap (Figure 13). Another case of 'occlusive' intricacy occurs in the most monumental of Giotto's enthroned virgins, the huge Ognissanti Altarpiece in the Uffizi (Figure 14). Although it contains far fewer flanking figures than the Berlin panel — fourteen against thirty-six — and correspondingly 'only' five or six thinly populated occlusion-layers, they are nonetheless developed with uncanny symmetry around the Virgin, including on either side near-matching faces whose frontal features are almost completely masked — even the *near* eye.

Another compositional element compensates for the lower count of attending figures in the Ognisanti panel by sharpening the subtlety of occlusion design. Because the figures are grouped tautly around the Virgin's throne, any irregularity in the mirror-imaging of their spatial disposition registers as if magnified. Moreover, occlusion becomes entangled with the implicit lines of perspective, which embody a slight off-axis viewpoint, shifted to the left of center. All elements of the scene, from the steps of the throne to the gable, are made to appear as if they are viewed from slightly off center. Displacements from perfect symmetry are consistent. The artifice is highlighted by the 'twin' saint's heads seen through the 'windows' in the sides of the throne, differentially occluded left and right to a slight but noticeable degree, as if seen from a slightly off-center viewpoint. It is this precision, rather than the excess of sightlines as in the Berlin painting, that heightens the viewer's awareness of herself an observing 'eye'.

Giotto's most intricate manipulation of these subtle asymmetrical effects, perspective sightlines, and studied occlusion occurs in Santa Croce, in the *Apparition of Saint Francis at Arles* fresco in the Bardi Chapel (Figure 15, roughly contemporaneous with the Uffizi panel). In this case the entirety of an intricate, multilayered configuration of figures and full-scale architecture is differentially half-revealed and half-obscured by the apparent

Fig. 12 Giotto, *Death of the Virgin*, Gemäldegalerie, Berlin.

Fig. 13 Detail of Figure 12.

'displacement' of the viewpoint, a shifting of the 'eye' that draws attention to the perspectival artifice of the architectural/pictorial structure of the cloister and adjacent chapterhouse into which we can 'see'. Because of the density of occlusion, the viewer is encouraged to self-consciously trace the light rays running between her eye point and various parts of the painting. These lines produce a spatio-visual structure so elaborate yet consistent in

Fig. 14 Giotto, *Ognissanti Madonna*, Uffizi.

its slight off-center visual axis that it is possible to reconstruct a credible 'groundplan' of Giotto's stage-set, including the position of the imaginary viewing eye (Figure 16). Such

Fig. 15 Giotto, *Apparation of Saint Francis at Arles*, Bardi Chapel, Santa Croce.

is the intricacy and consistency of the virtual spatial configuration and the visibility or occlusion of all of its parts that no other plan would fit the visual data of the fresco — suggesting the possibility that Giotto drew a plan of the architectural/sightline scheme before establishing its pictorial form (perhaps taking a page from the proliferation of ray-tracing in the illustrations of optical texts).

The density of sightlines in all three of Giotto's paintings, their intricate positioning of the ideal viewpoint, and their *excessive presence* relative to the nominal subject seem to invite a further conceptual move on the part of the beholder as a way of assimilating the complexity. One seems to *step back* from the viewpoint of seeing the painting, *behind* the former position of one's own eye. The sightlines themselves are self-consciously sensed as if they were highlighted rays. With a quasi-Dantesque self-consciousness of perception, one *sees oneself seeing the painting*, which for the informed trecento viewer would have meant 'seeing' the rays of light/sight (the 'species') that connect the pictorial object and the eye.

That such an unfolding of self-consciously layered perception might have been the intended reception of these paintings is supported by a factor already mentioned: the currency of the ancient debate about the directionality of vision, a controversy to which the works in question would plausibly have alluded. Regarding the point in question, the key aspect of this controversy was not the validity of either position — that modernity affirms intromission is irrelevant here — but rather the emphasis of both schools on the active power of vision (which involved cognitive powers of processing raw visual information) and the physicality of the rays of light ('species') that are either emitted or received by the organ of sight. In fact, so much close analysis in optical theory explicitly

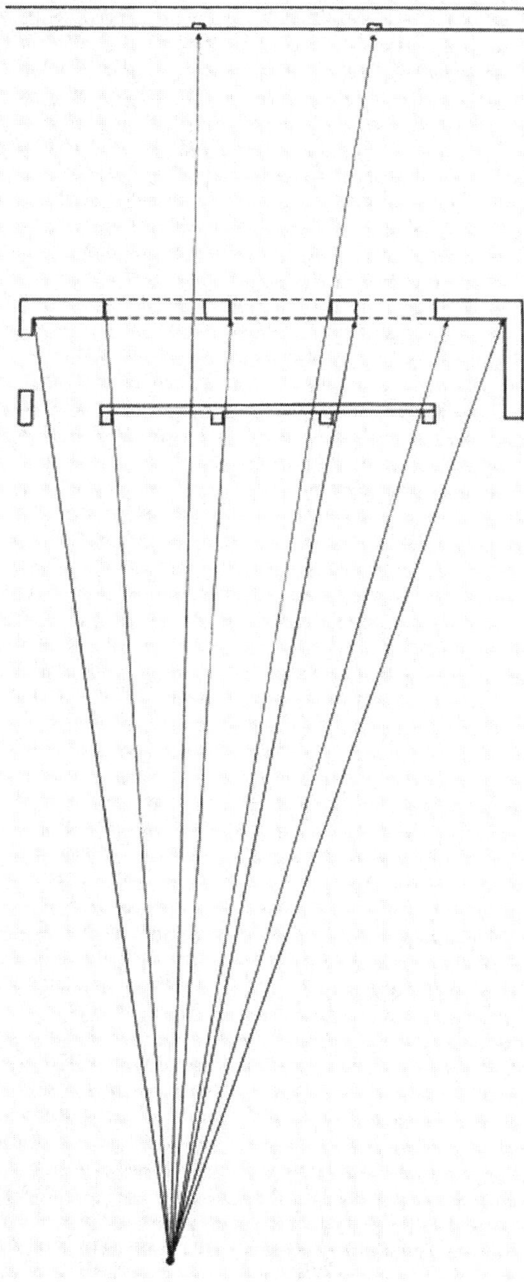

Fig. 16 Plan of sightlines/visual rays of Figure 15.

treats such rays with respect to their nature, direction, refracted ocular pathways, etc., that the texts sometimes seem almost visibly suffused by such 'radii' of light/vision — a gestalt that may have conditioned the imaging of 'real' rays of the observer linking the pictorial object and the eye, as suggested above regarding Giotto. This reading is sustained by the way Giotto himself actually depicted a figure whose eyes visibly emit such corporeal rays in the Arena Chapel.[37]

Yet it is crucial to the present argument that the debate over radiant directionality was not resolved in the period. Even theorists who favoured intromission — most participants of the late medieval period including Grosseteste, Bacon, and Pecham — deemed it necessary to accommodate the ancient, psychologically resonant theory of extramission if for no other reason than its origins and support by numerous voices of ancient authority still requiring deference (including Euclid, Plato, Ptolemy and Augustine, as against figures such as Aristotle and Alhacen on the other side). Moreover, the unity of knowledge was an enduring ideal.[38] Strange as it may seem to moderns, the issue thereby remained alive despite the preponderance of observation-based arguments favoring intromission. Because the nature of sight-rays was so critical to understanding the ever mysterious phenomenon of vision, the ability of humans to perceive light, one of God's first creations, and the world it illuminated (and may have even created according to Grosseteste),[39] its directionality persisted as a topic that continued to stimulate argument during the Age of Giotto (as well as that of Brunelleschi and Alberti).[40]

Ultimately, it was the perceived *power of the eye* — or of the 'eye-rays', running one way or the other — that was at stake in this discursive sensory event, and in this respect the ancient/medieval doctrine retains a certain validity today. Even in modern science, vision is understood, *mutatis mutandis*, as being far from passive. Never a mere receptor of information, the eye is 'active' as we feel ourselves sending out quanta of 'focused' attention in the direction of the object in view (not to mention the complex processing of the information sent to the brain from the retina).[41] The visio-projectile process that moderns construe metaphorically, premodern optical science understood literally and physically within a geometricised framework that was at the forefront of medieval intellectual life.

That the 'extramission/intromission' controversy raged during the Age of Giotto was not unrelated to the late-medieval debate over the truth-value of sight (and the 'lesser' senses); if vision were to be construed as capable of producing legitimate knowledge, the process of seeing needed to understood in terms that enabled its legitimacy by confirming

37 See Eva Frojmovič, 'Giotto's Circumspection', *Art Bulletin*, 89 (2007), 195–210, Figure 2: The eye-rays '[…] reference the medieval optical theory of vision by means of corporeal visual rays' (p. 200).

38 David C. Lindberg observes how Roger Bacon himself 'nicely succeeded in agreeing with everybody. He agrees with Alhacen and the Aristotelians that vision occurs through an impression made on the eye [intromission] […] But this does not forbid the existence of visual rays [extramission] that perform some other function […] a beautiful compromise is possible', see Lindberg, *Theories of Vision*, 115–16. Cf. David C. Lindberg and Katherine Tachau, 'The Science of Light, Seeing and Knowing', in *The Cambridge History of Science: Medieval Science*, ed. by David C. Lindberg and Michael H. Shank, 7 vols (Cambridge: Cambridge University Press, 2013), II, p. 505.

39 Following Plotinus, 'Grosseteste believed that God created an original point of light, which diffused itself in all directions, generating matter and giving rise to the cosmos as we know it', see Lindberg and Tachau, 'Science of Light', p. 498 n. 17.

40 The debate over directionality overlaid a deeper controversy: see note 24.

41 Except for bright sources of light, such as electric lamps and, of course, the sun. Cf. the broadened interpretation of perception of J. J. Gibson and his followers, see J. J. Gibson, *The Senses Considered as Perceptual Systems* (Boston: Houghton Mifflin, 1966).

its accuracy as well as defining its limitations.[42] That this burning set of issues would have been reflected in the most advanced visual culture of the period — and in the work of an artist officially praised by the city of Florence for his 'scientia et doctrina' — seems a plausible hypothesis given the visual evidence discussed here, and extended below.

Occlusion and the Other Optical Devices in Trecento Sculpture

The adoption of optical-theoretical and Vitruvian concepts by trecento pictorial practice is paralleled in sculpture and architecture. The modulation of normative physical form to situated perception was developed imaginatively by the Pisani between the mid-thirteenth and early fourteenth centuries. In his prophets and sybils on the facade of Siena Cathedral (1284–1300), for example, Giovanni Pisano twisted forward the necks of several of the figures, thrusting the enlarged, powerful heads out and turning their agitated faces toward the viewer down in the piazza (or toward other figures who look back). His half-length features in the exterior gallery of Pisa Baptistery are even more distorted. Similarly, at a smaller scale in the work of his father, Nicola Pisano, the relief sculptures of the Pisa and Siena cathedral pulpits (*c.* 1250–70) are often calculated for the eye of the viewer standing on the pavement, i.e., located below the relief level.[43] Often the composition as a whole falls into place only when viewed from below. Seen frontally, for example, the Pisan *Adoration of the Magi* scatters attention from figure to figure, whereas as actually experienced the Virgin dominates a scene that sweeps in rhythmic movement toward her (Figure 17). This perspective also resolves such oddities as the distorted horses.

There is often a spatial dimension to this treatment, which brings us back to occlusion and the doctrine of the linearity of light rays. Just as in trecento painting the illusionistic overlap of figures is an important compositional technique, so in the relief sculpture of the Pisani the actual overlap of figures creates much the same effect. In sculpture, however, unlike in painting this impression depends on the *real* angle of view. As is suggested in a generalised cross section of a typical Pisani relief panel — and clearly visible in the rows of figures in the Siena *Last Judgement*, which are angled forward — the effect obtains only as seen from a position below the scene (Figure 18). Viewed frontally — that is, abnormally, from scaffolding — the spatial effect fades, and anomalies such as gaps appear. The forward-leaning figures in the 'Saved' panel look particularly unstable.

Two cross-sectional diagrams of a typical relief panel of the pulpits, indicating the sightlines, make it obvious that this sculptural device depends on the linearity of light-rays (Figure 19). In this case, they are real rays rather than the imagined penetration of the eye into the composition in painting. That occlusion and the resulting compositional effects are produced only at the proper viewpoint, and that this viewpoint is the 'natural' position with the beholder standing on the same ground as the work itself, somatically unifies the beholder with the entire work.[44]

---

42 See note 30.

43 Dominique Raynaud observes an enlargement of details towards top of panels, especially the heads, in Giovanni Pisano. See Raynaud, 'Optique et perspective', p. 171.

44 See Trachtenberg, *Dominion*, 183–91.

Fig. 17 Nicola Pisano, *Adoration of the Magi*, Pisa Baptistery pulpit, from ground level and scaffolding, c. 1260.

Optical-Theoretical and Vitruvian Factors in Architectural Site Planning

The expansion of this interpretation to include monumental architecture is enabled by the intermedial practice of the period, when few architects went without experience in at least one of the other media. Reflections of Vitruvianism and surveying theory are found in their architectural designs, while the penetration of optical theory extended beyond the general bonding of spectator and architectural spectacle discussed earlier. The hide-and-seek occlusion/presentation dialectic, the alignment of light-rays and sight-lines, and

Fig. 18 Nicola Pisano, *Last Judgement*, Siena Duomo pulpit, from ground level and scaffolding.

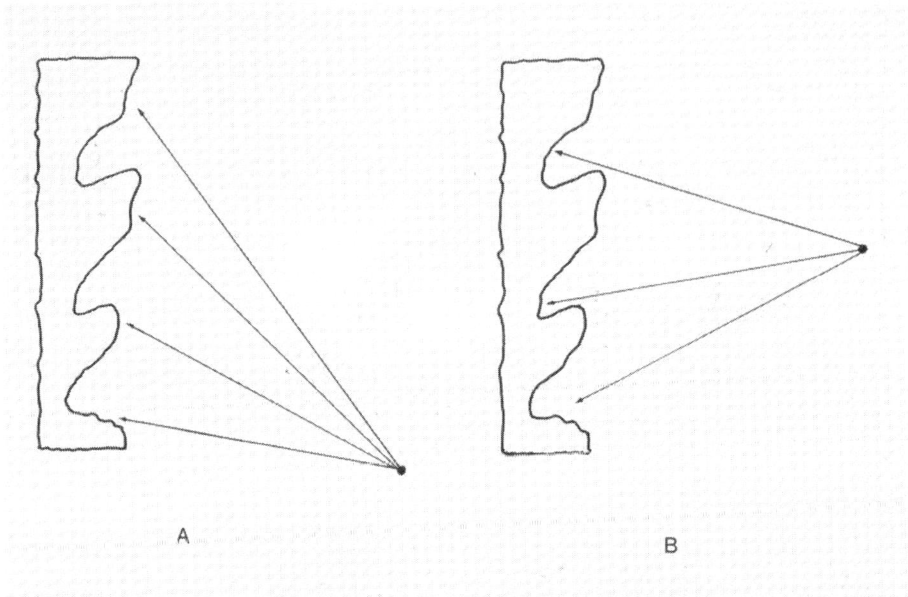

Fig. 19 Generalized cross section of Pisani pulpit relief, sight-lines from ground level and from scaffolding.

other optically-informed devices furthered the spread of the theoretical package studied here into architectural and urban planning. This dynamic appears to have enabled the achievement of precise, theoretically sustained solutions to key architectural problems — in many cases 'problems' produced by the methodology itself.

The role of the occlusion/presentation dynamic as studied above in painting and sculpture emerges in architectural planning in the closing years of the thirteenth century at the Florentine duomo, baptistery, and the space between them (Figure 20). This moment witnessed the embellishment, replacement, or restructuring of all three components of the complex. The baptistery incrustation was upgraded, a giant new cathedral was begun, with Giotto's colossal new bell tower following three decades later, and the interrelationship of the buildings was reformulated. The latter initiative was the key element with respect to the present discussion.[45]

Previously the old cathedral of Santa Reparata had stood closer to the Baptistery. It was, in fact, so close that from the primary viewpoint at the cathedral portal, the superstructure of the Baptistery could not be seen (Figure 21). The marble roof and lantern, of which the city was proud — displayed in all early views of Florence — were invisible from this primary viewpoint, or more precisely, from the position that the planners now *desired to be the main viewpoint*. The crown of the Baptistery was totally *occluded by the sightlines*. In the expansion of the piazza, the builders widened the space between the two buildings

45 Cf. Trachtenberg, *Dominion*, 42–55 for an expanded discussion of this site.

Fig. 20 Airview of Duomo group.

just enough so that occlusion was voided and from the critical viewpoint the Baptistery roof and lantern could be seen (Figures 22 and 23).[46]

That the new facade was designed by none other than a leading member of the Pisani school, Arnolfo di Cambio, who had apprenticed with Nicola at the Siena pulpit and, now a famous artist himself, was called to Florence explicitly for the duomo commission, reinforces this line of interpretation, as does the fact that his own sculptures in general are often formally inflected to the position of the observer.[47]

Vitruvianist Management of Shape and Proportions in Painting and Architecture

Taddeo Gaddi, the leading student of Giotto, was called 'another Vitruvius' by the trecento Florentine chronicler, Filippo Villani.[48] This enigmatic passage has been interpreted as a reference to the lavish architectural backgrounds in the artist's frescoes, in particular those in the Baroncelli Chapel in Santa Croce (*c.* 1330). An alternative explanation suggests itself when one considers that, quite apart from the issues discussed in this essay, painting and architecture were closely allied. Architecture was an 'imagistic' medium in Italy (rather than being more 'technically' oriented, as in Gothic France), with painters (along with sculptors) often serving as architects or on building committees.[49] Gaddi himself was a

46 Brunelleschi evidently found the view of the roof too steep/foreshortened for his demonstration panel, and moved the viewpoint inside the duomo a few steps.

47 A. M. Romanini, 'Nuove ipotesi su Arnolfo di Cambio', *Arte medievale*, 1 (1983), 157–202.

48 Trachtenberg, *Dominion*, p. 315 n. 420.

49 Trachtenberg, *Building-in-Time*, 261–83.

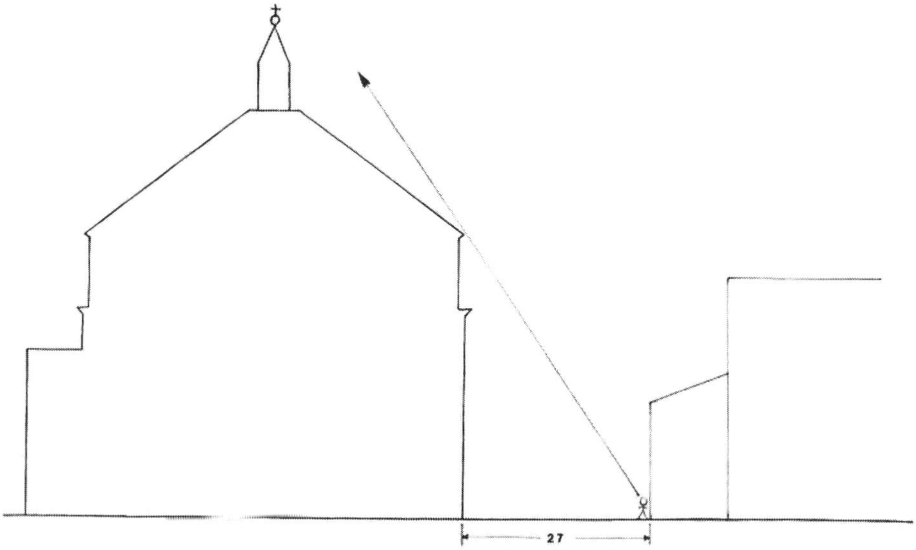

Fig. 21 Cross-sectional analysis, view of Baptistery from S. Reparata facade before 1296.

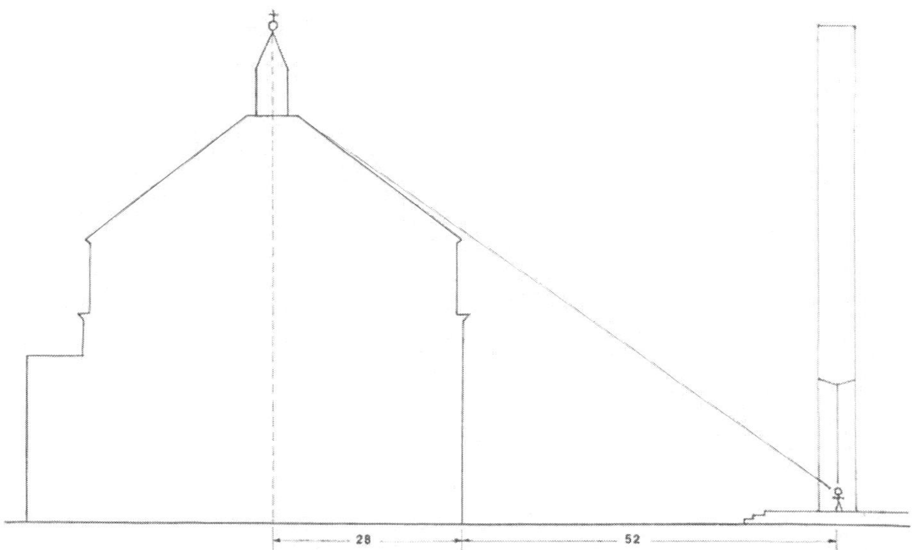

Fig. 22 Cross-sectional analysis, post 1296 view of Baptistery from Duomo.

leading member of the commission of artists who in the 1360s devised the executed plan for the duomo. In at least one of his pictorial stage sets he seems to have practiced a pictorial mode of Vitruvian optical refinements, while paying acute attention to the linearity of sight/light-rays and the phenomenon of occlusion. What is remarkable in this example is that the artist gave his pictorial buildings 'Vitruvian corrections' as if they were real structures.

Fig. 23 Baptistery from center portal of Duomo.

On the main wall of the Baroncelli Chapel one can observe the anamorphic inflection of architectural details in the painting to the observer, ideally standing on the chapel floor in front of the fresco (Figure 24). The same 'Temple' structure appears twice on this wall, below in the *Presentation of the Virgin in the Temple,* and again above in the *Expulsion of Joachim* (Figure 25). In the upper scene, however, not only is the temple two and a half times as big as its lower depiction, but the impost blocks and other vertical forms (including windows and buttresses) of the depicted structure are elongated with respect to the corresponding details in the lower panel, 'as if' stretched in anamorphic compensation for foreshortening. In fact, the entire upper building is stretched, in particular the piers. Even more remarkably, the clerestory windows of the Temple visible in the lower scene almost disappear behind the roof and buttresses, as if occluded by the steeper theoretical angle of view, as would have occurred with real buildings in real three-dimensional architectural space.[50]

No less than in the orchestration of overlapping heads and bodies observed in Giotto's painted scenes, in Gaddi's 'Vitruvian' architecture the simulation of the conditions of viewing real buildings from different vantage points is dependent on both the conjured linearity of light rays within the illusory spatial layers of the fresco *and* on the rays between the object seen in the painting and the viewer's eye in real space, all of which are folded together in a multi-modal illusionistic perceptual knot. No less than in Giotto's Berlin panel, Gaddi's fresco seems to illustrate the axiomatic optical precept of ray-linearity. As in the paintings of Giotto, the beholder of Gaddi's fresco cannot help but become aware of the linear rays of light running between her eye and the details of the painting. The attendant effects of faux-occlusion are perhaps *more* marked for Gaddi's viewer than with Giotto's overlapping bodies because the former *invenzioni* are so ingeniously contrived (hence, perhaps, the 'Vitruvian' identity of Gaddi).

What makes Gaddi's simulation of the conditions of visibility of real monumental buildings yet more telling is that in real buildings of the time such Vitruvian corrections occasionally echo the artist — and vice versa. In such monuments, although the linear-ray-occlusion device of the Baroncelli scenography sometimes plays a role, more common is the *actual enlargement of forms* placed high on the building, in particular the vertical elongation of key details to compensate for foreshortening in the eye of the beholder standing down in the piazza — in other words, corresponding exactly to the primal Vitruvian scene of visual mis-information advocated in the passages in *De architectura* cited earlier, a desire realised with inventiveness and clarity by the architects (and painter-architects) of the trecento.[51]

One of the most striking examples of this anamorphic mode of Vitruvianism is seen at the Palazzo Vecchio (1299–1315), whose unprecedented height and articulate form may have sensitised its builders to the perception theory of the ancient writer (Figure 26). They seem to have taken to heart his observation that 'the look of a building when seen close at hand is one thing, on a height it is another' and his advice that 'when the height is great [...] there must always be a corresponding increase in the proportions of the members [...]'.[52] Such adjustments are evident in two prominent sets of corbels that encircle the building. Those sustaining the huge ballatoio, or galleried-battlements, that crown the main, rusticated block are of a 'normative' shape, often seen on military and para-military buildings of the period. The corbels sited far above, however, sustaining the watch-box atop the closed tower shaft, are vertically stretched, drawn out to elongated, pyramidal forms. This double set of details provides an uncanny parallel in 'real' forms to the simulation of compensatory anamorphosis in the Taddeo Gaddi 'temples' painted a few years later.

Other features of the Palazzo are implicated in the Vitruvian dynamic of 'increase in the proportions of the members' far above the eye. The detailing of the main palace block is everywhere fine-grained and articulate, including the bifora windows and the

50 Trachtenberg, *Dominion*, 180 ff.

51 Trachtenberg, *Dominion*, 194–213.

52 Vitruvius, *Ten Books on Architecture*, 94.

Fig. 24 Taddeo Gaddi, fresco cycle of the *Life of the Virgin*, Baroncelli Chapel, Santa Croce, 1328–38.

Fig. 25 Baroncelli Chapel, *Expulsion of Joachim and Presentation of the Virgin* (detail of Figure 24).

classicizing stringcourses separating the stories. But the articulated columns of the ranks of bifora windows become gargantuan shafts with rough-leafed capitals in the belfry. In the superstructure all components are thus enlarged and simplified, appearing to belong to a different building in a detail photograph.

Similar observations can be made about another gigantic Florentine monument, the Campanile, whose cliff-like walls present a wealth of elaborate detailing ascending to a staggering height (144 *braccia*, almost ninety meters). Difference in scale is evident in all elements of lower and upper levels, including the size of windows, the relief of cornice work and moldings, and especially the tricolour marble inlay, which is extremely fine-grained in the lower stories but coarsely-sized in the upper levels. These distinctions are marked when one's attention is alerted to them, yet in 'normal' viewing they present a smooth, 'natural' effect. Either way, every element is clearly seen by the naked eye.

Like the optical theorists, the builders of these monumental works understood that an object needs to occupy a certain angle of view within the optical system of the eye to register 'forcefully'. Like Vitruvius, they were of the belief that distance of detail from the observer's eye demands a 'corresponding increase in the proportions of the members', and they applied these rules to the entire architectural colossus under construction. Such treatment, and its Vitruvian origins, implied a contingent set of aesthetic rules that valued clarity of perception over uniformity of detail.

The Eye of the Piazza (The Piazza as Eye)

In studying altarpieces, frescoes, sculptural ensembles, and monumental buildings, we have observed how Vitruvian and Euclidean theory figured in inflecting the work toward the viewer's eye, which was present as a dynamic, vital factor in the composition. In most painting and sculpture, however, the actual space in which this ideal viewer was situated was an area within a preexisting church or other monumental building. Although examples come to mind in which architectural space and the painting on its walls and vaults were coordinated — the Arena Chapel, notably — in general the actual spatial environment of the pictorial or sculptural image was not revised.

This was not the case with monumental architecture. The viewer was not left adrift on her own devices in the preexisting sea of streets into which a new or enlarged building was inserted, a condition that would have weakened the impact of the new structure. Whereas ecclesiastical and other interior spaces were colonised by the painted/sculpted image in a virtual reformation of visual dynamics, the space adjacent to a new architectural monument was usually opened up physically (through demolitions) and reshaped to provide not merely 'breathing space' for the building but a specific, framed viewpoint on it.

In the main Florentine sites the production of this viewpoint was not a haphazard process. Rather it was guided toward the instantiation of a scenographic model that was theoretically 'correct' according to the Vitruvian-Euclidean package of optical principles that we have examined. This design procedure was well-articulated in trecento Florence, and I have elsewhere given it a book-length analysis (*Dominion of the Eye*). Here I will recast several key points and introduce new observations that pertain to the theme of Vitruvian and Euclidean/medieval optical science.

My principal argument in this regard is that the configuration of the viewing space — the piazza — was not merely guided by optical-Vitruvian principles and models of surveying theory. Nor was the piazza merely 'like' a theoretically imagined eye. The relationship was

Fig. 26 Palazzo Vecchio, superstructure.

not instrumental or metaphorical but one of virtual *identity*: The *piazza itself was conceived as an 'eye' in a quasi-literal, active sense,* which was made possible by the fact that the human eye itself was construed not 'organically' but as a geometric, Euclidean abstraction, as noted. Thereby the piazza was imagined as and came to constitute a theoretically correct 'eye', complete with a pyramid of vision whose angular sweep comprised forty-five degrees above eye level and approached forty-five degrees to either side of a central visual axis that was focused on the primary features of the monument (Figures 10 and 11).[53] The former, vertical angle was the critical parameter given the extremely high buildings that were being inserted in the tight grid of streets inherited from Roman antiquity (especially the cupola, campanile, and Palazzo Vecchio). Within this matrix, optically correct perception of the entire vertical dimension of the structure was accommodated by carving out appropriately shaped and dimensioned viewing spaces, sometimes of great scale, as at the Signoria.

Yet without the viewer, the 'piazza-as-eye' was merely a potential visual mechanism. It was not yet a functioning organ of sight. It required the presence of the spectator at the spatio-visual fulcrum of the piazza for its latent perceptual capability to be activated, which brought the vast tectonic 'eye' to life as both conceptual and living mechanisms of sight folded into a single spatio-visual experience. When the living eye of the spectator was brought by his or her body to the spatio-visual fulcrum of the piazza — its main entrance — that eye fused with the architectonic eye of the piazza, activating its architectural scenography, bringing it literally to life, and into the life of the spectator. Thereby the 'eye of the piazza' served as a crucial mediation between individual and society as reified by the monument in view. It was as well a final link in the interconnectivity of optical science, artistic practice, and their human lifeworld.

The Problem of the Campanile Proportions and the 'Eye of the Piazza'

My reading of the trecento Florentine square as a 'Euclidean piazza', a scenographic apparatus that produced the primary view of a major monument in conformity with the norms of Vitruvian aesthetics, medieval-optical theory, and the representational habits of contemporary surveying theory, leads to a fuller understanding of a number of sites that otherwise might seem formless and even inscrutable. The Euclidean 'Eye of the Piazza' seems to inhabit every important view on a major monument in the center of the city, each structure having been provided with at least one 'correctly' framed forty-five degree-vertical, optically-centered view from a major entrance to the piazza created around it (Figure 27). Over the course of the mid and late trecento, this scenographic spectacle was realised thrice along the south side of the cathedral, once for the cupola and twice for the Campanile (close-up and distant views). All are closely studied in *Dominion of the Eye*.[54] However, one prominent singularity embedded in these views has remained a mystery. It can now be resolved by pursuing certain implications of the concept of the 'Eye of the Piazza'.

The mystery concerns a small incongruity of the Campanile. We have noted the Vitruvian exaggeration of detailing in its upper parts. Yet the basic volumetric units of the tower are also enlarged, as the stories increase in height from bottom to top, from the small zones bearing relief sculpture to the ever taller stories and larger windows that culminate in the gigantic trifora zone at the summit (Figure 28). Appearing in the midst of this Vitruvian dynamic is the doubling of the middle stories, that is, the two units that each present two bifora windows per side. These two stories are identical except for subtle

53 Trachtenberg, *Dominion*, 241–43.

54 Trachtenberg, *Dominion*, 72–85.

simplifications of detailing in the upper bifora unit, together with one more prominent variation: the upper bifora section is noticeably taller than the lower unit, being twenty-six *braccia* in height versus twenty-two *braccia* below. This distinction was probably intended as Vitruvian compensation for foreshortening characteristic of the period, not unlike the Palazzo Vecchio corbels, as well as being part of the general expansion of the stories of the Campanile itself from bottom to top. The question is whether a plausible logic can be found in the particular numbers embedded in the bifora stories, twenty-six and twenty-two.

Reduction to prime numbers yields only a meaningless pair, 13:11, not one of the symbolic Pythagorean (or Boethean) sets. But perhaps the numbers can be explained in the context of the overall dimensions of the tower itself. These were developed from the cathedral group as a whole, whose major dimensions manifest a dynamic coherence through four centuries of planning. The domed, octagonal baptistery provided the root dimension. The twenty-four *braccia* external span of each octagonal facet presented a double-duodecimal unit that not only determined the height of the building, seventy-two *braccia* (twenty-four times three), but also the principal dimensions of the duomo. Its nave is seventy-two *braccia* in width and height, 144 *braccia* in length, numbers which were then rotated in the cupola, which is seventy-two *braccia* in diameter and 144 *braccia* in internal height. The Campanile was part of this matrix. In plan twenty-four *braccia* square, with a 144 *braccia* total height, it is subdivided into three forty-eight *braccia* units: the closed lower stories bearing relief and full-size sculpture, the trifora unit at the top, and the intermediate zone of the 'twin'-bifora stories.

The traditional interpretive approach to pre-Renaissance architecture, which is to see its architects working by rough approximation, might offer an explanation of the varying bifora story dimensions such as follows. Given a total of forty-eight *braccia* for the twin-bifora zone, there would have occurred a trial-and-error juggling of numbers that subdivided it into two stories, with the upper made somewhat larger than the lower unit to satisfy Vitruvian criteria of perception, using whole numbers for the dimensions. Thereby, one could have made the upper unit twenty-five *braccia*, leaving twenty-three for the lower: too small a difference. Conversely, the upper unit could have been twenty-seven *braccia*, leaving twenty-one for the lower zone: too great a difference. Like the story of 'Goldilocks and the Three Bears', twenty-six and twenty-two *braccia* would have seemed just right, in any case the least problematic solution.

Although this scenario may have been a factor contributing to the design, another planning logic may be imagined that provided theoretical certainty and legitimacy in place of guesswork and approximation, situating the optically curated visibility of this prime monument within the planning procedures of the time. In this scenario, the dimensional distinction in question would not have been determined mathematically by juggling numbers until one set seemed (more or less) to work, but instead reasoned *geometrically* and *optically* according to Euclidean-Vitruvian logic. Thus, the calculation would *not have concerned linear values but rather the visual angles* produced by the two bifora dimensions, that is, the angle that they would subtend in the eye of the observer standing at a predetermined viewpoint at a theoretically correct distance from the tower. In conformance to the period logic observed in the scenographic planning of the Piazza della Signoria and other Florentine sites, this distance would not have been determined on the basis of the linear dimensions of the bifora stories, but rather have been equivalent

STATISTICS OF PERSPECTIVAL STRUCTURES OF CENTER CITY PIAZZAS (IN BRACCIA)

PIAZZA	BUILDING	VIEWPOINT	BLDG HT.	DIST-ANCE	VERT. ANGLE OF VIEW	HORIZ. ANGLE OF VIEW
BAPTISTERY	BAPTISTERY	Volta dei Pecori	72	81	42°	c. 90°
		Canto alla Paglia	72	77	43°	c. 90°
		Via dei Martelli	72	72	45°	(open)
		Corso degli Adamari	72	77	43°	c. 90°
		Duomo portal	72	80	40° **	90° /open
	CAMPANILE	Volta dei Pecori	139	130	46.5°	(open)
		Via dei Martelli	139	130	46.5°	(open)
	DUOMO FACADE	Baptistery corner/line	72	72+	45°	45°
	DUOMO FACADE*	East Baptistery portal	c.50	52	45° -	(90°)
VIA S. REPARATA	DUOMO WALL*	Opposite side of the street	36	36	45°	(open)
	DUOMO TRIBUNE*		27	27	45°	(open)
OPERA	CUPOLA	Corner Via Buia & Via Proconsolo	170***	170	45°	84° (78°)
CAPITOLO	CAMPANILE	East Piazza entrance	c.140	139	45°	(tight)
CAMPANILE	CAMPANILE*	Corso degli Adamari	50	50	45°	90°
SIGNORIA	P. VECCHIO	Via delle Farine	160	160	45°	(tight)
		Via dei Calzaiuoli	160	160	45°	90°
SANT' APOLLINARE	BARGELLO	Via della Condotta	52	54	44°	65° (90°)
	BARGELLO TOWER		94	100	43°	------

* selective/partial ideal view of portals, windows, iconic zones, etc.
** calculated to eyepoint of viewer at portal
*** estimated height of original project, including lantern

Fig. 27 Statistical table of the perspectival structures of monumental views in Florence.

to the total height of the Campanile, making the tower occupy forty-five degrees in the eye of the ideal viewer, corresponding to the dictates of the 'Eye of the Piazza'.

With this possibility in mind, I asked my research assistant at the time of this analysis, Theresa Flanigan, an architect by training, to draw up a simple rendering: a measured elevation of the tower, with the indication of sightlines to the Bifora stories from a point 144

Fig. 28 View of Campanile.

braccia distant from the 144 *braccia* high building — i.e., the theoretically correct viewpoint for the Campanile's 'Eye of the Piazza'. I was stunned by the outcome (Figure 29). Both stories subtended 7.5 degrees of the visual field. They each *occupied the same angle in the eye of the ideal viewer* and 'theoretically' would have thus 'appeared' to be of the same height, standing before the viewer as 'true' architectural twins according to Vitruvian aesthetic precepts and Euclidean/optical doctrine. The scheme evokes Propositions 10 and 11 of the *Optics*, reconfigured as a Vitruvian solution (Figure 5). Moreover, in this instance the comprehensiveness and ingenuity of the planners was such that the entire double-Bifora zone simultaneously occupied fifteen degrees of the forty-five degree visual arc of tower (measured at the 'Eye of the Piazza') and comprised forty-eight *braccia* (or one-third) of the 144 *braccia* high tower. In other words, both linear and angular measure were brought to converge in this middle section of the tower (although they do not in the lower and upper parts of the building).[55]

Rendering the evidence of the Campanile bifora dimensions even more telling is the fact that although these dimensions appear to have been determined by the spatio-visual logic of the 'Eye of the Piazza', the piazza in which the Bifora scenography was to be realised *did not yet exist at the moment of planning this part of the tower.* Nor was it even laid out. The demolition of the area in which the principal view of the Campanile was created, south of the duomo, did not begin until the 1350s, well *after* the construction of the Bifora zones in the mid and late 1340s. Indeed, this now partly-demolished viewing space (Figure 30) — its southern wall brutally realigned in the nineteenth century — was not completed until many decades later (1418, see Figure 31).[56] Evidently, so firmly did the geometric/proportional model of the scenographic 'Eye of the Piazza' exist by the 1340s, that the dimensions of the 'twin' Campanile bifora stories could be precisely proportioned 'toward' an *imaginary* such 'Eye' of the future, which the builders believed would eventually be accommodated in the 'correct' position.

The 1340s was also the moment when the ideal plan of the Piazza della Signoria probably was being formulated, which determined the siting of its actual 'Eye' (with no more than the usual degree of adjustment for site contingencies in execution).[57] The architect of the Campanile stories was Francesco Talenti (*c.* 1300–70), also author of the duomo nave, and he may have been the architect of the Signoria plan as well, which would assign a degree of named authorship to the emergence of full-blown scenographic planning in the city. Although the process that installed the precepts of Euclid, Vitruvius, Alhacen, Bacon, and Pecham in the major public spaces of trecento Florence was collective and transgenerational, it was individual architects like Talenti who produced the shapes and ran the numbers 'correctly', even though their names are often lost.

---

55 From a 144 *braccia* distance, this convergence does not obtain for either the lower zone (built by Giotto and Pisano), whose forty-eight *braccia* occupy more than fifteen degrees, or the top level of the trifora which occupies less than fifteen degrees.

56 Trachtenberg, *Dominion*, 77–80. Brunelleschi may have been involved in the final stages of shaping the campanile viewing space.

57 Trachtenberg, *Dominion*, 114–24.

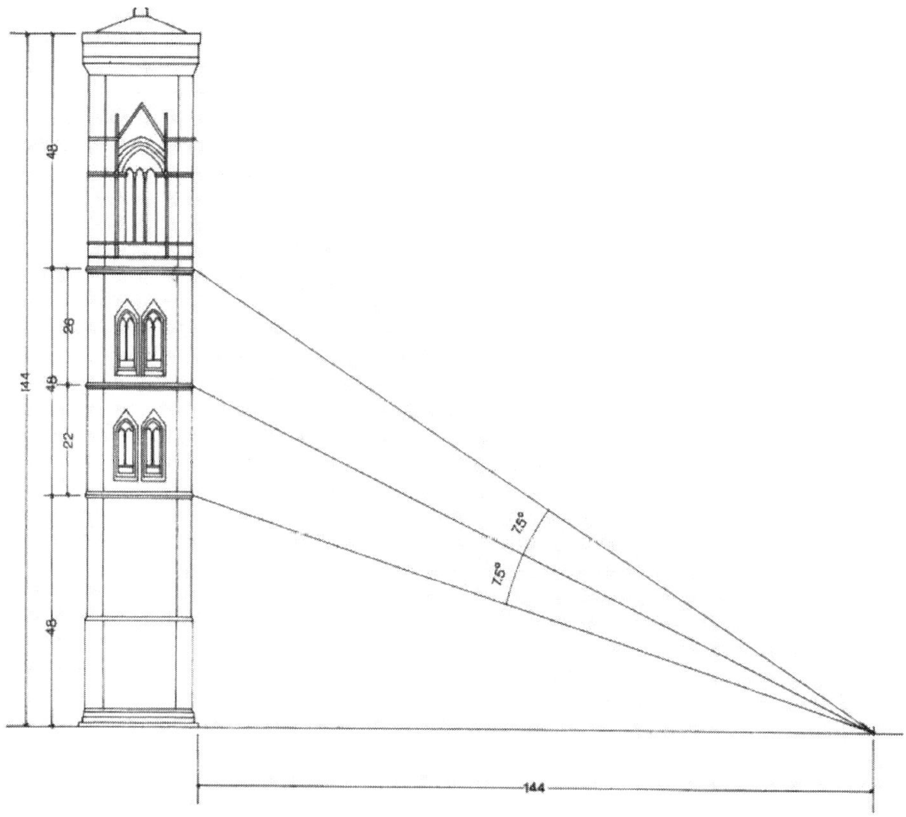

Fig. 29 Campanile, angular structure of view of whole at 144 *braccia* distance.

Conclusion

Brunelleschi's Perspective Panels: The 'Eye of the Piazza' Reified

Having traced the installation of the Eye of the Piazza at the Baptistery, Campanile, and Signoria, and studied its vivification in the living eye of the observer, I would briefly address the possible implications of this reading for the origin story of linear perspective. Although the literature is vast regarding Brunelleschi's lost demonstration panels — known to us only through their description by his biographer, Antonio Manetti — it is limited in methodological range, tending to be either technical calculations and conjectures, or speculation regarding ideas such as the Renaissance invention of modern visuality and universal proto-Cartesian space.[58] Such investigations are exemplified by the technical analyses of Richard Krautheimer (1956) and Martin Kemp (1990) on the one hand, and

58 Antonio Manetti, *The Life of Brunelleschi by Antonio di Tuccio Manetti*, ed. by Howard Saalman, trans. by Catherine Enggass (University Park: Penn State Press, 1970).

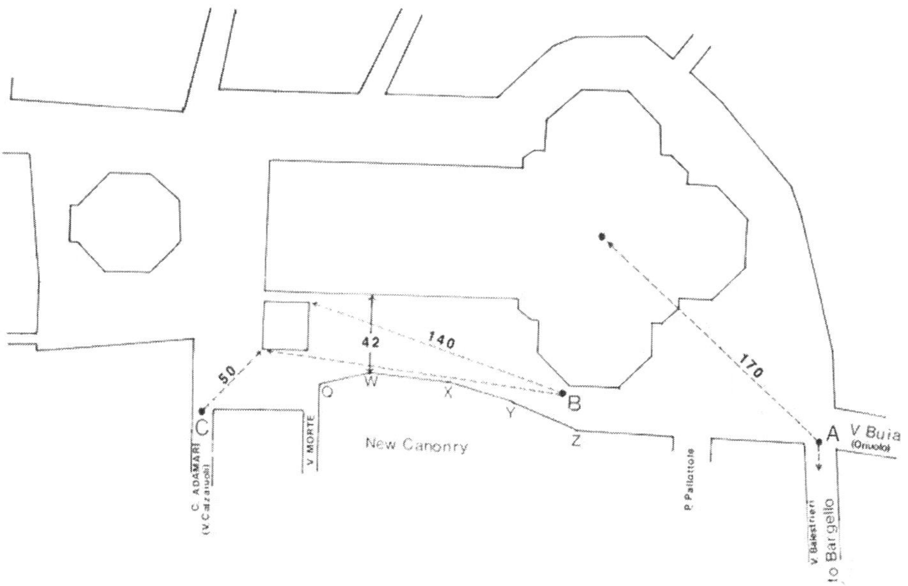

Fig. 30 Plan of zone to south of Duomo as planned and built in the trecento, with Campanile view at B.

on the other by Hubert Damish's Lacanian *The Origins of Perspective* (1994) and Erwin Panofsky's famous 'Perspective as Symbolic Form' (1927).[59]

Irrespective of method, however, such studies are teleological in orientation: Brunelleschi inevitably looks *forward* to Alberti and quattrocento pictorial practice; he looks out but he does not look back. The architect is not brought to engage the trecento and earlier Florentine past, even though it constituted his pictorial subject. Nor does he address optical theory in a meaningful way. At the same time, this critical discourse is highly formalist: the panels are treated as spatio-temporal abstraction and their subject — buildings and spaces alike — as empty form.

Thus, the question of Brunelleschi's choice of subject is not sufficiently addressed by this modernist critical school. Such problems as the fact that for his panels he did not select church interiors such as S.S. Apostoli and San Miniato, which he is often alleged to have studied as models for his own basilicas, or any of the other grand interiors of the Florentine past, not to mention his own projects, are not raised.[60] Nor are his actual selections interrogated, even though they present awkward panoramas for the supposed demonstration of proto-Albertian perspective, especially the asymmetrical, obliquely oriented Piazza della Signoria. Yet it is difficult to imagine that Brunelleschi's choice of subject was random. What, then, might have been its logic?

---

59 Richard Krautheimer and Trude Krautheimer-Hess, *Lorenzo Ghiberti* (Princeton: Princeton University Press), Chapter 16; Kemp, *Science of Art*, 11–21 and 344–45.

60 Cf. Marvin Trachtenberg, 'To Build Proportions in Time, or Tie Knots in Space? A Reassessment of the Renaissance Turn in Architectural Proportions', *Architectural Histories*, 2 (1) (2014) 1–8.

Fig. 31 View of Campanile in early nineteenth-century lithograph, prior to revision of the space.

The answer lies first in the very subjects of the panels, which often are misconstrued. That the buildings at their centers were key icons of Florentine civic identity is not the only point here. In both cases, Manetti's description takes pains to clarify that what was depicted was not just the central monument (as is often alleged) but the sweeping architectural spectacle of the monument in its setting, the *piazza* as a whole. Thus at the Signoria, Brunelleschi 'made a perspective of the piazza of the Palazzo dei Signori in Florence together with all that is in front of it and around it that is encompassed by the eye when one stands outside the piazza'. Similarly, in the 'Baptistery' panel, 'he painted that part of the piazza encompassed by the eye, that is to say, from the side facing the Misericordia up to the arch and corner of the sheep [market], and from the side with the column of the miracle of St Zenobius up to the corner of the straw [market], and all that is seen in that area for some distance'. Moreover, Manetti locates the viewpoint of the perspective. To paint the Baptistery-square panel, Brunelleschi 'stationed himself some three *braccia* inside the central portal of Santa Maria de Fiore'.[61] Although he writes with less exactitude about the Signoria panel, the viewing position clearly is the Via dei Calzaiuoli entrance, as is inferred in the passage. Thus, it was not the historical buildings in themselves that Brunelleschi depicted but the two instances of trecento architectural scenography, each represented at its visual fulcrum: the panel-images would thus have roughly corresponded to our figures 23 and 32. In effect, Brunelleschi transforms an already rationalised scenography constructed in real space and time by his forebears into an optically 'correct' pictorial *representation* of that scenography.

Brunelleschi's biographer provides the key insight into the nature of this representation. Manetti describes these scenographic vistas neither in angular terms nor as presented to a person standing at the viewpoint. His rendition is neither abstract nor fully embodied. Rather, Manetti depicts the visual reception of the trecento scenography much as John Pecham or Roger Bacon might have, that is, *explicitly in terms of the eye as optical organ.* The sentences in question sound not unlike 'propositions' inscripted in the optical treatises. Thus, the Baptistery panel displays '[...] that part of the piazza encompassed by the *eye*'. The Signoria panel replicates 'all that is in front of it and around it that is encompassed by the *eye*'.[62] It is not the observer that Manetti points to, or even the 'eyes' of the observer, in which the rest of the sentient body would be implicated, but the detached, clinical, theoretically reified eye-concept of the opticians. Yet paradoxically, at the same time it is the *living eye* of the viewer that provides the final agency in the scenographic production of the piazza, as it vivifies the *constructed* 'Eye of the Piazza', with the entire transaction now being reified by Brunelleschi's ingenious panels.

Brunelleschi's panels did not replicate the two main Florentine instantiations of the Eye of the Piazza as free-floating representations, however. They were instead site-specific in their intended *viewing*. Their singularities of shape and facture, as described by Manetti, implied that they were meant to be viewed not just anywhere, as imagined, but ideally at the spot where they were painted (or, more likely, drawn for later studio completion) — the actual locus of the Eye of the Piazza in question. This singularity did not involve their pictorial

61 Manetti, *The Life of Brunelleschi*, 52–56.
62 Manetti, *The Life of Brunelleschi*, 52–56.

Fig. 32 Reconstructed view of Piazza della Signoria in c. 1400 in principal perspective at Via dei Calzaiuoli entrance to the square.

representation of the scenography as such, presumably rendered in a tempera-on-wood technique like most panel paintings of the period. Rather the evidence resides in peculiar technical details of facture, including the shape and size of the two panels and the singular provisions for their viewing invented by the architect.

The representation of the Baptistery casts this process in clear relief. As Manetti informs us, the small panel was meant to be held in one's outstretched hand with the painted surface facing away (Figure 33). In the other hand, one held out a mirror further away but facing *back* toward the painting and the observer 'behind' it. To activate viewing, one brought one's eye — manifestly the *single* optical-theoretical'eye' — to the back of the panel, and looked out through a small conical opening sited at the level of the painting corresponding to the optically 'correct' viewing level. One now saw the Baptistery scene on the opposite side of the panel as reflected in the outstretched mirror, which one moved back and forth until the perspectivally correct distance-point was attained (an important criterion for Manetti). Moreover, one did not see a painted sky above the painted Baptistery, for this area was blank, substituted by mirroring, in which one glimpsed the reflection of the real sky relayed back to the eye zig-zag through its reflection in the main mirror.

Manetti neglected to mention several implicit points in his breathless description. One was that the to-be-mirrored image had necessarily been *painted in reverse*, which meant that it should *only be seen* in a mirror, which re-reversed the image back to its normal orientation (thereby behaving as optical theory described).[63] Although the Baptistery was

63 'In plane mirrors directly opposite, figures [faces] appear turned around, and right appears opposite left and vice versa'. Cf. Pecham, *Perspectiva communis*, ed. and trans. by Lindberg, Proposition II. 22.

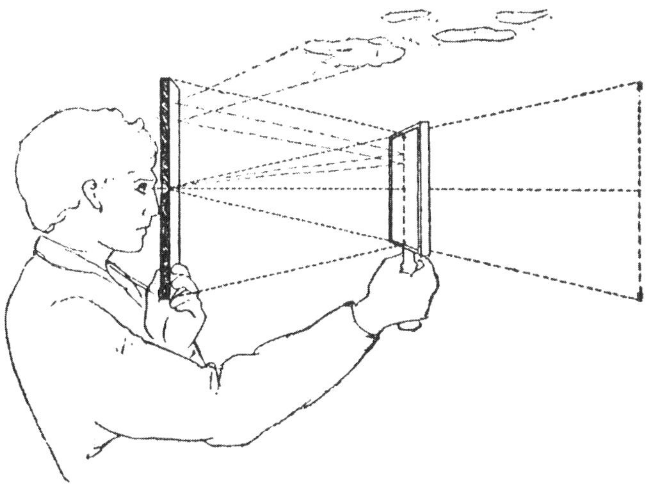

Fig. 33 Reconstruction of Brunelleschi's Baptistery panel, with indication of 'peep' hole and mirror(s).

symmetrical, the scene as a whole (with the Zenobius Column and Volta dei Pecori right and left) was not, and its reversal in direct viewing would have been contradictory to the veristic spirit of the panel. Mirrors so fascinated the optical theorists that a whole volume of Pecham, with some fifty-seven weighty propositions, is dedicated to the reflective behavior of various types of mirrors (including image-reversal). Perhaps to a degree Brunelleschi's double-mirrored optical-pictorial apparatus referred to that compendium of knowledge.

A crucial singularity of his specular machine indicates another of his likely intentions: that the reflective viewing was meant to take place at the very spot where the panel had been painted. There, just inside the duomo portal, the spectator held the apparatus up, mirror and all, and — as Manetti explains technically but not topographically — attained what seemed the perspectivally correct reflected view. Then, the description suggests, one lowered one's hand holding the mirror, and lo and behold the *real Baptistery* and its piazza slid 'up' into view substituting *precisely* the painted image seen in the mirror a few seconds previously (the experiment also works in reverse). This dazzling experiential validation of *theoretical* perspectival 'correctness' by correctness *seen* would have been, in effect, the main point of the intricate 'peep show' — not merely to render the 'first' illusion: *seeing* reality's *duplication* of the perspectival image was *believing* in the latter. As Manetti writes, 'the spectator felt he saw the actual scene when he looked at the painting'. In optical-theoretical terms, visual truth was 'certified'.[64]

The cumbersome to-and-fro construction of the apparatus and re/inversion of the mirror image are difficult to explain in any way other than as aimed towards this image-substitution/reification. The double distance from eye to painted image produced by the mirror

64 Manetti, *The Life of Brunelleschi*, 52–56.

was required by the specific perspectival construction of the panel — merely turning it around and holding it out facing the viewer would not allow sufficient distance for correct perspective (apart from the mirror reversal issue). The mirroring seems to have been the only way to confirm the distance-dependent perspectival 'correctness' that Manetti cites as the central criterion of its 'truth value'.

The panel's operations are entangled with yet another facet of the theoretical optical knowledge of the period. The back-and-forth oscillation of the Baptistery image, its turning, deflected path towards convergence in the 'eye' of the panel — which the peep-hole constituted — continuing to a second, living eye behind, evokes the complicated ray-trajectories of the image into and through the various complicated layers of the eye as construed by the medieval theorists. Only in post-Keplerian optics did vision come to be understood as a bundle of rays passing directly through a lens that projected them as an inverted image on the retina, the whole thereby constituting a simple chiasmatic double cone of light. Previously, not only was the function of the retina misconstrued and disabled, but inversion of the image by the lens of the eye (the 'anterior cristalline humor') was *excluded*. Instead there prevailed variations of a cumbersome hypothesis that conjured bent (but never inverted) rays of light following an intricate pathway through the various 'humors' of the eye: an overcomplicated, ultimately self-contradictory thesis eventually overturned by Kepler (in analogy to the Copernican reduction of ungainly Ptolemaic cosmology to elementary heliocentric logic).[65]

I suggest that the workings of the Baptistery panel might have been prompted by the pre-Keplerian model of vision as an intricate bending, refracted passage of visual rays through a complex pathway towards the production of an accurate and true, 'certified' image of the object seen. In this sense, Brunelleschi's panel figured as an uncanny token of the reception of high optical discourse in the imaginary of Florentine artisans (a mode of quirky reinterpretation that would later characterise Leonardo's optical speculations). His panel thereby would have been double-purposed: it registered the lived experience of a key work of Florentine architectural scenography, yet simultaneously it seemed to reflect, with singular inventiveness, the three-dimensional model of vision as theoretically (mis)construed by the opticians.[66]

Given this understanding of the Baptistery panel, and returning to the question of the situated perception of Brunelleschi's devices, one may surmise that on-site viewing might have been also intended for the Signoria panel, and that there the living eye of the viewer merged with the Eye of the Piazza both as represented pictorially and in 'real' monumental space: all would have been intended to telescope into a theoretically correct epiphany of

65 Lindberg's summary of Pecham's dense Propositions 31, 36, and 37 regarding the intricate path of light-rays within the eye reads: 'Vision is not completed in the glacial humor [the lens] [...] rays of the visual pyramid penetrate the various humors and tunics of the eye and reach the glacial humor without refraction. But then [...] before actually converging to a vertex at the center of the eye, they encounter the interface between the glacial humor and the vitreous humor, which is eccentric to all anterior tunics and humors. Here all the rays except the axis of the pyramid [...] are refracted away from the center of the eye and hence do not actually converge. This refraction prevents inversion of the visual impression, which would occur if the rays were to intersect and continue in inverted order', see Pecham, *Perspectiva communis*, ed. and trans. by Lindberg, 38.

66 Compare Leonardo's half-informed revisions of optical theory, as studied by Lindberg in his *Theories of Vision*, 154–68.

architectural 'reality' and 'illusion'. Although Manetti makes no explicit mention of this experiential dimension, it is implicit in his description.

He tells us that because the Signoria scene was so vast, the panel depicting it had to be much larger than the Baptistery panel. Thus it could not be held in one hand by the viewer, obviating the outstretched mirror arrangement. Instead, as he writes in the paragraph devoted to explaining this peculiarity, 'where in the San Giovanni panel he [Brunelleschi] had placed burnished silver, here he cut away the panel in the area above the buildings represented, and took it to a spot where he could observe it with the natural atmosphere above the buildings'. In other words, the top of the painting was cut in a zig-zag corresponding to the intricate roofline of the buildings in the square, presumably as seen from the ideal viewpoint (Figure 34).

The likely motivation of the zig-zag upper border would have been that the panel, like the Baptistery portrait, was meant to be viewed where it had been painted or at least composed, the Eye of the Piazza della Signoria, the visual fulcrum of the pictorial construction. Only at this point would real and pictorial skylines have been 'identical'. Standing at the piazza foyer, with both arms the viewer would have held the painting out toward the scene, adjusting its distance and height so that the pictorial roofline merged exactly with the *real* roofline, thereby certifying (to use the language of the theorists) the perspectival accuracy of the panel in a positive manner that a standard rectangular panel could not have achieved. And then, presumably, the panel (rather than a mirror, as at the Baptistery) was slowly lowered, and the real buildings and spaces rose into view.

Optics from Giotto to Brunelleschi and Alberti: A Common Ground

This excursus into the wilderness of the 'origins of perspective' has led to the realization that Brunelleschi was an engaged, albeit belated participant in the perspectivist visual culture of trecento Florence in his attentiveness to the hyper-rational, geometric spirit of the medieval perspectivists as well as their obsession with mirrors, visual pyramids, and ray-tracing the image through the multilayered 'humors' of the eye. Far from having been the first painter who ushered in Florentine perspectivist visual culture, now understood to be a symbiosis of high medieval theory and artisanal practice reaching back to Giotto, Brunelleschi was merely one of its most daring and imaginative protagonists. Indeed, the excessive optical complexity of his Baptistery panel, including its self-conscious ray-tracing, appears less surprising in light of Giotto's dazzling, virtuoso assemblages of occluded human and architectural form and Gaddi's mind-bending inversion of Vitruvian logic in his painted temples. With his two panels, Brunelleschi did not escape the past — of either the builders, the painters, or the theorists — but instead wrote another chapter into its brilliant history.

A decade later, the Brunelleschian representational turn became the springboard of Alberti's method, which in effect inverted the logic of his precursor's invention. Rather than Brunelleschi's rationalised transformation of real historical space into its pictorial simulacrum, with the 'eye' of the observer serving as the final hinge, in Alberti the observer's eye as construed in optical theory (and reified in the Baptistery panel) becomes a model instead for the construction of the virtual-illusionistic, *ideal* spatial tableau in which the *imaginary* sacred events of the *historia* are made to come to life *as if* 'rationally' distributed

Fig. 34 Reconstruction of Brunelleschi's Piazza della Signoria panel, with cut out roof silhouette.

in a perfect, *ex novo* spatial continuum in the world of the picture. In Brunelleschi's panels, a historical spectacle of the real world, full of irregularities, brought to focus at the Eye of the Piazza, is condensed on the panel and received by the eye. In Alberti, following the same optical geometries in reverse, an invented spectacle is projected out from the eye of the artist (and the viewer).

Thus, it could be said that, in effect, the optical models behind Brunelleschi's and Alberti's inventions were respectively those of intromission and extramission. Although the latter famously remarked that directionality didn't matter, it has been shown that in fact he favoured extramission theory.[67] Together, the two Florentines can be said to have imported the *controversy* regarding ray-direction into the visual arts.

Perhaps a more fundamental note on which to conclude this essay is that although Alberti may have been inspired by Brunelleschi's *invenzioni*, so different was his purpose that he was not guided by them. His synthesis of optical theory and pictorial space was original, as he claimed. Nevertheless, it was unimaginable outside the ancient-medieval, theoretical-practical discursive formation illuminated in this essay, running from Euclid, Vitruvius, and Al-Hazen to Pecham, Giotto, Talenti, Brunelleschi and beyond.

67 J. V. Field, *Piero della Francesca: A Mathematician's Art* (London: Yale University Press, 2005), p. 36.

Bibliography

Primary Sources

Alberti, Leon Battista, *Della pittura*, trans. by J. R. Spencer (New Haven: Yale University Press, 1954).

Manetti, Antonio, *The Life of Brunelleschi by Antonio di Tuccio Manetti*, ed. by Howard Saalman, trans. by Catherine Enggass (University Park: Penn State Press, 1970).

Parronchi, Alessandro, *Studi sulla dolce prospettiva* (Milan: Aldo Martello, 1964).

Pecham, John, *John Pecham and the Science of Optics: Perspectiva communis*, ed. and trans. by David C. Lindberg (Madison: University of Wisconsin Press, 1970).

Villani, Filippo, *Liber de origine civitatis Florentiae et euisdem famosis civibus*, ed. by G. C. Galleti (Florence, 1847).

Vitruvius, *Ten Books on Architecture*, ed. and trans. by Morris Hicky Morgan (New York: Dover Publications, 1960).

Secondary Works

Baxandall, Michael, *Painting and Experience in Fifteenth-Century Italy* (Oxford: Clarendon Press, 1972).

Carter, B. A. R., 'Perspective', in *The Oxford Companion to Art*, ed. by Harald Osborn (Oxford: Oxford University Press, 1970).

Conant, Kenneth J., 'The Afterlife of Vitruvius in the Middle Ages', *Journal of the Society of Architectural Historians*, 27 (1) (1968), 33–38.

Edgerton, Jr., Samuel Y., *The Renaissance Rediscovery of Linear Perspective* (New York: Harper and Row, 1975).

Elkins, James, *The Poetics of Perspective* (Ithaca: Cornell University Press, 1994).

Field, J. V., 'Alberti, the Abacus and Piero della Francesca's Proof of Perspective', *Renaissance Studies*, 2 (1997), 61–88.

Field, J. V., *The Invention of Infinity: Mathematics and Art in the Renaissance* (Oxford: Oxford University Press, 1997).

Field, J. V., *Piero della Francesca: A Mathematician's Art* (London: Yale University Press, 2005).

Frojmovič, Eva, 'Giotto's Circumspection', *Art Bulletin*, 89 (2007), 195–210.

Gibson, J. J., *The Senses Considered as Perceptual Systems* (Boston: Houghton Mifflin, 1966).

Gillam, Barbara, 'Occlusion Issues in Early Renaissance Art', *i-Perception*, 2 (9) (2011), 1076–97.

Hills, Paul, *The Light in Early Italian Painting* (New Haven: Yale University Press, 1987).

Johannsen, B. B., and M. Marcussen, 'A Critical Survey of the Theoretical and Practical Origins of Renaissance Linear Perspective', *Acta ad archaeologiam et artium historiam pertinentia* (1981), 191–227.

Kemp, Martin, *The Science of Art* (London: Yale University Press, 1990).

Krautheimer, Richard and Trude Krautheimer-Hess, *Lorenzo Ghiberti* (Princeton: Princeton University Press, 1956).

Krinsky, Carol Herselle, 'Seventy-Eight Vitruvius Manuscripts', *Journal of the Warburg and Courtauld Institutes*, 30 (1967), 36–70.

Kruft, Hanno-Walter, *History of Architectural Theory* (Princeton: Princeton Architectural Press, 1996).
Kuhn, Jehane R., 'Measured Appearances: Documentation and Design in Early Perspective Drawing', *Journal of the Warburg and Courtauld Institutes*, 53 (1990), 114–132.
Lindberg, David C., *Theories of Vision from Al-Kindi to Kepler* (Chicago: The University of Chicago Press, 1976).
Lindberg, David C., 'The Science of Optics', in *Science in the Middle Ages*, ed. by David C. Lindberg (Chicago: University of Chicago Press, 1978), pp. 338–68.
Lindberg, David C. and Katherine Tachau, 'The Science of Light, Seeing and Knowing', in *The Cambridge History of Science: Medieval Science*, ed. by David C. Lindberg and Michael H. Shank, 7 vols (Cambridge: Cambridge University Press, 2013).
Lubbock, Jules, *Storytelling in Christian Art from Giotto to Donatello* (London: Yale University Press, 2006).
Panofsky, Erwin, *Perspective as Symbolic Form*, trans. by C. S. Wood (New York: MIT Press, 1991).
Raynaud, Dominique, 'Geometrical and Arithmetical Methods in Early Medieval Practice', *Physis. Rivista in-ternazionale di storia della scienza*, 45 (2008), 29–55.
Raynaud, Dominique, 'Optique et perspective avant Alberti', in *Le printemps de la Renaissance. La sculpture et les arts à Florence*, ed. by B. Paolozzi Strozzi and M. Bormand (Paris: Musée du Louvre éditions, 2013), pp. 165–71.
Raynaud, Dominique, *Optics and the Rise of Perspective: A Study in Network Knowledge Diffusion* (Oxford: The Bardwell Press, 2013).
Romanini, A. M., 'Nuove ipotesi su Arnolfo di Cambio', *Arte medievale*, 1 (1983), 157–202.
Schild Bunim, Miriam, *Space in Medieval Painting and the Forerunners of Perspective* (New York: Columbia University Press, 1940).
Smith, A. Mark, *From Sight to Light: The Passage from Ancient to Modern Optics* (Chicago: University of Chicago Press, 2014).
Smith, Pamela, *The Body of the Artisan: Art and Experience in the Scientific Revolution* (Chicago: University of Chicago Press, 2004).
Tachau, Katherine H., *Vision and Certitude in the Age of Ockham: Optics, Epistemology, and the Foundation of Semantics, 1250–1345* (Leiden: Brill, 1988).
Trachtenberg, Marvin, *Dominion of the Eye: Urbanism, Art and Power in Early Modern Florence* (Cambridge: Cambridge University Press, 1997).
Trachtenberg, Marvin, *Building-in-Time: From Giotto to Alberti and Modern Oblivion* (London: Yale University Press, 2010).
Trachtenberg, Marvin, 'To Build Proportions in Time, or Tie Knots in Space? A Reassessment of the Renaissance Turn in Architectural Proportions', *Architectural Histories*, 2 (1) (2014) 1–8.
White, John, *The Birth and Rebirth of Pictorial Space* (New York: Thomas Yoseloff, 1958).

MARJOLIJN BOL

The Emerald and the Eye

On Sight and Light in the Artisan's Workshop and the Scholar's Study

Introduction

Through the case of the emerald, this chapter considers artisanal explorations of light and their influence on ideas about optics and vision as part of the history of perspective. Today's distinction between optics and perspective as projective geometry was not apparent in the pre- and early modern period.[1] Influential treatises on perspectivist optics, such as the Latin translation of Ibn al-Haytham's book of optics (*De aspectibus*, *c.* 1200) or those treatises that make heavy use of this work, including Roger Bacon's *Perspectiva* (*c.* 1265) Witelo's *Perspectiva* (*c.* 1275), and John Pecham's *Perspectiva communis* (*c.* 1280), typically focused on theories of light, colour and vision. Over the course of the fifteenth century however, mathematical perspective became an important tool for artisans to suggest the illusion of space on a two-dimensional surface. The history of this discovery and its subsequent development became a central subject in the history of art and science. Because of this emphasis on perspective as projective space, the artisan's creative play with and understanding of the optics of his materials has received far less attention.[2]

Yet, precisely this applied knowledge of how light interacts with the materials of art has found important resonance among those early writers theorizing the genesis and medicinal benefits of the materials of nature. To delve into this history of practical optics, this chapter therefore moves away from the above-mentioned corpus of perspectivist optics as it was constructed by for instance David Lindberg in *Theories of Vision* (1976), and, more recently, by A. Mark Smith in *From Sight to Light* (2014) and instead turns to pre- and early modern recipe collections and writings on natural history, such as encyclopedias and lapidaries.

More specifically, I will show how the special optics of the emerald stimulated not only its appreciation, but also encouraged interesting techniques for working this precious mineral and for its imitation. These emerald imitations, in turn, gave rise to theories about

1 See for instance Sven Dupré, 'The Historiography of Perspective and Reflexy-Const in Netherlandish Art', in *Art and Science in the Early Modern Netherlands* (= *Netherlands Yearbook for History of Art/ Nederlands Kunsthistorisch* Jaarboek 61), ed. by Eric Jorink and Bart Ramakers (Leiden: Koninklijke Brill, 2011), pp. 34–61.

2 Ernst H. Gombrich, *The Heritage of Apelles* (London: Cornell University Press, 1976).

Marjolijn Bol Utrecht University, m.a.h.bol@uu.nl

Perspective as Practice. Renaissance Cultures of Optics, ed. by Sven Dupré, Turnhout, 2019 (*Techne. Knowledge, Technique, and Material Culture*, 1), p. 71-101

DOI 10.1484/M.*Techne*-EB.5.117722

Fig. 1 Terminated raw emerald (var. beryl), Panjshir Valley, Afghanistan, Photo by Marjolijn Bol. [See colour plate 1]

the green gem's genesis in nature, its value and its use. Indeed, the emerald was not just considered intriguing and beautiful because of its special optics; its deep, translucent green colour had been long regarded a practical aid to the human eye. In this way, the emerald and the various practices devised for its imitation had a surprising and enduring impact on the history of furnishing both the scholar's studio and the artisan's workshop.

A Perfect and Soothing Green

The English term 'emerald' can be traced back to *smaragdos*, old Greek for 'green gem' which itself may go as far back as Sanskrit. Today, the emerald is known as the variety of beryl that receives its green colour from traces of chromium, and sometimes vanadium (Figure 1).[3] It is prized for its colour, clarity, rarity and hardness. In the premodern period, a number of green gemstones could be classified as *smaragdos*. These include today's emerald, but also less costly stones such as malachite and green turquoise. Despite the difficulties in establishing the exact mineral composition and origin of the emeralds described in historical sources, the most precious *smaragdi* have always been valued for the same qualities as our emeralds today. Highest value was placed on the rarest and hardest *smaragdi* of the most saturated lush green colour and greatest clarity. It was these precious emeralds that were the subject of a wide variety of artisanal practices attempting to enhance or imitate the gem's beautiful translucent green colour.

3 Not all gemologists recognise vanadium beryls as emeralds.

First, let us turn to the two earliest and most influential accounts of the emerald. It is here that we find the roots of the idea that the green stone is beneficial to sight. In *On Stones*, Theophrastus (*c.* 371–287 BCE), a pupil of Aristotle, provides us with the earliest surviving description of a green stone called *smaragdos*.[4] Theophrastus describes it as a small and rare stone that has the ability to imbue water with its own colour.[5] He also explains that the *smaragdos* is good for the eyes and that this is why people carry emerald seals; it helps them see better.[6] Approximately three centuries later, Pliny the Elder (23–79 CE), in book thirty-seven of his *Historia naturalis* (77–79 CE), lists no less than twelve kinds of *smaragdi*. The most precious are of an incomparable translucent green colour:

> Although we gaze eagerly at young plants and leaves, we look at *smaragdi* with all the more pleasure because, compared with them, there is nothing whatsoever that is more intensely green. [...] they remain the same in sunlight, shadow, or lamplight, always shining gently and allowing the vision to penetrate to their further extremity owing to the ease with which light passes through them, a property that pleases us in water.[7]

Pliny explains that because *smaragdi* are generally concave in shape, they have the ability to 'concentrate the vision' which is why 'a decree has forbidden them to be engraved'.[8] Echoing Theophrastus, to whom he also refers, Pliny points out that one of the most important qualities of the *smaragdus* is that it can help reduce eye strain by looking at it.[9] Moreover, 'they alone of gems, when we look at them intently, satisfy the eye without cloying it'. Indeed, even after straining our sight by looking at another object, we can restore it to its normal state by peering into a *smaragdus*: 'Engravers of gemstones find that this is the most agreeable means of refreshing their eyes, so soothing to fatigue is the mellow green colour of the stone'.[10]

---

4 Concerning the nature of the *smaragdos* Theophrastus is describing, see Theophrastus, *Theophrastus on Stones, Contributions in Physical Science*, ed. and trans. by Earle Radcliffe Caley and John F. C. Richards (Columbus, Ohio: Ohio State University, 1956), pp. 50 and p. 97 [nr. 23].

5 Theophrastus, *Theophrastus on Stones*, pp. 45–46, 50 [nrs. 4 and 23] and see also the commentary on pp. 66–67, 98.

6 Theophrastus, *Theophrastus on Stones*, p. 50 [nrs. 23–24]. Diane Morgan also pointed out that of all the stones that he describes, Theophrastus only recognised the emerald as having health benefits, see Morgan, *From Satan's Crown to the Holy Grail*, p. 55.

7 'Nam herbas quoque silentes frondesque avide spectamus, smaragdos vero tanto libentius, quoniam nihil omnino viridius comparatum illis viret. [...] non sole mutati, non umbra, non lucernis, semperque sensim radiantes et visum admittentes ad crassitudinem sui facilitate tralucida, quod etiam in aquis nos iuvat'. Pliny the Elder, *Pliny: Natural History. Books 36–37*, ed. and trans. by D. E. Eichholz, 10 vols (Cambridge and London: Harvard University Press, 1962), 10, XXXVII. 62–64.

8 'ut visum conligant'; 'quam ob rem decreto hominum iis parcitur scalpi vetitis'. Famously, the emperor Nero (AD reign 54–68) would have used such a concave *smaragdus*: 'Nero princeps gladiatorum pugnas spectabat in smaragdo' ('to watch the fight between gladiators'). See Pliny, *Natural History*, 10, XXXVII. 64–65. For a recent interpretation of this much debated passage see David Woods, 'Pliny, Nero, and the Emerald', *Acta philologica fennica*, 40 (2006), pp. 189–96.

9 Jean Trinquier, 'Les vertus du vert dans l'antiquité', in *Couleurs et vision dans l'Antiquité classique*, ed. by Laurence Villard (Rouen: L'université de Rouen, 2002), pp. 97–128. For the possible origin of the idea that *smaragdi* are beneficial to the eye, see also J. W. Meadows, 'Pliny on the Smaragdus', *The Classical Review*, 59 (1945), pp. 50–51.

10 'Scalpentibusque gemmas non alia gratior oculorum refectio est: ita viridi lenitate lassitudinem mulcent'. Pliny, *Natural History*, 10, XXXVII. 62–64.

These ideas that engravers of gemstones can relieve and restore their wearied eyes by looking at a *smaragdus* and that the emerald has a certain ability to magnify things, deeply influenced later writings about the precious green gem. The aforementioned physical characteristics of the emerald and its ophthalmic benefits are repeated, in more or less detail, from Isidore of Seville's (*c.* 560–636) *Etymologiae*, an important source for medieval encyclopedias, until far into the sixteenth century. Even the renowned first 'reformer' of the lapidary tradition, the German scholar Georgius Agricola (1494–1555), refers to Pliny as his authority when he writes about the emerald's ophthalmic benefits. In his mineralogical treatise *De natura fossilium* (1546) Agricola points out that emeralds *oculos implent* ('sooth eyes') and 'restore keen sight to eyes that see indistinctly as a result of strain'.[11] Used this way, the green colour of the emerald is most soothing to the eyes of the *scalpentibus gemmas* ('gem engraver').[12] Just like Pliny, Agricola explains that an emerald 'transmits images as if it were glass' and that 'the concave ones are best for focusing an image'.[13] Later on, I hope to show how these long-lasting ideas about the emerald's ophthalmic benefits not only become apparent in the lapidary tradition, but also inspired an intriguing material culture of sight. In order to make this point, I first have to turn to the practices devised to polish and imitate emeralds. The first, an ancient jeweller's technique, forms the basis of all others; the oiling of emeralds so as to enhance their colour and clarity.

Oiling Emeralds

When emeralds are found, their surface is rough and irregular, and as a result, the visible light is scattered in all directions. The admixture of this so-called white light with the emerald's green colour results in desaturation, and, as not all light is transmitted, decreased clarity (Figure 1). This fact is pointed out by Agricola when he writes that 'gems can occur in many ways, but are never as brilliant and transparent in the natural state as when polished'.[14] Through the act of polishing, jewellers create a smooth surface on the emerald that allows light to penetrate it without scattering.[15] As a result, only the pure wavelengths of the gem's beautiful green colour are reflected back to our eyes, while all the other colours are transmitted. This way, polishing brings out both the emerald's saturated

11 'quin & ab intentione alia obscurata, aspectu smaragdi di recreatur acies'. Georgius Agricola, *De natura fossilium (= Textbook of Mineralogy)*, ed. and trans. by Mark Chance Bandy and Jean. A. Bandy, The Geological Society of America, Special Paper 63 (New York: The Geological Society of America, 1955), p. 126. It should be said that, writing about the ophthalmic benefits of the emerald, Agricola adds one critical note when he points out that it soothes the eyes but *nec fatiant* ('does not cure them'). For the Latin text see: Georgius Agricola, *Georgii Agricolae de ortu & causis subterraneorum lib. V: De natura eorum quae effluunt ex terra lib. IIII: De natura fossilium lib. X: De veteribus & novis metallis lib. II: Bermannus, sive, De re metallica dialogus: Interpretatio germanica uocum rei metallicae: addito Indice foecundissimo* (Basel: H. Frobenium and N. Episcopium, 1546), p. 289.

12 Agricola, *Textbook of Mineralogy*, p. 126; Agricola, *De natura fossilium lib. X*, p. 289.

13 'autem tanquam specula imagines rerum reddunt'; 'sed concavi insuper visum colligunt'. Agricola, *Textbook of Mineralogy*, p. 126; Agricola, *De natura fossilium lib. X*, p. 289.

14 'gemmae quo tandem modo inventae fuerint, antequae poliuntur, minus pellucunt nitentque'. Agricola, *Textbook of Mineralogy*, p. 114, Agricola, *De natura fossilium lib. X*, p. 273.

15 For the late medieval and early modern history of polishing, see Marjolijn Bol, 'Polito et Claro. The Art and Knowledge of Polishing, 1200–1500', in *Gems in the Early Modern World*, ed. by Sven Dupré and Michael Bycroft (Cham: Palgrave Macmillan, 2019), pp. 223–57.

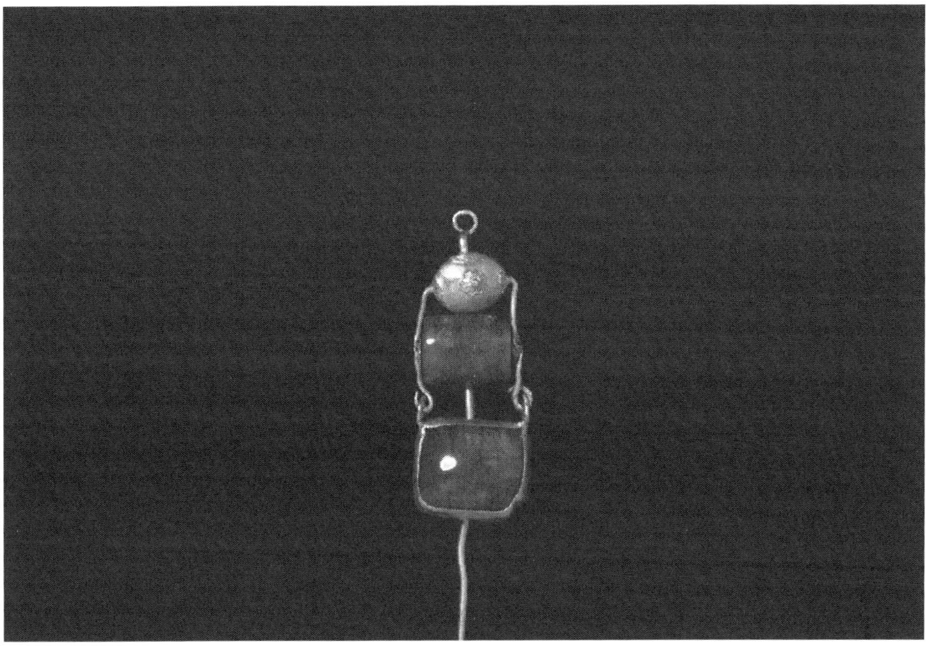

Fig. 2 'Pin or hair ornament, gold hung with emeralds', London, Victoria and Albert Museum, 1st century - 4th century (made), cat. nr. 8802-1863. © Victoria and Albert Museum, London. [See colour plate 2]

green colour and its clarity (Figure 2). But even with a perfectly polished surface, only minerals without inclusions, cracks or fissures have the ability to distribute light in such a manner that it results in the deeply saturated, glowing colour we admire so much in the most precious of gemstones. In this last respect, emeralds are not a perfect gem by any means. As they are naturally marked with cracks and fissures, emeralds need to be polished outside and inside to bring out their translucent green glow in the best possible manner. For this reason, jewellers treat emeralds with liquids that penetrate their crevices through the phenomenon of capillary action. When these liquids share a similar refractive index with that of the emerald, light can pass through uninterruptedly and the stone's colour and clarity improve. This method, nowadays known as fracture filling and still practiced on a large scale, was first documented in antiquity, and, significantly, would be discussed in increasing detail by almost every single lapidary in the ensuing centuries. Again, Pliny is the first writer who mentions the practice when he writes that even though emeralds appear naturally of green colour, their colour improves by the application of oil.[16] A few centuries later, Isidore of Seville provides us with additional details about the nature of the oil used in the practice of emerald oiling. He writes in his etymology of the emerald that '*smaragdi* improve from being treated with undiluted wine and green oil'.[17] In order to

16 Pliny, *Natural History*, 10, XXXVII. 71–72.

17 'Smaragdi autem mero et viridi proficiunt oleo, quamvis natura inbuantur'. Isidore of Seville, *The Etymologies of Isidore of Seville*, ed. by Stephen A. Barney, W. J. Lewis, J. A. Beach and O. Berghof (Cambridge: Cambridge University Press, 2006), XVI. vii. 2. For the Latin edition, see *Isidori Hispalensis episcopi Etymologiarum*

Fig. 3 Left: showing so-called early harvest oil ('Verde Esmeralda'). Right: regular cold-pressed olive oil. Photo by Marjolijn Bol. [See colour plate 3]

understand the nature of this 'green oil', we have to turn to Isidore's etymology of *oleum* ('olive oil') in which he explains that 'what is pressed from tawny, immature olives is called green oil'.[18] Interestingly, such so-called early harvest olive oils yield a greener type of oil compared to the golden yellow of olive oil pressed from later harvests (Figure 3). It is thus not hard to imagine why this particular olive oil would have been considered best suited for the polishing of a green precious stone.[19]

Outside the lapidary tradition, the practice of oiling emeralds also informed and inspired theological thought. In the earliest commentary on the *Book of Revelation* (variously dated to the second half of the sixth and the beginning of the seventh century[20]), the Greek bishop Andreas of Caesarea, writes about the symbolic meanings of the twelve stones that adorn the gates of the Heavenly Jerusalem. He connects the foundation stones of the holy city to the twelve apostles and explains that the emerald is connected to St John. The reason,

sive originum libri XX, ed. by W. M. Lindsay, *Scriptorum classicorum bibliotheca oxoniensis*, 2 vols (Oxonii: Clarendoniano, 1911), 2, XVI. vii. 2.

18 'Quod autem ex fulvis et nondum maturis fuerit expressum, viride appellatur'. Isidore of Seville, *Etymologies*, XVII. vii. 68 and Isidore of Seville, *Etymologiarum*, 2, XVII. vii. 68.

19 Today's retailers of early harvest olive oils describe its colour as 'emerald green' and one brand even named its olive oil after the emerald 'verde esmeralda' because of its associated history with the polishing of emeralds and packs it in a jewel box-type case, see: http://www.verdesmeraldaolive.com/tienda/en/history/ [accessed 05-11-2018].

20 For the later date, see *Commentary on the Apocalypse*, ed. and trans. by J. N. Suggit, The Fathers of the Church series, 127 vols (Washington DC: Catholic University of America Press, 2006), 112, p. 3.

so Andreas of Caesarea points out, is that St John's religious practice can be compared to that of the jeweller's practice of oiling emeralds:

> [...] it [the emerald] is nourished with oil, [so] that its translucency and beauty may not change; we conceive this stone to signify John the Evangelist. He indeed, soothed the souls dejected by sin with a divine oil [...].[21]

These accounts of the oiling of emeralds reveal how the jeweller's applied knowledge of the behaviour of light in this stone seeped into the lapidary tradition and theological thought. In the first case to explain how this practice could improve the stone's optical properties, and in the second to explain how, by analogy, the human soul could be similarly nourished with oil to treat its impurities. From the thirteenth century onwards, starting with Thomas of Cantimpré (1201–72) and Bartholomeus Anglicus (*c.* 1203–72), scholars start combining these ideas from the ancient lapidary tradition with more contemporary theological thought. Such later descriptions, although derivative, are of particular interest when they are translated into the vernacular. The need to find translations and equivalents for Latin technical terms and processes can often help understand aspects of craft practice that remain ambiguous in their Latin descriptions. More than mere iterations of the past, studying such vernacular translations and interpretations can also provide insight into the level of detail to which later scholars understood the crafts they were describing, and in what instances they were adding new knowledge. In the case of emerald oiling, a good example of this can be found in Konrad von Megenberg's (1309–74) *Das Buch der Natur* (*c.* 1349). Written in Middle High German, Von Megenberg's encyclopedia is a translation and adaptation of Cantimpré's popular *Liber de natura rerum* (1228–44). In the chapter *von dem smaragden*, Von Megenberg writes that the emerald 'strengthens the sight and clears the eyes'.[22] He also points out that St John is associated with the emerald because the gem symbolises *käusch* ('chastity').[23] Megenberg posits that the green colour of emeralds intensifies when they are washed and anointed with *paumöl*, the Middle High German term for olive oil.[24] The previous point reveals that Megenberg indeed identifies the *oleum* mentioned in the Latin sources with olive oil, but his translation lacks Isidore of Seville's specificity concerning the exact type of olive oil to be used.[25] The writings of the Liège author Jean d'Outremeuse (1338–1400) are another important source for our knowledge of

21 See Georg Frederick Kunz, *The Curious Lore of Precious Stones: Being a Description of Their Sentiments and Folk Lore, Superstitions, Symbolism, Mysticism, Use in Medicine, Protection, Prevention, Religion, and Divination: Crystal Gazing, Birthstones, Lucky Stones and Talismans, Astral, Zodiacal, and Planetary* (Philadelphia: Halcyon House, 1938), p. 312. For the entire passage in Greek with a Latin translation see: *Patrologiae cursus completus: seu bibliotheca universalis [...]*, ed. and trans. by Jacques-Paul Migne, 383 vols (Paris: Apud J. P. Migne, 1863), 106, pp. 435–43.

22 'sterkt daz gesiht und klaert diu augen'. Conrad von Megenberg, *Das Buch der Natur: Die erste Naturgeschichte in deutscher Sprache. Durch Conrad von Megenberg*, ed. by Franz Pfeiffer (Stuttgart: Verlag von Karl Aue, 1861), p. 459. There is no English edition, the translations are my own.

23 Von Megenberg, *Das Buch der Natur*, p. 459.

24 Von Megenberg, *Das Buch der Natur*, p. 459: 'Und wenn man in wescht und in salbt mit paumöl, sô erhoeht sich sein grüene'. It is the same as *Baumöl* ('tree oil') still used in the German language to describe olive oil.

25 Megenberg's most important source, Thomas of Cantimpré, mentions that the emerald improves from wine and oil (*vel vino vel oleo*), see *Thomas Cantimpratentsis liber de natura rerum: editio princeps secundum codices manuscriptos*, ed. by Helmut Boese, 2 vols (Berlin: Walter de Gruyter, 1973), I, p. 368. Another earlier source, Marbode of Rennes (1035–1123) mentions *olivo* instead of *oleo* for the oiling of emeralds in *De lapidibus* (*c.* 1090).

the history of emerald oiling. Not only does he reveal a deep understanding of the practice, Jean d'Outremeuse is also the first author to clearly state the role of wine in the process. In his *Trésorier de philosophie naturelle des pierres précieuses* (1382–90), d'Outremeuse explains that it is necessary to bathe the emerald in wine and rub it with a linen cloth so as to clean it. This way it is completely clean when it is oiled with *d'oile d'olive* to recover its clarity.[26] As a method to bring out the colour and clarity of emeralds, writers on stones thus consider the practice of treating them with naturally green oil in a positive light. Indeed, the green oils mentioned in all cases appear to refer to olive oil, but never to oils that have received their colour artificially with, for instance, dyes or pigments. It is therefore significant that when the same sources discuss the closely related practice of imitating emeralds by oiling transparent crystals with a green substance, it is condemned in all cases.

Verdigris, Emeralds and Imitation

Of all gemstones that historical sources discuss as having been the subject of imitation, emeralds are mentioned most often. In fact, the earliest surviving reference to the practice of gemstone imitation is about making an imitation emerald. The passage in question can be found in the *Epistles* of Seneca the Younger (1–65 CE). Seneca writes that the pre-Socratic philosopher Democritus (460–370 BCE), mainly known as the founder of ancient atomism, discovered how a *calculus* ('pebble') could be transformed into *zmaragdum* by boiling it.[27] A method, so Seneca points out, that was still practiced in his own time on coloured stones which are suitable for such treatment.[28] Pliny also frequently mentions emeralds when he discusses the practice of making and exposing gemstone imitations. He writes that the Indians have found ways to stain rock-crystals to make them look like more expensive jewels, and adds, in another passage, that he knows of treatises that describe how to imitate emeralds and other gems:

> There are treatises by authorities, whom I at least shall not deign to mention by name, describing how by means of dyestuffs emeralds and other transparent coloured gems

---

See Marbode of Rennes, *Marbode of Rennes' (1035–1123) De lapidibus Considered as a Medical Treatise with Text, Commentary and C. W. King's Translation Together with Text and Translation of Marbode's Minor Works on Stones)*, ed. and trans. by John M. Riddle and C. W. King (Wiesbaden: Franz Steiner, 1977), p. 45.

26 See Anne-Françoise Cannella, *Gemmes, verre coloré, fausses pierres précieuses au Moyen-Âge: Le quatrième livre du 'Trésorier de Philosophie naturelle des pierres précieuses' de Jean d'Outremeuse*, Bibliothèque de la Faculté de Philosophie et Lettres de l'Université de Liège (Geneva: Droz, 2006), Fascicule 288, p. 186: 'Se ceste pierre pert sa clarete si la baingnies en vin et la frotes bien de drap de lin et puis l'oindes d'oile d'olive et recouverra sa clarte'.

27 Ancient alchemists attributed four books on dyeing to Democritus, including one on the making of precious stones. As these attributions are unreliable, the author of these books has become known as pseudo-Democritus. Whereas we have fragments of some of these books, the book on precious stones unfortunately is lost, see *The Four Books of Pseudo-Democritus*, ed. and transl. by Matteo Martelli (New York: Routledge 2013), pp. 23–26. A later Byzantine recipe collection (thirteenth and fifteenth century), *The Deep Tincture of Stones, Emeralds, Rubies and Jacinths from the Book taken out from the Sancta Sanctorum of Temples* may bear some relation to the ancient treatise of Pseudo-Democritus. This text was edited by Marcelin Berthelot and Charles-Émile Ruelle, *Collection des anciens alchimistes grecs*. 3 vols (Paris: Georges Steinheil) 2, pp. 350–64.

28 Seneca the Younger, *Seneca. Epistles 1–65*, ed. and trans. by Richard M. Gummere, Loeb Classical Library 75, 3 vols (Cambridge, MA: Harvard University Press, 1917), 1, Epistle XC. 33.

> are made from rock-crystal [...] And there is no other trickery that is practised against society with greater profit.[29]

Isidore of Seville also stresses how convincing such fake *smaragdi* could be, and when he laments about the abundance of these practices of deception, Isidore, like Pliny, seems to suggest that the production of fake gems may indeed have been quite common:

> As a substitute for that most precious stone, the *smaragdus*, some people dye glass with skill, and its false greenness deceives the eyes with a certain subtlety, to the point that there is no one who may test it and demonstrate that it is false. It is the same with other matters in one way or another, for there is no mortal life free from deception.[30]

Considering this, it is perhaps not surprising that most lapidaries do not discuss specific details about the processes involved in making emerald imitations. Pliny does not even consider it appropriate to mention the names of the authorities on the matter.

It has been suggested that an echo of the contents of the treatises mentioned by Pliny and a few other ancient authors may have been preserved in the so-called *Stockholm Papyrus*.[31] This document, written in Greek, was possibly copied as a funerary gift around 200–300 CE, but is believed to have had a much older core.[32] The papyrus contains more than seventy recipes for the imitation of precious stones using less expensive materials, including ruby, beryl, amethyst and sunstone. It also includes more than twenty recipes for imitating emeralds, again making it the gem imitated most often. The recipes are various but always involve the 'corroding' or 'opening up' of a transparent mineral so that it becomes receptive to the colour that is to be added in the next step. In the case of the emerald, the colouring substance is always the pigment verdigris, usually suspended or dissolved in oil or resin.[33]

---

29 'Quin immo etiam exstant commentarii auctorum — quos non equidem demonstrabo — quibus modis ex crystallo smaragdum tinguant aliasque tralucentes, [...] neque enim est ulla fraus vitae lucrosior'. Pliny, *Natural History*, 10, XXXVII. 79–78 and 197–98.

30 'Nam et pro lapide pretiosissimo smaragdo quidam vitrum arte inficiunt, et fallit oculos subdole quadam falsa viriditas, quoadusque non est qui probet simulatum et arguat; sic et alia alio atque alio modo. Neque est sine fraude ulla vita mortalium'. Isidore of Seville, *Etymologies*, XVI. xv. 27 and Isidore of Seville, *Etymologiarum*, 2, XVI. xv. 27.

31 A related manuscript is kept in Leiden (*Leyden Papyrus X*). The most extensive critical edition is: *Les alchimistes grecs: Papyrus de Leyde – Papyrus de Stockholm – Recettes*, ed. and trans. by Robert Halleux, Collection des universités de *France* (Paris: Belles Lettres, 1981), I. For an English translation of the two papyri (based on Otto Lagercrantz's German translation), see Earle Radcliffe Caley, 'The Leyden Papyrus X. An English Translation with Brief Notes', *Journal of Chemical Education*, 3 (1926), pp. 1149–66 and Earle Radcliffe Caley, 'The Stockholm Papyrus: An English Translation with Brief Notes', *Journal of Chemical Education*, 4 (1927), pp. 979–1002.

32 Berthelot has suggested that the papyri were preserved in the mummy-case of an Egyptian chymist and Otto Lagercrantz similarly argued that the papyri were a luxury copy (internal evidence shows the manuscripts are copied from another source) made for the purpose of entombment. This would also explain why they have been preserved after Diocletan's 296 CE decree banning all treatises dealing alchemy. See Marcellin Berthelot, *Introduction à l'étude de la chimie des anciens et du Moyen-Âge* (Paris: G. Steinheil, 1889), p. 5 and *Papyrus Graecus Holmiensis (P. Holm.): Recepte für Silber, Steine und Purpur*, ed. and trans. by Otto Lagercrantz, (Uppsala: Arbeten Utgifna Med Understöd Af Vilhelm Ekmans Universitetsfond, 1913), 13, p. 55.

33 As mentioned before, I investigated the ideas and practices of counterfeiting precious stones in a previous publication. Here, I do not discuss the practice of quench cracking, but present another (related) method in which transparent minerals were given a translucent coating of drying oils ground with copper green (supported by practical reconstructions of the processes involved). See Bol, 'Coloring Topaz, Crystal and Moonstone',

The *Stockholm Papyrus* includes a separate instruction to make this colour. In a recipe entitled 'the preparation of verdigris for emerald', the green pigment is made by suspending clean, copper plates covered with oil above strong vinegar (Figure 4):

> Now if you put it in in the morning, then scrape off the verdigris carefully in the evening, but if you put it in in the evening, then scrape it off in the morning.[34]

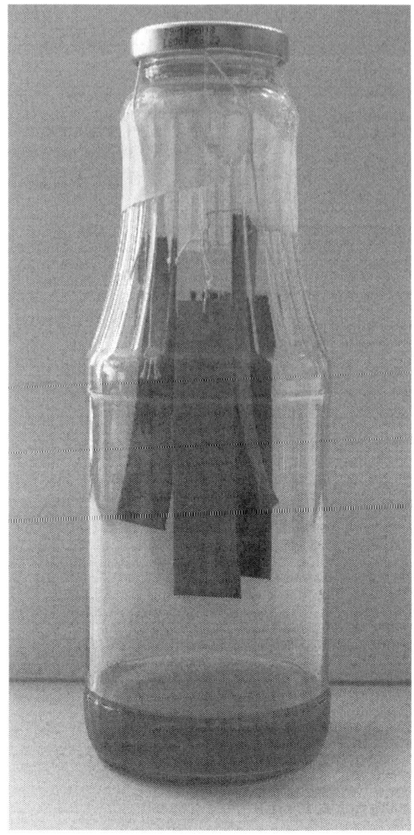

Fig. 4 Making 'verdigris' by suspending copper plates over vinegar. [See colour plate 4]

Verdigris, or the acetate of copper as it is known today, has special optical properties, which makes it particularly suitable for the imitation of a translucent green stone. Ground with oil or resin, this pigment not only has the potential to approach the deep green hues of the emerald, but also has the ability to match its translucency. Verdigris is the only relatively stable green pigment that has a refractive index closely matching that of oil, resin and crystal. This means that when verdigris is ground with oil or resin and then used to coat the crystal or to fill its fractures, light is allowed to pass through, while reflecting back a green hue similar to that of the emerald (compare Figures 1, 2 and 5.2-3). In a recipe for the 'preparation of emerald', the papyrus explains how a mixture of verdigris, celandine (a yellow dye also used to colour textiles) and 'scythian black', have to be pulverised and mixed with liquid resin:[35]

> Preparation of Emerald.
> Take pure pyrites or rock crystal and make the composition in the following way: Verdigris, 2 drachmas; celandine, 1 drachma; Scythian black, 3 drachmas; liquid resin, which one holds in the mouth, as much as necessary. Pulverize the dry materials, mix the

pp. 108–29.

34 Caley, 'The Stockholm Papyrus', [nr. 74] and Halleux, *Les alchimistes grecs*, [nr. 74]. The production of verdigris was also ancient, and similar methods are described by both Theophrastus and Pliny. The pigment is relatively easy to make and the reconstructions presented here have all been made with verdigris produced by the author.

35 If Lagercrantz's translation of the Greek text is correct, this may be mastic resin, which is known to have been used since antiquity as a type of 'chewing gum' for its beneficial impact on the teeth, see for instance *Pliny: Natural History. Books 24–27*, ed. and trans. by W. H. S. Jones and A. C. Andrews, 10 vols. (Cambridge and London: Harvard University Press, 1956) 8, XIV. 121–23. Halleux, however, points out that the passage should probably be understood as resin flowing directly from a cut in the tree: 'résine liquide, celle de l'ouverture'. See Halleux, *Les alchimistes grecs*, [nr. 76] and p. 196 n. 5.

> resin with them, and set it aside. Take liquid alum, pour water upon it so that it becomes very watery and preserve it in a clay vessel. Heat the stone in an earthen vessel and cool it off in the alum. Heat the stone and put it in the above-named composition. However, if you desire that it should be greener then again mix pulverized verdigris with it.[36]

The second mixture, made of 'liquid alum' with water, is used to quench the heated 'pyrite' or 'rock crystal'. The result is a quick change in temperature that would have cracked the stone without breaking it, causing internal fractures reaching up to its surface (Figure 6). According to the recipe, the stone should then be heated again and submerged in the first verdigris-resin substance. It is not entirely clear whether this green liquid is heated with the stones, or whether they are again quenched in it, but when we consider that the stones have to take up the resin, which is quite viscous when cold, it seems likely that it would have been heated.[37] Through the phenomenon of capillary action the green liquid can now penetrate the fractures created when the crystal was quenched. The reader may have noticed that this procedure to imitate emeralds is similar to that of oil polishing the stone, and it is indeed quite possible that these practices derived from one another. It is therefore significant that the papyrus also contains a recipe that describes a method for oiling emeralds. In fact, it is the only recipe I have been able to find that records this practice within the recipe tradition. The recipe in question explains how to 'polish' an emerald by putting it first in wax and garlic, after which it is cooked in an unspecified oil that is coloured green by the addition of the juice of leek:[38]

> Stick hard emerald into wax for 14 days. After this period, grate 'garlic' and make a cake out of it. Take the stone out (of the wax) and stick it into the cake of 'garlic' for 7 days. Take leek and extract the juice out of it. Mix with the leek juice an equal amount of oil, put this in a new pot, put the stones in it at the same time and boil 3 days (or) until they become satisfactory to you. The stones should be in a basket so that they do not touch the bottom of the pot.[39]

The use of artificially coloured oils for the enhancement of emeralds is nowadays considered a fraudulent practice, and, indeed, as I showed earlier, not mentioned in medieval lapidaries.[40] It is therefore not unthinkable that, because of its use of coloured oils, this

---

36 Caley, 'The Stockholm Papyrus', [nr. 76], and Halleux, *Les alchimistes grecs*, [nr. 76].

37 Robert C. Kammerling, John I. Kocvula, Robert E. Kane, Patricia Maddison, James Shighley, and Emmanuel Fritsch, 'Fracture Filling of Emeralds: Opticon and Traditional Oils', *Gems & Gemology*, 27 (2) (1991), pp. 70–85. The authors present a quench-cracked crystal filled with a synthetic resin that was dyed green to show how effective the resin could be (not dyed of course) for filling the fractures of emeralds.

38 In biblical scripture the emerald is often referred to as the 'leek-green' stone (*prasinus*), see for instance Robert Grosseteste's (*c.* 1175–1253) commentary of the biblical account of creation, Robert Grosseteste, *Robert Grosseteste on the Six Days of Creation: A Translation of the Hexaëmeron*, ed. and trans. by Christopher F. J. Martin, Auctores Britannici Medii Aevi (Oxford: Oxford University Press, 1996), p. 330. Robert Grosseteste also writes that 'Those who carve stones make no stone more graceful to the eyes than this. [the emerald] If its body is extended, it gives back images like a looking-glass'. (See p. 330).

39 Caley, 'The Stockholm Papyrus', [nr. 37] and Halleux, *Les alchimistes grecs*, [nr. 37].

40 Kurt Nassau, *Gemstone Enhancement. Heat, Irradiation, Impregnation, Dyeing, and Other Treatments Which Alter the Appearance of Gemstones, and the Detection of Such Treatments* (London: Butterworths, 1984), pp. 65–71, 101–02.

Fig. 5 From left to right: 1. Reconstruction of a copper green glaze ground with cold-pressed linseed oil on gold leaf; 2. small piece of emerald depicted in figure 1; 3. Three quench cracked crystals that have been "dyed" with copper green dissolved in larch balsam. Photo by Marjolijn Bol. [See colour plate 5]

particular recipe reflects a deceitful aspect of emerald oiling.[41] The possible deceit is made more apparent, because the recipe is presented alongside those recipes that, by means of a similar method, instruct how to transform transparent minerals into having the appearance of precious green emeralds.[42]

But the above methods of imitating emeralds were not always used to commit fraud. They were also adopted within various craft practices to beautify art objects, not to falsify them. Best-known perhaps, are the painters who have a long history of using verdigris ground with drying oils on their palette. Besides the many paint recipes that single out verdigris for the use with oil, often describing its colour as 'emerald green', scientific examination has revealed that panel painters used verdigris ground with drying oils and sometimes resins as early as the twelfth century for the purpose of evoking the saturated translucent green colour of emeralds.[43] Using a technique nowadays known as glazing, painters applied thin translucent layers of oil paint over polished gold, silver or tin (Figure 5.1). Similar to how light travels though crystals coloured green with oil and verdigris, the light is here transmitted through the green oil glazes. When light bounces back from the reflective

41 That this is the case is also suggested by Robert Grosseteste when he writes that 'put into oil or wine turns them green though they have their natural colour', see Grosseteste, *Robert Grosseteste*, p. 330.

42 Jean d'Otremeuse for instance, provides the reader with many recipes for the imitation of gems in book four of his treatise, but mentions the practice of emerald oiling in book two of his treatise discussing the properties of gems, see Cannella, *Gemmes*, p. 186.

43 In my dissertation I argue that the desire to create a material imitation of precious stones was an important factor in the adoption of drying oils in the workshop of the early medieval panel painter, see Marjolijn Bol, 'Oil and the Translucent. Varnishing and Glazing in Practice, Recipes and Historiography, 1100–1600' (doctoral dissertation, Utrecht University, 2012).

Fig. 6 Four quench-cracked crystals before colouring. [See colour plate 6]

metal leaf underneath, the green translucent paint layer is illuminated from within, creating a perfect material substitute of the optical effect and colour generated in gems set in the work of the goldsmith. Another important craft that used copper or verdigris to evoke the appearance of emeralds was that of glassmaking. When small particles of copper are added to quartz when it is fired in the glassmaker's kiln, they produce a green colour in the glass.[44] For the first written evidence of the use of copper as a colouring agent in glass, we have to turn to two early medieval recipe collections, the *Mappae clavicula* and *De coloribus et artibus Romanorum*. As was the case with the *Stockholm Papyrus*, the core of these recipes may have been composed several centuries (possibly the fourth century CE) before the earliest manuscripts that have come down to us (all dated between the eighth and twelfth century CE).[45] The manuscripts of the so-called *Mappae clavicula* family contain various recipes that describe how copper can be used to give glass a green colour:

44 Sandra Davison and R. G. Newton, *Conservation and Restoration of Glass* (Oxford: Butterworth Heinemann, 2008), pp. 76–77 and see also Robert Halleux and Anne-Françoise Cannella, 'Entre technologie et alchemie: De la teinture du verre à la fabrication des fauses pierres précieuses', in *Il colore nel medioevo: Arte, simbolo, tecnica: atti delle iornate di studi, Lucca, 2–3–4 maggio 1996* (Lucca: Istituto Storico Lucchese e Scuola Normale Superiore di Pisa, 1998), pp. 41–58.

45 Robert Halleux and Paul Meyvaert, 'Les origines de la mappae clavicula', *Archives d'histoire doctrinale et littéraire du Moyen-Âge*, 54 (1987), pp. 7–58. See also Francesca Tolaini, "De tinctio omnium musivorum". Technical Recipes for Glass in the So-Called "Mappae clavicula"', in *When Glass Matters*, ed. by Marco Beretta (Florence: Olschki, 2004), pp. 195–219. For an annotated edition, see Cyril Stanley Smith and John G. Hawthorne, '"Mappae clavicula": A Little Key to the World of Medieval Techniques', *Transactions of the American Philosophical Society*, 64 (1974), pp. 1–128 and for the Latin text, see Thomas Phillipps, 'Letter from Sir Thomas Phillipps, Bart. F. R. S., F. S. A., Addressed to Albert Way, Esq., Director, Communicating a Transcript of a Ms. Treatise on the Preparation of Pigments, and on Various Processes of the Decorative Arts Practiced During the Middle Ages, Written in the Twelfth Century, and Entitled "Mappae clavicula"', in *Archeologia: Or, Miscellaneous Tracts Relating to Antiquity* (London: Society of Antiquaries of London, 1770), pp. 183–244. Theophrastus may have been the

> Making a green color in glass.
> Grind glass well, and put 3 oz. of clean copper filings to a pound of glass, and cook it for 3 days.[46]

De coloribus et artibus Romanorum, a treatise attributed to a certain Eraclius, contains similar instructions that explain how one can make and colour glass green by adding copper filings. Eraclius delineates many uses of glass, from tableware to mirrors, and he adds that 'glass is often tinged in various colours to imitate gems, which many persons make of it'.[47] Naturally, such glass gems, similar to those produced by the methods described in the *Stockholm Papyrus*, could also be used to fool innocent buyers into thinking they are acquiring a real precious stone. But glassmaking of course did not become a popular craft to deceive *per se*, but precisely because it enabled medieval artisans to imitate the aesthetic of gems in church windows, reliquaries, enamels and so on.[48]

From oiling emeralds to imitating them on a base of transparent crystal, or substituting their optics on paintings and in glass; copper and verdigris were always used in one form or another. This close link between copper, verdigris and emeralds resulted in natural philosophers' explanations of the genesis of natural emeralds. The great medieval philosopher and theologian Albertus Magnus (*c.* 1200–80) for instance, writes in his *De mineralibus* that emeralds:

> occur in veins of copper which gives the gems their translucency because it has not yet actually become copper — for the rust of copper is green.[49]

The translucent emerald is thus thought to grow from verdigris, or, as Albertus Magnus describes the pigment, from the rust of copper. Approximately a century earlier, in the treatise on stones of the Persian Muslim scholar and polymath Al-Beruni (973–1048), we encounter a similar, yet more critical stance on the idea that emeralds and copper are

---

first to allude to this function of copper in the art of glassmaking when he writes: 'The most unusual earth is the one mixed with copper; for in addition to melting and mixing, it also has the remarkable power of improving the beauty of the colour', see: Theophrastus, *Theophrastus on Stones*, [nrs. 48 and 49]. This passage is difficult to translate and interpret. As Caley argues in his commentary, Theophrastus may have been speaking of copper as a colouring agent in glass but also of the manufacture of brass or a copper-arsenic alloy or a refining process in which bronze or copper was melted in the presence of an earthy substance (the latter, Caley considers most likely), pp. 162–67.

46 'Tinctio vitri prassina. Tere vitrum bene, et de limaturis eris mundi z. iij. mitte in libram vitri, et decoques per dies iij'. *Mappae clavicula*, [nr. 154], compare to [nrs. 155, 224 and 258] and for the Latin text of the recipe cited here see Phillipps, 'Letter from Sir Thomas Phillipps', p. 218 [nr. cliiij].

47 *Original Treatises Dating from the Xiith to Xviiith Centuries on the Arts of Painting*, ed. and trans. by Mary Philadelphia Merrifield, 2 vols (London: John Murray, 1849), I, pp. 208–12 [V, nr. 255].

48 Just one example of many, the *Schedula diversarum artium*, a recipe collection dating to the twelfth century, includes a recipe for the setting of gems without lead in painted glass. Theophilus explains that one uses *croceo vitro claro* ('clear saffron glass') to imitate the cross-nimbus of Christ, the Books, or the borders of draperies, all which in a painting are made of gold or orpiment. When you are done, you must decide on the places where you want to put 'the stones'; take some clear blue glass to make *iacinctos* ('jacinths') and green for *smaragdos* ('emeralds'). Theophilus, *The Various Arts: 'De diversis artibus', Nelson's Medieval Texts* (London: Thomas Nelson, 1961), pp. 57–58.

49 'rationale etiam est quod in venis aeris in perspicuo quod ad aeris substantiam non venit, nascitur: quia rubiginis aeris habet viriditatem quod in venis aeris'. Albertus Magnus, *Book of Minerals*, ed. by Dorothy Wyckoff (Oxford: Oxford University Press: Clarendon Press, 1967), p. 119. For the Latin text see: Albertus Magnus, *De mineralibus (Mineralium libri quinque)*, ed. by A. Borgnet, Opera omnia Alberti Magni, 38 vols (1890–99), V, pp. 45–46.

related substances. Al-Beruni explains that the Persian physician Rhazes (854–925 CE) considers emeralds and verdigris to be related because the colour of the emerald is like that of verdigris. Al-Beruni, however, adds the critical note that one can only accept this if the emerald is found in copper, not gold mines.[50] And, indeed, only gems that were considered the lesser species of *smaragdi*, malachite in particular, are found in copper mines.[51]

It appears that the theory that emeralds and copper were related substances was not based on knowledge of the geological origin of the gem. More likely, it became common belief because of the many artisanal practices that used verdigris to imitate the colour and clarity of emeralds. Yet, the idea that emeralds were found in copper mines was persistent and can be found in many lapidaries' accounts in subsequent centuries. Almost four hundred years later, the Italian metallurgist Vannoccio Biringuccio (*c.* 1480–*c.* 1539), while questioning various other theories about the emerald, including those about where it can be found, is still certain about its origin in copper:

> [...] it must be believed that all these strange things, wherever they may come from, are colored by the virtue and power of minerals of copper.[52]

The fact that the copper green pigment verdigris and the emerald came to be understood in close relation to one another, both optically and physically, may have resulted in the role the pigment came to play in the material culture of sight and light that I alluded to at the beginning of this chapter. Although natural philosophers did not necessarily believe that man-made imitations possessed the same 'magical' qualities as their natural counterparts, such a crossover appears to have happened in the material culture of the eye that I am about to discuss.[53]

---

50 Al-Beruni, *Al-Beruni's Book on Mineralogy. The Book Most Comprehensive in Knowledge on Precious Stones*, ed. and trans. by Hakim Mohammad Said, One Hundred Great Books of Islamic Civilization (Islamabad: Pakistan Hijra Council, 1989), 66, p. 141.

51 Based on similar ancient sources, Arab writings on stones appear to have developed slightly different ideas about the origin of emeralds and malachite. Compare for instance the entries on 'emerald' and 'malachite' in an Arab lapidary ascribed to Aristotle. Here emeralds are said to be found in gold mines, whereas malachite is found in copper mines where, similar to Albertus Magnus, it is said to grow from the green tarnish on copper, being verdigris, see: Pseudo-Aristotle, *Das Steinbuch des Pseudo-Aristotle nach der Arabische Handschrift*, ed. and trans. by Julius Ruska (Heidelberg: Carl Winter's Universitätsbuchhandlung, 1912), pp. 134, 146.

52 '[...] ma véghino la dove si voglino habian da credere che tutti Tebaidi sien tenti per virtu & potētia dele miniere del rame'. Vannoccio Biringuccio, *The Pirotechnia of Vannnoccio Biringuccio*, ed. and trans. by Cyril Stanley Smith and Martha Teach Gnudi (Cambridge: MIT Press, 1959; orig. publ. 1942), p. 12; Vannoccio Biringuccio, *De la pirotechnia libri X ... composti peril Vanoccio Biringuccio Sennese* (Venice: Venturino Roffinello, 1540), p. 41.

53 Compare for instance a passage from Albertus Magnus, *Commentarii in II. III, lib. sententiarum*, 15:86–87 in which he points out that alchemy does not give the things it produces the same properties as those things produced by nature: 'an alchemical sapphire does not cool off sexual ardor [...] nor does an alchemical carbuncle dispel a vaporous poison'. For the English translation, and an interesting analysis of this passage, see: William R. Newman, *Promethean Ambitions: Alchemy and the Quest to Perfect Nature* (Chicago: University of Chicago Press, 2004), pp. 48–49. See also: Robert Halleux, 'Albert le Grand et l'alchimie', *Revue des sciences philosophiques et théologiques*, 66 (1982), pp. 57–80.

Verdigris, Oil and the Eye

The roots of this peculiar tradition can be found with Galen (129–*c.* 200 CE / *c.* 216 CE). In *On the Usefulness of the Parts of the Body* Galen points out how the eyes of painters tire easily 'especially when they are working with white parchment'.[54] Galen goes on to say that 'to provide for this they place nearby gray or dark-colored objects, to which they keep looking away, resting their eyes'.[55] In his etymology of parchment, Isidore expresses a similar concern when he points out how the invention of white parchment in Rome negatively affected the reader's eyesight. In contrast to Galen's observation of the use of dark or grey objects to alleviate tired eyes, Isidore continues his account with the suggestion that the problem can be remedied with a variety of green materials; from marble and cloth to the backs of scarab beetles, all are of particular benefit to soothe the sight of scholars, money changers, gem carvers and painters:

> Parchment. They were made at first of a muddy color, that is, yellowish, but afterwards white parchment was invented at Rome. This appeared to be unsuitable, because it soils easily and harms the readers' eyesight – as the more experienced of architects would not think of putting gilt ceiling panels in libraries, or any paving stones other than of Carystean marble, because the glitter of gold wearies the eyes, and the green of the Carystean marble refreshes them. Likewise those who are learning money-changing put dark green cloths under the forms of the coins, and carvers of gems look repeatedly at the backs of scarab beetles, than which nothing is greener, and painters [do the same, in order that they may refresh the labor of their sight with the greenness of these scarabs].[56]

Certainly, to soothe your eyes with the perfect, translucent glow of a precious emerald was a luxury that most would not have been able to afford. For this reason, other green materials eventually replace the emerald in performing this function. This was especially the case for those who busied themselves with activities particularly straining to the sight. It is in this context that we try to understand an intriguing group of recipes in which the materials for emerald imitation — verdigris and oil — are used to transform parchment into having the appearance of green glass. The recipes in question can be divided into two groups that describe how oil and verdigris transform parchment into either a green 'screen' or into green 'window glass'. The earliest examples were recorded in various fifteenth century manuscripts now dispersed in libraries across Europe. Mostly written

54 Galen, *Galen on the Usefulness of the Parts of the Body, Translated from the Greek with an Introduction and Commentary*, ed. and trans. by Margaret Tallmadge May (Ithaca: Cornell University, 1968), p. 473.

55 Galen, *Galen on the Usefulness*, 473.

56 'DE PERGAMENIS. Pergameni reges cum carta indigerent, membrana primi excogitaverunt. Vnde et pergamenarum nomen hucusque tradente sibi posteritate servatum est. Haec et membrana dicuntur, quia ex membris pecudum detrahuntur. Fiebant autem primum coloris lutei, id est crocei, postea vero Romae candida membrana reperta sunt; quod apparuit inhabile esse, quod et facile sordescant, aciemque legentiuni laedant; cum peritiores architecti neque aurea lacunaria ponenda in bibliothecis putent neque pavimenta alia quam e Carysteo marmore, quod auri fulgor hebetat et Carystei viriditas reficiat oculos. Nam et qui nummulariam discunt, denariorum formis myrteos pannos subiciunt, et gemmarum sculptores scarabaeorum terga, quibus nihil est viridius, subinde respiciunt, et pictores [idem faciunt, ut laborem visus eorum viriditate recreent]'. Isidore of Seville, *Etymologies*, VI. xi. 1–3 and Isidore of Seville, *Etymologiarum*, 1, VI. xi. 1–3.

in the German vernacular, this recipe family is named after one of the manuscripts that was lost in the 1870 fire in the Strasbourg Library.[57] In the first group, which explains how to make a *schirmm* ('screen'), parchment is treated so you can see through it.[58] First, one should search for 'the clearest parchment'. If you then want to colour the parchment 'a beautiful fine green' so that 'you can see through it as you can through beautiful glass', you should grind verdigris and a dye used to colour leather with vinegar. The parchment has to be left in this liquid overnight to take on a green colour. It is then stretched on a frame and coated on both sides with a drying oil or a 'clear varnish made of mastic' until it looks quite *glantz* ('shiny').[59] Finally, the parchment is placed in the hot sunshine so it can dry. Once dry, the sheets can be removed from the frame, 'cut into three or four parts and kept in a book or press to keep them flat'.[60] Practical reconstructions of these parchment 'screens' have shown that green translucent parchment can indeed be produced by this method (Figure 7).[61] The mystery of what these pieces of see-through parchment might have been used for is solved by another group of recipes that, somewhat confusingly, are

57 This manuscript was destroyed in the 1870 fire in the Strasbourg Library, but preserved through a nineteenth century copy made by Charles Lock Eastlake, see the latter's *Methods and Materials of Painting of the Great Schools and Masters*, 2 vols (London: Longman, Brown, Green and Longman, 1847). For a recent critical edition with English translation, see: *The Strasbourg Manuscript: A Medieval Tradition of Artists' Recipe Collections*, ed. and transl. by Sylvie Neven (London: Archetype, 2016). See also the important work on this family of recipes by Emil Ernst Ploss, 'Das Amberger Malerbüchlein: Zur Verwandtschaft der Spätmhd. Farbrezepte', in *Festschrift Für Hermann Hempel Zum 70. Geburtstag [...]*, Hrsg. von den Mitarbeitern des Max-Planck-Instituts für Geschichte (Göttingen: Vandenhoeck & Ruprecht, 1972), pp. 693–703 and Anonymous, *Der 'Liber illuministrarum' aus Kloster Tegernsee. Edition, Übersetzung und Kommentar der Kunsttechnologischen Rezepte*, ed. and trans. by Anna Bartl, Christoph Krekel, Manfred Lautenschlager, and Doris Oltrogge (Stuttgart: Franz Steiner Verlag, 2005).

58 'wiltu nu daz bermit schön fin grun machen das man da dur sicht was man wil als du schön glas so nim spangrün als vil du wilt und rib das gar wol mit essich und müsche dar under des grünen da mit die sekler leder verwent und disu zwei temperer under enander das es weder ze dik noch ze dünne si dar nach so nim das geweschen bermit und netz das bermit in der varwe ze beiden sitten und las es also ligen in der varwe ein nacht dar noch so so swenke daz bermit in kalten wasser und spanne das berment stark uff ein ram und las es wol truken werden darnach so nim luter virnies das usser mastikel gemachet sÿ und mit dem selben virnis das berment ze beiden sitten bis daz es glantz wirt darnach so setz es in eine heisse sunne und las das bermit wol truken werden dar nach so schnid das bermit glich us der ramen und mach drie stuck oder 4 als breit als du es haben wilt und leg dis bermit in ein buch oder in ein presse das es schlecht belibe'. *The Strasbourg Manuscript*, [nr. 33, compare also nrs. 32, 95].

59 'des lutersten megdenbermenten'; 'schön fin grun'; 'man da dur sicht was man wil als du schön glas'; 'luter virnies das usser mastikel gemachet sig'. The mastic varnish described in the manuscript is made with linseed oil, hempseed oil or nut oil. It is praised for the fact that it is the best varnish possible and very good and clear, see *The Strasbourg Manuscript*, [nr. 33].

60 'mach drie stuck oder 4 als breit als du es haben wilt und leg dis bermit in ein buch oder in ein presse das es schlecht belibe'. *The Strasbourg Manuscript*, [nr. 33]. Many of the recipes I discuss have been published and transcribed in the ARTECHNE research database which contains fully searchable digitized sources on artisanal techniques, see: http://artechne.hum.uu.nl/home [accessed 10-01-2019]. Here, unless stated otherwise, translations are my own.

61 In collaboration with parchment specialist Henk de Groot and senior scientist, curator of paintings at the Rijksmuseum Arie Wallert, I published a first attempt at making such green translucent parchment in: Marjolijn Bol, Henk de Groot and Arie Wallert, 'Glass and Parchment with a View. Oil Paint and the Imitation of (Stained) Glass Windows, 1400–1600', in *Making and Transforming Art: Changes in Artists' Materials and Practice*, ed. by Hélène Dubois, Sigrid Eyb-Green and Joyce H. Townsend (London: Archetype Publishers, 2014), pp. 129–30.

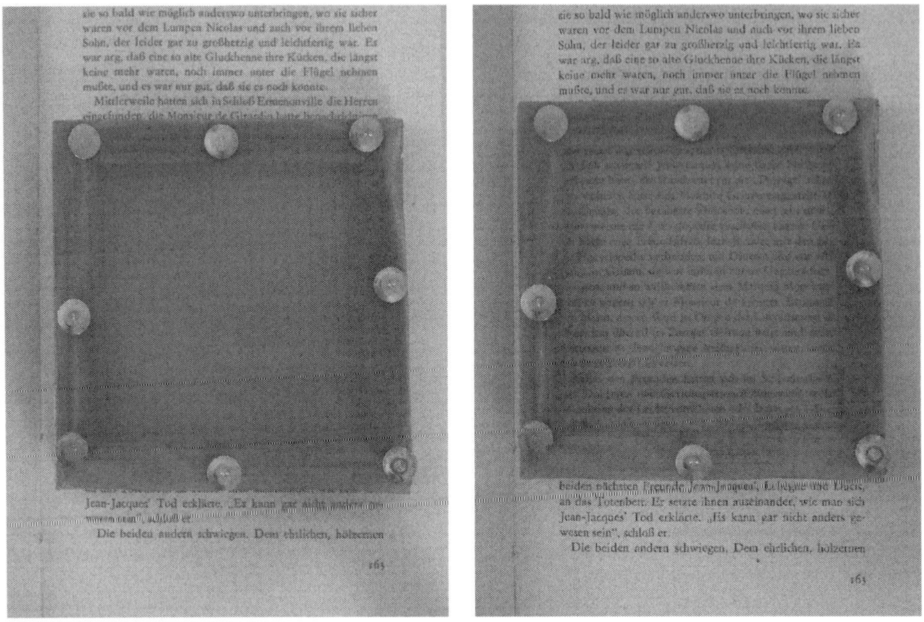

Fig. 7a-b Reconstruction of calf parchment coloured with copper green dissolved in vinegar. Left: shows the the colored parchment before oiling; Right: shows the parchment after having been oiled on both sides. Photo by Marjolijn Bol. [See colour plate 7 and 8]

not always included in the treatises that explain how to prepare the green parchment.[62] One of these recipes can be found in the *Amberger Malerbüchlein* (*c.* 1492). It clarifies that after having made parchment green and see-through according to the above methods it can be used as a reading device that soothes the eyes.[63] More spectacularly even, it was believed to magnify text:

62 Aside from the obvious fact that some recipes may not have come down to us, this could also imply that the practice was so common that it was not always considered necessary to explain what the translucent green parchment was used for.

63 This treatise contains the recipe that explains how to make the parchment green and translucent, compare: Amberger Malerbüchlein, Staatlichen Provinzial Bibliothek, Amberg, Cod. 77, fols 223^{v}–224^{r}, *Artechne database*, http://artechne.hum.uu.nl/node/86843 [accessed 05-12-2018]: 'Wil dw parmet durchleuchtig machen welcher varb dw (fol. 224^{r}) wilt als ein glas so nim des lautriste permacz den man vindet und wasch daz in lautterm wasser laugen gar wol pis dÿ laug gar chlar von dem permait get dar nach wasch es in lauterm wasser so ist daz permait lautter und chlar worden dar nach reib daz wasser auz dem permait Wil dw nun daz daz pirmait vein schon grun werd daz man dar durch sech waz man wil alz durch ein schon glaz So nim spon grun als vil dw wilt und reib daz grun iar wol mit ezzeich und mische dar under dez grun da man mit die seckkel verbt oder daz leder und dÿ zwaÿ gruen temper undereinander daz es weder ze dick noch ze duen werd dar nach nim dez gewaschen permacz und necz daz permait zw paider seÿten wol und laz es alzo ligen in der varb ein nacht dar nach so wasch daz permait auz chaltem wasser und spanne daz permait vast auff eÿn ram und laz es wol trucken werden Dar nach nim lautterm virnecz daz aus der mastikk gemacht seÿ und mit dem selben virneis ubervirneis daz permait ze peiden seÿten pis daz es glancz werd dar nach so secz es an ain haizzen sunne und laz daz permait wol trucken werden dar nach so schneid daz permait auz der ram und mach dar auz iii stuck oder vier alz dw es haben wilt und leg es danne in ein puch oder in ein prieff daz es schlecht werd'. For another recipe of the same kind see also: Amberger Malerbüchlein, Staatlichen Provinzial Bibliothek, Amberg, fol. 224^{rv}, *Artechne database*, http://artechne.hum.uu.nl/node/86845 [accessed 05-12-2018].

> When you lay the parchment over some writing, that which is small will appear to be large, and what one reads or studies through it does not harm the eyes and it keeps the sight fresh and secures the eyes.[64]

At the beginning of this chapter, I introduced Pliny's influential idea that emeralds not only have the ability to soothe the eyes of gem-cutters but can also 'concentrate the vision'. This final point has led to much speculation about whether lenses were known to the ancients. Unfortunately, there is no definitive evidence that supports this theory.[65] But whatever may have been true for the ancients, Pliny's idea did persist for many centuries and the fifteenth-century practice, in which green parchment was used to soothe the eyes of the scholar and magnify text, may have been used as a substitute for more precious reading devices such as the 'green stone for reading with' that is listed in the inventory of Henry VIII (1491–1547).[66] Whether Henry's stone was an emerald or not, a transparent green reading stone would have certainly cost a lot more than a piece of 'reading parchment'.

Reading through green parchment seems to have been an enduring practice, and a rather interesting recipe dating to the seventeenth century reveals a few more details about how and why it was thought useful. Similar to the German treatises, this recipe from an English alchemical treatise explains how to make a 'clear gren parchmente to read withall transparent', but, significantly, makes explicit that such parchment was used as an alternative reading device to glass, or see-through minerals: 'being laid on small writinge will mak yt showe better and bigger than any glasse, christall or berall'. The recipe concludes that the vellum thus prepared, when cut into pieces would be 'good for old men and such as have weke sighte'.[67]

Later recipes show that translucent green parchment was also used in a variety of other practices, significantly all related to sight. The seventeenth-century recipe collection compiled by the physician Sir Théodore Turquet de Mayerne (1573–1654 or 1655) for instance, contains several instructions for preparing green parchment using the above

64 'wer daz permait legt auff ein schrifft wÿ chlain sÿ ist so scheint sÿ groz und wÿ vil man dar durch list oder studirt so schat es den augen nicht und behat das gesicht in schoner frist und sichert dy augen'. Amberger Malerbüchlein, Staatlichen Provinzial Bibliothek, Amberg, fol. 224ʳ, *Artechne database*, http://artechne.hum.uu.nl/node/86844 [accessed 05-12-2018]. Compare with similar recipes in Bibliotheek der Rijksuniversiteit Leiden, Leiden, Voss. Chymicus octavo 6, fol. 23ᵛ, *Artechne database*, http://artechne.hum.uu.nl/node/89393 [accessed 05-12-2018] and Cod. XI D 10, Prager Malerbuch, Národní knihovna České republiky, Prague, fol. 84ᵛ, *Artechne database*, http://artechne.hum.uu.nl/node/85312 [accessed 05-12-2018].

65 There is a great deal of literature on this topic. A good discussion can be found in Edward Rosen, 'The Invention of Eyeglasses', *Journal of the History of Medicine and Allied Sciences*, 11 (1956), 13–46 and more recently: Marco Beretta, 'From the Eye to the Eye-Glass. A Pre-History of Spectacles', in *When Glass Matters: Studies in the History of Science and Art from Graeco-Roman Antiquity to Early Modern Era*, ed. by Marco Beretta (Florence: L. S. Olschki, 2004), pp. 249–82.

66 David Starkey, *The Inventory of King Henry VIII: Society of Antiquaries Ms 129 and British Library Ms Harley 1419, Reports of the Research Committee of the Society of Antiquaries of London* (London: Harvey Miller for the Society of Antiquaries of London, 1998). The green stone for reading is mentioned in the category 'spectacles and reading aids'.

67 I am grateful to Alexis Hagadorn for pointing me towards this recipe from MS Ashmole 1491, Bodleian Library, Oxford, p. 910. It was published in Ronald Reed, *Ancient Skins, Parchments, and Leathers* (London: Seminar Press, 1972), p. 144. It is perhaps significant that the recipe mentions to use abortive parchment, a very fine parchment probably made from stillborn animals, because it is the clearest as it 'hath no pores appearing therin'.

methods that is an 'excellent approximation of glass'.[68] According to De Mayerne, these pieces of glasslike parchment should be put on a metal frame and placed between a candle and the eye so that the eye is protected against the harsh flicker of the flames. A sixteenth century recipe collection, kept in the Bibliothèque nationale de France (BnF. Ms. 640[69]), contains another intriguing version of this practice. In this recipe it is not the weary sight of the reading scholar that is protected, instead it instructs how to make 'water that gives light to painters'.[70] In similar fashion to the green candle screen, distilled water of green vines is put in front of a candle to soothe the eyes of the painter:

> Distill some vine water and put it into a big bottle. Put your candle behind that, and it won't disturb your vision.[71]

Although dating much later, it is possible to understand Jean Baptiste Simon Chardin's 1775 self-portrait (Louvre) in this same context. As well as a pair of spectacles, Chardin portrayed himself wearing some type of green eyeshade or visor (Figure 8).[72] Chardin began to lose his eyesight after 1770 and as a remedy long known to painters, and likely practiced by many others, Chardin may have thus started wearing the green visors to help his troubled eyes.

Besides these interesting contrivances that use the translucent green parchment to magnify text and protect the sight of scholars and artisans, it also appears to have been used as window glass. These recipes represent a separate group that again first appears in the Strasbourg tradition. The procedures to make parchment windows are similar to the techniques used to make eye screens, and the fact that we are dealing with a different series of recipes only becomes clear from the terminology used to describe the end product. Instead of the word *schirmm* this green translucent parchment is consistently described as *venster glaz* ('window glass').[73] It is therefore notable that the recipes belonging to this

---

68 'excellent approche du verre'. Ernst Berger, *Beiträge zur Entwicklungs-Geschichte der Maltechnik* (München: Callwey, 1901), pp. 174–76 [nr. 82 and 83]. Recipe 83 is similar to recipe nr. 214 from the *Bolognese Manuscript*.

69 Bibliothèque nationale de France, Paris, Ms. Fr. 640. The *Making and Knowing Project*, directed by Pamela Smith, is currently working towards an open-access digital critical edition of this intriguing collection of recipes, see: http://www.makingandknowing.org/ [accessed 05-12-2018]. The manuscript has been digitised and can be accessed here: http://gallica.bnf.fr/ark:/12148/btv1b10500001g.r = .langEN [accessed 05-12-2018].

70 'Eau pour donner jour au painctre'. Bibliothèque nationale de France, Paris, Ms. Fr. 640, fol. 061v, transcription and translation (translation still in progress) supplied by *The Making and Knowing Project, A Digital Critical Edition of BnF Ms Fr. 640*, forthcoming.

71 'Distille leau de vigne & la mects dans une grande bouteille Et derriere icelle mects ta chandelle & elle ne nuyra point a ta veue'. Bibliothèque nationale de France, Paris, Ms. Fr. 640, fol. 061v, transcription and translation (translation still in progress) supplied by *The Making and Knowing Project, A Digital Critical Edition of BnF Ms Fr. 640*, forthcoming. This recipe appears right before an instruction that explains how 'German' miniature painters use a varnished canvas screen instead of glass to create a diffuse working light. *The English Housewife*, dating to the beginning of the seventeenth century, reveals that distilled vine waters were also used directly to soothe itching of the eye. Interestingly this treatise also explains how to give colour to such distillates by filling 'a glass of great strength' with the herbs or flowers and distill it again to give the liquid its colour, see: *The English Housewife by Gervase Markham*, ed. and trans. by Michael R. Best, (Kingston: McGill-Queen's University Press, 1986), pp. 125–26.

72 In 1813 the German painter Anton Graff (1736–1813) also depicted himself wearing a green eyeshade, Alte Nationalgalerie, Berlin, Ident.nr. A I 406.

73 2 Cod. 572, Staats- und Stadtbibliothek, Augsburg, fols 167–167rv, *Artechne database*, http://artechne.hum.uu.nl/node/39311 [accessed 05-12-2018]:'Von vensterglas Wiltu machen vensterglas von bermit [...] So hastu fenster glas das gu(o)t ist'. Compare with similar recipes from: Cod. Pal. Germ. 489, 'Ain gar schones unnd vast nutzliches handbuechlin von allerlaye farbenn *De coloribus*. Von den Farbenn- Von den Farben aus der federn

Fig. 8 Siméon Chardin, *Self-Portrait with Green Eye-Shade*, pastel on paper, Paris, Musée du Louvre, 1775. Photo © Musée du Louvre, Dist. RMN-Grand Palais / Martine Beck-Coppola. [See colour plate 9]

group also instruct to paint designs on the parchment windows and explain how to make it of colours other than green.

In his *Jewell House of Art and Nature* (1594), the Elizabethan scholar Hugh Platt (1552–*c.* 1611) provides us with a detailed account about the so-called parchment window. In a recipe entitled 'to make Parchment clear and transparent to serve for divers purposes', Platt explains how coloured parchment can be decorated before being made transparent

zu schreyben', Heidelberg, Universitätsbibliothek, 1562–63, fols 43^rv^, 177–78, 179^v^–180, *Artechne database*, http://artechne.hum.uu.nl/node/91857 [accessed 05-12-2018] and Bayerische Staatsbibliothek, München, Cgm 824, fols 8^v^–9^r^, *Artechne database*, http://artechne.hum.uu.nl/node/84837 [accessed 05-12-2018].

Fig. 9 *Saint John the Evangelist*, Tempera colors, gold, and silver on parchment, Ms. Ludwig II 3, fol. 127v, ca. 1120–1140, The J. Paul Getty Museum, Los Angeles. Photo courtesy of the Getty's Open Content Program. [See colour plate 10]

with oil or varnish so that '[…] it will shew very clear and serve in windows instead of glass, especially in such room as are subject to overseers. […]'.[74] According to Platt, these pieces of parchment stretched on little frames 'keep the room very warm' and 'will indure the blustring and stormy weather much better then [sic] paper'.[75] When coloured green, such parchment windows, like the green eye screens, appear to have also played a special

74 Hugh Platt, *The Jewell House of Art and Nature Conteining Divers Rare and Profitable Inventions, Together with Sundry New Experimentes in the Art of Husbandry, Distillation, and Moulding / Faithfully and Familiarly Set Downe, According to the Authors Owne Experience, by Hugh Platte*, (London: Peter Short, 1594), pp. 76–77.

75 Platt, *The Jewell House of Art and Nature*, pp. 76–77.

role in the scholar's study. The mathematician John Smith describes this particular use of green windows in *The Art of Painting in Oyl* (1687).[76] Smith explains how to make green translucent windows on parchment, paper or silk with verdigris and oil, much like the recipes dating to the fifteenth century.[77] He then accounts for the special ophthalmic properties of such green translucent windows saying that they are 'very comfortable to the sight', and 'of great benefit to those whose sight is impaired and weakened by poring too much upon their Books; the whiteness of the Paper being observed too often a great Enemy to the sight of some Men'.[78] Indeed, for John Smith, parchment coloured with verdigris and oil was the most suitable substitute or addition to green reading glasses or spectacles:

> [...] the inconveniences of which, such a green Light as this now mentioned, will infallibly prevent, beyond green reading Glass, Spectacles, or any other contrivance, yet found out; the like benefit some Tradesmen also receive from it.[79]

Tradesmen, it should be remembered, were also mentioned by Isidore of Seville to benefit from the colour green underneath their glistening coins. And, indeed, in addition to the fifteenth-century recipes discussed above, numerous paintings depicting scholars in their study and merchants at their dealing tables, dating variously to the fifteenth and sixteenth century, seem to suggest that the practice described by John Smith in the seventeenth century should be dated much earlier. The paintings in question show scholars and merchants surrounded by one or more green elements — green wall coverings, a green tablecloth, green and curtains. (Figures 9, 10 and 11).

Conclusion

Following the path of the emerald, this paper has taken us on a journey that connects the ideas and practices of lapidarists, encyclopaedists, theologians, jewellers, and painters to the history of gemstone imitation, parchment reading devices, eye protectors and window glass. As such, this largely unstudied history of how we have engaged with light and colour in our workspaces, connects a variety of practices moving between the artisan's studio and the scholar's study. The great variety of practices that revolve around imitating the green colour and clarity of the emerald became so closely intertwined with the gem itself that natural philosophers theorised a relationship between copper, verdigris and the genesis of the costly green stone in nature. What is more, this long history resulted in verdigris and oil

76 Smith's treatise was first published in 1676, but the relevant passage can be found from the second edition onwards (the work saw many editions and translations, well into the eighteenth century), see: John C. M. Smith, *The Art of Painting in Oyl. [...] the Second Impression with Some Alterations, and Many Useful Additions* (London: Printed for Samuel Crouch, at the corner of Popes-Head-Alley in Cornhill, 1687).

77 Smith, *The Art of Painting in Oyl*, 97–100: '[...] if any are troubled with weak Eyes and cannot indure a bright Light, this Varnish mixt with distilled Verdigrease and Paper Windows, or Sarsnet [a type of silk] ones done over with it, will make an incomparable green light, very comfortable to the sight, and of great benefit to such as love not too much brightness'.

78 Smith, *The Art of Painting in Oyl*, pp. 97–100.

79 Smith, *The Art of Painting in Oyl*, pp. 97–100.

Fig. 10 Jan van Eyck, *Saint Jerome in His Study*, oil on linen paper on oak panel, ca. 1435, Detroit Institute of the Arts. Photo courtesy of the Detroit Institute of the Arts. [See colour plate 11]

substituting emeralds to comfort and aid the weary eyes of those pouring over their books and making fine and detailed art.[80] This last fact seems to bear a particular relation to the idea that the invention of white parchment impaired the eyes of scholars and miniature

80 In Benito Daza Valdés's *Uso de los antoios* (*Use of Spectacles*, 1623), probably the first printed book on spectacles, Smith's green reading glass is indeed prescribed to a patient by the seventeenth century eye doctor who figures in the treatise. In a best practice dialogue concerning the eye, the doctor explains: 'lenses that have a little green

Fig. 11 Quentin Matsys, *The Money Changer and His Wife*, oil on panel, 1514, Paris, Musée du Louvre, Photo © RMN-Grand Palais (Musée du Louvre) / Gérard Blot. [See colour plate 12]

painters. In order to relieve eyes without using a costly emerald, the chosen methods for making a proper substitute were deeply embedded in the history of enhancing and imitating the optics and colour of the precious green gem. From parchment reading devices to green furnishings for the scholar's study and the artisan's studio, the impact of this history cannot be overestimated. Indeed, a green leather or baize cloth was a common feature of writing desks well into the twentieth century.[81] And with the increasing prevalence of electric light

are helpful to the vision, for green is a pleasant color. Green was given to delight our eyes, and all men enjoy the various green hues of herbs and plants in a valley'. See: *The Use of Eyeglasses. By Benito Daza De Valdés*, ed. and trans. by Paul B. Runge and Manuel Márquez (Oostende: J. P. Wayenborgh, 2004), p. 170.

81 Thomas Mortimer, *A General Dictionary of Commerce, Trade, and Manufactures: Exhibiting Their Present State in Every Part of the World; and Carefully Comp. From the Latest and Best Authorities* (London: Gillet and Son, 1810), np. [abbrev. lemma PAI]: 'Of painting the face in miniature. You are first to provide yourself with a mahogany desk for painting on, […] there is to be a lid covered with green cloth. […] The next thing you are to observe

NEW ENGLAND STATIONER AND PRINTER. 65

at the same distance from the eye; both of your hands are free and at rest and can hold a fan, or a field glass, or a handkerchief, or anything else that might be necessary to enhance the enjoyment of the hour.

EARLY DAYS OF JOURNALISM.

Mr. Higashi Kan-ichi contributes to the *Taiyo* a short story of journalism in early *Meiji* days, but unfortunately in most cases he has omitted to give dates. From the details supplied we cull the following: By some it is affirmed that the *Moshiogusa* was the first newspaper published in Japan, but this was rather a magazine than a newspaper, which first appeared in about 1868. The first actual newspaper published was Mr. Black's *Nisshin Shinjishi*, which was a daily consisting of four pages about the size of the present *Jiji Shimpo*. The next journal to appear was the *Yokohoma Shimbun*. This was mostly an advertising medium. As a newspaper it was in every way inferior to Mr. Black's organ. The *Nisshin Shinjishi* was first printed at a temple in Shiba called the Genkoji, but subsequently the office was moved to Ginza. Next came the *Nichi Nichi Shimbun*, which was first issued at a small office in Kawara-machi, Asakusa. It was a very poor affair at first, printed on Japanese paper. Then came the *Hochi Shimbun*, and what afterward became the *Choya Shimbun*.

It was toward the close of the year 1874 that the *Nichi Nichi Shimbun* enlarged its sheet to about its present size. But the facilities for printing available in this country at the time were so poor that even the following year it frequently happened that the paper appeared only every other day. These papers contained no leading articles at that time, the greatest reserve being necessary to avoid suspension. The writer who may be said to have set the ball of political discussion rolling was Mr. Fukuchi Genichiro. From the time of his connection with the *Nichi Nichi*, newspaper controversy became common and the prospect of opening a Diet soon furnished endless material for comment and criticism. The *Nisshin Shinjishi* first opened its columns to correspondents. At the time of which we write three magazines of considerable influence were in circulation, the *Hyoron Shimbun*, the *Kinji Hyoron* and the *Meiroku Zasshi*. The first of these used such violent language that the authorities found it necessary to issue press laws. These laws originated with the late Viscount Inoue Ki (minister of education for some little time) and Baron Ozaki Saburo. There is no doubt that in some respects newspapers have deteriorated in Japan. In early days their editors cared little what the public might think of their articles. They wrote from conviction and they made use of none of the modern artifices for pandering to popular taste; such as novelettes, spicy, suggestive paragraphs, slander, personalities, and the like.

FEATHERWEIGHT EYESHADES.

The days are getting shorter and the time when eyeshades are used most has arrived. In placing your orders we would advise that you lay in a stock of the Featherweight Eyeshades, manufactured by the Featherweight Eyeshade Company of Bayonne, N. J. The "Featherweight" has the following advantages: It is light in weight, weighing only about one-fourth ounce; it is adjustable to any size or position, it is durable, being made of material of great strength; it is hygienic, as

prespiration has no effect on it; it is healthful and is of a color peculiarly restful to the eye, it has no metal band to press the head or to tarnish, and it lies flat when not in use, thus preventing it being crushed out of shape by accident, and it can be carried in hat or rolled up in a small package to carry in the pocket. The visor is wider than other shades. The "Featherweight" is made in two styles, opaque for artificial light and transparent for natural light. The company recommends the opaque for those wearing glasses.

OHIO'S PAPER CONTRACTS.

The Oho State Printing Commission has awarded contracts for paper as follows:

The Diem & Wing Company of Cincinnati, 7,500 reams book paper, double super royal, at 2.97 cents per pound. The company had filed a letter agreeing to furnish the paper either white or in tints, as might be desired, at the same price. The Chatfield & Woods Company of Cincinnati, 40 reams of double medium, at 4 cents, and 100 reams of Crane's linen, at 19½ cents. The Brooks Company of Cleveland, 30 reams of linen ledger super royal, at 20¾ cents per pound. The Keough & Pike Company, Dayton, 500 reams of double flat cap, at 3¼ cents per pound.

The full details of the bidding were as follows:

Keough & Pike Company, Dayton, double super royal, 3.3 cents per pound; 500 reams double flat cap, 3.5 cents.

The Cincinnati Cordage and Paper Company, double super royal, 3.15; 40 reams double medium paper, 5 cents, and flat cap 4.75.

Diem & Wing Paper Company, Cincinnati, double super royal, 2.97; double medium, 4.49; 100 reams Crane's linen folio, 20; 30 reams linen ledger super royal, 23.75; flat cap 4.33.

The Brooks Company, Cleveland, double super royal, 2.98; double medium, 5.75; Crane's linen, 19.95; linen ledger, 20.25; double cap, 5.50.

Brown & Stuart, Cincinnati, double super royal, 3.03.

Central Ohio Paper Company, Columbus, double super royal, 3.11; double medium, 5.19; Crane's linen, 20.15; linen ledger, 20.49; flat cap, 3.9.

Ohio Paper Company, Miamisburg, double super royal, 3.21.

Chatfield & Woods Company, Cincinnati, double super royal, 3.04; double medium, 4; Crane's linen, 19.5; linen ledger, 24; flat cap, 4.125.

Mead Paper Company, Dayton, double super royal, 3.46.

The New York News Company's *Monthly Bulletin* for October is out, and its four pages are overflowing with attractive novelties for retailers. There doesn't seem to be any limit to this series, and each number of the *Bulletin* has more than the one preceding. Each retailer should have the *Bulletin* regularly for reference. Anything can be ordered by number. The subscription price is ten cents per year.

Fig. 12 Page from The New England Stationer and Printer. v. 15 (Apr. 1901-Mar. 1902) with an advertisement from the company Featherweight for its green visor.

by the end of the nineteenth century, writers, accountants, bankers and a whole range of other office clerks were using Chardin's green visor, now made of cellulose, to relieve their eyes from the associated strain (Figure 12). For the same reason, the famous banker's desk

is your choice of light [...] A north light must be attained. If there are more windows than one in the room, the second must be closed [...] and the one you sit at is to have a green baize curtain against the lower part of it, to reach about a foot higher than your head [...]'.

light also received a green, glass shade to shield the eyes from its bulb. Taking all of this into consideration, it is intriguing that science today has established that the best level of focus for the human eye is at peak spectral sensitivity (550 nm) — this, you may have guessed, is green.[82] Now that our eyes are suffering from a different kind of light, not from the whiteness of parchment, the flicker of candles or the harsh light of early incandescent bulbs, but from our computers, phones and other devices we may yet contrive, the history of emerald green as a soothing colour has become relevant once again.[83]

Bibliography

Manuscript and Archival Sources

Alte Nationalgalerie, Berlin, Ident.nr. A I 406.
Bayerische Staatsbibliothek, München, Cgm 824.
Bibliotheek der Rijksuniversiteit Leiden, Leiden, Voss. Chymicus, octavo 6.
Bibliothèque nationale de France, Paris, Ms. Fr. 640.
Bodleian Library, Oxford, MS Ashmole 1491.
Prager Malerbuch, Národní knihovna České republiky, Prague, Cod. XI D 10.
Staatlichen Provinzial Bibliothek, Amberg.
Staats- und Stadtbibliothek, Augsburg, 2 Cod. 572.
Universitätsbibliothek, Heidelberg, 1562–63.

Primary Sources

Agricola, Georgius, *Georgii Agricolae de ortu & causis subterraneorum lib. V: De natura eorum quae effluunt ex terra lib. IIII: De natura fossilium lib. X: De veteribus & novis metallis lib. II: Bermannus, sive, De re metallica dialogus: Interpretatio germanica uocum rei metallicae: addito Indice foecundissimo* (Basel: H. Frobenium and N. Episcopium, 1546).

Agricola, Georgius, *De natura fossilium (= Textbook of Mineralogy)*, ed. and trans. by Mark Chance Bandy and Jean. A. Bandy, The Geological Society of America, Special Paper 63 (New York: The Geological Society of America, 1955).

Al-Beruni, *Al-Beruni's Book on Mineralogy. The Book Most Comprehensive in Knowledge on Precious Stones*, ed. and trans. by Hakim Mohammad Said, One Hundred Great Books of Islamic Civilization (Islamabad: Pakistan Hijra Council, 1989).

Anonymous, *Der 'Liber illuministrarum' aus Kloster Tegernsee. Edition, Übersetzung und Kommentar der Kunsttechnologischen Rezepte*, ed. and trans. by Anna Bartl, Christoph Krekel, Manfred Lautenschlager, and Doris Oltrogge (Stuttgart: Franz Steiner Verlag, 2005).

82 Louise A. Bye, Neil. C. Modi, and Miles Stanford, *Basic Science for Ophthalmology* (Oxford: Oxford University Press, 2013), p. 216.

83 Much research is invested to study the presumed tiring effects of the light emitted by our computer screens. Software developers have already created specials overlay filters to make computer screens less straining to our eyes, and it was recently demonstrated that green light is soothing to people suffering from migraines, see: Rodrigo Noseda et al, 'Migraine Photophobia Originating in Cone-Driven Retinal Pathways', *Brain: A Journal of Neurology*, 139 (2016), pp. 1971–86.

Anonymous, *The Strasburg Manuscript: A Medieval Painter's Handbook/Das Strassburger Manuskript: Handbuch Für Maler Des Mittelalters*, ed. and trans. by Rosamund Borradaile and Viola Borradaile (München: Callwey 1976, 1st ed. 1966).

Biringuccio, Vannoccio, *De la Pirotechnia libri X … composti peril Vanoccio Biringuccio Sennese* (Venice: Venturino Roffinello, 1540).

Biringuccio, Vannoccio, *The Pirotechnia of Vannnoccio Biringuccio*, ed. and trans. by Cyril Stanley Smith and Martha Teach Gnudi (Cambridge: MIT Press, 1959; orig. publ. 1942).

Caley, Earle Radcliffe, 'The Leyden Papyrus X. An English Translation with Brief Notes', *Journal of Chemical Education*, 3 (1926), 1149–66.

Caley, Earle Radcliffe, 'The Stockholm Papyrus: An English Translation with Brief Notes', *Journal of Chemical Education*, 4 (1927), 979–1002.

Cannella, Anne-Françoise, *Gemmes, verre coloré, fausses pierres précieuses au Moyen-Âge: Le quatrième livre du 'Trésorier de Philosophie naturelle des pierres précieuses' de Jean d'Outremeuse*, Bibliothèque de la Faculté de Philosophie et Lettres de l'Université de Liège (Geneva: Droz, 2006).

Cantimpré, Thomas of, *Thomas Cantimpratentsis liber de natura rerum: editio princeps secundum codices manuscriptos*, ed. by Helmut Boese, 2 vols (Berlin: Walter de Gruyter, 1973).

Eastlake, Charles Lock, *Methods and Materials of Painting of the Great Schools and Masters*, 2 vols (London: Longman, Brown, Green and Longman, 1847).

Galen, *Galen on the Usefulness of the Parts of the Body, Translated from the Greek with an Introduction and Commentary*, ed. and trans. by Margaret Talladge May (Ithaca: Cornell University, 1968).

Grosseteste, Robert, *Robert Grosseteste on the Six Days of Creation: A Translation of the Hexaëmeron*, ed. and trans. by Christopher F. J. Martin, Auctores Britannici Medii Aevi (Oxford: Oxford University Press, 1996).

Halleux, Robert (ed.), *Les alchimistes grecs: Papyrus de Leyde – Papyrus de Stockholm – Recettes*, Collection des universités de France (Paris: Belles Lettres, 1981).

Jones, W. H. S., and A. C. Andrews (eds.), *Pliny: Natural History. Books 24–27*, 10 vols (Cambridge: Harvard University Press, 1956).

Kunz, Georg Frederick, *The Curious Lore of Precious Stones: Being a Description of Their Sentiments and Folk Lore, Superstitions, Symbolism, Mysticism, Use in Medicine, Protection, Prevention, Religion, and Divination: Crystal Gazing, Birthstones, Lucky Stones and Talismans, Astral, Zodiacal, and Planetary* (Philadelphia: Halcyon House, 1938).

Lagercrantz, Otto (ed. and tr.), *Papyrus Graecus Holmiensis (P. Holm.): Recepte für Silber, Steine und Purpur* (Uppsala: Arbeten Utgifna Med Understöd Af Vilhelm Ekmans Universitetsfond, 1913).

Magnus, Albertus, *De mineralibus (Mineralium libri quinque)*, ed. by A. Borgnet, Opera omnia Alberti Magni, 38 vols (1890–99).

Magnus, Albertus, *Book of Minerals*, ed. by Dorothy Wyckhoff (Oxford: Oxford University Press: Clarendon Press, 1967).

Markham, Gervase, *The English Housewife by Gervase Markham*, ed. and trans. by Michael R. Best, (Kingston: McGill-Queen's University Press, 1986).

Megenberg, Conrad von, *Das Buch der Natur: Die erste Naturgeschichte in Deutscher Sprache. Durch Conrad von Megenberg*, ed. by Franz Pfeiffer (Stuttgart: Verlag von Karl Aue, 1861).

Merrifield, Mary Philadelphia (ed. and trans.), *Original Treatises Dating from the Xiith to Xviiith Centuries on the Arts of Painting*, 2 vols (London: John Murray, 1849).

Migne, Jacques-Paul (ed.), *Patrologiae cursus completus: seu bibliotheca universalis [...]*, 383 vols (Paris: Apud J. P. Migne, 1863).

Phillipps, Thomas, 'Letter from Sir Thomas Phillipps, Bart. F. R. S., F. S. A., Addressed to Albert Way, Esq., Director, Communicating a Transcript of a Ms. Treatise on the Preparation of Pigments, and on Various Processes of the Decorative Arts Practiced During the Middle Ages, Written in the Twelfth Century, and Entitled "Mappae clavicular"', in *Archeologia: Or, Miscellaneous Tracts Relating to Antiquity* (London: Society of Antiquaries of London, 1770), pp. 183–244.

Platt, Hugh, *The Jewell House of Art and Nature Conteining Divers Rare and Profitable Inventions, Together with Sundry New Experimentes in the Art of Husbandry, Distillation, and Moulding / Faithfully and Familiarly Set Downe, According to the Authors Owne Experience, by Hugh Platte*, (London: Peter Short, 1594).

Pliny the Elder, *Pliny: Natural History. Books 36–37*, ed. and trans. by D. E. Eichholz, 10 vols (Cambridge and London: Harvard University Press, 1962).

Pseudo-Aristotle, *Das Steinbuch Des Pseudo-Aristotle nach der Arabische Handschrift*, ed. and trans. by Julius Ruska (Heidelberg: Carl Winter's Universitätsbuchhandlung, 1912).

Rennes, Marbode of, *Marbode of Rennes' (1035–1123) De lapidibus Considered as a Medical Treatise with Text, Commentary and C. W. King's Translation Together with Text and Translation of Marbode's Minor Works on Stones)*, ed. and trans. by John M. Riddle and C. W. King (Wiesbaden: Franz Steiner, 1977).

Seneca the Younger, *Seneca. Epistles 1–65*, ed. and trans. by Richard M. Gummere, Loeb Classical Library 75, 3 vols (Cambridge, MA: Harvard University Press, 1917).

Seville, Isidore of, *Isidori Hispalensis episcopi etymologiarum sive originum libri XX*, ed. by W. M. Lindsay, *Scriptorum classicorum bibliotheca oxoniensis*, 2 vols (Oxonii: Clarendoniano, 1911).

Seville, Isidore of, *The Etymologies of Isidore of Seville*, ed. by Stephen A. Barney, W. J. Lewis, J. A. Beach and O. Berghof (Cambridge: Cambridge University Press, 2006).

Smith, John C. M., *The Art of Painting in Oyl. [...] the Second Impression with Some Alterations, and Many Useful Additions* (London: Printed for Samuel Crouch, at the corner of Popes-Head-Alley in Cornhill, 1687).

Suggit, J. N., *Commentary on the Apocalypse*, The Fathers of the Church series, 127 vols (Washington DC: Catholic University of America Press, 2006).

Theophilus, *The Various Arts: 'De diversis artibus', Nelson's Medieval Texts* (London: Thomas Nelson, 1961).

Theophrastus, *Theophrastus on Stones, Contributions in Physical Science*, ed. and trans. by Earle Radcliffe Caley and John F. C. Richards (Columbus, Ohio: Ohio State University, 1956).

Valdés, Benito Daza De, *The Use of Eyeglasses. By Benito Daza De Valdés*, ed. and trans. by Paul B. Runge and Manuel Márquez (Oostende: J. P. Wayenborgh, 2004).

Secondary Works

Beretta, Marco, 'From the Eye to the Eye-Glass. A Pre-History of Spectacles', in *When Glass Matters: Studies in the History of Science and Art from Graeco-Roman Antiquity to Early Modern Era*, ed. by Marco Beretta (Florence: L. S. Olschki, 2004), pp. 249–82.

Berger, Ernst, *Beiträge zur Entwicklungs-Geschichte der Maltechnik* (München: Callwey, 1901).

Berthelot, Marcellin, *Introduction à l'étude de la chimie des anciens et du Moyen-Âge* (Paris: G. Steinheil, 1889).

Bol, Marjolijn, 'Oil and the Translucent. Varnishing and Glazing in Practice, Recipes and Historiography, 1100–1600' (unpublished doctoral dissertation, Utrecht University, 2012).

Bol, Marjolijn, 'Coloring Topaz, Crystal and Moonstone: Gems and the Imitation of Art and Nature, 300–1500', in *Fakes!?: Hoaxes, Counterfeits and Deception in Early Modern Science*, ed. by Marco Beretta and Maria Conforti (Sagamore Beach: Science History Publications, 2014), pp. 108–29.

Bol, Marjolijn, Henk de Groot and Arie Wallert, 'Glass and Parchment with a View. Oil Paint and the Imitation of (Stained) Glass Windows, 1400–1600', in *Making and Transforming Art: Changes in Artists' Materials and Practice*, ed. by Hélène Dubois, Sigrid Eyb-Green and Joyce H. Townsend (London: Archetype Publishers, 2014), pp. 129–30.

Bol, Marjolijn. 'Polito Et Claro. The Art and Knowledge of Polishing, 1200-1500', in *Gems in the Early Modern World*, ed. by Sven Dupré and Michael Bycroft (Cham: Palgrave Macmillan, 2019), pp. 223–57.

Bye, Louise A., Neil. C. Modi, and Miles Stanford, *Basic Science for Ophthalmology* (Oxford: Oxford University Press, 2013).

Davison, Sandra and R. G. Newton, *Conservation and Restoration of Glass* (Oxford: Butterworth Heinemann, 2008).

Dupré, Sven, 'The Historiography of Perspective and Reflexy-Const in Netherlandish Art' in *Art and Science in the Early Modern Netherlands* (= *Netherlands Yearbook for History of Art/ Nederlands Kunsthistorisch* Jaarboek 61), ed. by Eric Jorink and Bart Ramakers (Leiden: Koninklijke Brill, 2011), pp. 34-61.

Eastlake, Charles Lock, *Methods and Materials of Painting of the Great Schools and Masters*, 2 vols (London: Longman, Brown, Green and Longman, 1847).

Gage, John, *Colour and Culture: Practice and Meaning from Antiquity to Abstraction*, 60th edn. (London: Thames & Hudson, 2009; 1st edn. 1993).

Gombrich, Ernst H., *The Heritage of Apelles* (London: Cornell University Press, 1976).

Halleux, Robert, 'Albert le Grand et l'alchimie', *Revue des sciences philosophiques et théologiques*, 66 (1982), pp. 57–80.

Halleux, Robert, and Paul Meyvaert, 'Les origines de la Mappae clavicula', *Archives d'histoire doctrinale et littéraire du Moyen-Âge*, 54 (1987), pp. 7–58.

Halleux, Robert, and Anne-Françoise Cannella, 'Entre technologie et alchemie: de la teinture du verre à la fabrication des fauses pierres précieuses', in *Il colore nel medioevo: arte, simbolo, tecnica: Atti delle iornate di studi, Lucca, 2–3–4 maggio 1996* (Lucca: Istituto Storico Lucchese e Scuola Normale Superiore di Pisa, 1998).

Ilardi, Vincent, *Renaissance Vision from Spectacles to Telescopes, Memoirs of the American Philosophical Society* (Philadelphia, PA: American Philosophical Society, 2007).

Kammerling, Robert C., John I. Kocvula, Robert E. Kane, Patricia Maddison, James Shighley, and Emmanuel Fritsch, 'Fracture Filling of Emeralds: Opticon and Traditional Oils', *Gems & Gemology*, 27 (2) (1991), pp. 70–85.

Kunz, Georg Frederick, *The Curious Lore of Precious Stones: Being a Description of Their Sentiments and Folk Lore, Superstitions, Symbolism, Mysticism, Use in Medicine, Protection,*

Prevention, Religion, and Divination: Crystal Gazing, Birthstones, Lucky Stones and Talismans, Astral, Zodiacal, and Planetary (Philadelphia: Halcyon House, 1938).

Meadows, J. W., 'Pliny on the Smaragdus', *The Classical Review*, 59 (1945), pp. 50–51.

Morgan, Diane, *From Satan's Crown to the Holy Grail: Emeralds in Myth, Magic, and History* (Westport, Conn.: Praeger, 2007).

Mortimer, Thomas, *A General Dictionary of Commerce, Trade, and Manufactures: Exhibiting Their Present State in Every Part of the World; and Carefully Comp. From the Latest and Best Authorities* (London: Gillet and Son, 1810).

Nassau, Kurt, *Gemstone Enhancement. Heat, Irradiation, Impregnation, Dyeing, and Other Treatments Which Alter the Appearance of Gemstones, and the Detection of Such Treatments* (London: Butterworths, 1984).

Newman, William R., *Promethean Ambitions: Alchemy and the Quest to Perfect Nature* (Chicago: University of Chicago Press, 2004).

Noseda, Rodrigo, Carolyn A. Bernstein, Rony-Reuven Nir, Alice J. Lee, Anne B. Fulton, Suzanne M. Bertisch, Alexandra Hovaguimian, Dean M. Cestari, Rodrigo Saavedra-Walker, David Borsook, Bruce L. Doran, Catherine Buettner, and Rami Burstein, 'Migraine Photophobia Originating in Cone-Driven Retinal Pathways', *Brain: A Journal of Neurology*, 139 (2016), pp. 1971–86.

Pastoureau, Michel, *Green: The History of a Color* (Princeton, NJ: Princeton University Press, 2014).

Ploss, Emil Ernst, 'Das Amberger Malerbüchlein: Zur Verwandtschaft der Spätmhd. Farbrezepte', in *Festschrift Für Hermann Hempel Zum 70. Geburtstag […]*, Hrsg. von den Mitarbeitern des Max-Planck-Instituts für Geschichte (Göttingen: Vandenhoeck & Ruprecht, 1972), pp. 693–703.

Reed, Ronald, *Ancient Skins, Parchments, and Leathers* (London: Seminar Press, 1972).

Rosen, Edward, 'The Invention of Eyeglasses', *Journal of the History of Medicine and Allied Sciences*, 11 (1956), pp. 13–46.

Tolaini, Francesca, '"De tinctio omnium musivorum". Technical Recipes for Glass in the So-Called "Mappae clavicula"', in *When Glass Matters*, ed. by Marco Beretta (Florence: Olschki, 2004), pp. 195–219.

Trinquier, Jean, 'Les vertus du vert dans l'antiquité', in *Couleurs et vision dans l'Antiquité classique*, ed. by Laurence Villard (Rouen: L'université de Rouen, 2002), pp. 97–128.

Woods, David, 'Pliny, Nero, and the Emerald', *Acta philologica fennica*, 40 (2006), pp. 189–96.

SAMUEL GESSNER

The Perspective of the Instrument Maker

The Planispheric Projection with Gemma Frisius and the Arsenius Workshop at Louvain

Introduction

A set of mathematical notions, procedures, instruments, and practices were shared by a number of masters across various arts and disciplines in the fifteenth and sixteenth centuries. The development of perspective therefore needs to be considered in the light of the variety of the involved individuals: university masters of mathematics, private *abbaco* teachers or teachers at urban schools, painters, architects, engravers and goldsmiths, instrument makers, surveyors, and military engineers. This point of view has been eloquently developed by historians including Jane Andrews Aiken and Jim Bennett, who brought together the study of theoretical and practical mathematical cultures, most notably instrument making. The former forcefully argued that '[…] Alberti's perspective method and the theory supporting it may be regarded as an adaptation by artists of the values, vocabulary, and methods of graphic techniques already employed by astronomers and instrument makers'.[1] The latter showed, in a very broad sense, how those practitioners applied perspective in different areas of geometrical practice, while sharing the same projection constructions and recurring to the same mathematical instruments.[2] A case in point has also been provided by Sven Dupré's study concerning the important contribution made by mathematical instrument design to the development of orthographic projection in the sixteenth century.[3] The early modern mathematical practitioners thought of perspective in broader terms than that of 'mimetic representation' and understood it as optical theory. In the context of instrument workshops, perspective as a practice referred to the techniques of rendering spatial configurations, like those of the celestial circles, on a plane. In this chapter I will adopt their point of view to address a puzzling question that arose with the examination of

1 Jane Andrews Aiken, 'Truth in Images: From the Technical Drawings of Ibn al-Razzaz al-Jazari, Campanus of Novara, and Giovanni de'Dondi to the Perspective of Leon Battista Alberti', *Viator. Medieval and Renaissance Studies*, 25 (1994), 325–59 (p. 328).

2 Jim Bennett, 'Practical Geometry and Operative Knowledge', *Configurations*, 6 (2) (1998), 195–222.

3 Sven Dupré, 'Galileo, Mathematical Instruments and Orthographic Projection', *Bulletin of the Scientific Instrument Society*, 69 (2001), 10–20. On orthographic projection, see also Filippo Camerota, 'The Eye of the Sun: Galileo and Pietro Accolti on Orthographic Projection', in *Perspective, Projections & Design: Technologies of Architectural Representation*, ed. by Mario Carpo and Frédérique Lemerle (London: Routledge, 2008), pp. 115–25.

Samuel Gessner University of Lisbon, samuel.gessner@gmail.com

Perspective as Practice. Renaissance Cultures of Optics, ed. by Sven Dupré, Turnhout, 2019 (*Techne. Knowledge, Technique, and Material Culture*, 1), p. 103-122

DOI 10.1484/M.*Techne*-EB.5.117723

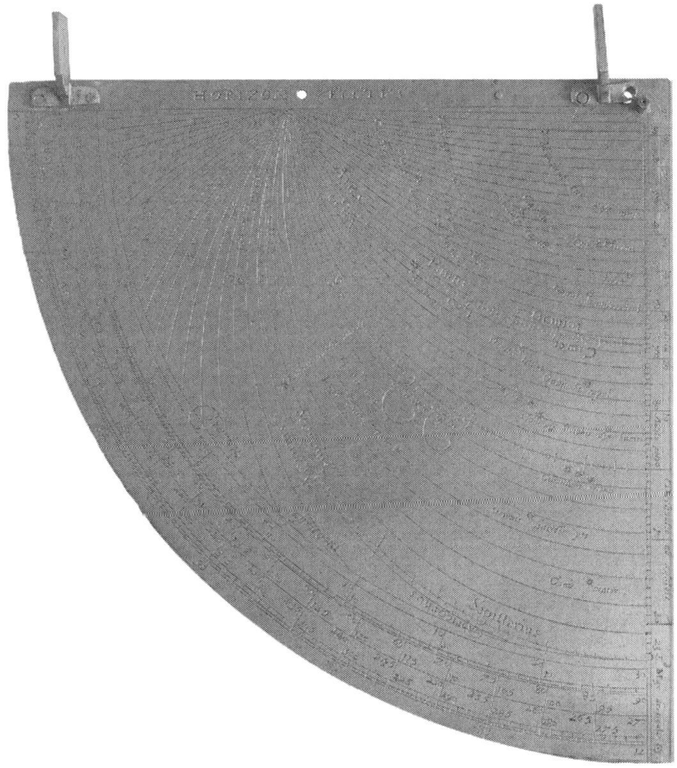

Fig. 1 Astrolabe quadrant by Ieremias Arscenius, 1573. Gilt brass, 170 mm. Photo by J.N. Lamas, MUHNAC, inv. nr. MUHNAC-UL-DEP262. Courtesy MUHNAC. [See colour plate 13]

a curious sixteenth-century instrument: a small 'astrolabe quadrant' by Ieremias Arsenius from 1573 (Figure 1). Its close examination will highlight the necessity for the historian to try and reconstitute workshop practice and the knowledge shared at the time to accurately interpret the instruments produced in the past.

The Arsenius workshop stood in the tradition of Gemma Frisius's and Gerardus Mercator's instrument making in Louvain.[4] It was mainly due to Gemma's activity that artisans, university students, and humanistic scholars with connections to courtiers were brought together here, precisely the wide spectrum of mathematical practitioners who would share knowledge and procedures, regardless of the differences between their fields. Gemma Frisius (1508–55), a physician by training, seems to have been mainly active in the edition of cosmographical textbooks, the teaching of mathematics, and the making of mathematical instruments including globes. He published the earliest known description of a particular surveying procedure by 'triangulation'.[5] Gemma collaborated with the

4 Koenraad van Cleempoel, *A Catalogue Raisonné of Scientific Instruments from the Louvain School, 1550–1600* (Turnhout: Brepols, 2002).

5 Fernand van Ortroy, *Bio-bibliographie de Gemma Frisius, fondateur de l'école belge de géographie, de son fils Corneille et de ses neveux les Arsenius*, Mémoire de l'académie royale de Belgique, 2nd series, 11 (2) (Brussels: M. Lamertin, 1920); Fernand Hallyn, *Gemma Frisius, arpenteur de la terre et du ciel* (Paris: Honoré Champion, 2008).

artist and engraver Gaspar van der Heyden (Casparus a Myrica) (1496–after 1549) and also with young Gerardus Mercator (1512–94) in the production of the first copper plate print terrestrial globe (in 1536) and a celestial globe (in 1537).[6] The first of these globes was dedicated to Maximilianus Transylvanus (*c.* 1490–*c.* 1538), author of the first published account of Magellan and Elcano's circumnavigation of the world (between 1519 and 1522), and a councillor to Emperor Charles V whose court resided often in Brussels. Gemma's home in Louvain could be seen as a centre that attracted many interesting visitors: John Dee (1527–1609) would visit Gemma and Mercator.[7] Juan de Rojas Sarmiento (fl. 1540–50) would also be deeply influenced by his time spent in Louvain. Mercator, following in Gemma's footsteps, brought instrument making to new heights, creating large and finely crafted gilt brass instruments.[8] Although he is remembered mainly as a cartographer today, he actually combined several skills, including calligraphy, copper engraving, astronomy, and geometry. Notably, Mercator designed the famous italic script type that we see used on the quadrant presented here, and that would be widely adopted for the labels on maps and instruments.[9] Later, according to Koenraad van Cleempoel, Gualterus Arsenius (d. 1580) seems to have taken over Mercator's workshop (and his patronage network) when the latter left for Duisburg in 1552, probably following being arrested upon religious charges. The association of the workshop with Gemma Frisius continued, too. Those working in that workshop often used the epithet *nepos Gemmae Frisii* ('nephew [or grandson] of Gemma Frisius') for themselves when signing their instruments, without it being quite clear in what sense they were Gemma's actual 'nephews'.[10] They included Gualterus Arsenius, Regnerus Arsenius (whose precise identity remains unclear), Ferdinand Arsenius (fl. 1573–1628), and as we have seen on our small quadrant: Ieremias Arsenius (fl. 1573). Cornelius Gemma (1535–79), Gemma's son, established himself as a mathematics teacher alongside Pierre Beausard (1535–77) in Louvain after Gemma's death in 1555. None of these individuals seem to share the same intellectual profile, but it is not surprising that they shared knowledge, problems, procedures, and concerns.

This milieu provides the cultural background against which the questions raised in our present case study — the astrolabe quadrant — can be adequately tackled. For this reason, before diving into the specific interpretation of the instrument, I will briefly address the following more general concerns: Firstly, the procedure to obtain a common planisphere is usually called 'stereographic projection' and not 'perspective', although both procedures are analogous. Both consist in the art of putting the eye of the beholder in one

6 The only surviving terrestrial globe (37 cm) belongs to a private collection, and is on loan to the Globenmuseum, Vienna. The only surviving celestial globe (37 cm) is conserved at the National Maritime Museum, Greenwich. See Elly Dekker, *Globes at Greenwich: A Catalogue of the Globes and Armillary Spheres in the National Maritime Museum, Greenwich* (Oxford: Oxford University Press, 1999), pp. 87–91, 341–42.

7 Bruno Almeida, 'On the Origins of Dee's Mathematical Programme: The John Dee-Pedro Nunes Connection', *Studies in History and Philosophy of Science*, Part A, 43 (3) (2012), 460–69.

8 Gerard L'E. Turner and Elly Dekker, 'An Astrolabe Attributed to Gerard Mercator, *c.* 1570', *Annals of Science*, 50 (5) (1993), 403–43; Gerard L'E. Turner, 'The Three Astrolabes of Gerard Mercator', *Annals of Science*, 51 (1994), 329–53.

9 Arthur S. Osley, *Mercator, A Monograph on the Lettering of Maps, etc. in the 16th Century Netherlands with a Facsimile and Translation of His Treatise on the Italic Hand and a Translation of Ghim's Vita Mercatoris* (London: Faber and Faber, 1969).

10 Van Cleempoel, *A Catalogue Raisonné*.

particular place and projecting an image onto a well-defined plane. Were they perceived as analogous procedures among scholars, 'connoisseurs', and instrument makers at the time? We will see that the answer to this question is yes, particularly if we take the example of the scholar Gemma Frisius. Secondly, ideas of perspective participated in the meaning of pictorial representations as a symbolic form. With a planisphere representing the vault of heaven, a unique vantage point for the whole world is given. Did this have implications with the symbolic use of planispheric instruments, such as astrolabes, in the Renaissance? And finally, a question central to this book as a whole concerns practice in workshops: to what extent do planispheric representations on astrolabes and other instruments enable us to learn about procedures, habits, and knowledge available in the workshops where they were produced?

The Whole World at a Glance: A Question of 'Perspective'

When the sky is represented on celestial globes or planispheres, the traditional forty-eight mythical figures surrounding the stars of the constellations are depicted. One realises that in most cases the figures of human shape are represented from the rear, turning their back to viewer. This derives from the fact that — in the night sky — all constellation figures are imagined as facing us (according to the so-called Hipparchos rule) and looking down onto the Earth.[11] From God's perspective — indeed, the way we ourselves look at a celestial globe — the constellations are to be seen from outside of the sphere of the fixed stars, with only their rear side discernible. This clearly very much becomes a question of 'perspective'.

One of the most influential star maps, and the first to appear in print, consisted of a pair of woodcuts by Albrecht Dürer (1471–1528) from 1515 (Figure 2).[12] For Dürer's *Imagines coeli*, according to the caption in the lower left corner, it was apparently Conrad Heinfogel (*c.* 1452–1517) who set the star positions, with Dürer surrounding them with the figures.[13] The arrangement of the overall design was made by Johannes Stabius (*c.* 1460–1522), who incidentally is known to have experimented with various types of 'projections' of the sphere, notably for representing the terrestrial globe.[14] Heinfogel presumably had a star list containing the degree values for the longitudes and latitudes that he used to fix

11 Elly Dekker, *Illustrating the Phaenomena: Celestial Cartography in Antiquity and the Middle Ages* (Oxford: Oxford University Press, 2013), p. 35.

12 Albrecht Dürer, *Imagines coeli septentrionales cum duodecim imaginibus zodiaci* and *Imagines coeli meridionales* (Nuremberg, 1515), woodcut prints, *c.* 45 × 43 cm (various extant copies: Metropolitan Museum of Art, New York, inv. 51.537.1, or Staatliche Graphische Sammlung, Munich, inv. 118930 and 118931, etc.); Walter L. Strauss, *Albrecht Durer, Woodcuts and Wood Blocks* (New York: Abaris, 1979), no. 171, 172. The wood blocks are conserved in the Kupferstichkabinett in Berlin. On the complex connections and various states of the surviving prints see: Hans Gaab, *Die Sterne über Nürnberg: Albrecht Dürer und seine Himmelskarten von 1515* (Petersberg: Michael Imhof, 2015).

13 'Ioann. Stabius ordinauit, Conradus Heinfogel stellas posuit, Albertus Durer imaginibus circumscripsit' ('Johannes Stabius ordained [it], Conrad Heinfogel set the stars, Alrecht Dürer surrounded them with images'). Dürer, *Imagines coeli.*

14 On Stabius, see Helmuth Grössing, 'Wiener Astronomen und Mathematiker des 15. und beginnenden 16. Jahrhunderts und ihre Instrumente', *Wiener Geschichtsblätter*, 38 (4) (1983), 149–62; on the so-called Stabius-Werner projections, see John P. Snyder, *Flattening the Earth: Two Thousand Years of Map Projections* (Chicago: University of Chicago Press, 1993), pp. 33–37.

Fig. 2 'Imagines coeli septentrionales', Albrecht Dürer, Johann Stabius, Conrad Heinfogel, 1515. Courtesy Kupferstichkabinett; Courtesy SKD. [See colour plate 14]

the positions of the stars on the planisphere. The one graduated circle, the ecliptic, on the Dürer *Imagines* allows determining one of the coordinates (ecliptic longitude) of each represented star. The second coordinate, the latitude, is not readily available, because it is presented in stereographic projection for northern stars, and, according to Elly Dekker, an approximation of this projection is used for stars south (i.e. outside) of the ecliptic. In other words, the mapping used for those stars uses a radial foreshortening.[15]

Several procedures, however, for setting the stars in their place on a planisphere were well known to those familiar with the tradition of astrolabe making. In contrast with Dürer's planispheres that imply a sort of projection centred at a pole of the *ecliptic*, the stereographic projection usually adopted for astrolabes used the southern pole of the *equator*. The treatise that would become most well-known during the sixteenth century to explain

15 Elly Dekker, 'Construction and Copy: Aspects of the Early History of Celestial Maps', Beiträge zur Astronomiegeschichte, Band 13, *Acta Historica Astronomiae*, 58 (2016), pp. 47–93.

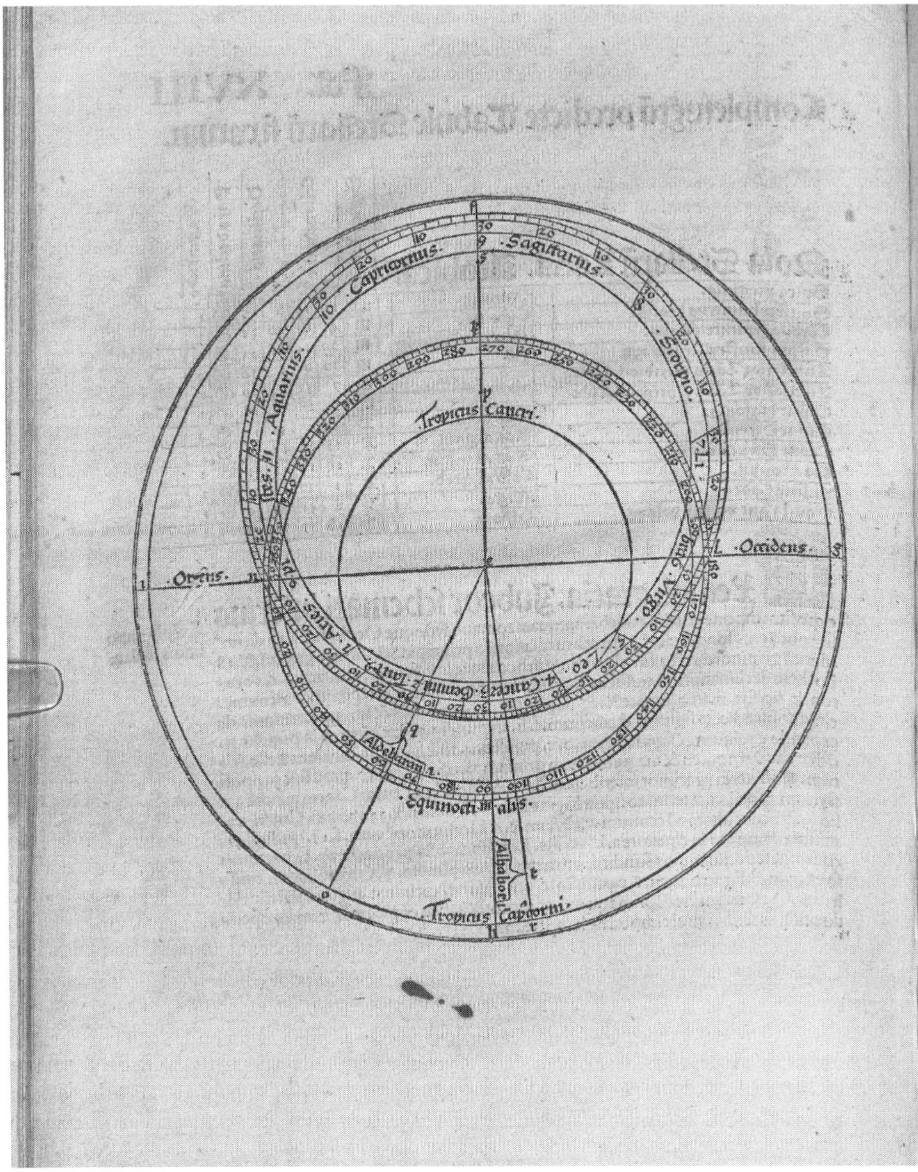

Fig. 3 Construction of the rete. Johann Stöffler, *Elucidatio fabricae ususque astrolabii*, (Oppenheim, 1513), fol. 18v. Bayerische Staatsbiblioth. Res/2 Math. a. 94. Courtesy Bayerische Staatsbibliothek.

the structure and use of the planispheric astrolabe was the *Elucidatio* by Johannes Stöffler (1452–1531), written around 1510 and first published in 1513.[16] For an astrolabe, all types of celestial circles and the positions of selected bright stars are represented on a plane parallel to the equatorial plane: the tropic circles, the celestial equator, the horizon, altitude and

16 Johannes Stöffler, *Elucidatio fabricae ususque astrolabii* (Oppenheim: Köbel, 1513).

azimuth circles of a given latitude, and also the line of twilight, the cusps (borders of the twelve celestial houses) and the hour lines of the unequal hours are all plotted onto a solid plate, held in the cavity of the astrolabe's *mater* in fixed position. As to the ecliptic and the star positions, they are represented by pointed tips on the *rete*, or spider, a rotating piece often produced as finely ramified strap work, superposed on the former.

Stöffler describes the way in which the circles are constructed and graduated, and the procedure of fixing the star positions in the correct place (Figure 3). For instance, the graduation of the ecliptic (twelve signs of thirty degrees each) can be done, according to Stöffler, in five different ways, two of which require tables, while three methods are purely geometrical. Among the geometrical methods Stöffler describes, one uses straight lines and another large circles — supposedly to obtain the same result.[17] Similarly with the positioning of the stars: Stöffler shows how to use star tables with celestial coordinates given either by declination and *mediatio coeli* ('mediation'), or then with ecliptic coordinates (latitude and longitude).[18] Never the word planisphere or perspective is used, at most the verbs *delineare* ('trace') or *pingere* ('paint') occur which, incidentally, establish a tenuous link with pictorial practice. What is important to remember here is that one and the same representation is said to be achievable by various modes: 'Five ways of dividing the rete'.[19] It is striking that this variety of methods to achieve the 'same' result were also admitted by perspectivist painters at the time, for instance in the treatise *Due regole* by Giacomo Barozzi da Vignola.[20] Today some of the geometric procedures may be considered correct and others wrong, but the historical actors would not use the same criteria as we do to make this distinction. This is consistent with the larger tradition of practical geometry in the early modern period.

Since the Middle Ages, astrolabe treatises have often offered variants for the way of dividing the ecliptic or setting the stars. The problem had been addressed *more geometrico* already in Jordanus Nemorarius's treatise *De plana spera* (before 1250), a work on the stereographic projection. As Aiken noted, the procedure we now call the 'planispheric projection' is described here as putting a *virtus visiva* ('capacity of seeing') at the pole.[21]

> For let us imagine that a flat surface touches the sphere at one of its poles. And we assume that the remaining pole has the capacity of seeing, and furthermore that the sides of the sphere are not able to limit a ray, but that [the ray] itself is carried on as far as the plane (which it is assumed that the sphere touches) and is shown on it; and that there, any point on the sphere is seen where the ray from the observing pole passing through that point intersects with the plane itself.[22]

17 Stöffler, 'De compositione retis vel arane[a]e. Propositio Undecima partis prim[a]e', in *Elucidatio*, fol. 13.

18 Stöffler, 'De descriptione et impositione stellarum. Propositio Duodecima partis prim[a]e. Stellas fixas reti via geometrica imponere', in *Elucidatio*, fol. 17.

19 'Quinque modi diuiendi rethe'. Stöffler, *Elucidatio*, fol. 13.

20 Jacopo Barozzi da Vignola, *Le due regole della prospettiva pratica con i commentarij del P. Egnatio Danti* (Rome: Zanetti, 1583).

21 Jane Andrews Aiken, *Renaissance Perspective: Its Mathematical Source and Sanction* (unpublished doctoral dissertation, Harvard University, 1986), p. 281.

22 'Ymaginemur enim quod plana superficies speram in alterutro polorum suorum contigat. Reliquumque polum virtutem ponemus habere visivam, partes autem spere non posse radium terminare sed ipsum usque ad planum (quod positum est speram contingere) deferri et ab eo ostendi, ibique quodlibet punctum spere videri ubi raius a

More widely available than Jordanus *De spera plana* was Pseudo-Masha'allah's *On the Astrolabe*. It already contained a variety of alternative methods of constructing the geometrical parts of the astrolabe. The variant in chapter seventeen reads:

> The flattening or extension, or more correctly the projection by sight, of a sphere onto a plane is effected in this manner: let the plane be line *MBN*, the axis of the sphere, line *AB*, standing perpendicular on the plane *MBN* so that the north pole touches plane *MBN* at point *B*. However, let the other one, that is the southern one, be at the greatest distance from the plane at point *A*, which is the eye of the observer.[23]

There seems to be an echo of these passages in the *Cosmographicus liber* of Petrus Apianus from 1524.[24] This book was later widely disseminated in many re-issues and re-workings by Gemma Frisius. Apianus defines 'cosmography' — in opposition to geography — as a *mundi descriptio* ('description of the world'): 'as it discerns the Earth by means of celestial circles, not by mountains, seas, and rivers'.[25] In cosmography, as Apianus explains, initially, the celestial circles are considered, and the situations of the Earth's parts are then determined according to the celestial circles they are 'subjacent' to: 'thereafter, using their distinction, [cosmography] shows by mathematical demonstrations the situation of landmasses subjacent to them [the celestial circles]'.[26]

As a didactic device to illustrate this definition the author inserts an extraordinary exploded view of the celestial circles on the eighth sphere, the Earth and the eye of the beholder at the centre of the world (Figure 4). This plate uses the concept of vision as an analogy to make this clear (a single eye emitting visual rays to the celestial circles through the surface of the Earth), to show how the place and proportions of geography are related to the celestial circles.

These examples from Jordanus, Dürer, Stöffler, and Apianus suggest that the process of representing the world as a whole was seen in similar terms as the process of vision, in particular, it is the vision of one single 'eye', and further, that various procedures lead to the same representation.

Gemma Frisius: The Astrolabe as an Effigy of the World

Gemma Frisius was a reader of the writings of Dürer, Stöffler, and Apianus and became one of the most active broadcasters of Apianus's cosmographical teaching. In Gemma

---

polo vidente per punctum illum transiens ad ipsum planum inciderit'. Ron B. Thomson, *Jordanus de Nemore and the Mathematics of Astrolabes: De plana spera*, Studies and Texts, 39 (Toronto: Pontifical Institute of Mediaeval Studies, 1978), pp. 146–47.

23 'Concussio sive extensio immo verius proiectio spere in planum per visum fit hoc modo. Sit planum linea MBN, axis spere linea AB stans ortogonaliter super planum MBN ita quod polus septentrionalis contigat planum MBN in puncto B. Alter vero scilicet meridionalis maxime distet a plano in puncto A qui est oculus videntis'. Pseudo-Masha'allah, *On the Astrolabe*, ed. by Ron B. Thomson, version 1.2 (Toronto, 2015). See: https://shareok.org/handle/11244/14221 [accessed 27-07-2017].

24 Petrus Apianus, *Cosmographicus liber* (Landshut: Weyssenburger, 1524).

25 'quia terram distinguit [tamen] per circulos coeli, non per montes, maria et flumina'. Apianus, *Cosmographicus liber*, 1.

26 'deinde ex ipsorum distinctione terrarum illis subjectarum situs [...] iuxta mathematicas ostensiones demonstrat'. Apianus, *Cosmographicus liber*, 1.

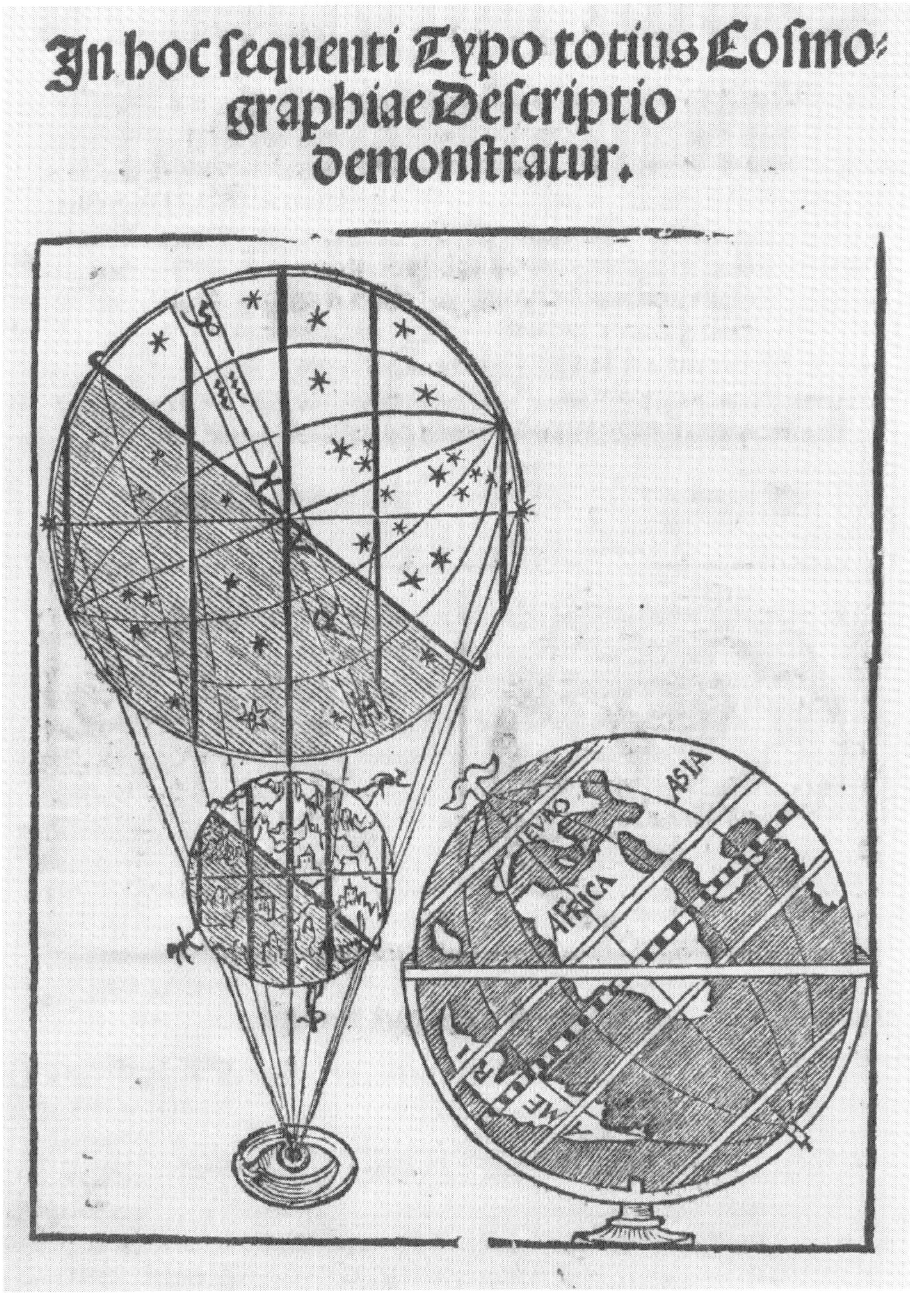

Fig. 4 Exploded view of the nested terrestrial and celestial spheres. Petrus Apianus, *Liber cosmographicus* (Oppenheim, 1524), p.2. Courtesy Dibner Library of History of Science.

we find a very clear expression of the analogy of optics and stereography, and also of the analogy of the methods of painters and the construction of a planisphere. This view is specifically expressed in his treatise on the universal astrolabe, *De astrolabo catholico*, posthumously published by his son Cornelius Gemma in 1556.[27] First of all, he uses the expression *proiectio* ('projection') in the title of the book's first chapter. It is this term 'projection' that will later be adopted in mathematics for precisely that kind of geometric procedure. The claim is furthermore made that this is exactly what the painters likewise do when they represent houses, theatres, and cities. The first chapter of the work is entitled 'Caput primum. De proiectione Sphaerae in planum, & de Astrolabi compositione' ('About the projection of the sphere on a plane, and about the constitution of the astrolabe'). The chapter begins:

> [...] the ancients have tried, and knew to transfer the circles of the globe by amazing ingenuity to a flat surface, through which we get usefulness of the greatest things, it is what Vitruvius calls Analemma as I understand it. But while it is impossible to perfectly adapt the surface of a sphere to a plane such that all elements conserve the same ratio to each other as they have on the curved surface of the globe, a different way has been invented by those industrious men, by which apparently all those things that appear to us set on the globe at a certain place, we perceive the same in the same way of vision on a plane, which is a trick of optics. By the same [trick] the painters graphically depict for us on a flat surface houses, theatres, even cities, or whatever you like, such that we seem to perceive these according to three dimensions, although they can only receive two [dimensions] on a flat surface.[28]

27 The analogy became so habitual that Commandino very naturally subsumed Ptolemy's and Jordanus's works on the planisphere under special cases of *scenographia* used by the architects. See Federico Commandino, *Ptolemaei planisphaerium* [...] (Venice: Aldo Manuzio, 1558). See also Sinisgalli's translation: Rocco Sinisgalli, *La prospettiva di Federico Commandino* (Florence: Cadmo, 1993). Later, Clavius would also refer to Gemma Frisius's concept of positioning an eye and producing a planisphere — here a universal one by using the vernal point as pole of projection — by the figure perceived on a plane: 'Quidam enim, inter quos est Gemma Frisius non ignobilis scriptor in Astrolabio suo universali, quod Catholicum appellat, oculum collocant in communi sectione Aequatoris atque Eclipticae, omnesque circulos caelestes in plano Coluri solstitiorum qui Meridianum circulum refert, ea forma describunt, qua eos oculus intuetur. [...] Ptolemaeus denique Astronomorum princeps constituit oculum in polo australi, circulosque omnes primi Mobilis, lineas, ac puncta in plano Aequatoris in infinitum extenso ea figura depingit, qua ex polo australi eo in plano cernuntur'. ('Now some, among whom there is Gemma Frisius a worthy writer in his *Universal astrolabe* that he calls *catholic*, who set the eye at the intersection of the equator and the ecliptic, and they describe all celestial circles on the plane of the solstitial colure — that corresponds to a meridian circle — by the shape in which the eye beholds them. [...] Ptolemy the Prince of the astronomers, finally, put the eye at the south pole, and depicts all circles of the prime mover, the lines and points on the infinitely extended equatorial plane by the shape in which they are seen on that plane from the southern pole'.) Christopherus Clavius, *Astrolabium* (Rome: Giunti, 1593), pp. 269–70.

28 '[...] veteres studuerunt, ac conati sunt miro ingenio globi circulos, per quos maximarum rerum commoditates accipimus, in planam superficiem traducere, id quod Analemma vocat Vitruuius ut ego [fol. 2^v] interpretor. At quandoquidem impossibile est globi superficiem plano prorsus adaptari, ita ut omnia eandem in plano seruent ad inuicem rationem, quam in gibba globi superficie habent, inuenta est industriis illis viris alia ratio, qua scilicet ea omnia quae nobis certo constitutis loco in globo videntur, eodem visus modo in plano conspiciamus, id quod της Ὀπτικῆς artificium est. Quo pictores in plana superficie nobis domos, theatra, imò urbes ac alia quaeuis ita γράφικη depingant, ut ea nobis cernere videamur secundum tres dimensiones, quae tamen in plana superficie non nisi duas obtinere possunt'. Gemma Frisius, *De astrolabo catholico liber: quo latissime patentis instrumenti multiplex usus explicatur, & quicquid uspiam rerum mathematicarum tradi possit continetur*, ed. by Cornelius Gemma (Antwerp: Johannes Steelsius, 1556), fols 1^r, 2^{r-v}. Translations of Gemma Frisius are provided by the author unless otherwise stated.

In the margin, Frisius adds the words: 'Short presentation of the optical theory which is amazingly useful for understanding the grounds of the planisphere'.[29] The text continues:

> However, by the gaze of sight or perception, a description on a plane can be made that exhibits the same for us to see as we behold otherwise in space. This trick the painters show us every day, and Albrecht Dürer, a noble painter as well as a mathematician, has given very beautiful illustrations of this. For he shows how on a flat surface that he applies to a window, any objects presented to the eye are thus described on a plane. By a second [example] he explains the same [thing] even more clearly: while a flat glass pane is set between the viewer and the object, whatever things are seen by a fixed eye through the glass he paints by hand on the glass surface. As this is primarily useful for those who learn graphic [arts], so it is for understanding our many rules of what is done. Actually, we too could trace, on a glass surface, the Sphere of the orbs with its circles that we see through the glass, following the gaze of our sight through all places of the sphere, and carefully sketching everything by hand.[30]

And Gemma illustrates this with a woodcut (Figure 5) that makes those procedures, already taught by Stöffler, easily intelligible as perspective practice:

> The description of which [of the astrolabe] has been extensively given by Johannes Stöffler, we put in front of the [reader's] eyes by just one diagram, however, the method of its construction, by which we will teach that this production of the sphere onto a plane arises from vision, i.e. by an eye's gaze, and we simultaneously offer the principles of the whole construction.[31]

It should be noted that the analogy of perspective representation with the intersection of the visual pyramid and a glass plane occurred before Dürer made reference to it, in the work of Leon Battista Alberti (1404–72), who explained the following in his *De pictura*:

> They [the painters] should understand that, when they draw lines around a surface, and fill the parts they have drawn with colours, their sole object is the representation on this surface of many different forms of surfaces, just as though this surface which they colour were so transparent and like glass, that the visual pyramid passed right

29 'Opticae doctrinae succincta traditio ad planae sphaerae rationem percipiendam mirè utilis'. Gemma Frisius, *De astrolabo*, fol. 2v.

30 '[...] Attamen per visus aspectum seu intuitum fieri potest descriptio in plano quae eadem nobis in plano videnda exhibeat quae alioqui in solido comprehendimus. Hoc artificium nobis quotidie pictores exhibent, & scribit de ea re pulcherrima exempla Albertus Durerus nobilis & pictor & mathematicus. Docet enim quomodo in plana superficie quam fenestrae applicat, quaecunque obiecta oculo obuertuntur ita in planum describantur. Secundo idem clarius innuens, vitream planam tabulam inter aspectum et rem obiectam collocans, aspectu fixo quaecunque videntur per vitrum manu depingit in vitrea superficie. Hoc ut primum γράφικημ discensibus utile est, ita ad nostrum institutum intelligendum plurimum facit. Nam et nos per vitrum aspicientes Sphaeram orbicularem cum suis circulis, in vitri superficie poterimus describere, sequentes intuitus aciem per omnia sphaerae loca decurrentem, manu omnia notantes diligenter'. Gemma Frisius, *De astrolabo*, fol. 3v.

31 'Huius descriptionem [astrolabii] Ioannes Stoflerus prolixè ac diligenter prosequitur, verùm nos uno tantum schemate methodum huius compositionis ob oculos ponemus, quo docebimus et hanc sphaerae in planum productionem ab intuitu, siue oculi inspectione [diagram 6v, fol. 7r] ortum habere, et simul totius compositionis fundamenta trademus'. Gemma Frisius, *De astrolabo*, fol. 6r.

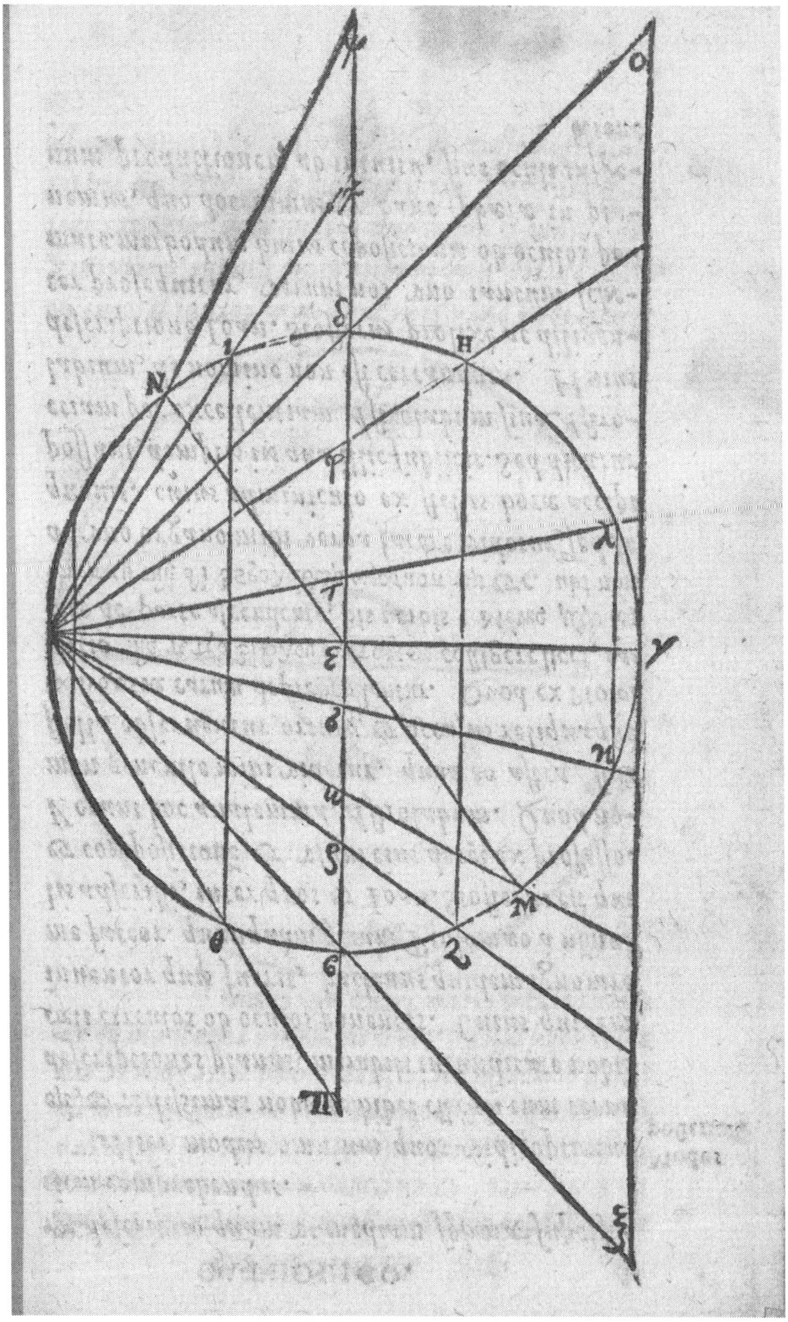

Fig. 5 Diagram explaining the stereographic projection underlying the flattening of the sphere. Rainerus Gemma Frisius, *De astrolabo catholico* (Antwerp, Joannes Steelsius, 1555), p. 6[v]. ETH-Bibliothek Zürich, Rar 4404. Courtesy ETH, Library, Zürich, http://www.e-rara.ch.

> through it from a certain distance and with a certain position of the centric ray and of the light, established at appropriate points nearby in space.[32]

Finally, given the analogy with depicting and pictorial representation, we now can more easily understand also the symbolic aspect of a planispheric projection on an astrolabe. The planisphere on an astrolabe is used as a metaphor by Gemma Frisius's son Cornelius. His long dedication poem to King Philip of Spain, the son of Charles V, dated 1555, in Gemma's astrolabe treatise, ends with four verses that suggest he was dedicating the book as well as the actual metallic instrument:

> Receive also the orbs with many orbs intertwined | an image of the Reign compressed into a mass not comparable in size, | but similar to the image sculpted by Vulcan on the shield, | that the Trojan Aeneas carried with himself, depicting the glory and fate of his descendants.[33]

To someone who is familiar with Virgil's long description of the round shield, *clipeus Volcani* ('Vulcan's shield'), used by Aeneas, the parallel with the astrolabe does not fail to conjure up a strong image. The legendary shield was covered with depictions in silver and gold of glorious events that lay ahead in Roman history (as can be seen in the *Aeneid* VIII, verses 626–731, that terminates with exactly the same expression: 'famamque et fata nepotum'). Aeneas carried the shield not knowing that the images predicted the glory of his descendants. If we consider the idea of prognostication by astrology, common in early modern Europe, the astrolabe could be seen as a forecaster of future events, similar to Aeneas's shield. Cornelius's poem plays with this similarity. It expresses a view of such instruments that was frequently held in the sixteenth century: the astronomical-mathematical features were meant not only for actual reckoning but also for providing a symbolic image and for reminding the beholder of the intricacies of the cosmic order.

We have now prepared the stage with notions of planispheric projection and its relationship to pictorial perspective. We have highlighted what Lucien Febvre has called the *outillage mental* typical of sixteenth century people, i.e. some linguistic, intellectual, and emotional elements that circumscribe the field of thoughts and values concerning the perception of the whole world at a glance. Now we may turn to look at the highly specific problem raised by the small brass quadrant mentioned at the beginning of this chapter.

---

32 'Ac discant quidem dum lineis circumeunt superficiem, dumque descriptos locos implent coloribus, nihil magis queri quam ut in hac una superficie plures superficierum formae repraesententur, non secus ac si superficies haec, quam coloribus operiunt, esset admodum vitrea et perlucida huiusmodi ut per eam tota pyramis visiva permearet certo intervallo certaque centrici radii et luminis positione cominus in aere suis locis constitutis'. Leon Battista Alberti, *De pictura praestantissima et nunquam satis laudata arte libri tres* (Basel: Th. Venatorius, 1540), p. 12. The English translation is from Leon Battista Alberti, *On Painting and on Sculpture, the Latin Texts of 'De pictura' and 'De statua'*, ed. by Cecil Grayson (London: Phaidon, 1972), pp. 48–49. While the text circulated in multiple manuscript versions, the *editio princeps* is *De pictura* (1540).

33 In the margin these last lines of the poem are referenced as *Dedicatio instrumenti* ('Dedication of the instrument'): '[…] Accipe & innumeris perplexos orbibus orbes, | Effigiem regni non aequa in mole coactam. | Sed qualem semet Vulcani pictus in armis | Tros tulit Æneas, famamque & fata nepotum'. Gemma Frisius, *De astrolabo*, [sig. † †4^{v}]. This is my translation and I am grateful for the help received from my colleague Bernardo Mota.

Thoughtlessness or Meaningful Change? The Astrolabe Quadrant by Ieremias Arsenius

In 1527, only three years after the first edition of Apianus's *Cosmographicus liber*, Oronce Fine's work on a *quadrans universalis* (or 'universal quadrant') is published.[34] The main underlying principle of this instrument, commonly called an astrolabe-quadrant, is the 'double-folding' of the planisphere, itself the result of a stereographic projection of the celestial circles onto a plane. In the Middle Ages, the quadrant with such a doubly-folded projection was called *quadrans novus*, as opposed to the well-known horary quadrant. Its inventor is usually thought to be Jacob ibn Tibbon (also known as Prophatius Judaeus, 1236–1305). When Oronce Fine (1494–1555) describes this type of instrument, he states in the preface that it is an adapted version from both a treatise by Petrus Dacus (or Peter Nightingale, fl. 1290–1300) and from a manuscript by his father from *c.* 1475. The title *quadrans universalis* ('universal quadrant') underlines the fact that it comprised several horizon circles for various geographic latitudes (Figure 6). This feature replaced the set of differently engraved plates, each one for a single given latitude, to be used with the typical astrolabe. Indeed, Fine's illustrating plate shows many different horizon circles, one every three degrees latitude.

The small brass instrument mentioned above features precisely such an astrolabe-quadrant. It is held at the Museu Nacional de História Natural e da Ciência (Lisbon, inv. nr. MUHNAC-UL-DEP262). It is inscribed with a signature that apparently connects it with the famous Arsenius workshop: 'Ieremias Arscenius nepos Gemmę Frisÿ | Louanÿ fecit anno | 1573 ·1· Augusti', ('Ieremias Arsenius, the nephew of Gemma Frisius, made [it], at Louvain, in the year 1573, 1st August'). The name Ieremias Arsenius, however, is not otherwise documented in the literature.[35] The quadrant has a radius measuring approximately seventeen centimetres; the plate that constitutes the main part is around half a millimetre thick. Two brass sights with pinholes are fixed to the top edge of the instrument. The instrument is however not quite complete any more: it certainly had been equipped once with a thread and sliding bead attached to the *brachiolus*, the hinged arm. An index ruler, like the one visible at the bottom of Fine's engraving (Figure 6), seems to also be missing. It would have born a scale for declination and zenithal distance and would have been indispensable for the astrolabe-quadrant's proper functioning, unless there was in its place another thread with a sliding bead.

The instrument bears a striking resemblance to several elements in Fine's sundial treatise. By carefully examining all inscribed features of this quadrant, it can be reasonably

34 Oronce Fine, *Descriptio partium, et succinta utilitatum elucidatio quadrantis cujusdam universalis* (Paris: Nicolaus Savetier, 1527). Oronce Fine includes an adapted version of this treatise in the part *De solaribus horologiis, et quadrantibus libri IIII*, on fols 157 ff. of his *Protomathesis opus uarium, ac scitu non minus utile quàm iucundum* (Paris: Gerard Morrhius and Joannes Petrus, 1532). See Books III and IV in particular: 'Liber tertius. De reductione planisphaerii seu uulgaris Astrolabij, in circuli quadrantem: eiusdem, uel aequè facilis cum ipso Planisphaerio usus atque commoditatis', fols 202r ff. and 'Liber quartus et ultimus, de praecipuis utilitatibus, usuque multiplici ipsius proximè descripti quadrantis universalis', fols 205r ff. In 1534 a new version of the quadrant treatise came out, this time under the title *Quadrans astrolabicus*. It contained no indications about the making but gave a brief description of the instrument and a list of over fifty of its possible uses.

35 Neither Van Ortroy nor Van Cleempoel seems to have encountered this name.

assumed that the instrument was made by using as a source some edition of that treatise and, in particular, the images on several adjacent pages.[36] It appears that the woodcuts of the book are used as general schemes determining the overall layout of the instrument. Fine's book served as a 'pattern book', from which Arsenius eclectically combined several instruments on the same piece. This combination of instruments on the quadrant enabled its user to perform cosmographical and time-keeping operations using the sun or stars, either by day or by night.

Particularly interesting, however, are the aspects where the maker chose to deviate from the model described in the book. The list of stars selected by Arsenius represents such a deviation. Only some stars coincide with those tabulated in Fine's treatise. And for all star coordinates an updated correction for the precession of the equinoxes was used. Moreover, the graphical design seems to have been adapted to aesthetic demands that diverged from those exemplified in Fine's book. This was achieved by simplifying and using the famous italic script designed by Mercator.

One of the changes in design concerns the labelling of the stars. This will now be examined in more detail. At first glance, it seems that Arsenius simply made the star labels uniform: all star and constellation names are engraved on concentric base lines, to be read in anticlockwise sense, with capitalised initials. In his book on the astrolabe-quadrant, however, Oronce Fine applied various distinct script types for the star labels and the zodiacal signs. The author explains this point clearly in Proposition VI: 'By a few [hints] to teach how to put the fixed stars on the quadrant'.[37] By folding the planisphere twice through the north pole, on a quadrant the four quarters of the sky appear superposed. Oronce insisted that one should keep track of this superposition by labelling the stars and the signs differently according to the celestial quadrant to which they belonged.

How to inscribe the names of the signs	Finally you inscribe the names of the signs: that is the boreal ones along the northern part *IN* of the ecliptic, the austral ones along its southern part *BI*, which you will distinguish as much by the proper order of the names with the corresponding symbols, as much as by the difference of the characters: such as clarified in the figure [...].
How to fix the stars	You will therefore label that star by its name, which is written so that the characters are similar, and towards the same side as the given sign of mediation [of that star] is labelled: just as you see is done in the figure with Oculus Tauri, Canis Maior and Vultur.[38]

36 Samuel Gessner, 'The Use of Printed Images for Instrument Making at the Arsenius Workshop', *Early Science and Medicine*, Special edition: 'Images in Comparative Perspective: Visual Forms in Astronomy, Medicine and Mathematics, 1470–1650', ed. by Nick Jardine and Isla Fay, 18 (1–2) (2013), 124–52.

37 'Ut stellae fixae ipsi quadranti ueniant imponendae, paucis edocere'. Fine, *Protomathesis*, fol. 203^{v}.

38 'Qualiter inscribenda signorum nomina | Inscribes tandem signorum nomina: borea scilicet in aquilonia Eclipticae parte IN, austrina uero in meridiana BI, quae tum proprio nominum cum eisdem signis ordine, tum characterum discrepantia inuicem separabis: uti figura dilucidat. [...]'; 'Qualiter imponenda sydera | Hanc

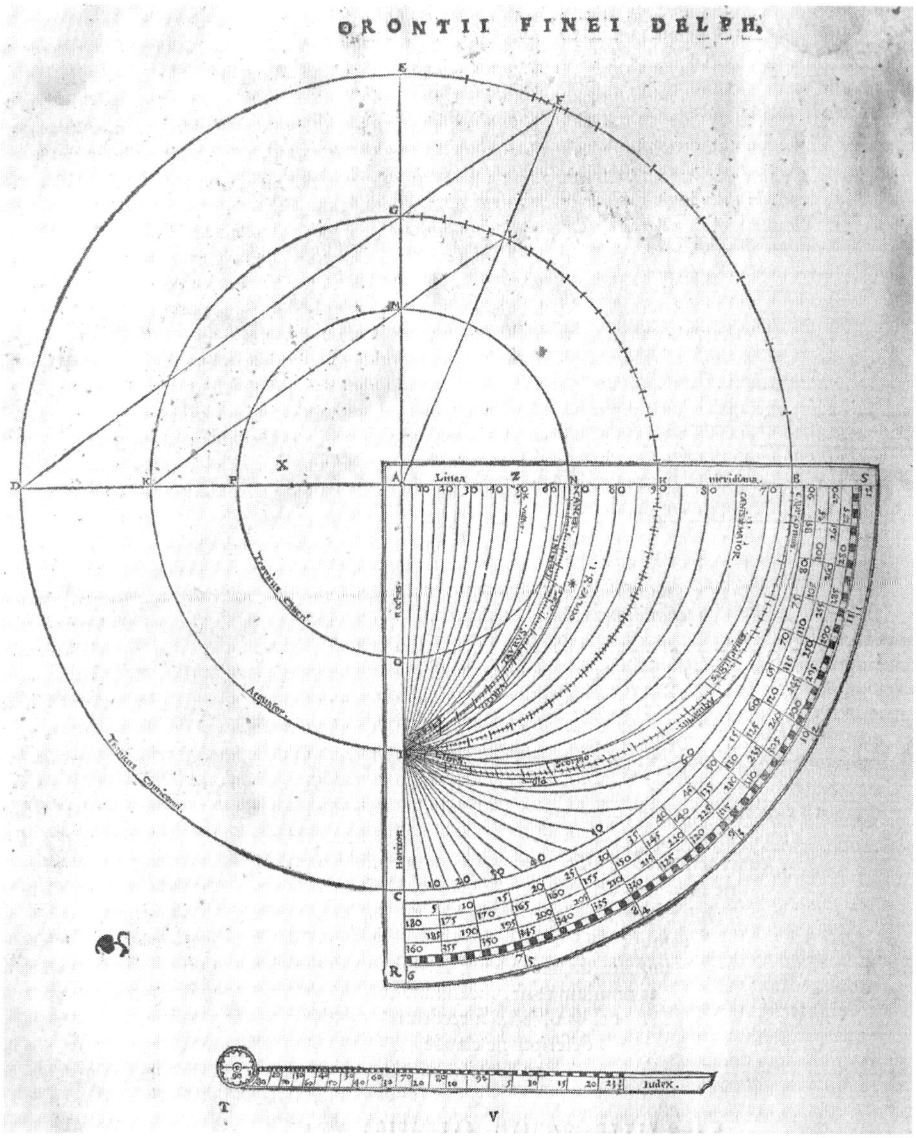

Fig. 6 Woodcut of the astrolabe-quadrant in Oronce Fine, *De solaribus horologiis* (Paris, 1532), 204$^{v^{\circ}}$. Cambridge University Library, Eee.84. Courtesy of the Syndics of Cambridge University Library.

The idea of identifying the corresponding quadrant (strictly speaking the range of right ascension to which it corresponds) of a given star by the script type and orientation of its label is exemplified on Fine's woodcut figure using three stars: 'OCVLVS ♉' (*oculus tauri*) is labelled in capital letters and written in the same sense as the three signs 'ARIES. TAVRVS.

[stellam] igitur proprio obsignabis nomine similibus quidem scriptum elementis, atque uersus eam partem, quibus signum medij Coeli datum annotatum est: quemadmodum uides obseruatum in figura, de oculo Tauri, Cane maiore, atque Vulture'. Fine, *Protomathesis*, fols 203^{v}–204^{r}.

GEMINI' (Figure 6). A table given in the text by Fine sets the mediation (*medium coeli*) of that star at Gemini 3°34'. 'CANIS ✶ MAIOR' is also labelled in capitals, but the orientation of the script is reversed, corresponding to the labels of the signs 'CANCER. LEO. VIRGO'. Indeed the value of mediation given in the table is Cancer 15°58'. Finally, the star Vega, here labelled by the name of the constellation 'Vultur' (Vultur cadens, i.e. Lyra) using lower case letters written in the same sense as the labels of the winter signs 'Capricornus. Aquarius. Pisces'. This is consistent with its value of mediation given in the table: Capricorn 3°51'.

The labels of Arsenius's brass quadrant contrast this arrangement, as we mentioned earlier, in so far as all labels are written with the same script orientation, and no distinctions with capital or lowercase letters are made. Was the introduction of uniform star labels an error? How can this discrepancy of source image and instrument be explained? Two mutually exclusive explanations come to mind. Firstly, Ieremias Arsenius thought that explicit attribution of the stars to one of the four quarters of the sky is not needed. The prospective user of the quadrant and he himself would be fully confident of their knowledge as to which sign each star belonged. After all, we are talking of twenty-three especially bright and well-known stars. Alternatively, we may assume that the maker did not realise that there arises this ambiguity by the double folding of the planisphere. The ambiguity would be problematic, for instance, when one wanted to reckon the time from the height of a star. One could not rely on the quadrant for knowing whether a star was below or above the horizon. It would have made using the quadrant more difficult for a beginner. On the other hand, if the instrument had a more symbolic use, and the planisphere was mainly there to signify an image of the cosmos, this distinction would perhaps not be so important.

It is an alternative proposition that makes the matter difficult to resolve. The dilemma shows how delicate an operation it is for the historian to deduce an instrument maker's knowledge from the inscriptions on an artefact alone. Let's keep in mind that it is characteristic of Renaissance culture that many instruments, and astronomical dials in particular, were intended to combine emblematic (symbolic) and operational (geometric) aspects. That means that the technical aspects were required to be correct, even when an instrument was conceived in a very symbolic way. The instruments could, at the same time, be used for teaching a student and for displaying the sophistication of their owner's astronomical learning. Moreover, it goes without mentioning that there always was some 'tacit knowledge' required for such gnomonic and astronomical instruments to be expertly used. Therefore, both possibilities remain equally probable.

The astrolabe-quadrant we have presented embodies a series of procedures and habits available in the Arsenius workshop: from using Mercator's italic script type to taking instrument treatises as a source for composing instruments. Concerning the knowledge about stereographic projection, we may only assume that the second or third generation of instrument makers of that workshop still treasured what Gemma Frisius once taught and wrote. If this assumption holds true, then Ieremias Arsenius knew about the fundamental analogy of the planisphere and the perspective representation of the sphere. Still the puzzle of the uniformity of the star labels cannot be explained.

Each Star in its Proper Place

In conclusion we must stress that the example of the Arsenius workshop with its connection to Gemma Frisius is exceptional in several regards. As we have shown above, the Louvain workshop represented an epicentre of quality instrument making with connections to many high-ranking personalities. Nevertheless, other instrument making workshops of the sixteenth century would share the particular practices and understand their connections to perspective and optics in the way Gemma Frisius articulated them.

In this chapter we have not addressed the mechanical devices used to produce images to which pictorial perspective practice had given rise. They often were based on Alberti's concept of the visual pyramid and provided a practical means to construct its intersections: e.g. the surface applied to a window mentioned by Gemma. Dürer was not the only proponent of such apparatuses, some versions of which have been preserved in instrument collections.[39] If we are interested in the connection of the practice of instrument workshops and perspective, however, it seems clear that the meaning of perspective must be extended beyond that of a technique for 'mimetic representation' of volumes on a flat surface. At least in Gemma's view, 'perspective' is taken in its general sense to comprise also the 'planispheric' projection. This realisation leads us to acknowledge that the ambiguity around the small brass quadrant by Ieremias Arsenius is not easily interpreted in terms of knowledge or ignorance, nor is it a question of abilities or shortcomings of the artisan. While it cannot be ruled out that the feature of uniform star labels is due to a negligent use of the source, it seems at least as probable that Ieremias brilliantly mastered the stereographic projection, and that this form of representation — even doubly folded — was so transparent to his eyes that he could dispense with the heavy handed mnemonic aid of diverse scripts.

For the sixteenth century we may assume with some certainty that a trained person would 'see' a sphere of three dimensions even when presented on a planisphere. To see three dimensions in only two always requires a previous knowledge of the objects that are projected. In those times, knowledge of the sky, of the place and order of the zodiacal signs, of the main celestial circles, and of the bright stars is inscribed in the *forma mentis* of a quite broad audience. Even a double folding of the planisphere, for an instructed contemporary, would not have meant a serious obstacle to 'seeing' the circles unfolded and each star in its proper place.

Bibliography

Woodcut Prints and Instrument Collections

Kupferstichkabinett, SKD, Dresden. 'Imagines coeli septentrionales', Albrecht Dürer, Johann Stabius, Conrad Heinfogel, Nuremberg, 1515, inv. A 1893-200.

---

39 Sven Hauschke, 'Albrecht Dürers Perspektivtische und ihre Nachfolger im 16. Jh.', in *Buchmalerei in der Dürerzeit. Dürer und die Mathematik*, ed. by Georg Ulrich Großmann (Nuremberg: Verlag des Germanischen Nationalmuseums, 2009), II, pp. 173–89. Baldassarre Lanci produced a device that, among other functions, was designed to capture perspective views on a surface: *Strumento topografico*, 1567, Museo Galileo, Florence, inv. nr. 152, 3165.

Metropolitan Museum of Art, New York. 'Imagines coeli', Albrecht Dürer, Johann Stabius, Conrad Heinfogel, 1515, inv. nr. 51.537.1.
Museo Galileo, Florence. '*Strumento topografico*', Baldassarre Lanci, 1567, inv. nr. 152, 3165.
Museu Nacional de História Natural e da Ciência, Lisbon, 'Astrolabe quadrant', Ieremias Arscenius, Leuven, 1573, inv. nr. MUHNAC-UL-DEP262.
Staatliche Graphische Sammlung, Munich. 'Imagines coeli', Albrecht Dürer, Johann Stabius, Conrad Heinfogel, Nuremberg, 1515, inv. nr. 118930 and 118931.

Primary Sources

Alberti, Leon Battista, *De pictura praestantissima et nunquam satis laudata arte libri tres* (Basel: Th. Venatorius, 1540).
Alberti, Leon Battista, *On Painting and on Sculpture, the Latin Texts of 'De pictura' and 'De statua'*, ed. by Cecil Grayson (London: Phaidon, 1972).
Apianus, Petrus, *Cosmographicus liber* (Landshut: Weyssenburger, 1524).
Clavius, Christopherus, *Astrolabium* (Rome: Giunti, 1593).
Commandino, Federico, *Ptolemaei planisphaerium* […] (Venice: Aldo Manuzio, 1558).
Commandino, Federico, *La prospettiva di Federico Commandino, tr. by* Rocco Sinisgalli (Florence: Cadmo, 1993).
Dürer, Albrecht, *Imagines coeli meridionales* (Nuremberg, 1515).
Dürer, Albrecht, *Imagines coeli septentrionales cum duodecim imaginibus zodiaci* (Nuremberg, 1515).
Fine, Oronce, *Descriptio partium, et succinta utilitatum elucidatio quadrantis cujusdam universalis* (Paris: Nicolaus Savetier, 1527).
Fine, Oronce, *Protomathesis opus uarium, ac scitu non minus utile quàm iucundum* (Paris: Gerard Morrhius and Joannes Petrus, 1532).
Gemma Frisius, Rainerus, *De astrolabo catholico liber: quo latissime patentis instrumenti multiplex usus explicatur, & quicquid uspiam rerum mathematicarum tradi possit continetur*, ed. by Cornelius Gemma (Antwerp: Johannes Steelsius, 1556).
Jordanus de Nemore, *Jordanus de Nemore and the Mathematics of Astrolabes: De plana spera*, ed. by Ron B. Thomson, Studies and Texts, 39 (Toronto: Pontifical Institute of Mediaeval Studies, 1978).
Pseudo-Masha'allah, *On the Astrolabe*, ed. by Ron B. Thomson, version 1.2 (Toronto, 2015), https://shareok.org/handle/11244/14221.
Stöffler, Johannes, *Elucidatio fabricae ususque astrolabii* (Oppenheim: Köbel, 1513).
Vignola, Jacopo Barozzi da, *Le due regole della prospettiva pratica con i commentarij del R.P.M. Egnatio Danti* (Rome: Zanetti, 1583).

Secondary Works

Aiken, Jane Andrews, *Renaissance Perspective: Its Mathematical Source and Sanction* (unpublished doctoral dissertation, Harvard University, 1986).
Aiken, Jane Andrews, 'Truth in Images: From the Technical Drawings of Ibn al-Razzaz al-Jazari, Campanus of Novara, and Giovanni de'Dondi to the Perspective of Leon Battista Alberti', *Viator. Medieval and Renaissance Studies*, 25 (1994), 325–59.

Almeida, Bruno, 'On the Origins of Dee's Mathematical Programme: The John Dee-Pedro Nunes Connection', *Studies in History and Philosophy of Science*, Part A, 43 (3) (2012), 460–69.

Bennett, Jim, 'Practical Geometry and Operative Knowledge', *Configurations*, 6 (2) (1998), 195–222.

Camerota, Filippo, 'The Eye of the Sun: Galileo and Pietro Accolti on Orthographic Projection', in *Perspective, Projections & Design: Technologies of Architectural Representation*, ed. by Mario Carpo and Frédérique Lemerle (London: Routledge, 2008), pp. 115–25.

Cleempoel, Koenraad van, *A Catalogue Raisonné of Scientific Instruments from the Louvain School, 1550–1600* (Turnhout: Brepols, 2002).

Dekker, Elly, *Globes at Greenwich: A Catalogue of the Globes and Armillary Spheres in the National Maritime Museum, Greenwich* (Oxford: Oxford University Press, 1999).

Dekker, Elly, *Illustrating the Phaenomena: Celestial Cartography in Antiquity and the Middle Ages* (Oxford: Oxford University Press, 2013).

Dekker, Elly, 'Construction and Copy: Aspects of the Early History of Celestial Maps', Beiträge zur Astronomicgeschichte, Band 13, *Acta Historica Astronomiae*, 58 (2016), pp. 47–93.

Dupré, Sven, 'Galileo, Mathematical Instruments and Orthographic Projection', *Bulletin of the Scientific Instrument Society*, 69 (2001), 10–20.

Gaab, Hans, *Die Sterne über Nürnberg: Albrecht Dürer und seine Himmelskarten von 1515* (Petersberg: Michael Imhof, 2015).

Gessner, Samuel, 'The Use of Printed Images for Instrument Making at the Arsenius Workshop', *Early Science and Medicine*, Special edition: 'Images in Comparative Perspective: Visual Forms in Astronomy, Medicine and Mathematics, 1470–1650', ed. by Nick Jardine and Isla Fay, 18 (1–2) (2013), 124–52.

Grössing, Helmuth, 'Wiener Astronomen und Mathematiker des 15. und beginnenden 16. Jahrhunderts und ihre Instrumente', *Wiener Geschichtsblätter*, 38 (4) (1983), 149–62.

Hallyn, Fernand, *Gemma Frisius, arpenteur de la terre et du ciel* (Paris: Honoré Champion, 2008).

Hauschke, Sven, 'Albrecht Dürers Perspektivtische und ihre Nachfolger im 16. Jh.', in *Buchmalerei in der Dürerzeit. Dürer und die Mathematik*, ed. by Georg Ulrich Großmann (Nuremberg: Verlag des Germanischen Nationalmuseums, 2009), II, pp. 173–89.

Ortroy, Fernand van, *Bio-bibliographie de Gemma Frisius, fondateur de l'école belge de géographie, de son fils Corneille et de ses neveux les Arsenius*, Mémoire de l'académie royale de Belgique, 2nd series, 11 (2) (Brussels: M. Lamertin, 1920).

Osley, Arthur S., *Mercator, A Monograph on the Lettering of Maps, etc. in the 16th Century Netherlands with a Facsimile and Translation of His Treatise on the Italic Hand and a Translation of Ghim's Vita Mercatoris* (London: Faber and Faber, 1969).

Snyder, John P., *Flattening the Earth: Two Thousand Years of Map Projections* (Chicago: University of Chicago Press, 1993).

Strauss, Walter L., *Albrecht Durer, Woodcuts and Wood Blocks* (New York: Abaris, 1979).

Turner, Gerard L'E., and Elly Dekker, 'An Astrolabe Attributed to Gerard Mercator, *c.* 1570', *Annals of Science*, 50 (5) (1993), 403–43.

Turner, Gerard L'E., 'The Three Astrolabes of Gerard Mercator', *Annals of Science*, 51 (1994), 329–53.

TAWRIN BAKER

Dissection, Instruction, and Debate

Visual Theory at the Anatomy Theatre in the Sixteenth Century

Introduction

In the history of science and medicine it is generally held that developments in anatomy had comparatively little impact on visual theory. After discussing the anatomists Realdo Colombo, Charles Estienne, Girolamo Fabrici d'Acquapendente (Hieronymus Fabricius ab Aquapendente), Johannes Jessenius, Costanzo Varolio, and André du Laurens on ocular anatomy, David Lindberg writes: 'None of the post-Vesalian authors that I have mentioned made significant alterations in visual theory'.[1] While somewhat more sympathetic to the anatomists, Alistair Crombie, after mentioning that 'an accurate general ocular anatomy was essential for a true optical analysis of its physiology', writes: 'It was the mathematicians who came to reform visual theory by proceeding through an optical analysis of ocular physiology, exploiting the models of eyeglasses and the camera obscura, and thus reformulating the problem itself'.[2] On this account anatomists provided empirical data about the structure of the eye, while mathematicians accomplished the more important theoretical work.

This picture rests upon several assumptions, two of which I will address in this paper. The first is that anatomists did not engage with theoretical issues regarding vision, and indeed that anatomists generated empirical data about the structure of the eye without interference from,

1 David C. Lindberg, *Theories of Vision from Al-Kindi to Kepler* (Chicago: University of Chicago Press, 1976), p. 175. Note that Lindberg only cites a 1614 reprint of Fabricius's *De visione*. Perhaps this is one reason why he did not examine Fabricius carefully in a work whose ultimate aim is to understand Kepler's *Paralipomena* (1604).

2 A. C. Crombie, 'Expectation, Modelling, and Assent in the History of Optics: Part I. Alhazen and the Medieval Tradition', *Studies in the History and Philosophy of Science*, 21 (4) (1990), 605–32 (p. 629). This is repeated in A. C. Crombie, *Styles of Scientific Thinking in the European Tradition: The History of Argument and Explanation Especially in the Mathematical and Biomedical Sciences and Arts*, 3 vols (London: Duckworth, 1994), p. 1120. A similar assessment can be found in Huldrych M. Koelbing, 'Anatomie de l'œil et perception visuelle, de Vésale à Kepler', in *Le corps à la Renaissance. Actes du XXXe colloque de Tours 1987* (Paris: Aux amateurs de livres, 1990), pp. 389–98. There is, notably, no mention of anatomy in Paolo Mancosu, 'Acoustics and Optics', in *Cambridge History of Science Volume 3: Early Modern Science*, ed. by Katharine Park and Lorraine Daston (Cambridge: Cambridge University Press, 2006), pp. 596–631. See also A. Mark Smith, *From Sight to Light: The Passage from Ancient to Modern Optics* (Chicago: University of Chicago Press, 2014), pp. 351–52. Smith mostly agrees with Lindberg's assessment on the role of anatomy in changes to visual theory, but contra Lindberg he considers Felix Platter's opinion that the retina is the site of visual sensation to be of great significance.

Tawrin Baker University of Pennsylvania, tawrin@gmail.com

Perspective as Practice. Renaissance Cultures of Optics, ed. by Sven Dupré, Turnhout, 2019 (*Techne. Knowledge, Technique, and Material Culture*, 1), p. 123-147

DOI 10.1484/M.*Techne*-EB.5.117724

or appeals to, theoretical concerns. On the contrary, in the sixteenth century, anatomy, and ocular anatomy especially, was inherently theoretical. Determining the structure of some body part through anatomical investigation always included an account of its temperament. Doing so required an account of the part's elementary qualities — hot or cold, wet or dry — and an analysis of its elemental composition, i.e., its respective ratio of earth, water, air, and fire; this knowledge was gathered through not just vision, but also touch (and perhaps smell and taste), observations that were inherently theory-laden. In the case of the eye, whether the temperament of the humours of the eye were fiery, as Plato and others insisted, pneumatic (i.e., composed of an ethereal animal spirit), as Galen of Pergamon argued, or watery, as Aristotle posited, was inextricably tied to both general theories of animal physiology as well as debates about extramission versus intromission theories of vision.[3] The latter debate raged throughout the sixteenth century among physicians, and accounts of the shape, size, colour, transparency, and consistency of the humours of the eye related to this debate in complex ways. One fundamental shift that occurred over the course of the sixteenth century was that the eye as revealed through dissection (as opposed to an eye constructed to accommodate *a priori* mathematical demands) became the basis for a mathematical analysis of vision. The eye that was bequeathed to seventeenth-century mathematicians was in part the resolution of intense debates among physicians concerning visual theory, and these arguments drove the precise determination of the composition of the humours (including their relative density), as well as the shape, size, and connection of the interior parts of the eye.

The second assumption is that intellectual developments in mathematical optics dwarfed other considerations. The diffusion of more basic knowledge about mathematical optics, the increasing engagement of philosophers and physicians with questions related to light, colour, and vision, and the development and spread of knowledge about ocular anatomy are, on most accounts, decidedly less significant.

But the diffusion of knowledge and cross-disciplinary interaction on matters related to vision were of great importance. One key sixteenth-century development in visual theory was the gradual integration of the three main traditions dealing with vision — mathematical optics, natural philosophy, and ocular anatomy. These were largely independent at the beginning of the century. By the end of the century the anatomical demonstration (both public and private) was an important occasion for the spread of the combined mathematical, philosophical, and anatomical approach to the problem of vision. Furthermore, anatomical texts were closely linked to demonstrations: as their prefaces make clear, such texts were not intended as repositories of information to be passively absorbed by the reader. Rather, they were framed as both guides for the reader's own dissections and a complement to a public dissection the reader (medical students in particular) would have attended.

The focus of this essay is the treatment of the eye and vision at the anatomy theatre itself, and this history follows a twisting path throughout the sixteenth century.[4] At the

3 Note that I use the term extramission in a broad sense to include any account that involves an effect (material, spiritual, qualitative, and so on) that propagates outwards from the eye.

4 On the anatomy theatre in Italy, see especially Giovanna Ferrari, 'Public Anatomy Lessons and the Carnival: The Anatomy Theatre of Bologna', *Past & Present*, 17 (1987), 50–106; Cynthia Klestinec, 'A History of Anatomy Theaters in Sixteenth-Century Padua', *Journal of the History of Medicine and Allied Sciences*, 59 (3) (2004), 375–412; Paul F. Grendler, *The Universities of the Italian Renaissance* (Baltimore: JHU Press, 2004), pp. 329, 340–41.

beginning of the century most anatomists did not emphasize public anatomical dissection of the eye, and although visual theory is often alluded to they did not seriously engage with the mathematical or philosophical traditions concerning vision. Furthermore, at the beginning of the century the practice of public anatomical demonstration, while common in Italy, was not ubiquitous, and even in major medical centres like Bologna and Padua some prominent anatomists questioned the usefulness of the public anatomy.[5]

Andreas Vesalius, who brought the public anatomy to prominence, is pivotal to this story. Although the notes we have to his 1540 Bologna anatomy reveal that he dissected the eye in public, they show him doing so rather sloppily. Moreover, the account of the structure of the eye in his *De humani corporis fabrica* in some ways assumes a Galenic extramissionist theory of vision. In part because of this, Vesalius's account of the shape and size of the humours departs from accounts from his predecessors. His idea of the eye, with a symmetrical crystalline humour placed directly in the centre, was explicitly compared to a geocentric model of the cosmos. Contrary to what has been repeated in the secondary literature, this was not typical of anatomical texts, and his contemporaries soon criticized him for it.[6] Two reasons for Vesalius's idiosyncratic eye are discussed. First is his emphasis on public dissection, for which this small, fluid-filled organ is not well suited. Second is the humanistic revival of Galenic anatomy, which reinvigorated the anatomical and philosophical grounds for extramissionist theories of vision.

In Padua towards the end of the sixteenth century we see that both the structure of the eye and visual theory were being taught at the anatomical theatre, and we have evidence that some optical experiments, on the crystalline humour especially, were performed there. This incorporation of visual theory into public (as well as private) anatomies was framed as an investigation of the action (*energeia* in Greek, *actio* in Latin) and usefulness (*chreia* in Greek, *usus* or *utilitas* in Latin) of the eye. These were teleological categories derived from Galen's works that addressed, respectively, what the eye and its parts *do* during the act of seeing (namely, how vision works) and *why* the eye and its parts perform their actions (that is, what *use* the parts and their actions contribute to the act of vision and, ultimately, to the life of the animal). According to Galen, only by accounting for the *chreia* or usefulness of the part does one obtain causal knowledge of it.[7] Vesalius framed *De fabrica* in part as a revival and perfection of ancient Greek anatomy, but many of his contemporaries and immediate successors found fault with Vesalius's attempt, and for investigating primarily the fabric of the body while often ignoring the action and use of its parts.[8] Yet, while attention to visual theory was stimulated in part by a rival of the

5 R. Allen Shotwell, 'The Revival of Anatomical Practices and Techniques in the Renaissance' (unpublished doctoral dissertation, Indiana University, 2013), pp. 155–56.

6 John B. de C. M. Saunders and Charles D. O'Malley, *The Illustrations from the Works of Andreas Vesalius of Brussels* (Cleveland: World Publishing Company, 1950), p. 200; Andreas Vesalius, *The Fabric of the Human Body: An Annotated Translation of the 1543 and 1555 Editions of 'De humani corporis fabrica libri septem'*, trans. by Daniel H. Garrison and Malcolm H. Hast, 2 vols (Basel: S. Karger Ag, 2014), pp. 1301 n. 1, 1306 n. 12; A. Mark Smith, 'Ptolemy, Alhazen, and Kepler and the Problem of Optical Images', *Arabic Sciences and Philosophy*, 8 (1) (1998), 9–44 (pp. 31–32).

7 Galen, *Galen on the Usefulness of the Parts of the Body. De usu partium*, trans. by Margaret Tallmadge, 2 vols (Ithaca: Cornell University Press, 1968); R. J. Hankinson, 'Philosophy of Nature', in *The Cambridge Companion to Galen*, ed. by R. J. Hankinson (Cambridge: Cambridge University Press, 2008), pp. 225–29.

8 On Vesalius's teleology and his relationship to Galen and ancient anatomy, see Nancy G. Siraisi, 'Vesalius and the Reading of Galen's Teleology', *Renaissance Quarterly*, 50 (1997), 1–37.

Galenic focus on final causes, eventually this Galenic method was used to reject Galenic extramission in favour of Aristotelian intromission accounts of vision.

With Hieronymus Fabricius ab Aquapendente we see an integration of the previously separate traditions of ocular anatomy, scholastic philosophical accounts of vision, and mathematical optics. In 1565 Fabricius succeeded Gabriele Falloppio as the chair of anatomy and surgery at the University of Padua, which was then the most well-regarded institution for anatomy and medicine in Europe. He is well known today for his innovations in surgery, for describing the *ostiola* ('little doors') of the veins, for developing a project of Aristotelian (or, more accurately, Galeno-Aristotelian) philosophical anatomy, and for teaching and influencing William Harvey.[9] Fabricius argued forcefully against the extramission theory of vision that seems to have been common, and perhaps dominant, among physicians in the sixteenth century. In response, he developed a new theory of vision that, while drawing heavily upon his reading of Aristotle and Risner's 1572 edition of Alhacen, departed from previous theories in key ways.[10] Although rarely endorsed by subsequent writers in its entirety, his theory of vision was well known in the seventeenth century, and Fabricius's ocular anatomy was in a sense re-performed in 1600 at Prague by Johannes Jessenius. Furthermore, Fabricius's attitude that visual theory must begin with a precise description of the eye as revealed by dissection — that is, that the foundation of optics is ocular anatomy — was a development of great significance and influence.

The present analysis is limited in some key respects. It focuses overwhelmingly on the University of Padua, and while Padua was undeniably important in both the history of the anatomy theatre and the integration of anatomy with philosophy and mathematics, it is hardly the only site worth studying. Furthermore, it will only roughly sketch the important controversy between Galenic extramission and Aristotelian intromission theories of vision, as well as the relatively less important role (for the sixteenth century at least) played by atomistic intromission theories and by Platonic and mathematical (particularly Euclidian) extramission theories.[11] Finally, further research into student notes and other historical evidence is necessary to determine more precisely what was taught and disputed at the anatomical theatre. To determine how exactly the ocular anatomy and visual theory encountered there migrated to other places in Europe, we would also need a more complete survey of students who attended major medical centres, particularly Bologna and Padua.

---

9 See Adelmann's introduction to Hieronymus Fabricius ab Aquapendente and Howard Bernhardt Adelmann, *The Embryological Treatises of Hieronymus Fabricius of Aquadenpente* (Ithaca: Cornell University Press, 1967); Andrew Cunningham, 'Fabricius and the "Aristotle Project" in Anatomical Teaching and Research at Padua', in *The Medical Renaissance of the Sixteenth Century*, ed. by Andrew Wear, Roger Kenneth French, and Iain M. Lonie (Cambridge: Cambridge University Press, 1985), pp. 195–222; Peter Distelzweig, 'Fabricius's Galeno-Aristotelian Teleomechanics of Muscle', in *The Life Sciences in Early Modern Philosophy*, ed. by Ohad Nachtomy and Justin E. H. Smith (New York: Oxford University Press, 2014), pp. 64–84.

10 I argue elsewhere that he likely developed this account in collaboration with others in either public or private anatomies, most notably with the scholastic natural philosopher Jacopo Zabarella who also presents the same account of vision as Fabricius in his natural philosophy textbook *De rebus naturalibus* (Venice: Paulus Meiettus, 1590). See Tawrin Baker, 'Color, Cosmos, Oculus: Vision, Color, and the Eye in Jacopo Zabarella and Hieronymus Fabricius Ab Aquapendente' (unpublished doctoral dissertation, Indiana University, Bloomington, 2014), pp. 219–82.

11 On the latter, see Sven Dupré, 'Kepler's Optics Without Hypotheses', *Synthese*, 185 (2012), 501–25.

Pre-Vesalian Ocular Anatomy

There are three clear humours that make up the interior of the eye: the aqueous humour is located towards the front of eye, followed by the crystalline humour (now called the crystalline lens or just the lens), with the vitreous humour at the rear. The *aranea*, which more or less corresponds to the lens capsule in modern anatomy, was described as a subtle, clear membrane covering at least the front end of the crystalline humour. Medieval perspectivists seem to have placed the crystalline humour towards the front of the eye, although this was not emphasized. In the texts of Alhacen, Witelo, and John Pecham the most important factor is that the cornea and the anterior surface of the crystalline humour form concentric spheres. At times Alhacen refers to the crystalline and the vitreous together as a single humour (usually called the *humor glacialis*, or 'ice-like humour'). Pecham follows Alhacen's suppositions, although the rear of the crystalline is rarely described with any precision by either. In sixteenth century print editions we see diagrams that present the crystalline humour in several ways: some give the same curvature in front as in the back, some show the crystalline-vitreous boundary to be flat, and some even depict the vitreous as spherical, giving the crystalline humour a crescent shape (Figure 1).[12] Furthermore, while the density of the humours was said to affect the refraction of rays of light and colour in the eye, the authors comprising the *Perspectivae* also appealed to the faculty of vision to account for the refraction of rays within the body. For them, the refraction at the crystalline-vitreous boundary was actively managed by the sensitive soul, and the twisting refraction of the visual image through the (supposedly) hollow optic nerve required the intervention of the sensitive soul as well. While anatomical details are included in their treatises, there is no evidence that any of the *Perspectivae* actually witnessed dissections of the eye. Notably, the treatises by Alhacen and Witelo each introduce the eye using the Galenic tripartite format of structure, action, and use. However, in their respective sections on the structure of the eye they construct a geometrical model eye based on the demands of the theory of vision to which they both subscribe. In their second sections on action, the core mathematical aspects of their account of vision are developed, while in their final sections on use these *Perspectivae* merely recast their accounts of vision teleologically.[13] This apriorism is a significant departure from Galen's method of determining how and why the parts of the body function.[14] For the perspectivists, the structure of the eye serves a geometric account of vision, and the shapes and sizes of the transparent tunics

---

12 Alhacen, *Alhacen's Theory of Visual Perception: A Critical Edition, with English Translation and Commentary, of the First Three Books of Alhacen's 'De aspectibus', the Medieval Latin Version of Ibn al-Haytham's 'Kitāb al-Manāẓir'*, ed. and trans. by A. Mark Smith, 2 vols (Philadelphia: Transactions of the American Philosophical Society, 2001), I, i–337 (p. 12); Witelo, *Witelonis perspectivae liber secundus et liber tertius*, ed. and trans. by Sabetai Unguru, Studia Copernicana, XXVIII (Warszawa: Ossolineum, 1991), pp. 105–11, 294–97; John Pecham, *John Pecham and the Science of Optics: Perspectiva communis*, ed. and trans. by David C. Lindberg (Madison: University of Wisconsin Press, 1970), pp. 114–15.

13 Alhacen, *Alhacen's Theory of Visual Perception*, ed. by Smith, I, pp. XLVII–XLIX, LX; Witelo, *Witelonis perspectivae*, ed. by Unguru, 107, 297. On the Galenic categories of structure, action, and use, see Nancy G. Siraisi, 'Historia, actio, utilitas: Fabrici e le scienze della vita nel Cinquecento', in *Il teatro dei corpi: Le pitture colorate d'anatomia di Girolamo Fabrici d'Acquapendente*, ed. by Maurizio Rippa Bonati and Jose Pardo-Tomas (Milano: Mediamed, 2004), pp. 63–73.

14 Alhacen, *Alhacen's Theory of Visual Perception*, ed. by Smith, I, p. XLIX.

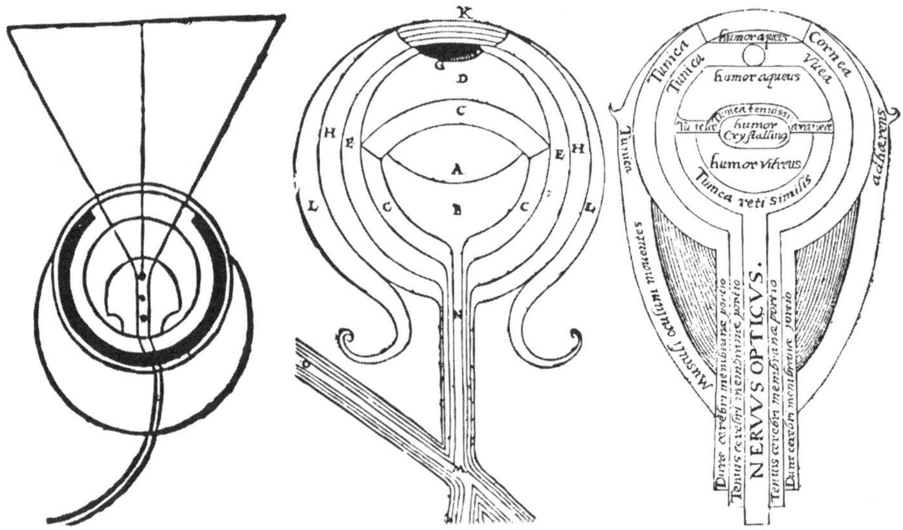

Fig. 1 Three cross-sections of the eye in sixteenth-century optical works. From left to right, based on: John Pechham, *Jo. archiepiscopi cantuariensis perspectiva communis* (Venice: Per Io. Baptistam Sessam, 1504), p. 6v; John Pechham, *Perspectiva communis*, ed. by George Hartman (Nuremburg: apud Iohan. Petreium, 1542); Alhacen and Witelo, *Opticae thesaurus*, ed. by Friedrich Risner (Basel: Per Episcopios, 1572), p. 6. For an image with a flat crystalline-vitreous boundary, based on diagrams in medieval manuscripts, see Witelo, *Vitellionis mathematici doctissimi Peri optikes* (Nuremburg: apud Ioan. Petreium, 1535; repr. 1551), p. 55r. Note that the left-most image was also reproduced in both the 1580 and 1592 Cologne editions of Peckham's *Perspectiva*, and that the upper-most dot represents the centre of curvature of both the cornea and the anterior surface of the crystalline humor.

and the interior humours were not derived from, or coordinated with, what is found in a dissected eye.[15] Moreover, because the *Perspectivae* required the active intervention of the visual faculty on the path of the rays (both in the eye at the crystalline-vitreous interface as well as within the twisting hollow of the optic nerve), the dissection of dead eyes and experiments with humoural refraction would appear to contribute little towards determining the precise path of rays in a living eye.[16] That the geometric depiction of the eye given by the *Perspectivae* is difficult to reconcile with anatomical experience does not appear as an important issue (indeed, doesn't seem to have been mentioned) before the middle of the sixteenth century, and is evidence that the domains of mathematical optics and anatomy were institutionally and conceptually isolated before then.

Most anatomists before Vesalius either placed the crystalline humour towards the front of the eye or were ambiguous about its precise location, but they gave a much clearer indication of the shapes of the vitreous humour and the posterior crystalline surface than

15 On whether the eye described by Perspectivists is real or theoretical, see Alhacen, *Alhacen's Theory of Visual Perception*, ed. by Smith, I, pp. xlix, lxxviii–lxxix; Pecham, *Perspectiva communis*, ed. and trans. by Lindberg, p. 247 n. 83; Witelo, *Witelonis perspectivae*, ed. by Unguru, p. 210 n. 2.

16 Alhacen, *Alhacen's Theory of Visual Perception*, ed. by Smith, I, 84–87; Witelo, *Witelonis perspectivae*, ed. by Unguru, 130, 319; Pecham, *Perspectiva communis*, ed. and trans. by Lindberg, 118–19.

the mathematicians. Mondino de Luzzi (*c.* 1270–1326), the medieval anatomist who was the primary authority before the works of Galen were revived during the late-fifteenth and sixteenth centuries, writes that the crystalline humour has 'a round or spherical figure, but with a certain flatness in the anterior part, and this humour is more toward the anterior than the vitreous in which it is placed'.[17] The anatomist Jacopo Berengario da Carpi (*c.* 1460–*c.* 1530) agrees with Mondino, but also says that 'some also add a fourth humour, which they call ethereal, and which they place immediately in front of the pupil';[18] Later, he writes 'There is another humor in the eye called the vitreous, which is greater in quantity than the other two', indicating that the crystalline is placed more towards the front of the eye.[19] Alessandro Benedetti (*c.* 1450–1512), writing somewhat earlier than Berengario, describes the front of the crystalline thus: 'The drop itself [i.e., the crystalline humour] lies more in the front part [of the eye]; it is rounded in its rear part, somewhat flattened in front'.[20] Two other prominent anatomists, Gabriele Zerbi (1445–1505) and Niccolò Massa (1485–1569) were less conclusive about the shape and location of the humours.[21]

Finally, Avicenna's *Canon of Medicine* — perhaps the most important book on medicine in the Middle Ages and the Renaissance — was widely cited in sixteenth-century anatomical treatises, especially prior to Vesalius. Book III, Fen 3, Tract 1, Chapter 1 is on the anatomy of the eye, and there we read of the *humor glacialis*:

> this humor is indeed clear, just like a hail-stone; the ice-like humor is indeed round, but its rotundity is diminished by a compression on its anterior part [...] and this humor is placed in the middle, since this is the most suitable place.[22]

'In the middle' here likely meant that the lens was the middle humor, but it could (though needn't) be interpreted as it being placed in the center of the eye. The primary anatomical

---

17 'rotundus sive figurae spericae, cum quadam planitie in parte anteriore, et hic humor vitreus in quo locatur, et ideo hic humor vitreus factus est ad christallinum locandum'. Jacopo Berengario da Carpi and Mondino, *Carpi commentaria cum additionibus super anatomia mundini vna cum textu eiusdem in pristinum et verum nitorem redacto* (Bologna: Hieronymus de Benedictis, 1521), p. 462^r.

18 'aliqui tamen addunt quartum humorem quem vocant etereum quem ponunt ante in directo pupille'. Berengario and Mondino, *Carpi commentaria*, p. 468^r.

19 'Est unus alter humor in oculo vitreus dictus, qui est in quantitate maior aliis duobus'. Berengario and Mondino, *Carpi commentaria*, p. 469^r; Cf. Jacopo Berengario da Carpi, *A Short Introduction to Anatomy (Isagogae Breves)* (Chicago: University of Chicago Press, 1959), p. 152.

20 'Gutta ipsa in partem priorem magis vergit, quae in parte postrema rotunda est, in priore leviter plana'. Alessandro Benedetti, *Historia corporis humani, sive anatomice*, ed. by Giovanna Ferrari (Florence: Giunti, 1998), p. 280.

21 Gabriele Zerbi, *Opus preclarum anathomie totius corporis humani et singulorum membrorum illius* (Venice: Scotus, 1533), pp. 122^r–33^r. The first edition was published in 1502. On the sizes of the humours, see especially p. 131^r. Note that the 1533 edition contains an error in the page numbering: after normal page numbering up to 134 r/v, the numbers return to 121. The page numbers cited above therefore refer to the second occurrence of the numbers 122^r–133^r. Nicolo Massa, *Liber introductorius anatomiae siue dissectionis corporis humani* (Venice: Franciscus Bindonus ac Mapheus Pasinus socios, 1536), pp. 92^r–93^v.

22 '& ipse quidem est humor clarus, sicut grando: glacialis vero est rotundus, cuius minuit rotunditatem compressio ipsius ab anteriore parte eius [...] & hic quidem humor positus est in medio, quoniam est dignior locis'. Avicenna, *Liber canonis: De medicinis cordialibus; et Cantica*, ed. by Benedetto Rinio (Basel: Ioannes Heruagius, 1556), p. 405A. The term *fen* is a transliteration of the Arabic *fann*, meaning part or section; it is traditionally used in references to Avicenna's *Canon*. On Avicenna's *Canon* in the Renaissance, see Nancy G. Siraisi, *Avicenna in Renaissance Italy: The Canon and Medical Teaching in Italian Universities after 1500* (Princeton, N. J.: Princeton University Press, 1987).

authors before Vesalius, then, describe the crystalline as having a flattened anterior, a rounded posterior, and they at times put the crystalline in the front part of the eye.[23] The humanistic revival of Galen's works, beginning towards the end of the fifteenth century, altered the anatomical context significantly, and set the stage for Vesalius. Galen's three main discussions of the eye are found in *On the Usefulness of the Parts, On the Doctrines of Hippocrates and Plato,* and *On Anatomical Procedures.* His section in *On Anatomical Procedures* where the anatomy of the eye is described, however, was not available to the Latin West in the sixteenth-century.[24] Thus in none of the texts available in Europe does Galen give a precise description of either the relative size of the three humours or the precise location of the crystalline humour within the eye.[25] Note that *On the Doctrines of Hippocrates and Plato* was largely unknown before the Greek edition of Galen in 1525; it is there that Galen harshly criticises Aristotle's intromission theory and offers his own synthesis of the extramission theories of Plato, the Stoics, and the mathematicians.[26] Its publication no doubt contributed to a revival of Galenic extramissionist theories towards the middle of the sixteenth century.

The pseudo-Galenic *Liber de oculis* ('Book on the Eye', sometimes also called *Galeni liber de oculis translatus a Demetrio*) was in fact a translation of a work by Hunain ibn Ishaq. Ibn Ishaq's work on the eye was published under Galen's name in many *Opera* early in the sixteenth century, and it was often believed to be genuine.[27] Here we read: 'The crystalline is a white, shining, and clear humour, not entirely round, but somewhat flattened, which is placed in the middle of the eye'.[28] Later this is reiterated: 'It is located

23 Although space limitations permit a full discussion, another important and contested issue in sixteenth-century anatomical works is the presence or absence of an foramen in the optic nerve, a crucial component of Perspectivist visual theory. The presence of a hole in the optic nerve was routinely denied by anatomists towards the end of the century.

24 It did influence Hunain ibn Ishaq, among others. Galen, *Galen on Anatomical Procedures: The Later Books*, trans. by Wynfrid Laurence Henry Duckworth, ed. by M. C. Lyons and B. Towers (Cambridge: Cambridge University Press, 2010), pp. XI–XIV. On Galen in the sixteenth-century, see: Andrew Wear, 'Medicine in Early Modern Europe, 1500–1700', in *The Western Medical Tradition: 800 BC–1800 AD*, ed. by Lawrence I. Conrad (New York: Cambridge University Press, 1995), pp. 250–64, 270–73. On the availability of Galen in the Renaissance, see Richard J. Durling, 'A Chronological Census of Renaissance Editions and Translations of Galen', *Journal of the Warburg and Courtauld Institutes*, 24 (3/4) (1961), 230–305; Stefania Fortuna, 'The Latin Editions of Galen's Opera Omnia (1490–1625) and their Prefaces', *Early Science and Medicine*, 17 (4) (2012), 391–412. On Galen and vision in the thirteenth and fourteenth centuries, see Fernando Salmon, 'The Many Galens of the Medieval Commentators on Vision', *Revue d'histoire des sciences*, 50 (1997), 397–420.

25 Galen, *Galen on the Usefulness*, 464–503; Galen, *On the Doctrines of Hippocrates and Plato: Books VI–IX*, trans. by Phillip De Lacy, Corpus Medicorum Graecorum (Berlin: Akademie-Verlag, 1980), p. 459.

26 Vivian Nutton, 'De placitis Hippocratis et Platonis in the Renaissance', in *Le Opere psicologiche di Galeno*, ed. by Paola Manuli and Mario Vegetti (Napoli: Bibliopolis, 1988), pp. 281–309. Prior to the sixteenth century, see Salmon, 'The Many Galens'.

27 Max Meyerhof, 'New Light on Hunain Ibn Ishaq and his Period', *Isis*, 8 (4) (1926), 685–724. The *Liber de oculis* was included in printings of Galen's *Opera* into the seventeenth century at least, although after some point was usually labelled spurious. It was also published as *Liber de oculis Constantini Africani* in *Omnia opera Ysaac* (Lyons: Trot, 1515).

28 'Crystallinus est humor albus, splendidus, lucidus, non omnino rotundus, sed aliquantulum planus, qui locatus est in medio oculorum'. Galen, *Galeni Opera ex octava Iuntarum editione* (Venice: Giunta, 1609), IX, p. 186r. Note that I have used the seventeenth-century Giunta edition of Galen's works, at which point the *Liber de oculis* was recognized as not authentic.

in the middle in order that the rest of the parts might serve it'.[29] The text does not mention the humour's asymmetrical flattening. As we saw in Avicenna, *loctus in medio oculorum* could be interpreted as 'the middle humour', although because the text does not give estimates of the relative sizes of the humours one can easily read the passage as locating the humour in the direct centre of the eye. Finally, in the pseudo-Galenic *Anatomia vivorum* ('Anatomy of the Living'), a very popular work which was also commonly believed to be genuine in the first half of the sixteenth century, we read that the crystalline humour is 'called *grando glacialis* by Aristotle, because it resembles a hailstone in form and colour' and that it is 'placed in the depth and in the middle'.[30] The text also does not refer to the relative sizes of the humours, but we do read that the crystalline is somewhat flattened in the front, but spherical behind.[31]

In short, with the exception of the pseudo-Galenic *Liber de oculis* and Massa's short introduction to anatomy, every anatomy text available to Vesalius described the crystalline humour as flattened in the front, while protruding and rounded in the back. While some placed it towards the front of the eye, some described it somewhat ambiguously as the middle humour, while two pseudo-Galenic works can easily be read as placing it in the direct centre. In most of these texts there is little connecting ocular anatomy with public dissection specifically. At the same time, the eye presented within mathematical optics did not draw upon experience with dissection: the *Perspectivae* constructed a geometrical eye according to the demands of visual theory, rather than abstracting the shapes, sizes, colours, densities, and other relevant properties of its parts from sensory experience.

Vesalius

Much of this changed with Vesalius. Vesalius put the crystalline humour in the direct centre of the eye and described it as symmetrical from front to back, with its shape being, as he says, 'as if you had used a saw to remove a thickish circle from a wooden globe along equidistant lines and then glued the two parts of the humour together' (Figure 2).[32] Vesalius twice draws a comparison between the eye and the cosmos.[33] Modern commentators sometimes say that he was continuing a medieval tradition of imagining the eye as a microcosm, but as we have seen this was not the case.[34] Instead, anatomists usually described the crystalline humour in relation to the vitreous as a precious stone set into a ring, conjuring up a far different sense of location, size, and relative shapes of the humours. Vesalius's departure is an important shift in sixteenth-century ocular anatomy. By the end of the century a general

---

29 'In medio locatus, ut caeterae partes sibi ministrent'. Galen, *Galeni Opera*, IX, p. 186^{r}.

30 George W. Corner, *Anatomical Texts of the Earlier Middle Ages: A Study in the Transmission of Culture* (Washington: Carnegie Institution of Washington, 1927), p. 98. On the history of this text, see pp. 35 ff. Note that I have been unable to find any passage in Aristotle referring to any part of the eye as a hail-stone; rather, this seems to be derived from Avicenna's *Canon*. See note 22.

31 Corner, *Anatomical Texts*, 98–99.

32 Vesalius, *The Fabric*, 1306. Unless noted, all translations are from this edition. Cf. Andreas Vesalius, *De humani corporis fabrica libri septem* (Basel: Johannes Oporinus, 1555), p. 801. Unless specified, all references to *De fabrica* are to the 1555 Basel edition.

33 Vesalius, *De fabrica*, 798, 801.

34 See note 6.

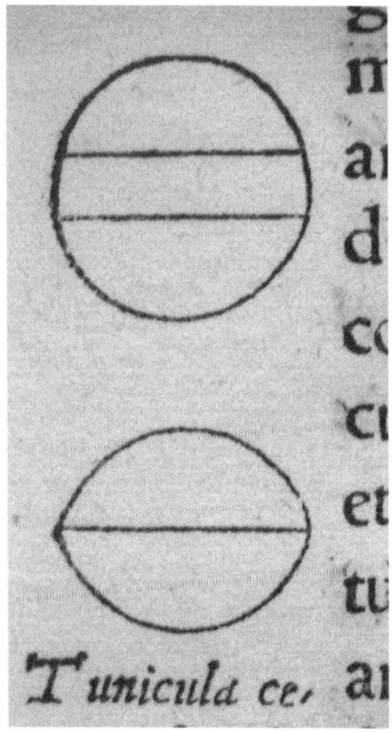

Fig. 2 Marginal image depicting how to generate the shape of the crystalline humor from a sphere. Andreas Vesalius, *De humani corporis fabrica libri VII* (Basel: Johannes Oporinus, 1543), p. 646. This is reproduced in the 1555 edition, p. 801. Courtesy Newberry Library.

consensus was reached on the size and shape of the humours, but many of the steps in the formation of this consensus were framed as a response to Vesalius — by Juan Valverde, Realdo Colombo, Gabriele Falloppio, and Hieronymus Fabricius ab Aquapendente, for example. On these questions, few if any authors after Vesalius referenced to earlier anatomists that Vesalius had eclipsed.[35] Also important is that Vesalius's depiction of the eye in cross-section — a sort of marriage of the optical and anatomical visual traditions — was highly imitated after *De fabrica*, even while the shapes and sizes of the humours were corrected (Figure 3).[36]

Compared with other sixteenth century anatomists, both before and after, Vesalius's conception of the eye was idiosyncratic. One possible cause for Vesalius's 'cosmic' eye is the role public dissection played in his career.[37] The eye, being small and fluid, is not an

35 Juan Valverde, *Historia de la composicion del cuerpo humano* (Rome: impressa por Antonio Salamanca y Antonio Lafrerij, 1556), p. 82^{v}; Realdo Colombo, *De re anatomica libri XV* (Venice: ex typographia Nicolai Builacquae, 1559), pp. 215–16, 218; Gabrielle Falloppio, 'Institutiones anathomica', in *Omnia, quae adhuc extant opera* (Venice: Felix Valgrisius, 1584), 282^{v} ff; Hieronymus Fabricius ab Aquapendente, *De visione, voce, auditu* (Venice: Franciscus Bolzetta, 1600), pp. 1–11.

36 Prominent examples of this are Valverde, *Historia*, p. 118^{r}; Felix Platter, *De corporis humani structura et vsu* (Basel: ex Officina Frobeniana, Ambrosius Frobenius, 1583), Plate 49.

37 For Vesalius's transformation of the public anatomical demonstration more generally, see R. Allen Shotwell, 'Animals, Pictures, and Skeletons: Andreas Vesalius's Reinvention of the Public Anatomy Lesson', *Journal of the History of Medicine and Allied Sciences*, 71 (1) (2016), 1–18.

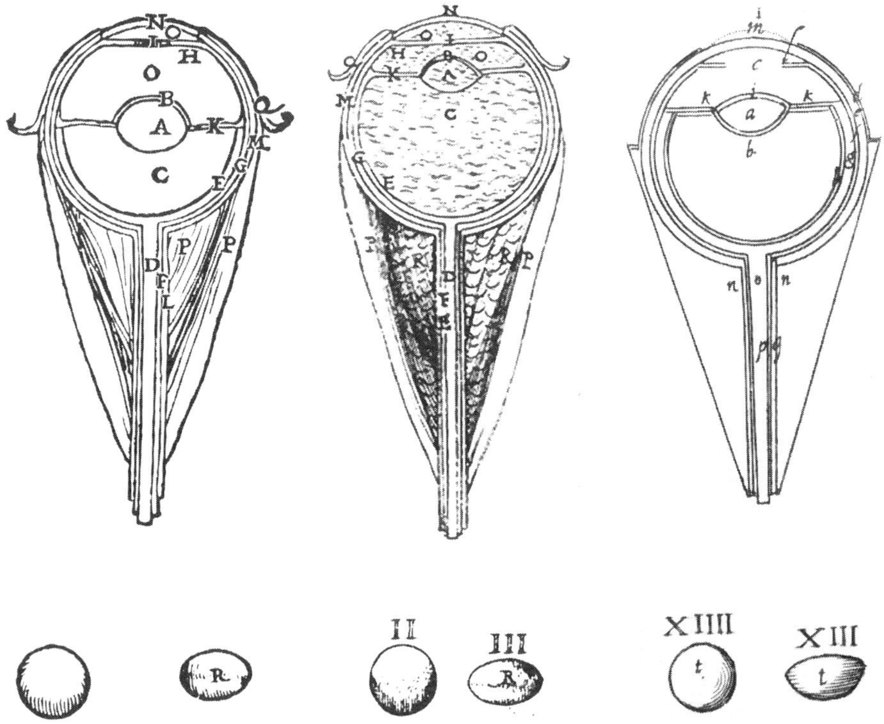

Fig. 3 Three cross-sectional images of the eye, along with the shapes of the crystalline humor, in sixteenth-century anatomical works. Left to right: Vesalius, *De fabrica* (1555), p. 798; Valverde, *Historia*, third table of the fifth chapter (following f. 82ᵛ); and Kepler, *Paralipomena*, plate 1 (following p. 164). Kepler's image is a faithful copy of Platter, *De structura*, table 49, but with the addition of a dotted line through *m*, indicating the corneal bulge.

ideal subject for public dissection, and the records we have of Vesalius's 1540 anatomy in Bologna show Vesalius making a rather careless spectacle of the organ. We read the following in the student Baldasar Heseler's notes:

> For there are, he said, four tunics of the eye, and three humours contained therein, which he named each one by turn and with their names, all of which you have in the text of Mundinus and in Galen. And he cut the eye through the middle with a razor, and he shook out into the hand the substance of the eye: the first humor, he said, is the albugineus one, the second is the vitreous and the third is the crystallinous humor, by which properly the vision occurs, and it is hard like a jewel. All this, he said, anybody can see for himself at home.[38]

Note that this was done at the beginning of Vesalius's very last demonstration, and that the eye dissected was that of a sheep.

38 Baldasar Heseler and Ruben Eriksson, *Andreas Vesalius' First Public Anatomy at Bologna: 1540* (Uppsala: Almqvist & Wiksell, 1959), pp. 290–91.

At the end of the 1555 edition of *De fabrica*, Vesalius says 'At present, I have described the eye using the account I give in the schools'.[39] But in the 1543 edition this sentence continues, explicating his classroom method in greater detail:

> [...] first explaining the construction of the eye and drawing a large picture on a sheet of paper as I go, like the one I have tried to show in the first figure placed at the beginning of this chapter. After my explanation, I perform my dissection as I shall add at the appropriate place in the final chapter of this book.[40]

That his diagram of the eye arose in a classroom context is significant, as is the fact that he taught ocular anatomy to his students by drawing a diagram prior to performing a dissection. His students would therefore have been disposed to import Vesalius's diagram into their observations — to use the diagram to bring order to the chaos that ocular dissections can become, particularly when performed as Vesalius describes.[41]

In *De fabrica* Vesalius advises dissecting the eye of a recently deceased human cadaver.[42] After denuding the eye of muscle and fat, he instructs the anatomist to hold the eyeball with the fingers of the left hand such that the pupil rests on the thumb and the nerve is held between the other fingers, thus securing the eye in place. Using an *acutior novacula* ('a well-sharpened razor') in the right hand, the anatomist should then make a long, transverse cut around the eyeball, 'so that all the humors of the eye can escape without squeezing'.[43] After this cut is made, one should pass the eye into the right hand, and then subsequently let the crystalline and vitreous fall out into the left together. (At this point the gelatinous vitreous and the wax-like crystalline might still be held together by their surrounding membranes). In sliding the two humours out together, one should take great care to ensure that the crystalline remains on top, otherwise 'you will spoil the job'.[44] That is to say, your audience will be confused. One should then pour out the remaining aqueous humour (until now resting in the front section of the eye) into the left hand also. Now the eye will be emptied of all three humours, and one can inspect its inner tunics.

These are instructions for giving a public anatomy demonstration, and they correspond exactly to Heseler's notes. This dissection technique shows Vesalius's great manual skill in cutting the tough, slippery eyeball and gracefully removing and displaying the delicate humours of the eye, but it is a poor method for understanding the precise size and location of the humours and inner tunics. (Having attempted, and failed, in front of an audience to imitate Vesalius's directions myself, I can attest to the skill required to pull it off.) Vesalius does, very briefly, give another method for dissecting the eye — the second eye in the

39 'Ego autem nunc ea oculum ennaratione persecutus sum, qua in scholis uti soleo'. Vesalius, *De fabrica*, 806; Vesalius, *The Fabric*, 1311.

40 Andreas Vesalius, *De humani corporis fabrica libri VII* (Basel: Johannes Oporinus, 1543), pp. 649. '[...] primum oculi constructionem explicans, & in charta maiusculam effigiem ita sensim delineans, atque in prima huic Capiti praeposita figura oculum exprimere conatus sum. dein post ennarationem ita sectionem obiens, quaeadmodum suo loco in ultimo huius libri capite subiiciam'. Vesalius, *The Fabric*, 1311.

41 On Vesalius's use of diagrams in the classroom and in public dissections see Shotwell, 'The Revival of Anatomical Practices', pp. 166–67.

42 Vesalius, *De fabrica*, 813. On Vesalius's dissection of human eyes, see note 53.

43 'ut citra compressionem omnes oculi humores excidere possint'. Vesalius, *De fabrica*, 813; Vesalius, *The Fabric*, 1323.

44 'negocium enim confuderis, si vitreus super crystallinum revoluatur'. Vesalius, *De fabrica*, 813; Vesalius, *The Fabric*, 1323.

animal under dissection — that is less well-suited to a public demonstration. Rather than executing one decisive cut through all of the tunics, one should make many softer cuts with the razor until one has come just through the sclera, thus revealing the uvea, at which point the sclera can be sufficiently detached from the uvea so that scissors can be inserted between the two to cut through the hard sclera. In this way, he says, one may see the veins passing from the sclera to the uvea, and then after cutting into the uvea, 'This will make it a little easier to see [...] the site of the humors'.[45]

Apart from the statement that vision occurs in the crystalline humour, there is no mention of visual theory in Heseler's notes. In *De fabrica*, apart from a statement that one should consider how the crystalline humour, when placed over text, magnifies it like a *specillum* ('glass lens'),[46] Vesalius only says the following:

> But just as I am accustomed to show the fabric of the eye so carefully that no viewer will have any questions about it, so I am quite silent here in explaining the function of its parts because I cannot satisfy myself about the primary instrument of vision. In this part of the body I cannot report what is sound in all respects. It is not as if I cannot say that the crystalline humor is the principal instrument of vision, and that the vitreous is made so the crystalline may be nourished, and that the other parts of the eye are formed for the sake of this humor, as Galen states at length in the tenth book of *De usu partium*, or that I am unable to write a great deal as well. But because I am not certain how vision takes place, because the remaining construction of the eye depends thereon, and especially because I believe I must sometime write separately about such scientific and medical controversies, in the task at hand I am keeping to a proper length.[47]

Vesalius never did write at length about this, although his successors certainly did. Nevertheless, from Vesalius's treatment of the eye we can tease out elements corresponding to a Galenic theory of vision — if not extramission *per se*, he at least retains much of Galen's ocular physiology. One reason Vesalius gives the aqueous humour a much greater bulk than is revealed during dissection stems from Galen's statement in *De usu partium* that 'if you care to dissect a dead diligently considered animal, you will see even before you do so that the eye is already somehow more shrivelled than it is in its normal state'.[48] Galen held that, in addition to the aqueous humour, a subtle pneuma also resides in the front aqueous

45 'humorum situm paulo meliùs contempleris'. Vesalius, *De fabrica*, 814; Vesalius, *The Fabric*, 1324.

46 Vesalius, *De fabrica* (1543), 655. Note that for the 1555 edition, Vesalius changed 'specillum' to 'specillum non excavatum', or 'a non-concave lens'. Vesalius, *De fabrica* (1555), 814. Cf. Vesalius, *The Fabric*, 1323, n. 42.

47 'Verùm ut oculi fabricam adeò studiose demonstrare consueui, ut spectantium nemo in ipsa quicquam desideraret: ita in ipsius partium usu explicando hîc prorsus obmutesco, quòd mihi ipsi de primario visus instrumento non satisfaciam, atque hac in parte aliquid quod sanum undique sit, à me non adferri posse, mihi persuadeam. Non sanè quasi crystallinum humorem, praecipuum visus instrumentum esse, & ut hic alatur, vitreum produci, & huius gratia oculi reliquas partes efformatas esse, quemadmodum Galenus in decimo de Partium usu diffuse persequitur, commemorare, aut hîc prolixè etiam scribere nequirem: sed quòd mihi certò, quî visus fiat, non constet, hincque reliqua oculi constructionis ratio dependeat, quodque potissimùm de huiusmodi philosophorum & medicorum controversis privatim quaedam aliquando mihi scribenda duxerim, hoc labore ad iustam nunc molem perducto'. Vesalius, *De fabrica*, 806. Translation based on: Vesalius, *The Fabric*, 1311. Note that the translators of this edition have mistakenly rendered '& huius gratia' (i.e., 'for the sake of this humour') as 'for the sake of the vitreous humour'. Vesalius meant 'for the sake of the crystalline humour', and this would have been obvious to his readers. I have corrected this in the translation above.

48 Galen, *Galen on the Usefulness*, 476.

chamber in living animals. In addition to the evidence from the laxity of the eyes of dead animals, which he twice makes reference to in *De usu*, Galen also says that this pneuma is proved by the fact that, if one closes one eye, the pupil of the other dilates.[49] This pneuma supposedly had its origin in the brain and issued forth through the optical nerves; because of the optic chiasma (so-called because the nerves cross, as in the Greek letter *chi*), if one eye were closed the spirits would push through to the other eye with greater force. For Galen this pneuma was an essential component for vision, although it was itself insensible.[50]

Vesalius followed Galen in the belief that living eyes contain a subtle pneuma or spirit in the anterior chamber. This is highlighted in his surviving notes to the 1555 *Fabrica,* written in preparation for a revised third edition that never appeared. Although criticism of Vesalius's description of the eye was beginning to circulate, Vesalius did not alter the shape or size of the crystalline or vitreous humour in his notes. He did, however, add that, compared to the vitreous, only a small amount of aqueous humour comes out of the eye upon dissection, and he noted two possible causes for this. The first is that in a living eye 'one must conclude that it [the aqueous] is largely composed of a sort of spirit and aerial substance', which supposedly dissipates after death.[51] This is Galen's position. After this Vesalius writes: 'perhaps someone [might say] that the vitreous humour occupies a larger space in the eye than the rear portion and thus that the lens along with the vitreous humour [is placed] off-centre in the front part of the eye'.[52] Vesalius's language gives the impression that he prefers the first explanation: 'one must conclude' the former, while 'perhaps' the latter is the case. Corroborating this reading is that, in these notes, he modifies neither of his other descriptions in the text nor his diagram showing that the aqueous and vitreous are of equal bulk in a living eye. Finally, in both editions Vesalius discusses the colour of the uvea (which includes the iris), and he writes that the colour of this tunic does not depend on spirits or humours: because the colour of the uvea is the same in a living and dead eye, spirits in the eye that dissipate upon death cannot be responsible for the tunic's colour.[53]

According to Vesalius some quantity of subtle spirit was present in the aqueous humour, and all evidence points to his having believed that this spirit accounted for the discrepancy between his depiction of a living eye compared to a dead, dissected eye. In all of this he appears to assume aspects of a Galenic theory of vision, and Galenic extramission theories were commonly held by sixteenth-century physicians.[54] The role

49 Galen, *Galen on the Usefulness*, 476–77.

50 See also Véronique Boudon-Millot, 'Vision and Vision Disorders: Galen's Physiology of Sight', in *Blood, Sweat and Tears: The Changing Concepts of Physiology from Antiquity into Early Modern Europe*, ed. by Manfred Horstmanshoff, Helen King and Claus Zittel (Leiden: Brill, 2012), pp. 549–67.

51 Vivian Nutton, 'Vesalius Revised. His Annotations to the 1555 Fabrica', *Medical History*, 56 (2012), 415–43 (p. 435). Translation by Nutton.

52 Nutton, 'Vesalius Revised', p. 435.

53 Vesalius, *De fabrica*, 804. Vesalius, *The Fabric*, 1309. Note that in this section Vesalius explicitly attributes the *tapetum lucidum* — a multi-coloured iridescence in the uvea behind the retina in some animals that aids in night vision — to humans. This is one indication that he did not carefully examine human eyes.

54 A prominent example of a physician arguing for Galenic extramission is Francisco Vallés, *Controversiarum medicarum et philosophicarum libri decem* (Frankfurt am Main: Andreas Wechelus, 1582), pp. 104–07. Originally published in 1556. On Vallés influence, see Nancy G. Siraisi, *The Clock and the Mirror: Girolamo Cardano and Renaissance Medicine* (Princeton: Princeton University Press, 1997), p. 55; Craig Martin, 'Francisco Vallés and the Renaissance Reinterpretation of Aristotle's "Meteorologica" IV as a Medical Text', *Early Science and Medicine*, 7, (2002), 1–30.

of public dissection in Vesalius's career also played a role in his idiosyncratic account of the relative shapes and sizes of the humours. Finally, Vesalius's image of the eye combined the cross-sectional diagram of the mathematical tradition with the notion — foreign to that tradition — that such diagrams represent what could be seen upon dissection. After Vesalius, cross-sectional diagrams of the eye, usually based upon Vesalius's illustrations, became ubiquitous in anatomical texts. (Whether such diagrams were also used in public anatomies, or in classrooms apart from Vesalius's, remains to be investigated.)

Padua and the Synthesis of Anatomy, Optics, and Natural Philosophy

Visual theory was addressed directly in anatomical (and medical) works increasingly after Vesalius. Furthermore, by the end of the century accounts of the substance of the humours become a major factor in the intromission-extramission controversy. In the second half of the century vision was usually addressed within a Galenic framework of structure, action, and use (*usus* or *utilitas*), and the reasons for the various temperaments of the parts of the eye noted in the structure sections of works were accounted for teleologically in the use sections. Moreover, Galen's brief treatment of mathematical optics in book ten of *On the Usefulness of the Parts of the Body* encouraged sixteenth-century authors to introduce optics in their sections on *usus* or *utilitas*, even when they failed to follow Galen's visual theory itself.[55] For example, in his *De corporis humani structura et usu* Felix Platter famously posited (without explanation or argument) that the retina was the primary author of vision, and took the crystalline lens to be a sort of *perspicillum* or looking glass that enlarged the visual species entering the eye.[56]

From the records from the German Nation of Artists (that is, the association of transalpine medical students) at the University of Padua we know that Hieronymus Fabricius ab Aquapendente, who at that time held the chair in anatomy, dissected the ear and eye during the annual public anatomy in February 1586.[57] Then, in June of that year, the physician and anatomist Paolo Galeotto performed a private anatomy at the request of the German Nation. We read:

> And so that nothing could be left wanting, at the close of his lectures in the month of June he added a thoroughly beautiful and complete demonstration of the eyes, as original as it was artful, in which he also found and noted certain things [that were] clearly new and extremely pleasing.[58]

55 Nancy G. Siraisi, 'Historia, actio, utilitas', pp. 63–73. See also Distelzweig, 'Fabricius's Galeno-Aristotelian Teleomechanics of Muscle', pp. 64–84.

56 Platter, *De corporis humani structura et vsu*. Lindberg writes that this development was 'modest', while more recently A. Mark Smith writes: 'Neither the radical nature nor the historical importance of this idea can be overstated'. Lindberg, *Theories of Vision*, 176; Smith, *From Sight to Light*, 352.

57 Antonio Favaro, *Atti della nazione Germanica artisti nello studio di Padova*, 2 vols (Venice: Prem. Tipografia Emiliana, 1911–12), I, p. 210.

58 'Utque nihil fere desiderari posset, etiam sub finem lectionum mense Innio adiecit perpulchram et absolutissimam oculorum tam nativorum quam factitiorum demonstrationem, in eisque quaedam plane nova et perquam iucunda adinvenit et annotavit'. Favaro, *Atti*, I, p. 211.

Later, in a private anatomy given in the winter of 1586–87, after describing the external features of the body, we read that Galeotto 'immediately seized upon the eye, and gave many lessons on its action and fabric so clearly and so skilfully that he conducted himself as one most expert in things anatomical and optical'.[59] Soon after in late January Fabricius dissected the eye and ear in his public anatomy, although because he both lacked a cadaver and had lost his voice he had to suspend his plan to treat the larynx as well.[60] The records of the German Nation also tell us that Fabricius dissected the eye and lectured on vision in a public anatomy in 1592.[61]

To some degree Galeotto's private anatomies competed with those given by Fabricius. In 1584, at the same time that Fabricius was accorded all the privileges given to ordinary professors of the first rank in medicine at Padua, the Venetian Senate gave Fabricius and whoever succeeded him in the chair of anatomy the sole privilege of performing the annual public anatomy during the winter months, while the surgeons were allowed to perform anatomies the rest of the year.[62] Although Galeotto's demonstrations were technically private — meaning that his anatomy, performed in a temporary anatomical theatre built especially for the purpose, was only open to paying students — in 1587 Fabricius complained to the *riformatori dello studio*, a small group of magistrates appointed by the Venetian senate to oversee the University, that Galeotto's private anatomy began before his own public anatomy had ended that year.[63] This competition to provide better accounts of the eye by incorporating philosophical and mathematical accounts of vision can also be seen in the nearly simultaneous publication of treatises on vision, voice, and hearing by Fabricius and the Paduan anatomist, surgeon, and physician Julius Casserius.[64]

The topics treated in Fabricius's public anatomy in 1587 correspond to his first published anatomical treatise, *De visione, voce, auditu*, first published in 1600 in Venice. In this text, Fabricius argues for a novel intromission theory: while he draws heavily from Aristotle and the *Perspectivae* (citing Alhacen, Witelo, and Pecham), his account differs from his sources in some important ways. In addition, Fabricius explicitly rejects Galen's account of vision, and moreover announces that he will build upon Galen, and indeed surpass him, by providing a thorough integration of anatomy and mathematical optics, as well as a detailed and accurate mathematical account based on a complete determination of the fabric of the eye.[65] Contra Aristotle, Fabricius argues that light is the vehicle for colour rather than, as the former held, the actuality of a transparent medium; that is, Fabricius argues that we see the colour of light which has become tinged by coloured bodies, rather than seeing

59 'protinus ad oculum aggressus Anatomicus, de eius actione et fabrica ita plane ita erudite multis disseruit lectionibus, ut virum sese gesserit anatomicarum rerum et opticae peritissimum'. Favaro, *Atti*, I, pp. 225–26.

60 Favaro, *Atti*, I, p. 227. See also Giuseppe Favaro, 'L'insegnamento anatomico di Girolamo Fabrici d'Acquapendente', in *Monografie storiche sullo studio di Padova, contributo del R. Instituto Veneto di Scienze, lettere ed arti alla celebrazione del VII centenario della Università di Padova* (Venice: Ferrari, 1922), p. 116.

61 Favaro, *Atti*, II, p. 32.

62 Jacopo Facciolati, *Fasti gymnasii Patavini* (Padua: Joannes Manfré, 1757), p. 388.

63 Klestinec, 'History of Anatomy Theaters', p. 387; Cynthia Klestinec, *Theaters of Anatomy: Students, Teachers, and Traditions of Dissection in Renaissance Venice* (Baltimore: Johns Hopkins University Press, 2011), pp. 79–88.

64 Julius Casserius, *De vocis auditusque* (Ferrara: Victorius Baldinus, 1601); Julius Casserius, *Pentaestheseion, hoc est de quinque sensibus liber* (Venice: Nicolaus Misserinus, 1609).

65 Fabricius, *De visione*, 55.

the colour of bodies themselves because light has actualized the transparent medium.[66] This became the dominant position over the course of the seventeenth century. Contra the *Perspectivae*, Fabricius argues, based upon evidence from dissection, that the cornea and the anterior surface of the crystalline humour are not concentric. He also denies that there is an optical foramen along whose twisting path tiny images are carried (a fact that nearly every sixteenth-century anatomist, including Vesalius, agreed upon), and he denies that light refracts differently in a living eye compared to a dead eye.[67] Because of the latter, Fabricius holds that the rays of light come to a focus just behind the crystalline humour.[68] In place of the scheme given by the *Perspectivae*, Fabricius argues that the faculty of sight makes an active judgement, within the eye itself, of the coloured images that show up in or on the crystalline humour.[69] Apart from this act of judgement (and the local motion of the eye) a living eye and a dead eye are largely identical. Thus in *De visione* Fabricius says that if you place the crystalline humour it in front of a candle it acts to gather the light together into a point, just as a burning lens.[70] The reason that Nature placed a burning lens in the eye was to focus the light passing through the crystalline into a point within the vitreous, thereby exhausting the light and preventing it from rebounding off the rear tunics, which would otherwise carry this interior image of the eye back to the crystalline humour.[71]

One crucial reason why the eye is watery, and especially why it is not airy (or filled with an airy or fiery spirit), is to prevent a flame from being kindled in the eye. Since air is hot and wet, it can be converted into flame via the action of a burning lens or mirror, whereas water, being cold and wet, cannot. (Note that it was commonly believed that sunlight was not itself hot, and that sunlight heats bodies only by rarefying the air near its surface and thereby causing the air, by virtue of its innate elemental quality of heat, to become more fiery). Another crucial reason Fabricius gives for why the eye is watery is that, unlike air, water can be condensed in various ways in order to produce the different refractive powers of the humours. The crystalline humour, being the most condensed, refracts light the most. The vitreous, being less condensed, refracts light less than the crystalline. Finally, the most tenuous humour, the aqueous, refracts light the least.[72] While Fabricius only gives a qualitative account of their refractive powers, because of the absence of pneuma in the eye and the fact that the visual faculty does not interfere with refraction in the eye, a precise determination of the refractive powers of the humours based upon dissection is possible in principle, and there is no reason why artificial models of the eye (as Vopiscus Fortunatus Plempius would later produce) cannot accurately replicate the path of rays in the eye. Indeed, Fabricius presents actual-size diagrams of both a human and a sheep

66 Fabricius, *De visione*, 39–42.

67 Baker, 'Color, Cosmos, Oculus', pp. 249–77.

68 Fabricius, *De visione*, 107.

69 Fabricius, *De visione*, 51–54.

70 Fabricius, *De visione*, 103.

71 For more detail, see Tawrin Baker, 'Why All This Jelly? Jacopo Zabarella and Hieronymus Fabricius Ab Aquapendente on the Usefulness of the Vitreous Humor', in *Early Modern Medicine and Natural Philosophy*, ed. by Peter Distelzweig, Benjamin Goldberg and Evan Ragland (Dordrecht: Springer, 2015); Baker, 'Color, Cosmos, Oculus', pp. 232–77.

72 Fabricius, *De visione*, 57–60.

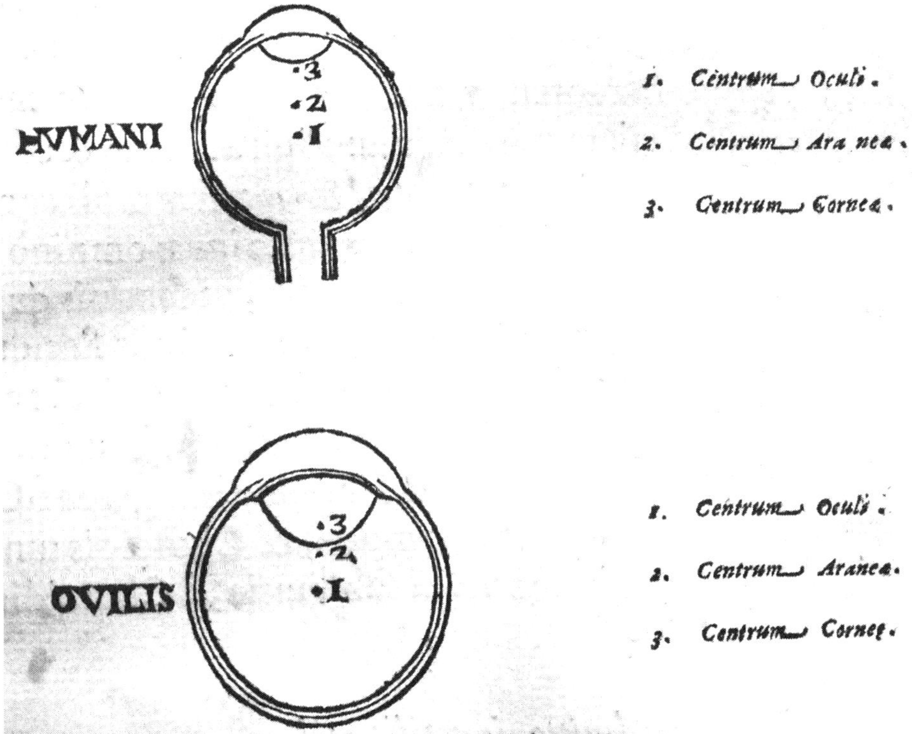

Fig. 4 Geometrical diagram of the eye abstracted from anatomical experience. Fabricius, *De visione*, 105. The top image depicts a human eye, the bottom a sheep. The dots represent the centres of curvature of: (1) the whole eye (i.e., the sclera), (2) the aranea (and thus also the anterior of the crystalline humor), and (3) the cornea. Note that the centres of the aranea and the cornea are not identical, and that the geometrical configuration of the eyes of the two animals are not identical. Courtesy Newberry Library.

eye, abstracted from anatomical experience, explicitly so that the mathematicians can determine the path of the rays in the eye (Figure 4).

> But so that those who produce works of optical science can accurately observe the diverse progression of rays, which are called visual, while they cross over from one humour into another; and [so that] they can accurately measure off the angles of refraction, and thence grasp the innumerable uses [*utilitates*] of the parts: we provide, with the most exact care, human and sheep eyes divided through the middle. And the whole magnitude and that of the individual parts, including their situations and figures, are described, and the place that each of their centres occupy is revealed, and everything is sketched out in the tables below.[73]

73 'Ut autem qui Opticae scientiae operam dant, accuratè obervare possint, progressum varium radiorum, quos visuales appellant, dum ab uno in alium humorem transeunt; atque angulos refractionis dimetiri, & inde innumeras utilitates partium excepere: curavimus exactissima diligentia, oculum humanum & ovilem per

We have indirect evidence that Fabricius taught this account of vision at the anatomy theatre. One source is the Paduan natural philosopher Jacopo Zabarella (1533–90) who in his natural philosophy textbook gives an account of vision that is nearly identical to Fabricius's — although Zabarella published his account ten years before Fabricius. Most importantly, Zabarella mentions attending a dissection of the eye during which he observed the very experiments on the crystalline humour that Fabricius describes in his text.[74]

Another source for Fabricius's ocular anatomies is the work of one of his students, the 1601 *Anatomiae Pragae* by Johannes Jessenius (1566–1621). Fabricius was Jessenius's teacher when he studied medicine at Padua, likely between 1588 and 1593. The *Anatomiae* were written as an account of a series of celebrated public anatomies that Jessenius gave in Prague in 1600. At the beginning of his section on the eye Jessenius writes 'My teacher, the Paduan anatomist D. Hieronymus Aquapendente, always seemed to me to reign supreme in the anatomy of the eye'.[75] Although much condensed, lacking images, and containing fewer references to mathematical optics, Jessenius ocular anatomy has nearly all the other distinctive features of Fabricius's *De visione* — including the presence of a burning lens in the eye and Fabricius's precise reasons why the eye has a watery temperament. Indeed, in his section of the eye Jessenius says nothing that is not included in *De visione*. We can conclude that, in his dissection of the eye, Jessenius relied only on notes taken from Fabricius in Padua (and not from the text of *De visione*) from the following. First is the timing: Fabricius dates his dedicatory remarks in *De visione* as being written in the month of December 1600, while Jessenius's anatomies in Prague occurred in June of that year. Furthermore, Kepler said he did not have access to Fabricius's *De visione* in Prague;[76] if his friend Jessenius bought a copy and cribbed from it for the published version of the *Anatomiae*, he would certainly have lent his copy of *De visione* to Kepler. On the other hand, whether Jessenius's ocular anatomy was derived from Fabricius's classroom lectures or from a public anatomy is more difficult to determine. Regardless, Jessenius's *Anatomiae Pragae* show that the international students at Padua disseminated the Paduan approach to vision to other parts of Europe. Indeed, we might say that Fabricius's philosophical anatomy of the eye was re-enacted in Prague. Just as a play might be performed in different locations by different actors, a specific public anatomy had the potential to be replicated — whether in temporary constructions or in one of the growing number of permanent anatomy theatres — at distant places by those who had learned the 'script', i.e., the actions and words constituting the anatomy, together with the meaning behind them.

This also reveals an important (and overlooked) translation of knowledge from the anatomy theatre to mathematical optics. In his ground-breaking *Ad Vitellionem paralipomena*

---

medium secari, & magnitudinem totius, ac singularum partium, nec non earundem situs, & figuras describi, & loca qua eorum centra obtinent inveniri, & omnia in subiecta tabella delineari'. Fabricius, *De visione*, 105.

74 Jacopo Zabarella, *De rebus naturalibus libri XXX, quibus quaestiones, quae ab Aristotelis interpretibus hodie tractari solent, accurate discutiuntur* (Venice: Paulus Meiettus, 1590), pp. 632–33. Corresponding passages can be found in Fabricius, *De visione*, 97, 102–03, 107, 110. For a more rigorous argument and more detail of this complex relationship, see Baker, 'Why All This Jelly?', pp. 59–88; Baker, 'Color, Cosmos, Oculus', pp. 277–82.

75 'Praeceptor meus, D. Hieronymus Aquapendentius, Anatomicus Patavinus, in oculorum anatome, regnum tenere mihi semper visus est'. Johannes Jessenius, *Anatomiae Pragae* (Prague: Laurentius Seuberlich, 1601), fol. 113^{v}.

76 Johannes Kepler, *Ad Vitellionem paralipomena, quibus astronomiae pars optica traditur* (Frankfurt: Claudius Marnius & haeredes Ioannis Aubrii, 1604), p. 159.

of 1604, Kepler emphasized that he had never performed or witnessed a dissection of an eye, and instead relied on the anatomical authorities of Felix Platter and, through Jessenius's text, Fabricius ab Aquapendente. Through this, Kepler was exposed to up-to-date accounts of ocular anatomical structure, such as the shapes, location, relative sizes, and relative refractive powers of the humours and transparent tunics. But he was also introduced to the Paduan notion that incoming rays come to a point of burning within the eye.

Kepler coined the term 'focus' in its modern sense: 'focus' in Latin literally meant hearth, but Kepler brought this term into the study of conics and optics, referring to the points (two determinate foci for the hyperbola and ellipse) at which, if rays are drawn from each focus to meet at a single point on the conic section, they form equal angles with the tangent to the section at that point. Kepler uses the term 'focus,' he says, 'for the sake of light'.[77] That is, if we imagine a source of light at one focus of an ellipse, parabola (whose one focus is at infinity), or circle (whose two foci are identical), and we imagine the interior of the conic section to be a perfectly smooth mirror, then all the rays emanating from one focus will collect at the other focus, creating a 'hearth' or burning point. The mirror analogy breaks down in the case of the hyperbola (where the rays meeting at equal angles to the tangent are on opposite sides of the conic section), but it is exceedingly practical for the case of the parabola, which forms a perfect burning instrument using parallel rays of the sun.

Identifying the point of burning with the point of inversion was a major development in sixteenth century optics,[78] and it has been argued that one of Kepler's key innovations was his using the same mathematical analysis for both visual phenomena in optics and the behaviour of light rays in burning mirrors and lenses.[79] Prior to Kepler, *Perspectivae* analysed just a single ray leading from every point of an object to the eye. Kepler transformed optics by (among other things) solving the problem of pinhole images, and in so doing he was able to take into account all the rays issuing from every point of an object that reach the eye. That is, Kepler posited that the rays from every point at the surface of a visible object form a cone of rays on the eye (or any lens), and that all these rays are refracted again onto a single point on the retina (or any screen). For our story, the key is that Kepler was the first to take a toolkit designed to handle the mathematics of burning (i.e., identifying loci where many rays come together) and apply it to the analysis of images, including the eye. But a burning lens was already, for teleological reasons, placed in the eye by anatomists and philosophers in Padua, although this was separate from their account of images. One might say, then, that Kepler revolutionized optics by combining his experiences with the *camera obscura* at court with Jessenius's written version of a particular anatomical performance (which, I have suggested, we can see as a restaging of Fabricius's performance). There is more to it than this simplified story, of course, but the point is that, in addition to the crucial structural details of ocular anatomy, there were important optical ideas stemming from anatomy and the anatomical theatre that very likely influenced Kepler's optics, not to mention others.

77 Johannes Kepler, *Optics: Paralipomena to Witelo & Optical Part of Astronomy*, ed. William H. Donahue (Santa Fe, New Mexico: Green Lion Press, 2000), pp. 107–8.

78 Dupré, 'Kepler's Optics Without Hypotheses'.

79 Smith, *From Sight to Light*.

Conclusion

Cynthia Klestinec has investigated the anatomy theatre as a public space in great detail. It was not only, or even primarily, a place for seeing, but for listening and instruction, for disputation, and for disciplining the behaviour of potentially unruly students. In contrast to the private anatomical demonstration, she argues, the public demonstration in Padua at the end of the sixteenth century was used to present and discuss anatomy considered as natural philosophy, rather than to teach the manual skills that students desired in order to become practising physicians and surgeons.[80] When the eye was the subject of dissection, instruction, and disputation, the focus was ultimately on visual theory.

Controversies among anatomists and physicians, especially Galenic versus Aristotelian accounts of vision, affected how one reconstructed a living eye from the evidence of dissection. Because of this controversy, along with responses to Vesalius and increasing competition among anatomists, anatomies became important sites for disputation about visual theory and the dissemination of optical ideas. Importantly, the visual theory presented at the anatomy theatre combined the previously isolated traditions of anatomy, natural philosophy, and mathematical optics. Debates between followers of Galen and Aristotle were a driving force for the gradual consensus, achieved at the end of the sixteenth century, on issues such as the material composition of the eye, the relative density of its humours, and the size, shape, and connection of its parts. The eye that emerged from this anatomical and medical context was the one bequeathed to both the physicians and the mathematicians of the first half of the seventeenth century, most of whom held that any theory of vision must be grounded in an accurate knowledge of the eye, gathered — ideally first-hand — via repeated, meticulous ocular dissection and experimentation.[81]

Bibliography

Primary Sources

Aguilon, François d', *Francisci Aguilonii e Societate Jesu opticorum libri sex philosophis juxtà ac mathematicis utiles* (Antwerp: Ex Officina Plantiniana, 1613).

Alhacen, *Alhacen's Theory of Visual Perception: A Critical Edition, with English Translation and Commentary, of the First Three Books of Alhacen's 'De aspectibus', the Medieval Latin Version of Ibn al-Haytham's 'Kitāb al-Manāẓir'*, ed. and trans. by A. Mark Smith, 2 vols (Philadelphia: Transactions of the American Philosophical Society, 2001).

80 Klestinec, *Theaters of Anatomy*, 14–15, 67–70, 89, 95–123.

81 See especially François d'Aguilon, *Francisci Aguilonii e Societate Jesu opticorum libri sex philosophis juxtà ac mathematicis utiles* (Antwerp: Ex Officina Plantiniana, 1613), pp. 1–15 (p. 11); Christoph Scheiner, *Oculus, hoc est, fundamentum opticum* (Innsbruck: Agricolum, 1619); Vopiscus Fortunatus Plempius, *Ophthalmographia, sive Tractatio de oculi fabrica, actione, & usu praeter vulgatas hactenus philosophorum ac medicorum opiniones* (Amsterdam: Henricus Laurentius, 1632). Kepler is a partial exception, in that he chose to trust his anatomical authorities and confessed that he had never performed or witnessed an ocular dissection. Kepler, *Ad Vitellionem paralipomena*, 158–59.

Avicenna, *Liber canonis: De medicinis cordialibus; et Cantica*, ed. by Benedetto Rinio (Basel: Ioannes Heruagius, 1556).

Benedetti, Alessandro, *Historia corporis humani, sive anatomice*, ed. by Giovanna Ferrari (Florence: Giunti, 1998).

Berengario da Carpi, Jacopo and Mondino, *Carpi commentaria cum additionibus super anatomia mundini vna cum textu eiusdem in pristinum et verum nitorem redacto* (Bologna: Hieronymus de Benedictis, 1521).

Berengario da Carpi, Jacopo, *A Short Introduction to Anatomy (Isagogae Breves)* (Chicago: University of Chicago Press, 1959).

Casserius, Julius, *De vocis auditusque* (Ferrara: Victorius Baldinus, 1601).

Casserius, Julius, *Pentaestheseion, hoc est de quinque sensibus liber* (Venice: Nicolaus Misserinus, 1609).

Colombo, Realdo, *De re anatomica libri XV* (Venice: ex typographia Nicolai Builacquae, 1559).

Fabricius ab Aquapendente, Hieronymus and Howard Bernhardt Adelmann, *The Embryological Treatises of Hieronymus Fabricius of Aquadenpente* (Ithaca: Cornell University Press, 1967).

Fabricius ab Aquapendente, Hieronymus, *De visione, voce, auditu* (Venice: Franciscus Bolzetta, 1600).

Facciolati, Jacopo, *Fasti gymnasii Patavini* (Padua: Joannes Manfré, 1757).

Falloppio, Gabrielle, 'Institutiones anathomica', in *Omnia, quae adhuc extant opera* (Venice: Felix Valgrisius, 1584).

Favaro, Antonio, *Atti della nazione Germanica artisti nello studio di Padova*, 2 vols (Venice: Prem. Tipografia Emiliana, 1911–12).

Galen, *Galeni Opera ex octaua Iuntarum editione. Quae, quid superioribus praestet, pagina versa ostendit. Ad amplissimum Venetorum medicorum Collegium: Galeni librorum septima classis curatiuam methodum tum diffuse tum breuiter descriptam, … tractatum continet* (Venice: apud Iuntas, 1609).

Galen, *Galeni Opera ex octava Iuntarum editione* (Venice: Giunta, 1609).

Galen, *Galen on the Usefulness of the Parts of the Body. De usu partium*, trans. by Margaret Tallmadge, 2 vols (Ithaca: Cornell University Press, 1968).

Galen, *On the Doctrines of Hippocrates and Plato: Books VI–IX*, trans. by Phillip De Lacy, Corpus Medicorum Graecorum (Berlin: Akademie-Verlag, 1980).

Galen, *Galen on Anatomical Procedures: The Later Books*, trans. by Wynfrid Laurence Henry Duckworth, ed. by M. C. Lyons and B. Towers (Cambridge: Cambridge University Press, 2010).

Israeli, Isaac, and Constantine the African, *Omnia opera Ysaac in hoc volumine contenta* (Lyons: Barthelemy Trot, 1515).

Jessenius, Johannes, *Anatomiae Pragae* (Prague: Laurentius Seuberlich, 1601).

Kepler, Johannes, *Ad Vitellionem paralipomena, quibus astronomiae pars optica traditur* (Frankfurt: Claudius Marnius & haeredes Ioannis Aubrii, 1604).

Massa, Nicolo, *Liber introductorius anatomiae siue dissectionis corporis humani* (Venice: Franciscus Bindonus ac Mapheus Pasinus socios, 1536).

Pecham, John, *John Pecham and the Science of Optics: Perspectiva communis*, ed. and trans. by David C. Lindberg (Madison: University of Wisconsin Press, 1970).

Platter, Felix, *De corporis humani structura et vsu* (Basel: ex Officina Frobeniana, Ambrosius Frobenius, 1583).

Plempius, Vopiscus Fortunatus, *Ophthalmographia, sive Tractatio de oculi fabrica, actione, & usu praeter vulgatas hactenus philosophorum ac medicorum opiniones* (Amsterdam: Henricus Laurentius, 1632).
Scheiner, Christoph, *Oculus, hoc est, fundamentum opticum* (Innsbruck: Agricolum, 1619).
Vallés, Francisco, *Controversiarum medicarum et philosophicarum libri decem* (Frankfurt am Main: Andreas Wechelus, 1582).
Valverde, Juan, *Historia de la composicion del cuerpo humano* (Rome: impressa por Antonio Salamanca y Antonio Lafrerij, 1556).
Vesalius, Andreas, *De humani corporis fabrica libri VII* (Basel: Johannes Oporinus, 1543).
Vesalius, Andreas, *De humani corporis fabrica libri septem* (Basel: Johannes Oporinus, 1555).
Vesalius, Andreas, *The Fabric of the Human Body: An Annotated Translation of the 1543 and 1555 Editions of 'De humani corporis fabrica libri septem'*, trans. by Daniel H. Garrison and Malcolm H. Hast, 2 vols (Basel: S. Karger Ag, 2014).
Witelo, *Witelonis perspectivae liber secundus et liber tertius*, ed. and trans. by Sabetai Unguru, Studia Copernicana, XXVIII (Warszawa: Ossolineum, 1991).
Zabarella, Jacopo, *De rebus naturalibus libri XXX, quibus quaestiones, quae ab Aristotelis interpretibus hodie tractari solent, accurate discutiuntur* (Venice: Paulus Meiettus, 1590).
Zerbi, Gabriele, *Opus preclarum anathomie totius corporis humani et singulorum membrorum illius* (Venice: Scotus, 1533).

Secondary Works

Baker, Tawrin, 'Color, Cosmos, Oculus: Vision, Color, and the Eye in Jacopo Zabarella and Hieronymus Fabricius Ab Aquapendente' (unpublished doctoral dissertation, Indiana University, Bloomington, 2014).
Baker, Tawrin, 'Why All This Jelly? Jacopo Zabarella and Hieronymus Fabricius Ab Aquapendente on the Usefulness of the Vitreous Humor', in *Early Modern Medicine and Natural Philosophy*, ed. by Peter Distelzweig, Benjamin Goldberg and Evan Ragland (Dordrecht: Springer, 2015), pp. 59–88.
Boudon-Millot, Véronique, 'Vision and Vision Disorders: Galen's Physiology of Sight', in *Blood, Sweat and Tears: The Changing Concepts of Physiology from Antiquity into Early Modern Europe*, ed. by Manfred Horstmanshoff, Helen King and Claus Zittel (Leiden: Brill, 2012), pp. 549–67.
Corner, George W., *Anatomical Texts of the Earlier Middle Ages: A Study in the Transmission of Culture* (Washington: Carnegie Institution of Washington, 1927).
Crombie, A. C., 'Expectation, Modelling, and Assent in the History of Optics: Part I. Alhazen and the Medieval Tradition', *Studies in the History and Philosophy of Science*, 21 (4) (1990), 605–32.
Crombie, A. C., *Styles of Scientific Thinking in the European Tradition: The History of Argument and Explanation Especially in the Mathematical and Biomedical Sciences and Arts*, 3 vols (London: Duckworth, 1994).
Cunningham, Andrew, 'Fabricius and the "Aristotle Project" in Anatomical Teaching and Research at Padua', in *The Medical Renaissance of the Sixteenth Century*, ed. by Andrew Wear, Roger Kenneth French, and Iain M. Lonie (Cambridge: Cambridge University Press, 1985), pp. 195–222.

Distelzweig, Peter, 'Fabricius's Galeno-Aristotelian Teleomechanics of Muscle', in *The Life Sciences in Early Modern Philosophy*, ed. by Ohad Nachtomy and Justin E. H. Smith (New York: Oxford University Press, 2014), pp. 64–84.

Dupré, Sven, 'Kepler's Optics Without Hypotheses', *Synthese*, 185 (2012), 501–25.

Durling, Richard J., 'A Chronological Census of Renaissance Editions and Translations of Galen', *Journal of the Warburg and Courtauld Institutes*, 24 (3/4) (1961), 230–305.

Favaro, Giuseppe, 'L'insegnamento anatomico di Girolamo Fabrici d'Acquapendente', in *Monografie storiche sullo studio di Padova, contributo del R. Instituto Veneto di Scienze, lettere ed arti alla celebrazione del VII centenario della Università di Padova* (Venice: Ferrari, 1922).

Ferrari, Giovanna, 'Public Anatomy Lessons and the Carnival: The Anatomy Theatre of Bologna', *Past & Present*, 17 (1987), 50–106.

Fortuna, Stefania, 'The Latin Editions of Galen's Opera Omnia (1490–1625) and their Prefaces', *Early Science and Medicine*, 17 (4) (2012), 391–412.

Grendler, Paul F., *The Universities of the Italian Renaissance* (Baltimore: JHU Press, 2004).

Hankinson, R. J., 'Philosophy of Nature', in *The Cambridge Companion to Galen*, ed. by R. J. Hankinson (Cambridge: Cambridge University Press, 2008), pp. 225–29.

Heseler, Baldasar and Ruben Eriksson, *Andreas Vesalius' First Public Anatomy at Bologna: 1540* (Uppsala: Almqvist & Wiksell, 1959).

Klestinec, Cynthia, 'A History of Anatomy Theaters in Sixteenth-Century Padua', *Journal of the History of Medicine and Allied Sciences*, 59 (3) (2004), 375–412.

Klestinec, Cynthia, *Theaters of Anatomy: Students, Teachers, and Traditions of Dissection in Renaissance Venice* (Baltimore: Johns Hopkins University Press, 2011).

Koelbing, Huldrych M., 'Anatomie de l'œil et perception visuelle, de Vésale à Kepler', in *Le corps à la Renaissance. Actes du XXXe colloque de Tours 1987* (Paris: Aux amateurs de livres, 1990), pp. 389–98.

Lindberg, David C., *Theories of Vision from Al-Kindi to Kepler* (Chicago: University of Chicago Press, 1976).

Mancosu, Paolo, 'Acoustics and Optics', in *Cambridge History of Science Volume 3: Early Modern Science*, ed. by Katharine Park and Lorraine Daston (Cambridge: Cambridge University Press, 2006), pp. 596–631.

Martin, Craig, 'Francisco Vallés and the Renaissance Reinterpretation of Aristotle's "Meteorologica" IV as a Medical Text', *Early Science and Medicine*, 7, (2002), 1–30.

Meyerhof, Max, 'New Light on Hunain Ibn Ishaq and his Period', *Isis*, 8 (4) (1926), 685–724.

Nutton, Vivian, 'De placitis Hippocratis et Platonis in the Renaissance', in *Le Opere psicologiche di Galeno*, ed. by Paola Manuli and Mario Vegetti (Napoli: Bibliopolis, 1988), pp. 281–309.

Nutton, Vivian, 'Vesalius Revised. His Annotations to the 1555 Fabrica', *Medical History*, 56 (2012), 415–43.

Salmon, Fernando, 'The Many Galens of the Medieval Commentators on Vision', *Revue d'histoire des sciences*, 50 (1997), 397–420.

Saunders, John B. de C. M. and Charles D. O'Malley, *The Illustrations from the Works of Andreas Vesalius of Brussels* (Cleveland: World Publishing Company, 1950).

Shotwell, R. Allen, 'The Revival of Anatomical Practices and Techniques in the Renaissance' (unpublished doctoral dissertation, Indiana University, 2013).

Shotwell, R. Allen, 'The Revival of Vivisection in the Sixteenth Century', *Journal of the History of Biology*, 46 (2013), 171–97.

Shotwell, R. Allen, 'Animals, Pictures, and Skeletons: Andreas Vesalius's Reinvention of the Public Anatomy Lesson', *Journal of the History of Medicine and Allied Sciences*, 71 (1) (2016), 1–18.

Siraisi, Nancy G., *Avicenna in Renaissance Italy: The Canon and Medical Teaching in Italian Universities after 1500* (Princeton, NJ: Princeton University Press, 1987).

Siraisi, Nancy G., 'Vesalius and the Reading of Galen's Teleology', *Renaissance Quarterly*, 50 (1997), 1–37.

Siraisi, Nancy G., *The Clock and the Mirror: Girolamo Cardano and Renaissance Medicine* (Princeton: Princeton University Press, 1997).

Siraisi, Nancy G., 'Historia, actio, utilitas: Fabrici e le scienze della vita nel Cinquecento', in *Il teatro dei corpi: Le pitture colorate d'anatomia di Girolamo Fabrici d'Acquapendente*, ed. by Maurizio Rippa Bonati and Jose Pardo-Tomas (Milano: Mediamed, 2004).

Smith, A. Mark, 'Ptolemy, Alhazen, and Kepler and the Problem of Optical Images', *Arabic Sciences and Philosophy*, 8 (1) (1998), 9–44.

Smith, A. Mark, *From Sight to Light: The Passage from Ancient to Modern Optics* (Chicago: University of Chicago Press, 2014).

Wear, Andrew, 'Medicine in Early Modern Europe, 1500–1700', in *The Western Medical Tradition: 800 BC–1800 AD*, ed. by Lawrence I. Conrad (New York: Cambridge University Press, 1995), pp. 250–64, 270–73.

JAIME CUENCA

The Princely Viewpoint

Perspectival Scenery and its Political Meaning in Early Modern Courts

Introduction

The use of illusionistic scenery was a distinctive feature of Western modern theatre culture until well into the twentieth century, distinguishing it from Antique and Medieval performances, as well as from theatrical traditions from other parts of the world.[1] This aspect of dramatic production has significantly influenced the evolution of theatre architecture, stage machinery, audience-stage relationships, and performance conventions in the Western dramatic tradition. The introduction of the rules of perspective in stage design, which brought a convincing illusion of depth, must be seen as a trigger for some of the processes that shaped Western theatre in the long term: from the sixteenth century on, theatre not only became a site of specific perspective knowledge and practices in Europe, but was also deeply transformed by them. The application of perspective stage scenery developed initially in Italian courts and remained largely a courtly matter as it spread throughout Europe. For most of the seventeenth century, in England and Spain, for example, perspectival settings distinguished aristocratic theatre from its popular counterpart.[2] Perspective was not only used to delight aristocrats in early modern courts; perspectival settings betray a distinctive performative function, where matters of perspective became politically relevant.

In court theatres, perspective was always part of the multisensorial, complex performances that were offered to an audience gathered around the prince. These events, which relied on the work of a highly specialised staff of artists and engineers, provided the ruler with a prime opportunity to show off the dynasty's wealth. But they also fulfilled a much more precise political function by letting perspective act as an ordering principle for a seating

1 This chapter was partly written during a stay as a visiting scholar at the Max Planck Institute for the History of Science in Berlin. I would like to thank the German Academic Exchange Service (DAAD) for funding my stay, as well as the members of the Max Planck Research Group Art and Knowledge in Pre-Modern Europe for their support and our discussions.

2 See Andrew Gurr, *The Shakespearean Stage, 1574–1642* (Cambridge: Cambridge University Press, 1992) and José María Ruano de la Haza, 'Hacia una metodología para la reconstrucción de la puesta en escena de la comedia en teatros comerciales del siglo XVII', *Criticón*, 42 (1988), 81–102.

Jaime Cuenca University of Deusto, Institute of Leisure Studies, jaime.cuenca@deusto.es

Perspective as Practice. Renaissance Cultures of Optics, ed. by Sven Dupré, Turnhout, 2019 (*Techne. Knowledge, Technique, and Material Culture*, 1), p. 149-172

DOI 10.1484/M.*Techne*-EB.5.117725

distribution that embodied the hierarchy of the court. While questions of precedence were a vital issue in everyday courtly life, only here were they combined with contemporary optical knowledge. If central perspective is strictly applied to a given scenery, the perfect illusion of depth will only be achieved from a particular spot in the auditorium: the specific position directly mirroring the vanishing point of the perspectival setting. At some point in the evolution of early modern court theatre, this place began to be reserved for the prince. The rest of those in attendance would be seated around it according to rank. In this way, as Orgel and Strong explain referring to the Stuarts' masques, the audience was 'transformed […] into a living and visible emblem of the aristocratic hierarchy: the closer one sat to the King, the 'better' one's place was, and only the King's seat was perfect'.[3] These descriptions of the audience have become common place among scholars when exploring the role of perspective in early modern courts, which has helped visualise its political dimension, but has led, too, to some confusing generalizations.

One such generalization is to be found in Roy Strong's influential essay on Renaissance festivals. Strong attaches great relevance to perspectival settings in his exploration of the expression of power in the early modern era. He stresses the privileged point of view obtained from the prince's seat, and the political message this carries. He traces from this a continuum between Italian festivals of the *cinquecento* and a fully-fledged Baroque court culture in the following century. At the beginning of this continuum, as far as perspective is concerned, Strong cites the celebrations held at the Palazzo Medici in 1539 for the wedding between Eleonora di Toledo and Cosimo I. The second courtyard of the palace was transformed by Bastiano da Sangallo into a richly decorated, temporary theatre, where *Il commodo*, by Antonio Landi, was performed. Strong writes:

> The setting was a *prospettiva* of Pisa with its campanile and baptistery cupola towering above palace facades alternating with street vistas. Significantly, its lines of perspective radiated from the ducal throne, foreshadowing the alliance of perspective and absolutism that was to be the potent reason for its adoption later by the Bourbon kings of France and the Stuart kings of England. Vitruvius wrote of the centre of the circle of his amphitheatre where all lines of vision met but the place was left unoccupied. In both temporary and permanent renaissance re-creations of antique theatres for courts this inevitably was appropriated for the royal box, an innovation that had no archaeological precedent.[4]

Strong is far from ambiguous here, as he locates the Duke's seat at the privileged viewpoint where all lines of perspective meet. It is thus surprising that the scarce contemporary sources that do exist fail to confirm this. Neither Pierfrancesco Giambullari, in his account published that year, or Vasari, in the short description included in his later biography of Bastiano da Sangallo — these being the only extant written sources — mention that Cosimo's seat was

3 Stephen Orgel and Roy Strong, *Inigo Jones: The Theatre of the Stuart Court. Including the Complete Designs for Productions at Court for the Most Part in the Collection of the Duke of Devonshire Together with their Texts and Historical Documentation*, 2 vols (London: Sotheby Parke-Bernet; Berkeley: University of California Press, 1973), I, p. 7.

4 Roy Strong, *Art and Power: Renaissance Festivals, 1450–1650* (Berkeley: University of California Press, 1984), p. 35.

placed at the point where the perspectival scenery was best seen.[5] Giambullari describes the decorations of the courtyard for the dinner of the day before the comedy, and he simply states that the table of the spouses was under the loggia opposite the stage with long table extensions on each side, running the length of the courtyard.[6] If they occupied the same position the following evening when the comedy was performed — a detail not mentioned by Giambullari — they were sitting at the bottom of a U-shaped seating arrangement, and not in the centre of any recreated amphitheatre, as Strong suggests. The positioning was thus probably more a question of feasting etiquette than of optics. There is little doubt that Strong takes his continuum too far back.

This kind of backward projection can be found, too, in an essay from Dorothée Marciak exploring the political implications of the identification between the prince's seat and the privileged viewpoint on the stage. She constantly refers to the Medici court theatre as a 'laboratory' where the technologies of power later applied at the court of Louis XIV were tested.[7] For Marciak, as for Strong, the first station of this long-standing alliance between central perspective and absolute power is those very wedding celebrations of 1539.

If we want to assess the political implications of central perspective applied to theatre, such projections and generalizations must be carefully avoided by closely following contemporaneous accounts of performances and taking into consideration the reality of the scenographers' knowledge and practices at each historical moment. This is precisely what I will do in this chapter, while tracing the history of how the privileged point of view on the stage came to be identified with the prince's seat, how this identification evolved and how it was finally discarded.

The Emergent Awareness of the Viewpoint

The first extant description of perspective being incorporated into stage design can be found in a letter from Bernardino Prospero written in 1508, where a 'perspective of a land with houses, churches, bell towers, and gardens' is mentioned appreciatively as forming the scenery for a performance of Ariosto's *La cassaria* in Ferrara.[8] While in this case the wording leads us to think of a backdrop painted with a landscape, sceneries soon became three-dimensional, as is clearly suggested by Vasari's description of the stage created by Baldassare Peruzzi for Bernardo Dovizi da Bibbiena's *La calandria* in Rome in 1514. He mentions 'the lights inside the scenery used for enhancing the illusion of depth'.[9] By then, the Italian Renaissance stage was already a combination of three-dimensional foreshortened

5 See Pierfrancesco Giambullari, *Apparato et feste nelle noze dello Illustrissimo Signor Duca di Firenze, et della Duchessa sua Consorte, con le sue Stanze, Madriali, Comedia, & Intermedii, in quelle recitati* (Florence: Benedetto Giunta, 1539) and Giorgio Vasari, *Le vite de' più eccellenti pittori, scultori e architettori nelle redazioni del 1550 e 1568*, ed. by Rosanna Bettarini and Paola Barocchi, 6 vols (Florence: Sansoni, 1966–87), V, pp. 398–400.

6 Giambullari, *Apparato*, 30.

7 Dorothée Marciak, *La place du prince. Perspective et pouvoir dans le théâtre de cour des Médicis, Florence (1539–1600)*, Études et Essais sur la Renaissance, 50 (Paris: Honoré Champion, 2005), pp. 345–46.

8 'prospetiva di una terra cum case, chiesie, campanili e zardini'. Letter of Bernardino Prospero to Isabella d'Este, quoted after Giuseppe Campori, *Notizie per la vita di Ludovico Ariosto tratte da documenti inediti* (Modena: Carlo Vincenzi, 1871), p. 69.

9 'i lumi di dentro che servono alla prospettiva'. Vasari, *Le vite de' più eccellenti pittori*, IV, p. 323.

buildings erected over an elevated platform (which sloped away from the audience), receding towards a painted backdrop in perspective. The central perspective formed by the side walls stretched often approximately ten metres back towards the backdrop, but the actors could not penetrate into this deep area without breaking the effect of the sharp foreshortening, and thus performed on a narrow proscenium before it.[10]

We have abundant information concerning one stage of this kind, designed by Peruzzi for the wedding of Giuliano Cesarini and Giulia Colonna in Rome in 1531. The plan and elevation of the stage kept in the Uffizi give a very precise idea of the general measurements and an account by Marco da Lodi informs us about the seating distribution:

> On the opposite side of the stage there were tiered wooden steps for seating, and on the other side, close to the wall of the courtyard of the palace, there were like three terraces, one above, another one in the middle, and the last one at the bottom, and there one could see cardinals, ambassadors, prelates, and other noble courtiers of quality and high condition.[11]

While the exact distribution cannot be determined with absolute accuracy, it is obvious that spectators were seated according to their rank. There is no extant plan of how Peruzzi designed the auditorium, but an alternative project for the same performance by Antonio da Sangallo the Younger, kept in the Uffizi, shows a Vitruvian orchestra.[12] If that was the inspiration for Peruzzi, too (which is rather probable, since his disciple Serlio also followed the Vitruvian model), the audience was seated in a semicircle. Since no special place is mentioned for the nuptial couple or for anyone else in Da Lodi's account, we should rather imagine an empty orchestra between the audience and the stage. And, perhaps more importantly, while Marco da Lodi comments extensively on the perspectival scenery, he does not mention a privileged viewpoint at all. It is probable that no one was thinking of such a thing. Nobuhide Nao has proved that Peruzzi's plan and elevation of the scenery imply an ideal viewpoint located at a distance of eighty metres from the front of the stage.[13] Considering that the square where the theatre was built was seventeen metres wide, we must conclude that Peruzzi did not work at all with the idea of a physical viewpoint where someone would have been able to sit.[14] It is difficult to believe that only eight years after this, anyone at the Medici wedding of 1539 would think of the lines of perspective as 'radiat[ing] from the ducal throne', as Strong puts it.[15]

10 See Götz Pochat, *Theater und bildende Kunst im Mittelalter und in der Renaissance in Italien* (Graz: Akademische Druck- u. Verlagsanstalt, 1990), pp. 279–80.

11 'Da l'altro capo del apparato vi erano e gradi di tavole da sedere, et da l'altra parte vicina al muro della corte del palazzo erano a guisa di tre loggie, una di sovra, l'altra in mezzo et l'ultima da basso, et qui stavano a vedere cardinali, ambasciatori, prelati et altri nobili cortegiani di qualità et conditione'. Marco Cademosto da Lodi, *Le splendissime et signorili nozze de li magnanimi Cesarini con li illustrissimi Colonnesi fatte a dì XXVIII di maggio MDXXXI*: reproduced in Appendix X of Pochat, *Theater und bildende Kunst*, 384. For a very precise reconstruction of the measures of the stage see Pochat, *Theater und bildende Kunst*, 291 ff.

12 See Pochat, *Theater und bildende Kunst*, 300 (Figure 211).

13 See Nobuhide Nao, 'A Study of "The Stage Set of *Le Bacchide*" by Baldassarre Peruzzi: The Development of "Prospettiva Solida" in Early Sixteenth Century Italy', *Journal of Graphic Science of Japan*, 39 (2005), 11–18 (pp. 16–17).

14 For the measures of the square, see Pochat, *Theater und bildende Kunst*, 301.

15 Strong, *Art and Power*, 35.

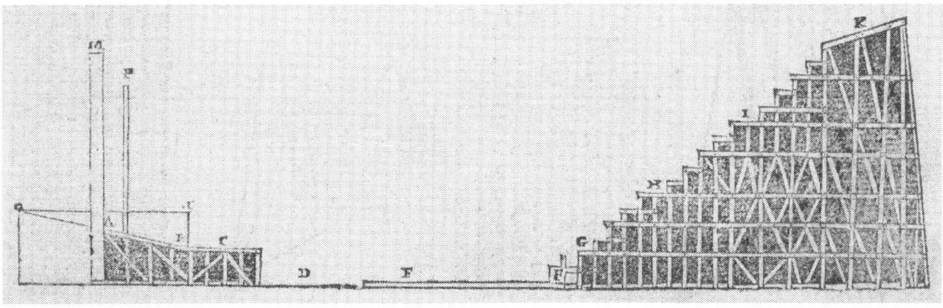

Fig. 1 Section view of a temporary theater. From Sebastiano Serlio, *Tutte l'opere d'architettura* (Venice: Senese, 1584), fol. 47^{v}. © Max Planck Institute for the History of Science.

Sebastiano Serlio, who worked at Peruzzi's Roman workshop for more than ten years, published the second of his *Seven Books on Architecture*, dealing with perspective, in Paris in 1545.[16] In its final pages, he codifies the scenographic knowledge of his master and thus extensively contributed to its dispersal. Serlio offers a complete, though brief, explanation of how to build a temporary theatre, relying heavily on his experience at the Palazzo Porto, in Vicenza, in 1539.[17] For our purposes, it is remarkable to note that, before starting to deal with the building of sceneries, Serlio feels the need to set up a distinction between the perspective rules he has explained before and the system he shall apply to the stage, 'for those are imagined on flat walls, and this is material and in relief'.[18] This distinction refers mainly to the way in which the *orizonte* ('vanishing point') is positioned.[19] What he means is best understood by looking at the theatre's profile he presents (Figure 1). The wall of the room where the temporary theatre is erected is marked with an M, while a P marks the painted backdrop (leaving some distance to the wall for the actors to cross unseen from one side of the stage to the other). C is the proscenium where the actors actually performed, before the three-dimensional decorations mounted on the inclined plane B-A. As can easily be imagined, the most difficult problem for a scenographer designing such a stage was to be sure that the foreshortened buildings on the B-A plane matched seamlessly — when viewed from the audience — to the urban perspective painted on P. For this illusion to work effectively, Serlio explains that more than one vanishing point must be included. The lines of the squares on the inclined ground as well as those that should appear as horizontal on the foreshortened facades of both sides of the stage do not converge into the bottom of the backdrop, but into a point beyond the wall of the room

16 Sebastiano Serlio, *Tutte l' opere d' architettura. Dove si trattano in disegno, quelle cose, che sono più necessarie all'architetto: et hora di nuovo aggiunto, oltre il libro delle porte, gran numero di case private nella città, & in villa, et un indice copiosissimo* (Venice: Senese, 1584), in *ECHO. Cultural Heritage Online:* http://echo.mpiwg-berlin.mpg.de/MPIWG:EAQFSZDY [accessed 27-07-2017].

17 See Carla Bino and Ilaria Tameni, 'Il teatro umanistico e rinascimentale', in *Storia essenziale del teatro*, ed. by Claudio Bernardi and Carlo Susa (Milan: Vita e Pensiero, 2005), pp. 121–64 (p. 136).

18 'per essere quelle imaginate sopra le mura piane, & questa per essere materiale & di rilievo'. See Serlio, *Tutte l' opere d' architettura*, fol. 48^{r}.

19 I want to warmly thank Pietro Roccasecca for pointing this out to me and being of great help in understanding Serlio's text.

(O). A stake (L) must be erected at the limit between the proscenium and the sloping platform, and from there a horizontal line should be traced to the vanishing point:

> where this [line] ends on the last wall of the stage, there will the horizon be, which will only serve for that wall though, and this line will be the one that will always be the horizon for the facades of the houses in front view.[20]

Serlio is not very clear here, but he probably means that the line must be used to find the vanishing points of the facades that are facing the audience. This line represents the view axis and is not just an imaginary geometrical aid on the plan, but actually a string that crosses the stage from its front to the backdrop: 'But the said line should be a stable thing, because it will serve all those canvasses that are in front view for finding the thicknesses of some things'.[21] The vanishing points are placed where the string is cut by the extensions of the planes of the frontal facades. The lines representing the 'thicknesses' on these facades, i.e., those lines appearing to recede into the painted surfaces (like those of a windowsill or a balcony), will converge into these vanishing points.[22] Although today we would likely describe all these vanishing points as projections of one point (O) hitting different sections of the view axis, Serlio clearly interprets each of them as an *orizonte* ('vanishing point') in its own right. If the sense of unity a single vanishing point provides is missing, then it is unlikely that a single viewpoint was relevant to Serlio. It was not some viewpoint in the auditorium that mattered for the design and construction of the perspectival stage, but a point placed at the limit between the proscenium and the inclined part of the stage. It is the string starting from the stake erected there that was essential to 'find the thicknesses' of the frontal facades (and probably to project the receding lines on the foreshortened ones too).

Now, if we turn to Serlio's description of the auditorium, we can confirm that a privileged viewpoint does not play a role here either. While he distributes the spectators into the tiered seats from the bottom up, according to their social rank, he provides no optical reason for this, and it rather must be understood as an attempt at reconstructing the uses in antique theatres. Serlio merely states that 'where F is seen, are the seats of the most noble', without mentioning the fact that those sitting in the middle of the half circumference would enjoy a much better view of the perspectival stage than those at both lateral extremes (Figure 2).[23] In fact, if the view axis L-O is extended towards the auditorium, it hits the tiers at some point between G and H, which shows that the first row, reserved for the most noble spectators, is placed too low.

The next relevant source for scenographical knowledge of the time is *La pratica della perspettiva* by Daniele Barbaro, published in Venice in 1569.[24] In contrast to Serlio, Barbaro

---

20 'dove essa finirà nel muro ultimo della Scena, ivi sarà l'Orizonte, il qual però servirà solamente per quel muro, & questa linea sarà quella che sarà sempre Orizonte alle faccie de' casamenti che saranno in maestà'. Serlio, *Tutte l' opere d' architettura*, fol. 47ᵛ.

21 'ma la detta linea sia una cosa stabile, perche questa servirà à tutti quei telari, che saranno in maestà, per trovare le grossezze di alcune cose'. Serlio, *Tutte l' opere d' architettura*, fol. 48ᵛ.

22 See Günter Schöne, *Die Entwicklung der Perspektivbühne von Serlio bis Galli-Bibiena nach den Perspektivbüchern*, Theatergeschichtliche Forschungen, 43 (Leipzig: Leopold Voss, 1933), p. 12.

23 'dove si vede F, sono le sedie de più nobile'. Serlio, *Tutte l' opere d' architettura*, fol. 47ᵛ.

24 Daniele Barbaro, *La pratica della perspettiva. Opera molto profittevole a' pittori, scultori, et architetti* (Venice: Borgominieri, 1569), in *ECHO. Cultural Heritage Online:* http://echo.mpiwg-berlin.mpg.de/MPIWG:CM4MKEMB [accessed 27-07-2017].

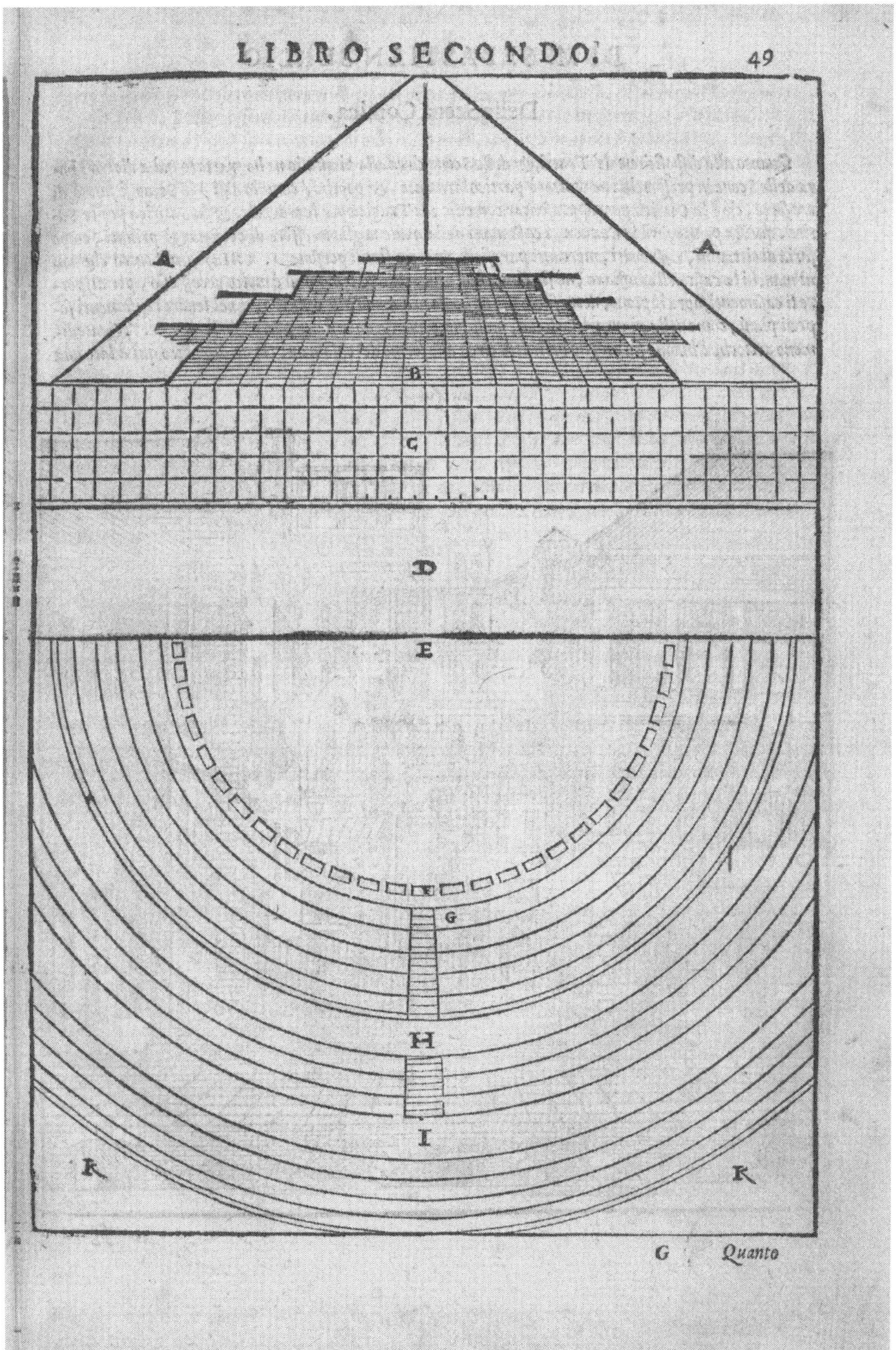

Fig. 2 Plan of a temporary theater. From Sebastiano Serlio, *Tutte l'opere d'architettura* (Venice: Senese, 1584), fol. 49r. © Max Planck Institute for the History of Science.

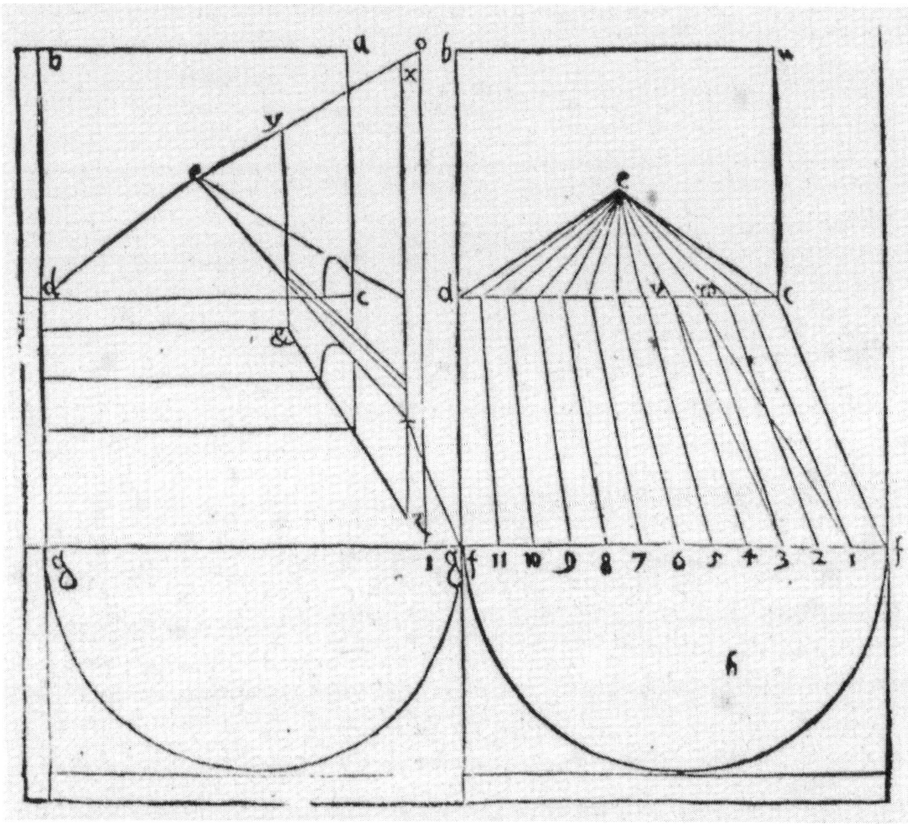

Fig. 3 Diagram showing the construction of a perspectival stage. From Daniele Barbaro, *La pratica della perspettiva* (Venice: Borgominieri, 1569), p. 155. © Max Planck Institute for the History of Science.

was above all a humanist scholar, and his knowledge of the stage did not come from any experience as a scenographer, but rather from his study of Vitruvius. He was acquainted with Andrea Palladio, who contributed the illustrations for Barbaro's Latin edition of Vitruvius's work. It was precisely Palladio who designed the plans of the Teatro Olimpico for the Academy of Vicenza, to which both belonged; this being one of the most exhaustive attempts of the Italian Renaissance to reconstruct an antique theatre,[25] it is not surprising that Barbaro, as far as the auditorium is concerned, does not deviate from the Vitruvian model in his book any more than Serlio does.[26] His (rather clumsy) diagram shows a semi-circular orchestra, with no hint that he would seat the most noble spectators in any place other than the first row (Figure 3). He mentions that the scenographer should take into account the distance from the audience when painting the scenery, but he refers

25 See Robert Klein, 'Vitruve et le théâtre de la Renaissance italienne', in *La forme et l'intelligible* (Paris: Gallimard, 1970), pp. 294–309.

26 See Ian F. Verstegen, 'Tacit Skills in the Perspective Treatise of the Late Renaissance', in *Cognition and the Book: Typologies of Formal Organisation of Knowledge in the Printed Book of the Early Modern Period*, ed. by Karl A. E. Enenkel and Wolfgang Neubner (Leiden: Brill, 2005), pp. 187–213 (pp. 203–06).

only to choosing a distance from which the spectators are able to both see the stage and hear the actors.[27] The only truly new contribution beyond what Serlio had already written is a method for tracing the orthogonals as projections of a string, an idea Barbaro acknowledges is taken from Pompeo Pedemonte. A string is attached to the vanishing point of the backdrop and stretched to one of the marks that divide the front of the stage in equal parts. 'Once the string is fixed, it is necessary to withdraw to the middle of the theatre, to the point H, and look at the stretched string'.[28] The receding line is traced on the floor and the backdrop so as to appear at one with the string when seen from the centre of the theatre. In comparison to Serlio's stake, which was on the stage itself, this procedure shows progress in the direction of taking one viewpoint as the basis for the perspectival scenery. Yet, even if that is the case, the method is still quite imprecise and the point from which the string is looked at is not defined with too much specificity. More importantly, as already seen, Barbaro's classically oriented mind-set prevented him from seating anyone at the centre of his semi-circular orchestra.

The tendency towards considering a given viewpoint in the auditorium as the basis for the scenery (both theoretically and practically) can be even more clearly discerned in what became one of the most prestigious and most frequently reprinted Renaissance treatises on perspective: *Le due regole della prospettiva pratica* (1583).[29] The core of this text was written by Giacomo Barozzi da Vignola, but it was not until ten years after his death that the mathematician Ignazio Danti published the treatise with numerous comments. It is among those comments that we find Danti's considerations on the perspectival stage. Even if he was not a scenographer, but a cosmographer and mathematician, he could well have acquired some very direct knowledge on the construction of sceneries, since he spent over a decade in Florence in close relationship with the ducal court. Firstly, Danti explains that Serlio's method of representing perspective on stage using two points, so that the painted and the three-dimensional parts match, is incorrect.[30] This disagreement is rather a question of wording; while Serlio is inclined to speak of several *orizonti* ('horizons') in the scenery, Danti writes about one *punto principale* ('principal point'), into which the lines of the foreshortened facades should converge, and about some other points for the frontal facades of the houses. These latter points are projections of the real vanishing point along the view axis and, when viewed by the audience, 'these said points become one with the principal point'.[31] Both mean the same thing, using different words; what is significant is that Danti, strictly following Vignola's spirit, refuses to speak of several points, but rather tends to unify all of them under the *punto principale* ('principal point') just as

27 Barbaro, *La pratica della perspettiva*, 155.

28 'Fermata la corda bisogna retirarsi al mezzo del theatro, come nel punto h, & guardare la corda tirata'. Barbaro, *La pratica della perspettiva*, 155.

29 Iacomo Barozzi da Vignola and Egnatio Danti, *Le due regole della prospettiva prattica di M. Iacomo Barozzi da Vignola con i commentari del Reverendo Padre Maestro Egnatio Danti* (Bologna: Longhi, 1682), in *ECHO. Cultural Heritage Online*: http://echo.mpiwg-berlin.mpg.de/MPIWG:TMVX65MM [accessed 31-07-2017). For more information on the relevance of the book, see Timothy K. Kitao, 'Prejudice in Perspective: A Study of Vignola's Perspective Treatise', *The Art Bulletin*, 44 (1962), 173–94 (pp. 186–87).

30 Vignola and Danti, *Le due regole della prospettiva prattica*, 90.

31 'li quali punti faranno nondimeno con il punto principale tutt'uno'. Vignola and Danti, *Le due regole della prospettiva prattica*, 90.

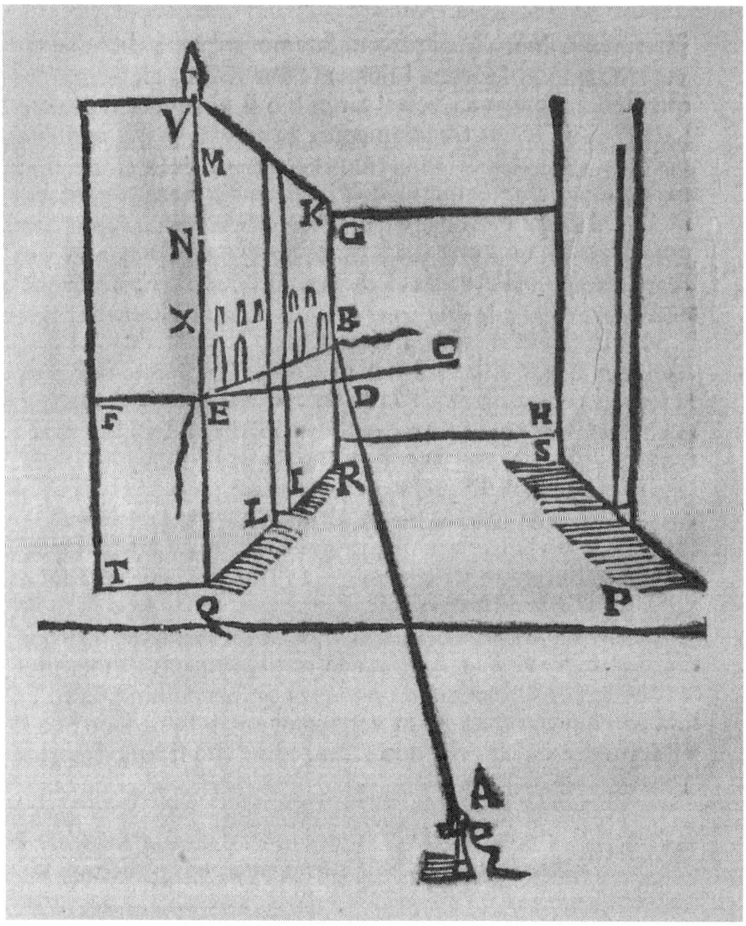

Fig. 4 Diagram showing the construction of a perspectival stage. From Iacomo Barozzi da Vignola and Egnatio Danti, *Le due regole della prospettiva prattica* (Bologna: Longhi, 1682), p. 91. © Max Planck Institute for the History of Science.

Vignola himself does in his second rule.[32] This reveals a more unitarian comprehension of the stage centred on only one vanishing point.

Danti consistently gives one single viewpoint more importance in the actual construction of the scenery than Serlio does. Strings and threads are also involved, but a specific viewpoint is much more precisely used as the basis for tracing all the receding lines of the scenery (Figure 4):

> And to make that the facades of the houses M L, and I K converge into the point C, and match with the houses painted on the wall G H, so that the eye, which is at the point A, the distance point, sees everything converging into the point C, it will be done as follows. A ruler will be placed in plumb at the point A, the distance point, as high as the

32 See Kitao, 'Prejudice in Perspective', p. 183.

> eye of the one who is looking, or a bit more, so that when stretching a thread from point A to point C, the principal point of perspective, it is level: then, another thread will be linked to the point C, and wanting to mark on the facades M L and I K, for instance, the cornice E B, for mounting the windows on it and to find also the heights of the windows and any other thing that we would like to draw in perspective, they must be first marked perfectly at the front of the perspective T V, after the measures that seem better, and then stretching the thread from point C to the angle of the front V Q, as the thread C D that goes to the point E, to touch the cornice F E, marked at the front T V, and from point A a thread is stretched to the angle of the house K R, as high or low as to touch the thread C E at point D, and making a point at the said angle on B, a line E B will be stretched, which corresponds to F E and will converge into point C.[33]

What the mathematician Danti offers here is a method to construct an imaginary plane by connecting three points with threads: the viewpoint, the vanishing point, and any other point on the foreshortened sides of the stage (here, E). The desired foreshortened line E-B is in fact the intersection between this imaginary plane and the side wall, which Danti finds by extending the plane into the corresponding back corner of the stage by means of a thread, and then connecting that point (B) with its counterpart at the front of the stage (E). While Danti does not express his ideas in such geometrical terms (as Guidobaldo del Monte later does), the procedure has some mathematical elegance to it and is far more precise than Barbaro's. In any case, scenographers did not care much about mathematical beauty and we will soon find simplified versions of this suggestion which were much easier to implement. What matters here is that Danti had a more unitarian understanding of the perspectival stage than his predecessors and succeeded in conveying this understanding through a precise method for building scenery that attaches the same importance to the vanishing point as to a given point of view.

Danti says nothing about the seating, but in this respect, there is a relevant account of those years. In 1589 the Duke Ferdinando de' Medici married Christina of Lorraine. The celebrations included the staging of a comedy in the brand-new court theatre built by Bernardo Buontalenti in the Uffizi. It had been erected in 1586, for the wedding of Virginia de'Medici, but was entirely remodelled three years later.[34] It seems that Vasari had already provided his temporary theatre in the Palazzo Vecchio with a seat for the Duke on a dais in the middle of the auditorium.[35] By 1589 the custom was firmly established: an elevated

33 'E per fare, che le facciate delle case M L, e I K, corrino al punto C, e s'accordino con le case finte nella parete G H, acciò l'occhio, che sta nel punto A, della distanza, vegga andare ogni cosa ad unirsi al punto C, si opererà in questa maniera. Si pianterà nel punto A, della distanza vn regolo à piombo tanto alto, quanto è l'occhio di chi mira, ò poco più, acciò tirando un filo dal punto A, al punto C, principale della Prospettiva, stia á livello: dipoi al punto C, si leguerà un altro filo, & volendo segnare nelle facciate M L, & I K, ponian caso, la cornice E B, per piantarvi sopra le finestre, e trovare anco l'altezze delle finestre, & ogn'altra cosa, che ci vorremo disegnare in Prospettiva, si segneranno la prima cosa perfette nella fronte della Prospettiva T V, secondo la misura che ci parrà, e poi tirando il filo dal punto C, all'angolo della fronte V Q, come é il filo C D, che và al punto E, à toccare la cornice F E, segnata nella fronte T V, e dal punto A, si tiri il filo all'angolo della casa K R, tanto alto ò basso, fin che tocchi il filo C E, nel punto D, e facendo nell'angolo detto un punto al legno B, si tirerà la linea E B, la quale corrisponderà alla F E, e correrà al punto C'. Vignola and Danti, *Le due regole della prospettiva prattica*, 90.

34 See James M. Saslow, *The Medici Wedding of 1589. Florentine Festival as 'Theatrum Mundi'* (New Haven: Yale University Press, 1996), p. 79.

35 Marciak, *La place du prince*, 173–74.

platform was reserved for the ducal couple in the centre of the squared-off U-shaped auditorium, separated from their guests by a balustrade.[36] Bastiano de' Rossi describes the dais in full detail, although he does not say anything about it occupying a particularly privileged viewpoint.[37] It cannot be ruled out, of course, that both Buontalenti and the duke, and perhaps also part of the audience, were aware of the use of the privileged point of view. But it seems that de' Rossi was not, which at least shows that this was not yet a part of the political message transmitted at the celebrations that diligent chroniclers would register. Nevertheless, the central position of the prince here is very relevant to this history, since the wedding of 1589 was circulated to all European courts by means of written accounts and etchings. Buontalenti's designs for the *intermedii* were reproduced by many scenographers in the following decades, and so too his arrangement of the auditorium.

Il luogo per il Prencipe

At the end of the sixteenth century in Italy we can already discern the implementation of two elements essential to the theatre model that will spread among European courts over the following decades. The first of these is the idea of perspectival scenery, constructed in relation to one single viewpoint. The second is the seating of the prince at the centre of the audience, which had become the custom in Florence at least. Both rulers and scenographers soon connected these two elements into one single device, indebted both to ideas of etiquette and optics, and imbued with profound political meaning. But there are still certain elements missing before we can find the fully developed and self-aware model in the work of Sabbatini.

The architect Lorenzo Sirigatti published *La pratica della prospettiva* in 1596.[38] Only in the final, short chapter of his first book does he deal with the building of sceneries, or, more precisely, with the problem of how to make the painted backdrop and the foreshortened side walls match. The chapter's title runs 'how to draw the painted scene so that it matches the houses of the stage'.[39] This concrete problem, as stated, was in fact one of the biggest challenges facing scenographers before side wings became commonly used. Of course, it was also a very relevant question for Serlio, Barbaro, and Danti, but there is a crucial difference between how the three of them addressed it, and how Sirigatti did. Serlio justifies locating the vanishing point 'beyond' the stage walls, arguing that 'in this way, all houses and other things will have more sweetness in the foreshortening', whereas, otherwise, 'all houses fall together'.[40] While he explains how to improve the way the scenery can be seen,

36 Saslow, *The Medici Wedding*, 81.

37 See Bastiano de' Rossi, *Descrizione dell'apparato e degl'intermedi fatti per la commedia rappresentata in Firenze nelle nozze de' Serenissimi Don Ferdinando Medici e Madama Cristina De Loreno, Gran Duchi di Toscana* (Florence: Padovani, 1589), p. 15.

38 Lorenzo Sirigatti, *La pratica della prospettiva* (Venice: Girolamo Franceschi Sanese, 1596), in *Internet Archive*: http://archive.org/details/gri_33125008485225 [accessed 31-07-2017].

39 'per disegnare il finto della Scena talmente che unifica con le case del palco'. Sirigatti, *La pratica della prospettiva*, fol. 42^{v}.

40 'cosi tutti li casamenti, & altre cose havranno più dolcezza ne gli scorci [...] tutti li casamenti se adunano'. Serlio, *Tutte l' opere d' architettura*, fol. 48^{r}.

he does not refer to any particular point of view and simply speaks as if the appearance of the scenery would improve generally if his rules are followed, regardless of the position of the viewer. Danti writes in similar abstract terms: Before explaining his aforementioned thread method, he states:

> And to make that the facades of the houses M L, and I K converge into the point C, and match with the houses painted on the wall G H, so that the eye, which is at the point A, the distance point, sees everything converging into the point C, it will be done as follows.[41]

The appearance of the scenery is now seemingly referred to as an exact viewpoint (A), but after explaining his method, Danti concludes:

> Now marking exactly this way all other things on the receding facades of the houses in relief, everything will converge into C, the principal point, and so the painted houses of the wall G H will match exactly with those in relief, and it will be proceeded with only one point, according to the true rules and to how Nature works in our seeing.[42]

The mention of a particular viewpoint has disappeared and, instead, a reference to Nature should convince the reader that, following Danti's rules, the scenery will be rightly built in itself, and according to Nature. Even if the viewpoint has such relevance for Danti's method, he does not acknowledge the obvious conclusion: that the perspectival construction will be contingent and only function properly when viewed from this point, while it will look strange from any other point in the auditorium. Sirigatti is the first to show a strong awareness of the contingency of the perspectival effect of the scenery. Not that his predecessors were unaware of this important element, of course, but they did not consider it relevant to include this awareness, in any sense, into their working procedures. Sirigatti, on the contrary, is very concerned with the different qualities of views among spectators. And he is the first to write such words as,

> all those who will have the sight in the line P.E.X. will judge the relief match the painting, but those that will have the sight outside of this said line will know the deception, and more so the farther they are from it.[43]

With a growing understanding of perspectival stages, some kind of innocence necessarily got lost: scenographers were increasingly aware that, under a central perspective system, the stronger the illusion of the scenery, the more the success of this illusion becomes

---

41 'E per fare, che le facciate delle case M L, e I K, corrino al punto C, e s'accordino con le case finte nella parete G H, acciò l'occhio, che sta nel punto A, della distanza, vegga andare ogni cosa ad unirsi al punto C, si opererà in questa maniera'. Vignola and Danti, *Le due regole della prospettiva prattica*, 90.

42 'Hora segnandosi cosi sattamente ogn'altra cosa nelle facciate digradate delle case di rilievo, correrà ogni cosa al punto C, principale, e cosi le case finte della parete G H, accorderanno giustamente con quelle di rilievo, e si opererà con un sol punto, conforme alle regole vere, & à quello che la Natura opera nel veder nostro'. Vignola and Danti, *Le due regole della prospettiva prattica*, 91.

43 'tutti quelli che haranno la vista nella linea P.E.X. guidicheranno il rilievo unire con il finto, ma quelli che saranno con la vista fuori di detta linea conosceranno l'inganno, e tanto più quanto si allonteranno da essa'. Sirigatti, *La pratica della prospettiva*, fol. 42^{v}.

restricted to a particular viewpoint. As Robert Klein puts it, the perspectival stage ended up becoming simply a painting for one spectator.[44] By 1625, this process had reached a stage of completion, made very clear when Pietro Accolti explicitly corrected Sirigatti in *Lo inganno de gl'occhi*, stating:

> And firstly, in these endeavours no one should be persuaded, as the Cavaliere Lorenzo Sirigatti has believed [...], that a scene can be drawn so exactly that it wholly obeys to a number of spectators, disposed one after the other in a straight line, and will coincide in all its parts to the eyes of many [...] thus it will follow that it can seem more or less discordant to some among them in diverse distances; therefore, the concordance of the painted things does not happen in the form of a line, but of a point.[45]

But before that, in 1600, the mathematician Guidobaldo del Monte published his six books on perspective, the last of which is devoted to the stage.[46] His is the first successful attempt to find the mathematical foundation of perspective, although in the developments we are dealing with here, he does not play a very relevant role, other than conveying in the most precise geometrical terms what his predecessors had already said.[47] While speaking of lines instead of threads, he explains a method similar to Danti's. Even if his method to connect viewpoint, vanishing point, and any point at the front of the stage is not identical to Danti's, he does refer to Danti's choice (without mentioning him) and acknowledges its accuracy. He makes explicit the geometrical reasoning behind this method, i.e., understanding the orthogonals on the stage as intersections between the side walls and the imaginary planes that rotate around the axis connecting viewpoint and vanishing point.[48] It was probably his clear geometrical understanding of the perspectival stage that allowed him to find a simplified, but not very intuitive, method for obtaining the same results. The orthogonal (PQ) must be found on the side wall along which a particular point (P) at the front of the stage projects into the vanishing point. As Del Monte explains, it can be determined by looking up and down at the line connecting the viewpoint (A) and the vanishing point (G) from the opposite side of the stage until P is obscured. From this exact point, 'and with one eye closed', PQ can be traced on the wall so that it appears to the viewer to be in accordance with AG.[49] This method can be

44 See Robert Klein, 'Vitruve et le théâtre de la Renaissance italienne'.

45 'Et prima, in questi propositi, nessuno si persuada, come ha creduto il Cavaliere Lorenzo Sirigatti [...] poter disegnarsi una Scena si sattamente, che à più numero di Spettatori, per una certa data linea, & dirittura l'un dopo l'altro disposti, onninamente obbedisca, & à gl'occhi di molti concordi nelle sue totali parti [...] ne seguirà, che in diverse distanze possino discordare più, e meno tra di loro; adunque, non si dà in processo di linea, ma in punto la concordanza delle cose finte'. Pietro Accolti, *Lo inganno de gl'occhi. Prospettiva pratica* (Florence: Cecconcelli, 1625), pp. 91–92, in *ECHO. Cultural Heritage Online*: http://echo.mpiwg-berlin.mpg.de/MPIWG:KSKH7YK2 [accessed 31-07-2017].

46 Guidobaldo del Monte, *I sei libri della prospettiva*, trans. by Rocco Sinisgalli (Rome: L'Erma di Bretschneider, 1984).

47 See Kirsti Andersen, 'Guidobaldo: The Father of the Mathematical Theory of Perspective', in *Guidobaldo del Monte, 1545–1607: Theory and Practice of the Mathematical Disciplines from Urbino to Europe*, ed. by Antonio Becchi, Domenico B. Meli, and Enrico Gamba (Berlin: Max Planck Research Library for the History and Development of Knowledge, 2013), pp. 145–66.

48 See Del Monte, *I sei libri della prospettiva*, 221–22.

49 'e, con l'occhio fermo'. Del Monte, *I sei libri della prospettiva*, 222.

implemented more swiftly than Danti's, but as it seems to introduce a new point of view onto the stage (although it does not do so) it is not very intuitive and was probably only made possible by Del Monte's understanding of the lines of an image in perspective as intersections of planes.[50]

Finally, it was Nicola Sabbatini who clearly expressed the conjunction of the privileged viewpoint and the seat for the prince. This connection is outlined explicitly in the thirty-fourth chapter of his treatise *Pratica di fabricare scene e machine ne' teatri* (Ravenna, 1638).[51] The custom of placing the ruler's seat in the middle of the auditorium may by then have been firmly established in Italian courts (probably following the Florence model), but Sabbatini was the first scenographer to connect it with the fully developed method for building the scenery with the point of view in mind (in Del Monte's version). He represents the bridge between a courtly custom and the optical knowledge that had been developing around the stage since the first half of the sixteenth century. After determining the limits of the stage, placing the vanishing point, and marking on the front of the stage the beginning of the first two houses' frontal facades, the next step is the position of the viewpoint. It is taken here quite literally as a viewing point. Its location is determined by an optical procedure: taking a square, holding it horizontally before the eye and looking toward the stage from the middle of the auditorium and going back and forth until the place is found where the two arms of the square meet the two marks on the front of the stage.[52] Sabbatini does not explain why he chooses this unique method, but it is clearly intended as a way of finding a position where the entire stage can be seen from one view. The choice of the right angle as the angle of vision seems to be a matter of what is practical: while it had been a traditionally extended answer for the amplitude of the view angle, by that time it was rather old-fashioned. Danti writes that two-thirds of a right angle is a preferable measure, and Del Monte himself discusses the matter extensively to conclude that the perfect viewpoint cannot be determined solely by the angle; the distance to the object and its relative position must also be considered.[53] After locating this point on the floor of the auditorium and erecting a stake there, a horizontal string is stretched between the stake and the vanishing point, and it is exactly where the string hits the stake that Sabbatini establishes the *punto della distanza* ('viewpoint').[54] Once the viewpoint is fixed and connected to the vanishing point by a horizontal string, Sabbatini proceeds to trace the receding orthogonals on the stage with Del Monte's simplified method. He then locates the prince's seat:

> As we are already finished dealing with how the scene must be made, it seems reasonable to me to also say how and where should the place be set for the prince or any other personage that should attend. It must be taken into account to choose the closest possible place to the distance point, and it should be so high from the floor of the hall

50 See Del Monte, *I sei libri della prospettiva*, 41, 57.

51 Nicola Sabbatini, *Pratica di fabricare scene e machine ne' teatri* (Ravenna: de' Paoli and Giovanelli, 1638), in *Internet Archive*: http://www.archive.org/details/praticadifabricaoosabb [accessed 31-07-2017].

52 Sabbatini, *Pratica di fabricare*, 9.

53 See Vignola and Danti, *Le due regole della prospettiva prattica*, 69 and Del Monte, *I sei libri della prospettiva*, 42–46.

54 Sabbatini, *Pratica di fabricare*, 10.

> that, when sitting, the sight is on the very same plane than the vanishing point, for in this way all things marked on the scene will appear better than in any other place.[55]

The viewpoint is now determined so precisely that the prince should not only occupy one particular position on the floor of the auditorium, but also be sitting as close as possible to the actual height to which the scenery has been built. The ruler's eye should identify with the mathematical point that operates as the basis of the whole perspectival construction. Given such an awareness of the desirable coincidence between the optical relation to the stage and the prevailing social hierarchy, it is not surprising that the whole seating distribution, not only the prince's throne, is ordered according to the same criteria. Sabbatini recommends that the person in charge of seating the spectators should know personally as many of them as possible, since he should try to arrange things so 'that the idiot and common people sit on the tiers and at the sides'.[56] The reason given for this is that some imperfections will be visible from those positions, imperfections to which these kinds of people will not be paying much attention. 'But the industrious and well-mannered people must be accommodated on the ground of the hall, as close to the middle as possible'.[57] In this way, the seating distribution, with the varying quality of views of the stage, is brought to correspond to a social hierarchy that is equated with an allegedly natural scale of intelligence. On the peak of this pyramid sits the prince, who embodies the apex of the visual pyramid for which the stage has been built. Everything fits together, and it is not surprising that such a clear ideological model of the social body of the time successfully spread over Europe within a matter of years.

Spreading and the End of the Model

During the first half of the seventeenth century, and while the final stages of its development were still taking place, the perspectival stage spread from Italy outward over Europe. The forms and techniques of Italian scenography had a profound impact on the masques staged by Inigo Jones at the English court between 1605 and 1640, as well as on the work of Joseph Furttenbach at his Stadttheater in Ulm, built in 1641. Both visited Italy at some point in their careers, and their designs show the enduring influence of Giulio Parigi, scenographer of the Medici. A disciple of Parigi, Giacomo Lotti, was called to the Spanish court, where he built the Coliseo of the Palacio del Buen Retiro for Philip IV in 1632. And in 1645 Cardinal Mazarin hired Giacomo Torelli to repeat in Paris the formidable successes he had enjoyed in Venice. All these scenographers contributed in different ways to the general trend towards a fully changeable stage, but it was Torelli who made the most important

---

55 'Mi pare ragionevole, essendosi di già finito di trattare come si debba fare la Scena, di dire anco come, & in qual sito si debba accomodare il luogo per il Prencipe od altro Personaggio, che vi doverà intervenire. Si haverà per tanto in consideratione di far' elettione di luogo piu vicino, che sia possibile al Punto della distanza, e che sia tanto alto dal piano della Sala, che stando à sedere, la vista sia nel medesimo piano del Punto del concorso, che cosi tutte le cose segnate nella Scena appariranno meglio, che in alcuno altro luogo'. Sabbatini, *Pratica di fabricare*, 55.

56 'che le persone idiote, e plebee si accommodino ne gli Scaloni, e dalle bande'. Sabbatini, *Pratica di fabricare*, 66.

57 'Ma le persone faccenti, e di garbo, si debbano accomodare nel piano della Sala, più nel mezo che sia possibile'. Sabbatini, *Pratica di fabricare*, 66.

step forward: He improved the system of sliding flat wings invented by Giovanni Battista Aleotti in 1606 by mounting them on so-called chariots under the stage, which allowed them to be changed mechanically, quickly, and simultaneously.[58]

This made scene changes an almost instantaneous affair and brought about a radical development of scenic machines and wonders. But as far as our purposes are concerned, none of this altered the conjunction of the prince's seat and the privileged viewpoint as formulated by Sabbatini. Nevertheless, the central problem for scenographers did change: it was no longer the case that one (foreshortened) real wall needed be matched with a painted one, the problem was rather matching a whole succession of parallel planes. If anything, this led to a deepening of the stage space: 'With scenic vistae painted on two-dimensional surfaces instead of constructed in three, stage depth had to be increased to make the two-dimensional illusion as effective as the built-up one'.[59] Courts began to compete on the depth of their respective stages. In Dresden, for instance, the *Comödienhaus* ('court theatre') was built by Wolf Caspar von Klengel between 1664 and 1667 with a stage equipped with six sets of wings. Only one year later, the building was reformed to add three more pairs of wings and a new final prospect. The stage reached back thirty metres, while the auditorium was only fifteen metres long.[60] The stage of the Salle des Machines at Les Tuileries in Paris was even deeper, with a reach of forty-four metres. This size seems however to have caused horrible problems with the acoustics which prevented it from being used after its opening.[61]

As the vanishing point on the backdrop receded more and more into the back of the stage, the less spectators in the auditorium were able to see it, although it must be remembered, of course, that the vast majority of them were not able to see the perspective functioning properly in any case. The close relationship between the vanishing point and the princely viewpoint reached its apotheosis with these deep stages and led to ways of conveying the central position of the prince through perspectival means only. While in London and Paris the king himself performed in front of his courtiers, wherever etiquette did not approve of the royal persons' personal involvement in plays and ballets, perspective allowed for other more formal solutions. In 1668, during the celebrations of the wedding between Emperor Leopold I and Margaret Theresa of Spain, the whole court, with the Imperial couple sitting at the privileged viewpoint, looked at a scene where an equestrian statue of Leopold I himself, surrounded by ten emperors of the past, occupied the vanishing point (Figures 5 and 6). And something similar happened at the performance of Pedro Calderón de la Barca's piece *Hado y divisa de Leonido y Marfisa* in the Coliseo of the Palacio del Buen Retiro in Madrid on the occasion of the wedding between Charles II and Marie Louise of Orléans in 1680.

---

58 See Schöne, *Die Entwicklung der Perspektivbühne*, 51 and Donald C. Mullin, *The Development of the Playhouse: A Survey of Theatre Architecture from the Renaissance to the Present* (Berkeley: University of California Press, 1970), pp. 25, 43.

59 Mullin, *The Development of the Playhouse*, 24.

60 See Uta Deppe, *Die Festkultur am Dresdner Hofe Johann Georgs II. von Sachsen (1660–1679)* (Kiel: Ludwig, 2006), pp. 63–64.

61 See Hans-Georg Hofmann, 'Das Dresdner Planetenballett 1678/79: Aspekte Einer Inszenierung', *Basler Jahrbuch Für Historische Musikpraxis*, 23 (1999), 75–100 (p. 89) and Marie-Claude Canova-Green, 'Le ballet de cour en France', in *Spectaculum Europaeum. Theatre and Spectacle in Europe. Histoire Du Spectacle En Europe (1580–1750)*, ed. by Pierre Béhar and Helen Watanabe-O'Kelly (Wiesbaden: Harrassowitz, 1999), pp. 485–512 (p. 503).

Fig. 5 The opera *Il pomo d'oro* being performed before the court of Leopold I in Vienna in 1668. Drawing by Lodovico Ottavio Burnacini and print by Frans Geffels. © Trustees of the British Museum. [See colour plate 15]

When the curtain rose, seven sculptures of past kings of Spain appeared on each side of the stage and at the vanishing point, was a picture of the King and Queen that mirrored their position in the middle of the auditorium.[62] As these examples show, in the court theatre of the seventeenth century the sitting prince was as essential a part of the spectacle as what happened on the stage. Viewpoint and vanishing point became the two equal foci of shows offered to courtly audiences.[63] In this context, the visibility of the prince as the ideal spectator, seated at the privileged viewpoint, was crucial, which explains the fact that at some court theatres of the time the raised royal boxes that theatres were equipped with were ignored and temporary daises on the parterre continued to be used (as in Madrid and Dresden).[64]

The general trend that developed over the 1600s towards a deepening and sharpening of central perspective of theatrical stages was suddenly interrupted at the turn of the century when the *scena per angolo* appeared. Allegedly invented by Ferdinando Galli Bibiena,

62 See Biblioteca Nacional de España, Madrid, Anonymous, 'Descripción de la comedia yntitulada Ado y divisa de Leonido y de Marfisa que se hizo a SS.MM. Don Carlos Segundo y Doña María Luysa en el Coliseo del Retiro el día 3 de março del año de 1680', MSS/9373, in *Biblioteca Digital Hispánica*: http://bdh-rd.bne.es/viewer.vm?id=0000139724&page=1 [accessed 27-07-2017], fols 110^{r}–24^{v}.

63 See Sebastian Neumeister, *Mythos und Representation. Die mythologischen Festspiele Calderons* (Munich: Fink, 1978), p. 281, and John E. Varey, 'The Audience and the Play at Court Spectacles: The Role of the King', *Bulletin of Hispanic Studies*, 61 (3) (1984), 399–406 (pp. 403–04). I would like to thank Sebastian Neumeister for his kind explanations about the use of perspective in the theatrical representations at the Spanish court.

64 See Biblioteca Nacional de España, Madrid, MSS/9373, fol. 110^{v} and Deppe, *Die Festkultur*, 65.

Fig. 6 The apotheosis of Leopold I in the first of the stage sets designed by Lodovico Ottavio Burnacini for the opera *Il pomo d'oro* (Vienna, 1668). Print by Frans Geffels. © Trustees of the British Museum. [See colour plate 16]

the *scena per angolo* was a form of perspectival scenery that showed an oblique view on a building or a part of a building. The centre of the scene was now a corner, from which walls, corridors, or stairs laterally receded into two vanishing points that could not be seen by the audience but could be imagined somewhere at both sides of the stage. This simple forty-five-degree twist of the geometrical main structure of the scenery paradoxically both privileged interior settings (where crossings of corridors and colonnades could likely be located) and brought about a new sense of infinite space, by hiding the end of the perspectival image from the view of the audience. The first datable *scena per angolo* was produced by Ferdinando Galli Bibiena in 1687 for the staging of *Didio Giuliano*, by Lotto Lotti and Bernardo Sabadini, at the Teatro Ducale in Piacenza.[65] After almost thirty years serving the Farnese family as a painter, architect, and scenographer, in 1711 Ferdinando Galli Bibiena wrote *L'architettura civile preparata su la geometria, e ridotta alle prospettive*, in which he explains his invention. He was then called by the Emperor Charles VI to Vienna, where he worked as court scenographer with great success between 1712 and 1717. Both the publication and the impact of his Viennese designs, which circulated in the form of illustrated librettos, helped the fast spreading of the *scena per angolo* throughout Europe. Of course, the novelty in itself was very appealing for audiences so used to central perspective sceneries. But there was also another, optical, reason for the popularity:

> The immense success of *la scena per angolo* [...] must have been due to its power of illusionism that was less artificial; for its viewpoint is not so restricted, its picture

65 See Carroll Durand, 'The Evolution and Actualization of the "Scena Per Angolo": High Baroque Scenography of the Galli Bibienas (Italy)' (unpublished doctoral dissertation, Tufts University, 1983), p. 83.

> plane not so obtrusively established, its distortion not so emphasized, and its proper visibility more universal.[66]

This new kind of oblique view of the stage was certainly far less dependent on a particular privileged viewpoint than the central perspective. It would have been a more reasonable choice for theatrical sceneries from the very beginning, as the illusionism works when viewed from almost any point in the auditorium. In fact, the question is why it was kept outside the stage for so long. These sorts of constructions *per angolo* were well-known to painters in the seventeenth century, since they 'could be readily found in almost any of the numerous books [on perspective] available to them'.[67] However, according to Timothy Kitao, it was precisely one of these books — and one with particular influence — that contributed to preventing the use of oblique perspectival constructions in the first instance: Vignola's. As stated before, Vignola presented two rules of perspective. The second involved not only a vanishing point, but a distance point too, which in fact worked exactly as a second lateral vanishing point. Although the distance point was for the method 'identical in nature and function with the principal point and just as essential', in his book Vignola constantly refuses to put it in those terms and subordinates the former to the latter.[68] Throughout his explanations he is strictly committed to a one-point-only type of construction and he emphatically states: 'One should work with a single point, and not with two'.[69] As Kitao explains:

> The feasibility of a system with two or more points is not merely questioned; it is altogether rejected, and at the same time the 'oblique' representation, which the author proves is as rational as any other perspective representation, is — however implicitly — discouraged, because it suggests the use and existence of *more than one point*.[70]

By doing so, Vignola expressed a deeply rooted preconception among Italian Renaissance painters who identified spatial illusionism with a centralised, unifocal perspective. He was able to 'distil the predilection of the century into a coherent theory' and, soon considered an authority, he prevented the use of bifocal perspectival representations for more than a century.[71] As Kitao notes, the pervasive impact of this prejudice is particularly striking in *quadratura* painting and theatrical settings. Since both involve the creation of large illusionistic images that are in the first case to be looked at by moving spectators, or, in the second, from several positions at the same time, a central perspective dependent on a single viewpoint would not be the most suitable option for any of them. And, in fact, while *quadratura* painters continued to avoid using oblique views in their paintings, they did turn away from strict one-point ideas of perspective, not least because such formations do not work well on the ceilings of large spaces. Both Carlo Cesare Malvasia and Filippo Baldinucci credited the Bolognese fresco painter Agostino Mitelli (1609–60) with the invention of a kind of large perspective that allowed a more flexible viewpoint, while a

66 Kitao, 'Prejudice in Perspective', p. 190.
67 Kitao, 'Prejudice in Perspective', p. 188.
68 Kitao, 'Prejudice in Perspective', p. 181.
69 'che si operi con un sol punto, e non con due'. Vignola and Danti, *Le due regole della prospettiva prattica*, 53.
70 Kitao, 'Prejudice in Perspective', p. 185.
71 Kitao, 'Prejudice in Perspective', p. 186.

geometrical method for softening the result by manipulating the focal point was advocated in print by Gioseffe Viola Zanini in 1629.[72]

The fact that no similar type of 'softened' system was applied to stage design before the irruption of the *scena per angolo* at the end of the seventeenth century — despite its obvious usefulness — is very telling. It stresses the background political reasons that were at play in the theatricalised appearances of the prince amid his court but did not apply to *quadratura* painting. This has to be taken into consideration when it comes to explaining why this enduring resistance to everything other than the strict central perspective system suddenly disappeared and gave way to the spectacular success of the *scena per angolo*. Some reasons for this sudden change have been suggested, like the emphasis on unity of place brought about by the neo-classical reform of the opera or, more generally, the 'baroque fascination with limitless space'.[73] But these explanations do not take into account the *scena per angolo*'s political dimension and fail to explain why princes would be prepared to leave aside such an exquisite optico-political device that had proven itself so effective as an emblem of prestige. In other words, if the new stage designs were going to make a prominent viewpoint useless, then the princes must have been thinking of something better than a throne in the middle of the parterre. And indeed, they were. It was the prominent raised royal box.

Furthermore, it was precisely the Galli Bibiena family that introduced the raised and fully decorated form of royal box in the Viennese court, along with the *scena per angolo*. The court theatre designed by Francesco Galli Bibiena (Ferdinando's brother) in 1704 included a magnificent double box in front of the stage for the Imperial family. Of course, we should be cautious when emphasizing the link between the *scena per angolo* and the royal box. In 1685, a couple of years before the first oblique stage design by Ferdinando Galli Bibiena, Gasparo and Domenico Mauro remodelled the theatre of the electoral court in Munich, the Opernhaus am Salvatorplatz, and added a spectacular box for the elector.[74] Such developments occurred the other way around too: the use of temporary daises in the parterre is attested at the Viennese court theatre many years after the construction of a raised royal box, as late as 1744 (which can probably be attributed to the very conservative ceremonial at the Habsburg court).[75] While a causal relationship between oblique settings and raised boxes cannot be drawn, it is true that both demonstrate a shift away from the optical alignment of central perspective and the prince's seat. Given the pervasive resistance to oblique perspectival constructions in theatre and the coincidence of their irruption with the spreading of prominent raised royal boxes, it is not too much of an assumption to suggest that only after the first of these royal boxes appeared, the previous model based on central perspective lost its attraction as a vehicle of ostensive representation. After all, a temporary dais in the parterre cannot offer the same opportunities for duly signifying

---

72 Filippo Baldinucci, *Notizie de' professori del disegno da Cimabue in qua* (Florence: Santi Franchi, 1681–1728); Carlo Cesare Malvasia, *Felsina pittrice. Vite de pittori bolognesi. Tomo primo* (Bologna: Per l'Erede di Domenico Barbieri, 1678); Gioseffe Viola Zanini, *Della architettura* (Padua: Francesco Bolzetta, 1629).

73 Durand, 'The Evolution and Actualization of the "Scena per angolo"', pp. 47, 82.

74 See Simon Tidworth, *Theatres: An Illustrated History* (London: Pall Mall Press, 1973), p. 74.

75 See Andrea Sommer-Mathis, 'Opera and Ceremonial at the Imperial Court of Vienna', in *Italian Opera in Central Europe. Volume One: Institutions and Ceremonies*, ed. by Melania Bucciarelli, Norbert Duwoby and Reinhard Strohm (Berlin: Berliner Wissenschafts-Verlag, 2006), pp. 178–91 (pp. 186–87).

the presence of the prince. Once this choice was made, the resistance towards anything other than central perspective disappeared, and with it came the end of the model that had shaped court theatre for almost two centuries. Perspective would never play such a crucial role in the European political arena again.

Bibliography

Manuscript and Archival Sources

Biblioteca Nacional de España, Madrid, Anonymous, 'Descripción de la comedia yntitulada Ado y divisa de Leonido y de Marfisa que se hizo a SS.MM. Don Carlos Segundo y Doña María Luysa en el Coliseo del Retiro el día 3 de março del año de 1680', MSS/9373, fols 110r–24v.

Primary Sources

Accolti, Pietro, *Lo inganno de gl'occhi. Prospettiva pratica* (Florence: Cecconcelli, 1625).

Baldinucci, Filippo, *Notizie de' professori del disegno da Cimabue in qua* (Florence: Santi Franchi, 1681–1728).

Barbaro, Daniele, *La pratica della perspettiva. Opera molto profittevole a' pittori, scultori, et architetti* (Venice: Borgominieri, 1569).

Campori, Giuseppe, *Notizie per la vita di Ludovico Ariosto tratte da documenti inediti* (Modena: Carlo Vincenzi, 1871).

Giambullari, Pierfrancesco, *Apparato et feste nelle noze dello Illustrissimo Signor Duca di Firenze, et della Duchessa sua Consorte, con le sue Stanze, Madriali, Comedia, & Intermedii, in quelle recitati* (Florence: Benedetto Giunta, 1539).

Malvasia, Carlo Cesare, *Felsina pittrice. Vite de pittori bolognesi. Tomo primo* (Bologna: Per l'Erede di Domenico Barbieri, 1678).

Monte, Guidobaldo del, *I sei libri della prospettiva*, trans. by Rocco Sinisgalli (Rome: L'Erma di Bretschneider, 1984).

Rossi, Bastiano de', *Descrizione dell'apparato e degl'intermedi fatti per la commedia rappresentata in Firenze nelle nozze de' Serenissimi Don Ferdinando Medici e Madama Cristina De Loreno, Gran Duchi di Toscana* (Florence: Padovani, 1589).

Sabbatini, Nicola, *Pratica di fabricare scene e machine ne' teatri* (Ravenna: de' Paoli and Giovanelli, 1638).

Serlio, Sebastiano, *Tutte l' opere d' architettura. Dove si trattano in disegno, quelle cose, che sono più necessarie all'architetto: et hora di nuovo aggiunto, oltre il libro delle porte, gran numero di case private nella città, & in villa, et un indice copiosissimo* (Venice: Senese, 1584).

Sirigatti, Lorenzo, *La pratica della prospettiva* (Venice: Girolamo Franceschi Sanese, 1596).

Vasari, Giorgio, *Le vite de' più eccellenti pittori, scultori e architettori nelle redazioni del 1550 e 1568*, ed. by Rosanna Bettarini and Paola Barocchi, 6 vols (Florence: Sansoni, 1966–87).

Vignola, Iacomo Barozzi da, and Egnatio Danti, *Le due regole della prospettiva prattica di M. Iacomo Barozzi da Vignola con i commentari del Reverendo Padre Maestro Egnatio Danti* (Bologna: Longhi, 1682).

Zanini, Gioseffe Viola, *Della architettura* (Padua: Francesco Bolzetta, 1629).

Secondary Works

Andersen, Kirsti, 'Guidobaldo: The Father of the Mathematical Theory of Perspective', in *Guidobaldo del Monte, 1545–1607: Theory and Practice of the Mathematical Disciplines from Urbino to Europe*, ed. by Antonio Becchi, Domenico B. Meli, and Enrico Gamba (Berlin: Max Planck Research Library for the History and Development of Knowledge, 2013).

Bino, Carla, and Ilaria Tameni, 'Il teatro umanistico e rinascimentale', in *Storia essenziale del teatro*, ed. by Claudio Bernardi and Carlo Susa (Milan: Vita e Pensiero, 2005), pp. 121–64.

Canova-Green, Marie-Claude, 'Le ballet de cour en France', in *Spectaculum Europaeum. Theatre and Spectacle in Europe. Histoire Du Spectacle En Europe (1580–1750)*, ed. by Pierre Béhar and Helen Watanabe-O'Kelly (Wiesbaden: Harrassowitz, 1999), pp. 485–512.

Deppe, Uta, *Die Festkultur am Dresdner Hofe Johann Georgs II. von Sachsen (1660–1679)* (Kiel: Ludwig, 2006).

Durand, Carroll, 'The Evolution and Actualization of the "Scena Per Angolo": High Baroque Scenography of the Galli Bibienas (Italy)' (unpublished doctoral dissertation, Tufts University, 1983).

Gurr, Andrew, *The Shakespearean Stage, 1574–1642* (Cambridge: Cambridge University Press, 1992).

Haza, José María Ruano de la, 'Hacia una metodología para la reconstrucción de la puesta en escena de la comedia en teatros comerciales del siglo XVII', *Criticón*, 42 (1988), 81–102.

Hofmann, Hans-Georg, 'Das Dresdner Planetenballett 1678/79: Aspekte Einer Inszenierung', *Basler Jahrbuch Für Historische Musikpraxis*, 23 (1999), 75–100.

Kitao, Timothy K., 'Prejudice in Perspective: A Study of Vignola's Perspective Treatise', *The Art Bulletin*, 44 (1962), 173–94.

Klein, Robert, 'Vitruve et le théâtre de la Renaissance italienne', in *La forme et l'intelligible* (Paris: Gallimard, 1970), pp. 294–309.

Marciak, Dorothée, *La place du prince. Perspective et pouvoir dans le théâtre de cour des Médicis, Florence (1539–1600)*, Études et Essais sur la Renaissance, 50 (Paris: Honoré Champion, 2005).

Mullin, Donald C., *The Development of the Playhouse: A Survey of Theatre Architecture from the Renaissance to the Present* (Berkeley: University of California Press, 1970).

Nao, Nobuhide, 'A Study of "The Stage Set of *Le Bacchide*" by Baldassarre Peruzzi: The Development of "Prospettiva Solida" in Early Sixteenth Century Italy', *Journal of Graphic Science of Japan*, 39 (2005), 11–18.

Neumeister, Sebastian, *Mythos und Representation. Die mythologischen Festspiele Calderons* (Munich: Fink, 1978).

Orgel, Stephen, and Roy Strong, *Inigo Jones: The Theatre of the Stuart Court. Including the Complete Designs for Productions at Court for the Most Part in the Collection of the Duke of Devonshire Together with their Texts and Historical Documentation*, 2 vols (London: Sotheby Parke-Bernet; Berkeley: University of California Press, 1973).

Pochat, Götz, *Theater und bildende Kunst im Mittelalter und in der Renaissance in Italien* (Graz: Akademische Druck- u. Verlagsanstalt, 1990).

Saslow, James M., *The Medici Wedding of 1589. Florentine Festival as 'Theatrum Mundi'* (New Haven: Yale University Press, 1996).

Schöne, Günter, *Die Entwicklung der Perspektivbühne von Serlio bis Galli-Bibiena nach den Perspektivbüchern*, Theatergeschichtliche Forschungen, 43 (Leipzig: Leopold Voss, 1933).

Sommer-Mathis, Andrea, 'Opera and Ceremonial at the Imperial Court of Vienna', in *Italian Opera in Central Europe. Volume One: Institutions and Ceremonies*, ed. by Melania Bucciarelli, Norbert Duwoby and Reinhard Strohm (Berlin: Berliner Wissenschafts-Verlag, 2006), pp. 178–91.

Strong, Roy, *Art and Power: Renaissance Festivals, 1450–1650* (Berkeley: University of California Press, 1984).

Tidworth, Simon, *Theatres: An Illustrated History* (London: Pall Mall Press, 1973).

Varey, John E., 'The Audience and the Play at Court Spectacles: The Role of the King', *Bulletin of Hispanic Studies*, 61 (3) (1984), 399–406.

Verstegen, Ian F., 'Tacit Skills in the Perspective Treatise of the Late Renaissance', in *Cognition and the Book: Typologies of Formal Organisation of Knowledge in the Printed Book of the Early Modern Period*, ed. by Karl A. E. Enenkel and Wolfgang Neubner (Leiden: Brill, 2005), pp. 187–213.

JULIET ODGERS

The Optical Construction of John Evelyn's 'Dyall Garden' at Sayes Court

Introduction

In 1652, the virtuoso, John Evelyn (1620–1706), returned to England to settle at Sayes Court in Deptford, a suburb of London near Greenwich. He had been abroad for a substantial part of the 1640s, travelling in Europe and latterly living in Paris where he met his wife, Mary Browne, the daughter of King Charles II's official Resident, Sir Richard Browne. Evelyn was a Royalist and returning to England to set up his marital home, he found himself excluded from public life under the hostile Cromwellian regime. He consequently had time to study and to write, to plant and to tend his garden at Sayes Court — a garden which, for all its relatively modest size, was to become internationally famous.[1] The garden was destroyed long ago and is known to us now principally through an annotated drawing by Evelyn himself, made at the time it was first laid out (Figure 1). In this paper I bring together the design shown in this drawing and a text drafted by Evelyn during the 1650s — his unfinished gardening book, the *Elysium Britannicum, or the Royal Gardens*, a part of which survives in the British Library in manuscript form.[2] In the *Elysium Britannicum* Evelyn aspired to provide a 'universal' treatment of the theory and practice of garden design, and included amongst his topics the proper use of hortulan perspective. The *Elysium* provides

1 Michael Hunter, 'John Evelyn in the 1650s: A Virtuoso in Quest of a Role', in *John Evelyn's 'Elysium Britannicum' and European Gardening*, ed. by Therese O'Malley and Joachim Wolschke-Bulmahn (Washington, DC: Dumbarton Oaks Research Library and Collection, 1998), pp. 79–106. On Sayes Court, see Mark Laird, 'Parterre, Grove, and Flower Garden: European Horticulture and Planting Design in John Evelyn's Time', in *John Evelyn's 'Elysium Britannicum'*, ed. by O'Malley and Wolschke-Bulmahn, pp. 171–221; Prudence Leith-Ross, 'The Garden of John Evelyn at Deptford', *Garden History*, 25 (1997), 138–52; Prudence Leith-Ross, 'A Seventeenth-Century Paris Garden', *Garden History*, 21 (1993), 50–57.

2 British Library, London, Evelyn Papers, Additional MS 78342. Page numbers refer to the modern transcription: John Evelyn, *Elysium Britannicum, or the Royal Gardens*, ed. by John E. Ingram (Philadelphia: University of Pennsylvania Press, 2001); this version of *Elysium Britannicum* will be referenced throughout the text unless otherwise stated and referred to as *Elysium Britannicum* or *Elysium*. I follow Ingram's transcription, which includes cancellations and shows insertions by including them in curly brackets thus: {insertion}. For state of manuscript see Frances Harris, 'The Manuscripts of the "Elysium Britannicum"', in Evelyn, *Elysium Britannicum*, ed. by Ingram, 13–21. For summary of contents see Therese O'Malley, 'Introduction', in *John Evelyn's 'Elysium Britannicum'*, ed. by O'Malley and Wolschke-Bulmahn, pp. 9–34.

Juliet Odgers Newcastle University, Juliet.Odgers@newcastle.ac.uk

Perspective as Practice. Renaissance Cultures of Optics, ed. by Sven Dupré, Turnhout, 2019 (*Techne. Knowledge, Technique, and Material Culture*, 1), p. 173-201

DOI 10.1484/M.*Techne*-EB.5.117726

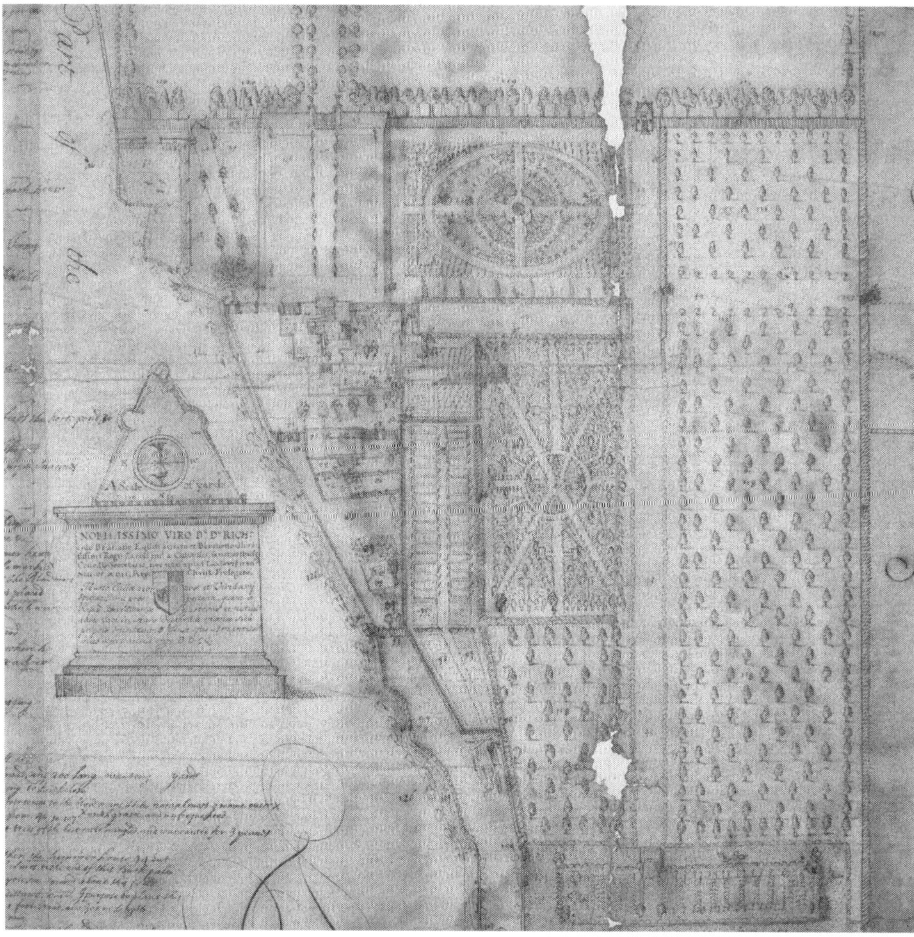

Fig. 1 John Evelyn, Detail of Sayes Court Plan, *c.* 1653. © The British Library Board, Add 78628, fol. A. [See colour plate 17]

the theoretical material against which I interpret his design for Sayes Court garden and, more particularly, the oval parterre, or 'Dyall garden', one of the two main ornamental plots within the larger enclosure.

Evelyn was an important figure in the cultural landscape of Early Modern England. He was not primarily a garden designer but, independently wealthy, found a role for himself principally as an author.[3] Within the field of art history he is known both for his book on engraving and print collecting, *Sculptura*, and for his translation into English of Fréart de

3 General account: Gillian Darley, *John Evelyn: Living for Ingenuity* (New Haven: Yale University Press, 2006). For Evelyn's writings, see Geoffrey Keynes, *John Evelyn: A Study in Bibliophily with a Bibliography of his Writings* (Oxford: Clarendon Press, 1968).

Chambray's treatise on the classical orders.[4] He is also well known as a diarist.[5] During his own lifetime his fame rested on his translations of French gardening books and on his hugely successful book on forestry, *Sylva: or a Discourse of Forest-Trees*, which was the first book to be published under the auspices of the Royal Society.[6] Evelyn was a founder member and an active participant of this institution for many decades. The *Elysium*, as Evelyn first conceived and drafted it was, however, a fruit of the decade preceding the foundation of the Royal Society, the 1650s.[7]

Evelyn structured the *Elysium Britannicum* into three books. The first book is theoretical, covering issues such as plant generation and general natural philosophical principles. Here Evelyn's thinking exhibits the strong influence of Hermetic ideas, transmitted, not least, through his study of Paracelsian chemistry during the late 1640s and early 50s at the Jardin du Roi, in Paris coupled to a dawning appreciation of ideas derived from the Classical School of Epicurean atomism.[8] The second book is practical, covering issues such as garden layout, garden 'ornaments', and horticultural practice. The final, third book is now lost. Perspective emerges from the pages of the second book as a 'rare *Arte*',[9] which serves two purposes — it is used to correct and order the view of the garden, rendering it 'harmonious'; and it is used to construct 'wonders' of vision, distinct quasi-theatrical episodes within the wider garden. When discussing the visual experience of the garden, Evelyn invariably refers to the one who views as the 'spectator', a register of the common Baroque conception of a garden as a theatrical domain.[10] Amongst Evelyn's hortulan wonders we find 'mathematical' ornaments such as rainbow making fountains, hydraulic statues that sound instruments or utter noises, music making machines, sundials and illusionistic trompe l'œil — the latter as theatrically engaging as any of the more obviously performative devices.

For Evelyn, perspective practice was embedded in the broader field of optically oriented, projective mathematical arts, as was typical for the Renaissance and Early Modern period — a field which included surveying, observational astronomy, and

4 John Evelyn, *Sculptura: Or the History, and Art of Chalcography and Engraving in Copper [...]* (London: J. C. for G. Beedle, T. Collins and J. Crook, 1662); Roland Fréart, sieur de Chambray, *A Parallel of the Antient Architecture with the Modern; [...] With L. B. Alberti's Treatise of Statues*, trans. by John Evelyn (London: John Place, 1664).

5 John Evelyn, *The Diary of John Evelyn. Now First Printed in Full from the Manuscripts [...]*, ed. by E. S. de Beer (Oxford: Clarendon Press, 1955).

6 John Evelyn, *Sylva, or a Discourse of Forest-Trees, and the Propagation of Timber in His Majesties Dominions [...]* (London: Jo. Martyn and Ja. Allestry, 1664; further editions 1670 and 1679).

7 See Hunter, 'John Evelyn in the 1650s'.

8 Juliet Odgers, 'Water in Use and Philosophy at Wotton House: John Evelyn and the History of the Trades', *arq: Architectural Research Quarterly*, 15 (2011), 237–47; Michael M. Repetzki, 'John Evelyn: Virtuoso and the Venture of Atomism', in *John Evelyn's Translation of Titus Lucretius Carus 'De rerum natura'* (Frankfurt: Peter Lang, 2000), pp. XI–LII, XX–XXVIII; Juliet Odgers, 'Resemblance and Figure in Garden and Laboratory: Gaffarel's Influence on John Evelyn', in *Jacques Gaffarel: Between Magic and Science*, ed. by Hiro Hirai (Rome: Serra, 2014), pp. 85–109; F. Sherwood Taylor, 'The Chemical Studies of John Evelyn', *Annals of Science*, 8 (1952), 285–92; Hunter, 'John Evelyn in the 1650s', pp. 79–106 (pp. 99–100).

9 Evelyn, *Elysium Britannicum*, ed. by Ingram, 215.

10 Evelyn, *Elysium Britannicum*, ed. by Ingram, 128, 216, 128; On garden as theater see John Dixon Hunt, *Garden and Grove: The Italian Renaissance Garden and the English Imagination 1600–1750* (London: Dent, 1986), pp. 59–72; William Howard Adams, *The French Garden, 1500–1800* (London: Scolar Press, 1979), pp. 63–73.

(particularly relevant here) dialling.[11] Recognising these continuities facilitates a richer and more accurate understanding of the garden and, when applied to the Dyall garden at Sayes Court, provides crucial interpretational clues. It prompts a recognition that the perspectival alignments and elongations that Evelyn uses in his design are not solely the result of his concern with orchestrating the visual spectacle of the garden, but rather arise from what I see as the central conceit of the Dyall garden — that the garden itself is analogous to a dial, an idea that informs the emblematics of the design and its spatial relationships with both the surrounding fabric of the garden and larger spatio-temporal order of the cosmos. As Evelyn writes, 'Dyalls are an ornament of necessary use in our Gardens', all the more so if, like Evelyn, the gardener follows an astrological routine of cultivation, courting both the visible and invisible celestial radiations which he understood to emanate from the 'stars' — a term that includes the planets, the sun, and the moon in seventeenth-century usage.[12] This point informs the final discussion of the design of Sayes Court below, but first we set out the context of Evelyn's endeavours and examine his presentation of perspective in the *Elysium*, whilst giving some attention to the current historiographical context.

Context

The surviving part of the *Elysium* is lavishly illustrated with Evelyn's drawings of the individual ornaments of the garden, but any illustration of the garden as a whole, which may originally have been produced for the book, is now lost. This want may be supplied in part by reference to some relevant precedents. Evelyn was omnivorous in his consumption of exemplary gardens, but judging from the text of the *Elysium*, on the issue of hortulan perspective he takes the French tradition as his guide.[13] He frequently lapses into French when discussing the topic; and depends on the work of Jacques Boyceau and Claude Mollet for his discussion of the setting out of avenues, one of the principle sites of perspective manipulation in the garden; he also makes repeated reference to the exemplary perspectival construction of the Tuileries and the Jardin du Luxembourg.[14] We must imagine these as they appear on Gomboust's 'Plan de Paris' of 1652, the year that Evelyn left Paris for good (Figure 2, 3). Both gardens eventually came under the care of André le Nôtre but, as Evelyn

11 Jim Bennett, 'Projection and the Ubiquitous Virtue of Geometry in the Renaissance', in *Making Space for Science: Territorial Themes in the Shaping of Knowledge*, ed. by Jon Agar, Crosbie Smith, and Gerald Schmidt (Basingstoke: Macmillan, 1998), pp. 27–38. See also Alberto Pérez-Gómez, *Architecture and the Crisis of Modern Science* (Cambridge, MA: MIT Press, 1983), pp. 91–94, 99–103.

12 Evelyn, *Elysium Britannicum*, ed. by Ingram, 213. For Evelyn's astrological gardening see Odgers, 'Resemblance and Figure in Garden and Laboratory'.

13 On Evelyn's dependence on European precedent see: Hunt, *Garden and Grove*, 143–79; Sally Jeffery, 'The Way of Italian Gardens', in *A Celebration of John Evelyn: Proceedings of a Conference to Mark the Tercentenary of his Death*, ed. by Mavis Batey (Surrey: The Surrey Gardens Trust, 2007); On Evelyn's dependence on the French tradition see Laird, 'Parterre, Grove, and Flower Garden…' in *John Evelyn's 'Elysium Britannicum'*, ed. by O'Malley and Wolschke-Bulmahn, 171–221.

14 Jacques Boyceau, *Traité du jardinage, selon les raisons de la nature et de l'art […]* (Paris: M. Vanlochom, 1638); Claude Mollet, *Théâtre des plans et jardinages […]* (Paris: Charles de Sercy, 1652).

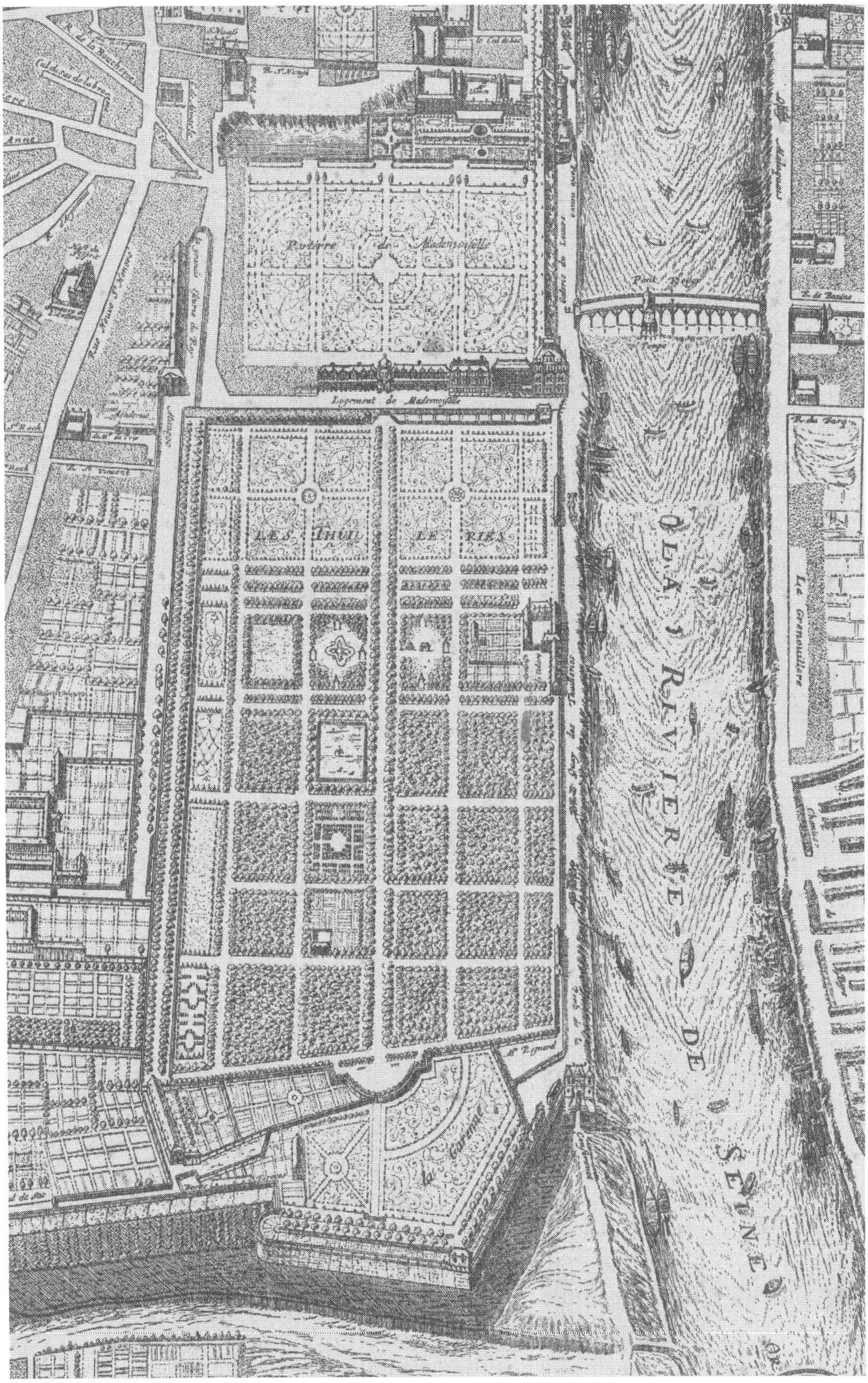

Fig. 2 Jacques Gomboust, *Plan de Paris* (1652), detail showing the Jardin des Tuileries. © Cambridge University Library.

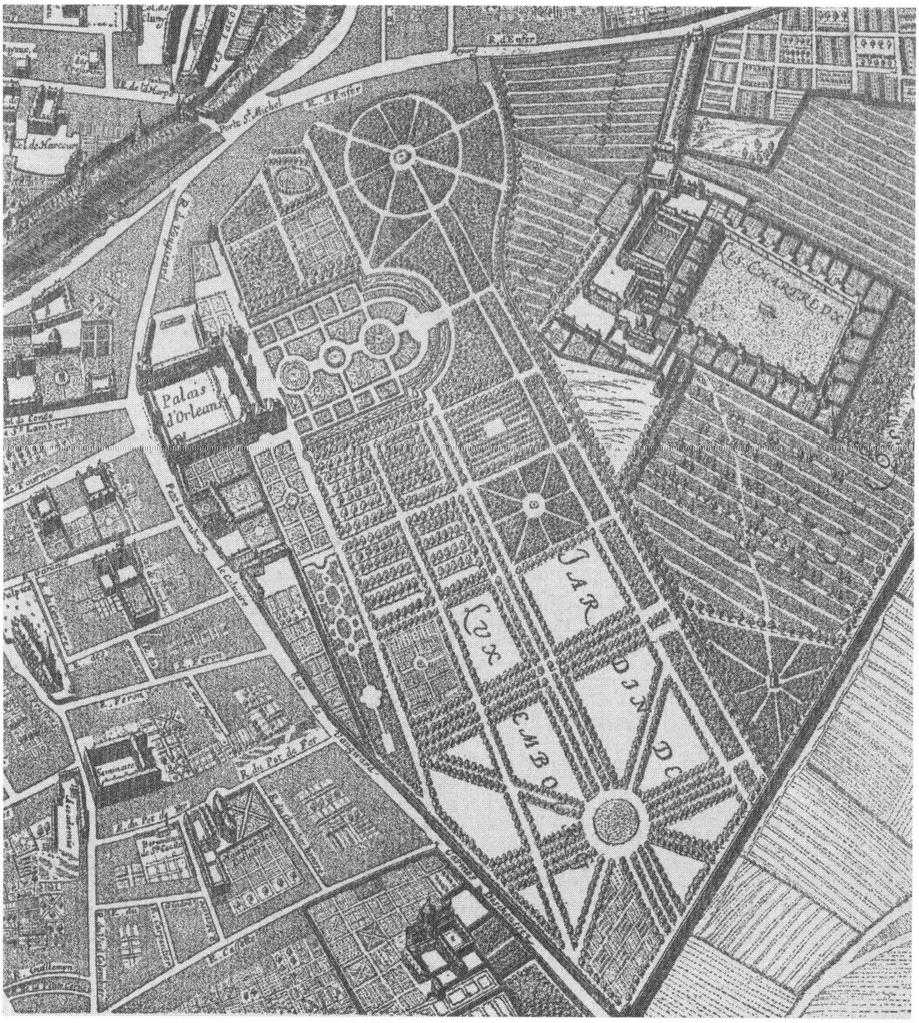

Fig. 3 Jacques Gomboust, *Plan de Paris* (1652), detail showing the Jardin du Luxembourg. © Cambridge University Library.

knew them they were the work of an earlier generation — le Nôtre's own forebears, the Mollet and Desgots families, and Boyceau.[15]

In this French gardening context, perspective is the historiographical topic *par excellence*.[16] A prominent strand of these discussions presents the perspectival constructions

15 Kenneth Woodbridge, *Princely Gardens: The Origins and Development of the French Formal Style* (New York: Rizzoli, 1986), pp. 117–18, 134–38; Franklin Hamilton Hazlehurst, *Gardens of Illusion: The Genius of Andre Le Nostre* (Nashville: Vanderbilt University Press, 1980), pp. 167–85.

16 Allen S. Weiss, *Mirrors of Infinity: The French Formal Garden and 17th-Century Metaphysics* (New York: Princeton Architectural Press, 1995), pp. 33–66; Allen S. Weiss, *Unnatural Horizons: Paradox and Contradiction in Landscape Architecture* (New York: Princeton Architectural Press, 1998), pp. 45–63; Frank Hamilton Hazlehurst, *Jacques Boyceau and the French Formal Garden* (Athens: University of Georgia Press, 1966), pp. 33–38; Hazlehurst,

of the seventeenth-century 'French formal garden' in terms of a, variously defined, 'Cartesian' spatiality. We find this, for example, in the work of Alberto Pérez-Gómez and of Allen Weiss, who establish such garden perspectives as metaphors for both 'Cartesian' epistemology and 'Cartesian space'.[17] I cannot hope to offer any adequate summary here of the nuanced work of these authors, or similar approaches from beyond the boundaries of garden history, a task which has already been admirably performed by others.[18] But leaving Weiss's elliptical meditations aside, and turning to Pérez-Gómez's narrative, we find perspective 'corresponding [...] to Cartesianism and implying the imposition of a geometrical scheme on reality in order to establish a relation between *res cogitans* and *res extensa*'.[19] He tells us that the perspectival figuration of seventeen-century gardens, 'implicitly demonstrated the belief in the transcendent nature of the new geometrical knowledge'.[20] In short, the perspectivally ordered spaces of a seventeenth-century garden with their grid-like, optically corrected spatial structuring, are interpreted as representing a Cartesian epistemology in which the disembodied eye of the subject is imagined to be positioned at the centre of a mathematically defined spatial extensivity — an emblem of the radical separation of the subjective viewer from the world.[21]

The Cartesian nature of garden perspective is presented by Pérez-Gómez (and Weiss) as a broad cultural metaphor, rather than a specific and individual response to Descartes's thought. This is a problematic proposition for some, for whom the idea that a French formal garden, any garden in fact, might appropriately be presented as a register of novel philosophical ideas coeval with the garden itself represents an uncomfortable anachronism.[22] In Evelyn, however, such an interpretative direction has its merits, for Evelyn had 'philosophical' aspirations for his garden. His philosophy of nature occupies much of the first book of the *Elysium*, and he repeatedly cajoles his imagined noble readership into taking an interest in philosophical 'Speculations and Subjects [...] so copious and withall so usefull', whilst entertaining them with accounts of hortulan 'experiments', not

Gardens of Illusion, 17–45, 136–47; Pérez-Gómez, *Architecture*, 174–75; William Howard Adams, *The French Garden, 1500–1800* (London: Scolar Press, 1979), pp. 63–73; Thierry Mariage, *The World of André Le Nôtre* (Philadelphia: University of Pennsylvania Press, 1999), pp. 42–43; George Farhat, 'Le Nôtre and the Quarrel of the Ancients and Moderns', in *Andre Le Notre in Perspective*, ed. by Georges Farhat and Patricia Bouchenot-Dechin (New Haven: Yale University Press, 2014).

17 Weiss, *Mirrors of Infinity*, 33–34; Pérez-Gómez, *Architecture*, 174–75.

18 On Weiss's Cartesian approach, see Brigitte Weltman-Aron, *On Other Grounds: Landscape Gardening and Nationalism in Eighteenth-Century England and France* (Albany: State University of New York Press, 2001), pp. 117–23; Lyle Massey, *Picturing Space, Displacing Bodies: Anamorphosis in Early Modern Theories of Perspective* (University Park: Pennsylvania State University Press, 2007), pp. 23–35; the history of the 'Cartesian' metaphor and its philosophical foundations in the work of Martin Heidegger and Henri Bergson, together with a review of recent literature: James Elkins, *The Poetics of Perspective* (Ithaca and London: Cornell University Press, 1994), gives both a detailed presentation of the history of the Cartesian metaphor and a detailed critique of Panofsky, see pp. 1–44, 190–205.

19 Pérez-Gómez, *Architecture*, 174–75.

20 Pérez-Gómez, *Architecture*, 174–75.

21 Pérez-Gómez's point ultimately is to critique this spatial reductivism in an attempt to reinstate the full phenomenological richness of the world to our modern understanding of space. See Pérez-Gómez, *Architecture*, 174–81.

22 Michel Conan, 'The New Horizons of Baroque Garen Cultures' in *Baroque Garden Cultures: Emulation, Sublimation, Subversion*, ed. by Michel Conan (Washington, D.C.: Dumbarton Oaks Research Library and Collection, 2005), pp. 1–36.

least those in perspective.[23] Furthermore, reference to Evelyn's commonplace books from the 1650s show him to have been well acquainted with Descartes's work on optics and on metaphysics. So, are his perspectivally constructed garden spaces also 'Cartesian' in some sense and what exactly does he have to say about perspective in the *Elysium*?

Perspective in the *Elysium Britannicum*

Evelyn's discussion of perspective is dispersed over several chapters of the *Elysium*, and this is perhaps why the topic has, as yet, received such scant consideration.[24] As it concerns layout and view, historiographical commentary on the *Elysium* has typically centred on the priority which Evelyn gives to the natural potentials of site — the variety of terrain and its affective potentials.[25] These are important considerations, but we would be mistaken in dismissing the equally prominent issue of the garden's subsequent geometric ordering, which in Evelyn's descriptions proves to be a perspectival concern. I order the discussion here around three distinct though related types of view: the overall layout of the garden seen as a static 'prospect' from a specific vantage point; the composition of individual scenes viewed in dynamic sequence as the visitor wanders through the garden; and finally, the dramatic visual spectacle provided by wonder-working illusionistic perspectives, or trompe l'œil.

Prospect

In his discussion of the garden spectacle, Evelyn distinguishes between 'prospects' and 'perspectives'. By 'prospect' he means a grand overview, while 'a perspective' is a view clearly delineated by parallel lines, seen diminishing as they recede into the distance. Evelyn also uses 'a perspective' to denote a trompe l'œil painting.[26] A prospect may include elements of perspective, but it signifies something grander — the perspective emerges within it. When Evelyn uses the word 'perspective' his intent is, I believe, to direct the reader's attention towards the discourse of artificial perspective and geometric optics (which he sometimes refers to as 'natural perspective') or the 'mysterious' skill by which the perspective is constructed.[27]

He introduces the topic of perspective in chapter III, 'of Fencing, Enclosing, plotting and disposing the ground', where he announces that the layout of the Elysium should:

---

23 Evelyn, *Elysium Britannicum*, ed. by Ingram, 42. See also Chapter II, 'Of a Gardiner, and how he is to be qualified', pp. 33–35; for 'experiments', see for example, pp. 297–306.

24 Evelyn, *Elysium Britannicum*, ed. by Ingram, especially Chapters III, V, VI, VII, pp. 100–01, 123–25, 126–28,132–86.

25 Peter H. Goodchild, '"No Phantasticall Utopia, but a Real Place": John Evelyn, John Beale and Blackbury Hill, Herefordshire', *Garden History*, 19 (1991), 105–27; Douglas Chambers, '"Wild Pastoral Encounter": John Evelyn, John Beale and the Renegotiation of Pastoral in the Mid-Seventeenth Century', in *Culture and Cultivation in Early Modern England: Writing and the Land*, ed. by Michael Leslie and Timothy Raylor (Leicester: Leicester University Press, 1992), pp. 173–94; John Dixon Hunt, 'Evelyn's Idea of a Garden: A Theory for All Seasons', in *John Evelyn's 'Elysium Britannicum'*, ed. by O'Malley and Wolschke-Bulmahn, 269–87.

26 'Prospect': Evelyn, *Elysium Britannicum*, ed. by Ingram, 97, 131; 'Perspective': Evelyn, *Elysium Britannicum*, ed. by Ingram, 100, 127.

27 'Natural Perspective': Evelyn, *Elysium Britannicum*, ed. by Ingram, 100.

Fig. 4 Adam Perelle, views of the Château de Richelieu, showing the extended central axis: Grand Parterre de le Demi-Lune (above) general view (below), *Venues des belles maisons de France*, (c. 1650) from the Metropolitan Museum of Art, Bequest of Phyllis Massar, 2011. Copyright: CC0 1.0 Universal; Public Domain Dedication.

> be contrived so as a prospect being had of the whole from the first stage of the Mansion, there may result a sweete & agreable correspondency in the parts, though considered by themselves, they ~~are~~ {seeme} altogether irregular & heterogene: Such a plot has a perfect resemblance to the Universe it selfe, of which contemplative men & such as best skill how to enjoy the virtuous delights of Gardens are never sated ~~withall~~ , but find always something of new and extraordinary to enhance their thoughts withall.[28]

This paradigmatic view, which elsewhere he establishes as one which extends along the primary axis of symmetry coincident with the main walk, is intended as a spectacle that reveals the harmony of the microcosmic garden (Figure 4).[29] In this opening passage there is no mention of perspective as such, but this soon appears, for the harmony Evelyn seeks relies on a skilful manipulation of geometric figure, an essentially perspectival enterprise. He details how the whole garden should be longer than it is wide, in order to accommodate views, along:

> the principall walkes and Allys; which do […] with their length alone afford a pleasant and most gracious perspective, whilst they concurr to decline to a poynt, especially if planted with tall trees, then which nothing can be more ravishing and agreeable.[30]

28 Evelyn, *Elysium Britannicum*, ed. by Ingram, 99.
29 Evelyn, *Elysium Britannicum*, ed. by Ingram, 216.
30 Evelyn, *Elysium Britannicum*, ed. by Ingram, 100.

Fig. 5 A. Herrocks, Frontispiece to Nicholas de Bonnefons, *The French Gardiner*, trs. by John Evelyn (1658). © British Library Board. [See colour plate 18]

He also writes that such a primary walk should 'stretch itselfe {a perte di vieue} quite through the whole enclosure'.[31]

Visual variety is also at a premium, for, as Evelyn says, the pleasure of 'the Eye is not altogether in the Lontanace {distance}, & losse of its object; but in the diversity of the notices of it'.[32] Variety is embedded in the paradigmatic axial prospect of the garden as the principal walks lead into a network of subsidiary pathways, connecting individual plots, 'cast into almost all the Species of Squares, circles, crooked & oblique lines, […] round, Elliptic, polygone'.[33] Again perspectival techniques come into play, for Evelyn advises that in choosing a figure for an individual plot:

> incase you affect the quadrangular you give it some what more of longitude than of latitude; because of an effect of natural perspective will else too much contract and foreshorten it at the front, & from ~~any~~ {all the} more superiour views & avenues'.[34]

Such elongations can be seen in both the Tuileries and, more strikingly, in the Jardin du Luxembourg (Figure 2, 3).

For Evelyn the primary topic of the grand prospect is the harmony and variety of the world. Perspective is presented as a mediating structure used to correct apparent distortion. Similar concerns inform the peripatetic experience that comes into play as the spectator leaves the mansion and sets off into the interior spaces of the garden (Figure 5).

Sequential View

Evelyn would have his Elysium composed according to the principle of a sequential affective variety, an idea he expresses in the opening passage of his chapter 'Of Rocks, Grots, Crypta's […]', where he writes:

> Nor is there certainely anything more agreable then after the eye has bin entertained with the pleasure & refreshments of Verdures, the fragrant Flowers, the christall Fountaines and other delicious and sense-ravishing objects, to be unexpectedly surprised with the horror and confusion of naturall or artificiall Rocks, Grotts, Caverns, Mounts Precipices well reppresented.[35]

The tour of the garden takes the form of a sequence of well-structured contrasts, which, by analogy with a piece of music, accommodates passages of discord within the dynamic and affecting harmony of the whole. Within this structure, perspective plays a part, both in maintaining the harmony of the garden and in producing diversionary experiences.

One privileged site where perspective works to powerful effect is the tree-lined walk. Evelyn dwells at some length on the techniques needed to produce an avenue of suitably

31 Evelyn, *Elysium Britannicum*, ed. by Ingram, 126.

32 Evelyn, *Elysium Britannicum*, ed. by Ingram, 126, 195.

33 Evelyn, *Elysium Britannicum*, ed. by Ingram, 100.

34 Evelyn, *Elysium Britannicum*, ed. by Ingram, 100. On variety, see Hunt, *Garden and Grove*, 83–89; Hunt, 'Evelyn's Idea of a Garden', pp. 269–87 (p. 283).

35 Evelyn, *Elysium Britannicum*, ed. by Ingram, 187.

affecting magnificence, here relying on the writings of Jacques Boyceau and Claude Mollet.[36] Firstly, proportions must be varied according to whether a walk is 'open' at the top or 'closed', 'for Cover'd Walkes do make the Allee seeme broader to the Eye than the open and free'.[37] In all cases:

> length and breadth [should] hold an agreable correspondency; so as they neither seeme too narrow, broad, flat, ~~or~~ round or uneven, which are all of them vices {sedulously} to be prevented by the Artist.[38]

Evelyn details this principle further in relation to the layout of the 'more principal walks', here using the language of geometric perspective:

> Let the proportions of {long} Walkes be so ordered, ~~as~~ that what in Perspective we name the poynt principall, may as one walkes continue without straightening the area; for there is nothing more August then to behold a Walke so contracted after one is a little engaged {& that hides the gaping} ~~then to see it gape~~ at both the extreames.[39]

He clearly changed his mind on the question of 'gaping', but whichever way the question is settled, it is clear that Evelyn regards the necessary proportioning as a perspectival enterprise.[40] How exactly perspective would have been engaged in the design and planting of an avenue such as this is, however, by no means clear, for there would have been no need for the gardening team to have a technical grasp of either optics or perspective in order to achieve the proposed ends, though some accurate surveying techniques would certainly have been entailed. Nonetheless, Evelyn clearly wants his reader to identify with the field of optical erudition that informs his descriptions, and to think of the avenues as a site of perspectivally defined 'experiment'.

Moving from 'August' avenues to 'gracefull' parterres, Evelyn considers perspectival issues in the design of even the most placid and lowly elements of the garden.[41] He advises, for example, on the ideal contour of *Embossements* — profiled mounds, or piles of soil intended to give visual definition to a parterre pattern.[42] These, he says:

> are made with a gracefull swelling and *Relievo* [...] But they should by no meanes be layed too high, halfe a foote is sufficient in the very ridge, unlesse the Bordure, or Circle is for Cammomile, or Carpets of turfe; but for Flowers a lesser declivity [...]

36 Cf. Evelyn, *Elysium Britannicum*, ed. by Ingram, 127–28; Mollet, *Théâtre des plans et jardinages [...]*, 113–14; Boyceau, *Traité du jardinage*, Chapter 4: 'Des allées et longs promenoirs' (p. 72). A relevant passage from p. 72 of Boyceau is translated in Frank Hamilton Hazlehurst, *Jacques Boyceau and the French Formal Garden*, 26–47 (p. 36). [A toise = approx. 6ft].

37 Evelyn, *Elysium Britannicum*, ed. by Ingram, 126–27.

38 Evelyn, *Elysium Britannicum*, ed. by Ingram, 126.

39 'Straightening' means narrowing. Evelyn, *Elysium Britannicum*, ed. by Ingram, 127.

40 Evelyn, *Elysium Britannicum*, ed. by Ingram, 127.

41 Evelyn, *Elysium Britannicum*, ed. by Ingram, 124, 127.

42 Laird, 'Parterre, Grove, and Flower Garden...' in *John Evelyn's 'Elysium Britannicum'*, ed. by O'Malley and Wolschke-Bulmahn, 171–221 (p. 180).

> seing the designe is onely to bring such flowers in sight at once, as in *plano*, could not appeare to one that were walking at any reasonable distance.[43]

The spectator must be well above the parterre in order to enjoy the pattern, or presumably very near the planting in order to enjoy an individual flower. Walking at any 'reasonable distance' a parterre becomes background. For Evelyn, the peripatetic perspective emerges at quite specific moments when moving around the garden, only to dissolve again as the spectator moves away from the ideal viewpoint.

The expectation of discovering such fleeting perspectives would have been supported by contemporary interest in extreme forms of perspective distortion, or anamorphosis, with which Evelyn was certainly familiar, for he owned a substantial collection of perspective treatises. Alongside older works, he had two books by his friend Abraham Bosse, the result of a collaboration with the prodigiously talented mathematician Girard Desargues. He also owned copies of François Niceron's *Thaumaturgus opticus*, and Gaspar Schott's *Magia universalis natura et artis*, a book that brought together the treatment of curious acoustic musical phenomena with 'magical' optical techniques including Niceron-like 'Curious Perspectives'.[44] Evelyn could not have been ignorant of anamorphic wonders such as the picture of St John which Niceron illustrates in his book (Figure 6). Here the image of the saint is designed to appear only when the painting is viewed from the extremely oblique viewpoint of the doorway of the room. As the spectator moves to a more 'normal' viewing position in the centre of the room, the image disappears. I do not suggest that Evelyn proposed that a princely garden be designed around the sort of sophisticated and exacting perspectival deceptions characteristic of the mature 'anamorphic' gardens of Le Nôtre.[45] The similarity between Evelyn's use of perspective in the garden and Niceron's painting is rather in the peripatetic instability of the perspectival moment. Evelyn's 'August' perspective appears as the spectator moves down an avenue towards the centre and disappears as he or she approaches the other end.[46] Perspective is not a constant way of seeing, but rather a transient wonder that appears within privileged geometrically structured territories within the garden — a point most explicitly illustrated by the trompe l'œil.[47]

---

43 Evelyn, *Elysium Britannicum*, ed. by Ingram, 124.

44 Giacomo called Il Vignola Barozzi, *[...] Le due regole della prospettiva pratica [...]* (Roma: Nella Stamparia Camerale, 1611); Sebastiano Serlio, *The First Booke of Architecture, [...]*, trans. by Robert Sir Peake (London: Simon Stafford and Thomas Snodham, 1611); Gaspar Schott, *Magia universalis naturæ et artis ... opus quadripartitum. pars. I. continet optica. ii. acoustica. iii. mathematica. iv. physica [...] cum figuris, etc* (Herbipoli, 1657). See Christie, Manson and Woods Ltd, Sale catalogue of 'The Evelyn Library', 4 vols. (1977), Items 1521, 1349 and 1316. The British Library holds Evelyn's copies of Abraham Bosse and Girard Desargues, *Maniere vniuerselle de mr. Desargues, pour pratiquer la perspectiue par petit-pied, [...]* (Paris: P. Des-Hayes,1648); Abraham Bosse, *Moyen uniuersel de pratiquer la perspectiue sur les tableaux ou surfaces irregulieres. [...]* (Paris: Chez Bosse, 1653); Jean François Niceron, *Thaumaturgus opticus, seu admiranda optices [...]* (Lutetiæ Parisiorum: typis & formis Francisci Langlois, 1646). On Bosse and Evelyn, see Sheila McTighe, 'Abraham Bosse and the Language of Artisans: Genre and Perspective in the Academie royale de peinture et de sculpture, 1648–1670', *Oxford Art Journal*, 21 (1998), 1–26.

45 Weiss, *Unnatural Horizons*, 47; Weiss, *Mirrors of Infinity*, 33–46; George Farhat, 'Great Vistas in the Work of Le Nôtre' in *Andre Le Notre in Perspective*, ed. by Farhat and Bouchenot-Dechin, pp. 172–87.

46 Evelyn, *Elysium Britannicum*, ed. by Ingram, 127.

47 On the moving body in perspective see Massey, *Picturing Space*.

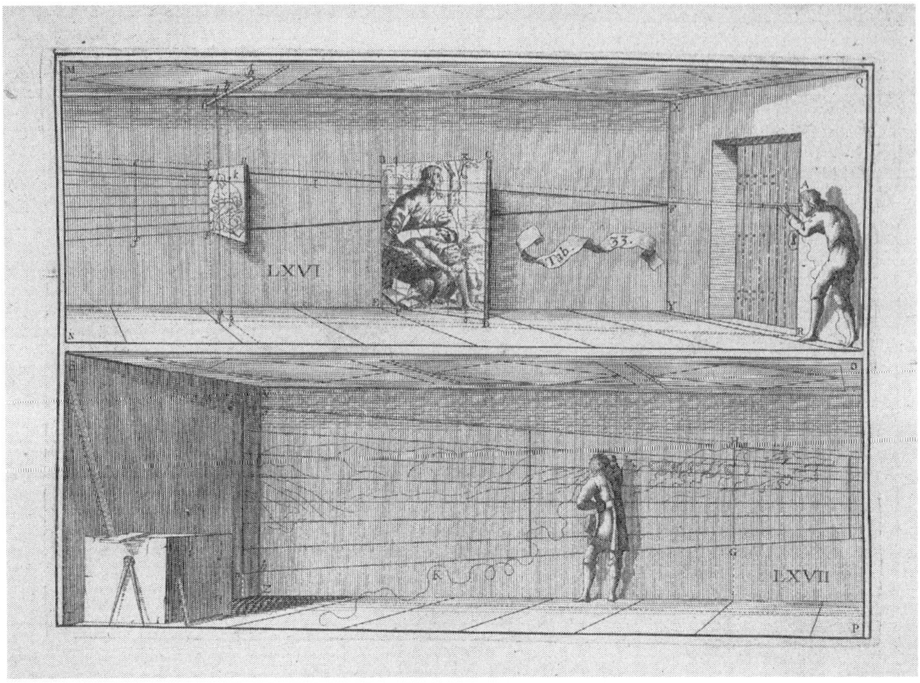

Fig. 6 Jean François Niceron, Anamorphic image projection explained in *La perspective curieuse [...] : Avec L'optique et la catoptrique du R. P. Mersenne* (Paris: Jean Du Puis, 1663), Plate 32. © British Library Board.

Trompe l'Œil

Evelyn describes trompe l'œil as 'Pleasant deceptions' designed to 'tempt & divertise the eye of the Spectator'.[48] He gives his *Elysium* readers some basic instructions on how to achieve a simple illusionistic painting on a garden wall, illustrating his discussion with three diagrams (Figure 7). Here, perspective is no longer achieved by trial and error, framed by a rhetoric of mathematical technique, it is rather presented through the description of geometric construction. The mathematical marvels of trompe l'œil bring to the fore the drama implicit in the ephemeral emergence of perspective in other parts of the garden — first deceiving and then undeceiving, they frame the drama of seeing sight itself. But whilst the trompe l'œil provide a stage for the dramatic enactment of seeing, Evelyn also points out that these perspective illusions may be used to maintain the visual harmony of the garden, in the face of a defect of site. He recommends that a trompe l'œil be placed where a walk is brought up short against a wall, writing: 'Perspective do can do wonders, & is able to give the Eye a {Lycean} passage {even} through a stone wall'.[49]

48 Evelyn, *Elysium Britannicum*, ed. by Ingram, 215. For Parisian context of trompe l'œil and Evelyn, see Luke Morgan, 'The Early Modern Trompe L'oeil Garden', *Garden History*, 33 (2005), 286–93.
49 Evelyn, *Elysium Britannicum*, ed. by Ingram, 215.

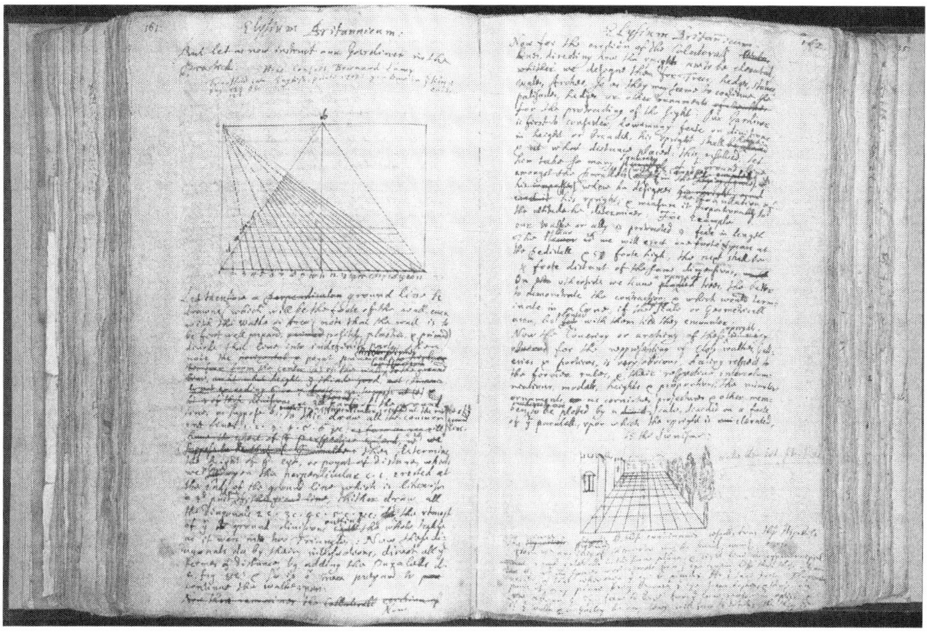

Fig. 7 John Evelyn, Diagrams for the perspective construction of trompe l'oeil wall paintings from the *Elysium Britannicum*. © The British Library Board, Add 78342, fols. 161-162.

To enliven his discussion, Evelyn gives several examples of trompe l'œil, including a painted '*Triumphal Arch of Constantine* in the *Cardinals Villa* at *Ruell*', later adding a construction that he saw in the gardens of Dr John Wilkins, at Wadham College in Oxford in 1654.[50] This piece appears to have been a hortulan version of Borromini's Palazzo Spada sequence with added mirrors:

> It was a close cradle walke short but much protracted by Art, here the ground rises & the sides or poling were contracted, at the extreame ~~places~~ was plac'd a looking glass set so declining as to take in the skie, having a landskip painted under it, which made a wonderfull effect.[51]

He also includes an unnamed 'stately instance', composed thus: a cloudy sky is painted on a high wall, which is screened by another wall of equal height standing in front of it.[52] The front wall is painted to resemble antique ruins and is pierced with windows and arches. The painted clouds and sky can be glimpsed through the openings and 'as the Spectators walked or changed their steps, reppresent the motions, or rack of the Clowdes, seeming to flye before the wind'.[53] Evelyn follows his description with an excerpt from his own translation of Lucretius's *De rerum natura*:

50 Evelyn, *Elysium Britannicum*, ed. by Ingram, 216.
51 Evelyn, *Elysium Britannicum*, ed. by Ingram, 218.
52 Evelyn, *Elysium Britannicum*, ed. by Ingram, 216.
53 Evelyn, *Elysium Britannicum*, ed. by Ingram, 216.

'[…] Ships with transport us move,| when fix't they seeme,| And those at Anker,| under Saile we deeme,| And ~~to~~ {towards} the Barke, Hills, & Fields seeme to Flie, Where as, in truth, we Saile & passe them bye'.[54]

Consideration of these trompe l'œil demonstrates that Evelyn's understanding of 'perspective' has porous boundaries, which admit a range of other optical and mathematical concerns beyond the strict confines of geometrically constructed illusionistic perspective. His 'perspectives' incorporate illusions dependent on parallax, on catoptrics, as well as on reflections and on the principle of relative motion.

In summary, the primary place of geometry in Evelyn's spatial economy, as far as it concerns visual perception, is to render proportions harmonious to the spectator — to represent divine order to flawed human perception. The secondary place is to open up occasional manipulable territories in which particular local mathematical marvels of sight can be staged. Perspective, then, appears to be not so much an all pervasive schematization of space as a locally applicable wonder-working art. With this in mind, we return to the question of the supposed 'Cartesian' character of 'French' perspectival gardens and the possibility of seeing Evelyn's Franco-British perspectives in these terms.

The Limits of Cartesianism

Evelyn was certainly acquainted with Descartes's work from as early as 1651 when he acquired a copy of *Discours de la méthode … plus la dioptrique*, as a gift from his father-in-law.[55] Furthermore, Descartes appears repeatedly in the extensive reading notes that Evelyn made in his vast commonplace book from the 1650s, the 'Tomus tertius', whilst his posthumously published *History of Religion* makes slight, though revealing, comment on Evelyn's attitude to 'the Cartesians' and to Descartes himself.[56] From these sources it is clear that whilst Evelyn gave some credit to Descartes's 'acute' observations on the 'mechanics' of physical phenomena — the 'screw, spring, counter-poise, or like mechanism' of the human body for example — he probably followed the Epicurean school of Pierre Gassendi and Walter Charleton more closely than Descartes in formulating his thoughts on optics.[57] The bulk of Evelyn's optical notes in the 'Tomus tertius' are taken from Walter Charleton's *Physiologia Epicuro-Gassendo-Charltoniana*, a defense of Pierre Gassendi's Epicurean physics (a version of mechanical atomism), and this seems to have been the proximate source for Evelyn's consideration of Descartes's 'supposed' optical principles, which appear

54 Evelyn, *Elysium Britannicum*, ed. by Ingram, 216.

55 Rene Descartes, *Discours de la méthode pour bien conduire sa raison et chercher la vérité dans les sciences, plus la dioptrique, les météores et la géométrie qui sont des essais de cette méthode* (Leyde: Jan Maire, 1637). Evelyn's copy of this work contains the inscription: 'Ex dono R. Bruni Equitis et Baronetti Parisiis. […]', i.e. Sir Richard Borwne, Evelyn's father-in-law, see Christie, Manson and Woods Ltd, Sale catalogue of 'The Evelyn Library', Item 456.

56 British Library, London, Evelyn Papers, Additional MS 78330, John Evelyn, 'Tomus tertius', fols 126^{r}–141^{v}; John Evelyn, *The History of Religion*, ed. by R. M. Evanson, 2 vols (London: Henry Colburn,1850), I, pp. 70, 283, this work is a transcription of the manuscript London, British Library, Additional MS 78367.

57 Evelyn, *The History of Religion*, 70.

in Charleton's book in digested form.[58] Descartes is not the source for Evelyn's record of a theoretical understanding of light in terms of material 'rays' or 'effluvias' — streams of particles which he shows being refracted in the course of their passage through lenses of various shapes. Similarly, although the diagram of the section through the human eye which Evelyn copied into this 'Tomus tertius' ultimately derives from Descartes's *Dioptriques* (Figure 8), the page numbers that Evelyn jotted down in his notes reveal that he copied the image from Charleton's book.[59] Evelyn was certainly an active reader, for he makes his own insertions into the flow of Charletonian thought — for example, he composed a small diagram showing diminishing visual pyramids to illustrate how 'as the object is farr or neare so dos the pyramid appear straighten'd or widen'd', thus delivering a small or large retinal image,[60] and concludes that the faintness of a distant object can be accounted for by considering the number of light particles entering the eye. In a distant object the visual pyramid is more 'straightened' (i.e. narrowed), consequently fewer 'rays' — that is to say fewer particles — enter the eye and the image is consequently dimmer.[61] Many of these principles are reflected in the optical effects that Evelyn writes about in his *Elysium* — progressive foreshortening; parallax, and so on — and, consequently, we might be justified in seeing the perspectivally structured spaces of Evelyn's imagined garden as a laboratory for testing optical principles, but there is nothing necessarily 'Cartesian' in this.[62]

Judging from the limited material available in his manuscripts and other writings, Evelyn was not 'Cartesian' in either his understanding of nature or of space. It is not that he could not follow Descartes in imagining an 'indefinite' spatial extensivity, geometrically represented, but rather that he disagreed with Descartes — or as he has it 'the Cartesians' — on the fundamental issue of the specific qualities, or lack of qualities, that they imputed to this spatial extension.[63] The evidence for this is found in *The History of Religion*, where Evelyn writes:

> The Cartesians tell us that there is no such thing as substantial life any where; and that even human volition is mechanically produced from certain effluvia and exurious membranes, as it were. They will not endure any scale or degree of entities lest they should find a link or chain which should bring them to a First Being.[64]

Though this passage is undatable, it is consistent with what we read in the *Elysium*, where Evelyn explains 'Nature' in idiosyncratic mechanico-Spiritual terms, fusing the 'well-restor'd doctrine of Epicurus' (that is to say Epicurean atomism) with 'Spiritual' ideas derived from Renaissance Hermetic Neoplatonism.[65] He equates 'Nature herselfe' with the 'Universal Spirit', an 'anima', or 'energie', which rotates from stars to earth and back

58 Walter Charleton and Epicurus, *Physiologia Epicuro-Gassendo-Charltoniana, or, a Fabrick of Science Natural, Upon the Hypothesis of Atoms Founded by Epicurus Repaired [by] Petrus Gassendus* … (London: Thomas Heath, 1654), pp. 137–207; British Library, London, Evelyn Papers, Additional MS 78330, fols 132v–133r.

59 Evelyn, 'Tomus tertius', fol. 132r.

60 British Library, London, Evelyn Papers, Additional MS 78330, fols 132v–133r.

61 British Library, London, Evelyn Papers, Additional MS 78330, fol. 132v.

62 Evelyn, *Elysium Britannicum*, ed. by Ingram, 100, 216.

63 Repetzki, 'John Evelyn', pp. XVI–XVII.

64 Evelyn, *The History of Religion*, ed. by Evanson, I, p. 283.

65 Evelyn, *Elysium Britannicum*, ed. by Ingram, 37.

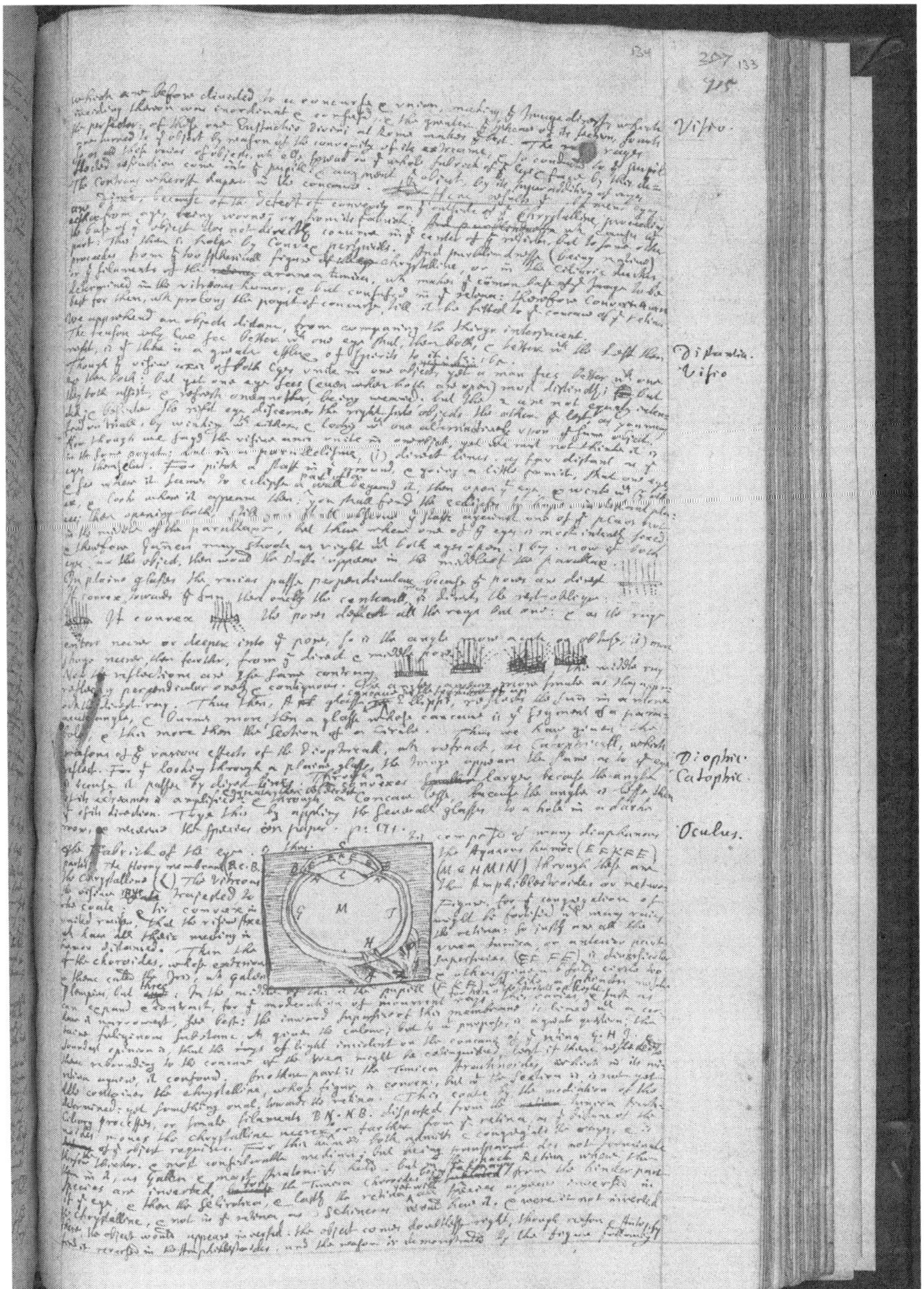

Fig. 8 John Evelyn's notes from Walter Charleton, *Physiologia Epicuro-Gassendo-Charltoniana*, in '*Tomus Tertius*' commonplace book. © The British Library Board, Add 78330, fol. 133.

again, vivifying the world as it interpenetrates and fructifies the material 'matrices' of the earth. He articulates this vision of the universal economy with recourse to the Hermetic dictum, '*nihil est inferius, quod non fit superius*, and *è contra*' [sic].[66] It is the banishment of 'substantial life' — a pervasive spiritual vitality that he saw as the fundamental stratum of created nature — and the levelling of space into the non-hierarchical homogeneity of material extension that Evelyn really objected to in Cartesianism. For him it posed a threat to religion. Consequently, the spatial grounding of Evelyn's *Elysium*, despite his very obvious interest in optics (Cartesian and otherwise), is not ultimately provided by the geometric spatial articulation of the garden. We have seen Evelyn casting perspective as a mathematically underpinned artistic technique which might be applied to the ordering of garden space, but it would appear that, for him, the geometric articulation of space does not touch the fundamental mysteries of vitality, of soul, or of the Universal Spirit — 'a powerfull emanation from the ~~first~~ {primarie} Cause, seene by few, but felt by every body, and flowing through all the workes of the creation.'[67] If we were to look for a metaphysical dimension to Evelyn's understanding of perspective, Descartes is not a promising direction. We would do better to look towards the Hermetic tradition, but Evelyn is silent on this topic and I will not consider it further here. We note, however, that it is Evelyn's Hermetic understanding of Nature as Spirit which provides the theoretical substrate for the astrologically nuanced gardening routine that he recommends – a routine intended to nurture plant life by capturing beneficial influences cast by the celestial bodies 'above' on the earth 'below', whilst avoiding the harmful effects of baleful constellations.[68] Given this interest in the order of the heavens, it is not surprising to find that Evelyn's reading notes show him to have taken a keen interest in the associated optical arts of astronomy and dialling, topics which inform our interpretation of Sayes Court garden.[69]

Sayes Court

When Evelyn took possession of Sayes Court, he decided, to his later regret, that he would not demolish and rebuild the house, but rather alter and extend, consequently eradicating almost every trace of any pre-existing garden in his new vastly expanded scheme. Referring to Evelyn's drawing of *c.* 1652, the house and garden occupied a plot of estuarine land, bounded on the east side by a stream that ran awkwardly close to the existing Tudor manor house (Figure 1 — south is at the top of the plan). Beyond the stream, to the north, lay the Thames naval dockyard, separating Sayes Court from the river front. By deciding to keep the house, Evelyn accepted at the outset the impossibility of establishing a garden

66 The account is dispersed across Evelyn, *Elysium Britannicum*, ed. by Ingram, Chapters III–XII, for a summary, see Odgers, 'Water in Use and Philosophy at Wotton House'. The Hermetic dictum may be translated as: 'that which is below, is as that which is above and the opposite', it appears in Evelyn, *Elysium Britannicum*, ed. by Ingram, 42.

67 Evelyn, *The History of Religion*, ed. by Evanson, I, p. 70; Evelyn, *Elysium Britannicum*, ed. by Ingram, 38.

68 For Evelyn's astrological gardening, see Juliet Odgers, 'Scales of Reference: John Evelyn and Caruso St John Architects', *Made at WSA* (2013), pp. 88–97; Evelyn, *Elysium Britannicum*, ed. by Ingram, 42.

69 British Library, London, Evelyn Papers, Additional MS 78330, CAP.II: MATHEMATICAE DISCIPLINAE. Arithmetica, Geometria, Geographia {ASTRONOMIA'} Cosmographia, Hydrographia, Optica, Musica, fols 88–95^{v}.

ordered along a primary central walk, aligned with the principal door of the house. Sayes Court's plan undoubtedly displays axiality, but it is a fractured and dispersed axiality — nowhere do we find the overarching perspectivally nuanced, harmonious prospect that Evelyn describes in the *Elysium*.[70]

Looking more closely at the layout of the garden, the axial approach to the house from the south is aligned with the entrance, but the axis terminates at the front door. In order to enter the garden, the visitor must pass through a door in the northwest corner of the entry court, which leads into the pathway bordering the northern edge of Dyall garden. Or, more familiar visitors might enter the garden through the house, by way of the parlour and Evelyn's private flower garden on the west front. This enclosure is overlooked by Evelyn's chemical laboratory pavilion, to the south of which a door opens onto a short flight of steps, which lead up to a raised walk, running between the Dyall garden to the south and the formal grove to the north. The main organising axis of the garden as a whole runs parallel to the entry pathway. It starts at the small banqueting house and runs northwards, to terminate in an island surrounded by a stream. The central walk is flanked by the parterre and grove on the right, and, on the left, the orchard. If we posit the banqueting house, rather than the mansion itself, as the organizing centre of this composition, the plan falls more into line with Evelyn's *Elysium* ideal, but there are still anomalies, the most puzzling of which is the orientation of the Dyall garden.

Evelyn's *Elysium* prescriptions for the layout of a garden dictate that plots should be laid out to align lengthwise with the main optical axis of the composition, thus helping to elongate the view down the primary central walk.[71] Supposing that the relatively modest elevation of the little banqueting house serves as the central viewing point for the primary prospect, the Dyall garden should be aligned north-south, parallel to the main axial walk, thus not only 'prolonging' the view of the walk, but also compensating for any foreshortening of the parterre pattern, when seen from this important angle.[72] But it is aligned east-west instead. We cannot attribute this seeming misalignment to site conditions, for there were no significant limitations on the size and shape of the Dyall garden since, when Evelyn began setting out this plot, the ground consisted of 'one intire fild of 100 Ackers, without any hedge: excepting the hither holly-hedge joyning to the bank of the mount walk'.[73] He could have shifted the location of the banqueting house southwards to allow his Dyall garden to be laid out along the long axis of the garden, but he chose otherwise. Why is this?

Perhaps, given the unfortunate placing of the house with respect to the boundaries of the land, Evelyn despaired of achieving a harmoniously balanced and symmetrical layout for the garden as a whole, and concentrated his perspectival artistry instead on the local and sequential experience of the garden. In this case, the elongation of the Dyall garden might be judged solely in terms of the immersive experience of the peripatetic spectator, rather than in terms of its place within an overall prospect. Thus, we may imagine the

70 This point is made in passing by Laird, 'Parterre, Grove, and Flower Garden …' in *John Evelyn's 'Elysium Britannicum'*, ed. by O'Malley, 171–221 (p. 217).

71 Evelyn, *Elysium Britannicum*, ed. by Ingram, 100.

72 Evelyn, *Elysium Britannicum*, ed. by Ingram, 124.

73 Evelyn, *The Diary of John Evelyn*, 17 January 1653. Quoted in Leith-Ross, 'The Garden of John Evelyn at Deptford', p. 138.

raised walk that stands between the parterre and the grove initially providing a fine vantage point from which to view the parterre pattern and indeed the grove, at least in the early years after planting. As the thickets that surround the central oval of his Dyall garden grew to maturity, however, (they were six feet high by 1658), this evergreen growth would have obscured the view of the parterre entirely.[74] The only enduring view point for the parterre pattern must have been the raised mount at the centre of the oval enclosure. From here the spectator could have walked around the sun dial that stood at the centre of the composition, and taken a giddy view of the radial beds and ornamental patterns below. Whilst it is clear that the staging of these experiences entailed perspectivally nuanced judgements, it is not clear that any of the considerations framed above gives a positive reason from the east-west alignment of the Dyall garden. This, I suggest, can be found by unfolding the central conceit of the garden: that it is a dial of sorts.

The Dyall Conceit

In the *Elysium*, Evelyn describes the parterre as a privileged representational domain, saying that here the gardener 'may be able to compose Impressees, Mottos, Dialls, Escutcheons, Cyphers and innumerable other devices with wonderfull felicity & effect', all arranged 'according to the fantsy, & judgement of the Gardiner Artist!'[75] Such a 'fantsy' should be 'neither too faint, nor too open'.[76] The Dyall garden is not a literal operational dial such as those Evelyn claims to have 'frequently seene planted in the Parters & traile works', but an elaboration of the primary metaphor of garden as microcosm unfolded through a calendrical theme.[77] This 'fantsy' informs the figural geometries and emblematic numbers that Evelyn embeds in his design, and it informs its orientation (Figure 9). The Dyall garden, like its context, is aligned exactly with the cardinal directions, establishing a basic microcosmic frame within which the numbers of the design unfold more specific meanings. Thus, the four quarters of the garden emblematize the four quarters of the world, while holding a temporal significance as a register of the four seasons. Similarly, the central bed, subdivided into twelve compartments of ornamental grotesque work, registers the twelve months; whilst the boss-like shrubs that occupy the band around the perimeter of the circle indicate the twenty-four hours of the day. Meanwhile the band around the edge of the oval is planted with thirty shrubs, which I propose be seen as a register the lunar month. This last detail is supported by another drawing found among a collection of Evelyn's manuscript notes held in archive in the British Library (Figure 10).[78] This sketch

74 Leith-Ross, 'The Garden of John Evelyn at Deptford', pp. 138–52 (p. 145).

75 Evelyn, *Elysium Britannicum*, ed. by Ingram, 123.

76 Evelyn, *Elysium Britannicum*, ed. by Ingram, 214.

77 Evelyn, *Elysium Britannicum*, ed. by Ingram, 213.

78 British Library, London, Evelyn Papers, Additional MS 15950, fol. 173: John Evelyn, 'MISCELLANEOUS Notes… including Notes and collections for Evelyn's intended work, entitled "Elysium Britannicum"'. See also a lunar parterre sketch, described as: 'Sketch plan of parterre, attributed to John Evelyn by William Upcott but of uncertain draftsmanship'at British Architectural Library/RIBA, London, Drawings Collection, reproduced in Laird, 'Parterre, Grove, and Flower Garden…' in *John Evelyn's 'Elysium Britannicum'*, ed. by O'Malley (2001), 171–221 (p. 172).

Fig. 9 John Evelyn, Detail of Sayes Court Plan, showing the Dyall Garden, *c.* 1652. © The British Library Board, Add 78628, fol. A.

shows a variant of the Sayes Court design in which the space between circle and oval is filled with thirty crescent moons. The ideas are to an extent commonplace, but the reading becomes more interesting when we consider Evelyn's astrological gardening routine and his concern with the minutiae of the fluctuating influences of the sun and the moon.

In the *Elysium* chapter 'Of Celestiall Influences, particularly, the Sun, and Moon; and of the Climates', Evelyn characterises the sun as 'a visible fire which we may call the Soule of our Gardens', of all 'Celestial inhabitants the most vigorous and active instrument'.[79] The sun is 'the life of the World; the gemme of heaven [...] the measure of Tyme'; 'the Celestial Genitor'; 'husband' of the earth.[80] The moon, characterised as feminine, is equally important, being 'of all the rest neerest the Earth; so hath she a very greate influence on the

79 Evelyn, *Elysium Britannicum*, ed. by Ingram, 55.
80 Evelyn, *Elysium Britannicum*, ed. by Ingram, 55–56.

Fig. 10 John Evelyn, Sketch of an alternative design for Sayes Court parterre. © The British Library Board, Add 15950, fol. 173.

Labours of our Gardiner, during the entire course of her periodic moneth.'[81] Her influence waxes and wanes with proximity and distance from the earth, 'in relation to Excentricitys

81 Evelyn, *Elysium Britannicum*, ed. by Ingram, 55.

and Epicycles in their [conjunction] and [opposition].'[82] Consequently the gardener should know the moon's course and consult her phases when he is deciding 'when to take up, cutt, Graffe, Transplant or Sow; for Seedes committed to the earth at the end or beginning of the Moone, produce lusty and goodly plants, those in the full Low & shrubby'.[83] Evelyn practiced what he preached and observed the phases of the moon when planting out the gardens and orchards of Sayes Court. The diary entry for March 19, 1653 reads: 'I planted the Ortchard at Says-Court, New Moone, wind West'.[84] A letter from his wife to a friend written in 1663/4, tells wryly 'of a New Almanack now under presse fortelling the disasters of plants if not sett just in such a face and minute of the Moone'.[85]

It appears that Evelyn's preoccupation with the minutiae of lunar planting increased over time, for he later starts adding lunar data in the margins of his *Elysium* manuscript, saying:

> the lunary yeare consists of 12 Synodical moneths or 354 days, some odd hours & Scruples, eleven dayes lesse than the Solar; & it is not rectified till the cycle of 19 yeares is Effluxed.[86]

and in another note:

> The illumination of the Moone begens from its first apparition, but the measure of it is uncertaine, because sometimes she appears from the 4th day after coition; sometimes from the 3d, yea & sometimes from the very first.[87]

Such material is also reflected in his somewhat earlier reading notes, particularly his study of Hevelius's *Selenographia*, of 1653, from which he recorded copious material in his commonplace book. He copied many of Hevelius's detailed observations, together with several of his diagrams, including two that show the aspects of the moon (Figure 11).[88] Evelyn was not alone in his concern for an accurate understanding of the cycles of the heavens, for his young wife, Mary, had a certain expertise in dialling. The manuscript collection of notes which contains the crescent moons parterre sketch, mentioned above, also contains a set of horological 'mathematical exercises' executed by Mary Evelyn in 1650 (after their marriage in 1647, but before she left Paris to join her husband in Deptford).[89] These exercises appear in the collection almost immediately after the parterre sketch, a proximity that suggests that the young couple's discussion of their garden included this more exacting mathematical register of the calendrical rhythms.[90] If the emblematic numbers embedded in the pattern of the Dyall garden articulate the temporal measure of the heavens at a general level, it is in the sundial, mounted at the centre of the garden,

82 Evelyn, *Elysium Britannicum*, ed. by Ingram, 57.
83 Evelyn, *Elysium Britannicum*, ed. by Ingram, 57.
84 Evelyn, *The Diary of John Evelyn*, III, p. 81.
85 Leith-Ross, 'The Garden of John Evelyn at Deptford', 147: Mary Evelyn Letter, Book 4, in which she writes in 1663/64 about her husband's book *Sylva*.
86 Evelyn, *Elysium Britannicum*, ed. by Ingram, p. 57 n. 4.
87 Evelyn, *Elysium Britannicum*, ed. by Ingram, p. 57 n. 3.
88 British Library, London, Evelyn Papers, Additional MS 78330, fols 89^{r}–91^{v}; Joannes Hevelius, *Selenographia: sive, lunæ descriptio. Addita est, lentes expoliendi nova ratio; [...]* (Gedani: Hünefeld, 1647).
89 British Library, London, Evelyn Papers, Additional MS 15950, fols 178–88.
90 British Library, London, Evelyn Papers, Additional MS 15950, fols 173–74; Laird, 'Parterre, Grove, and Flower Garden...' in *John Evelyn's 'Elysium Britannicum'*, ed. by O'Malley and Wolschke-Bulmahn, 171–221 (p. 185).

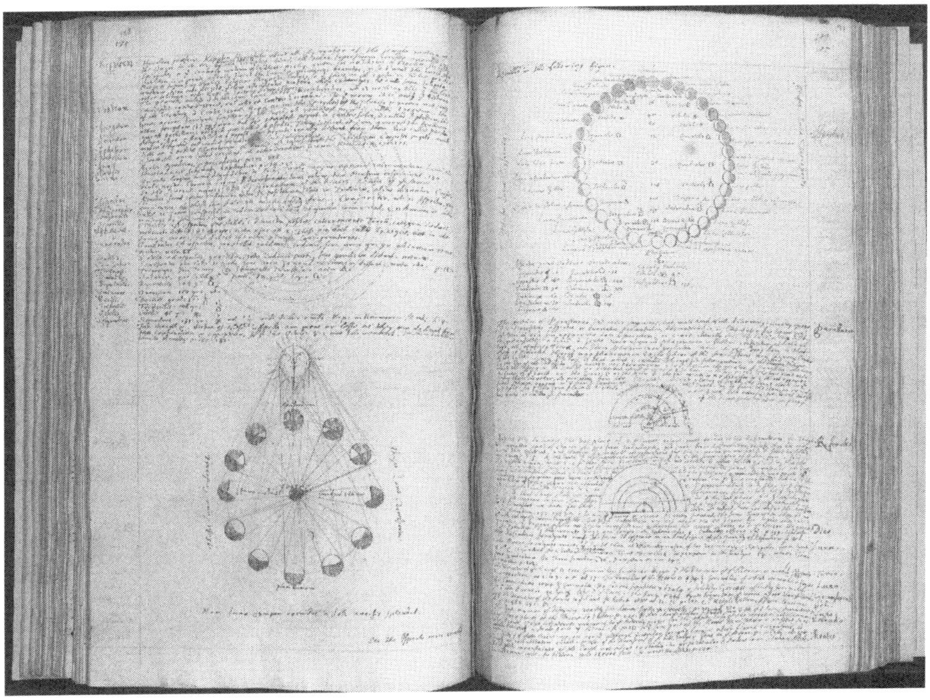

Fig. 11 John Evelyn, diagrams of lunar phases, after Hevelius *Selenography* (1653), drawn in the '*Tomus Tertius*' commonplace book. © The British Library Board, Add 78330, fols. 90^{v}, 91.

that we find the true register of the shifting geometries of the skies, played out in shadows across its face.

The topic of shadows returns us to the question of the orientation of Evelyn's Dyall garden. Without wishing to suggest that the complexities of any design decision can be deterministically pinned to one specific cause, the play of shadows does finally suggest a convincing positive reason for the east-west elongation of the enclosure. If the pattern of the parterre represents the stability of spatio-temporal rhythm, its orientation accommodates the play of shadows across the bright floral pattern of its floor, as they lengthen and shorten with the Sun's movement across the sky. At the junctions between the main axial paths of the parterre and the circular path at the base of the mount, Evelyn has posted eight slim cypresses. These would have cast their shadows, gnomon-like across the ground, long in the morning and evening, short at midday, just as Evelyn has drawn a long morning shadow against each of the trees in his orchard.

In 1647, whilst Evelyn was still resident in Paris, his friend, Abraham Bosse, produced a book on dialling, *La maniere vniuerselle [...] pour poser l'essieu & placer les heures & autres choses aux cadrans au soleil*, in collaboration with his usual partner Girard Desargues.[91]

91 Abraham Bosse and Girard Desargues, *La maniere vniuerselle de mr. Desargues [...] pour poser l'essieu & placer les heures & autres choses aux cadrans au soleil* (Paris: P. Des-hayes, 1643).

This appeared in English translation at the end of the 1650s, with a preface composed by the surveyor mathematician Jonas Moore, who writes:

> Dyalling I accompt one kind of Perspective, for that glorious body the Sun, the eye of the world, traceth out the lines and hour-points by his Diurnal Course, and upon subjected Plane by the laws of Picture, Scenographically deliniates the Dyal.[92]

Moore's text provides a pertinent example, form one in Evelyn's milieu, which places dialling in strict continuity with a larger field of perspectivally conceived geometrical projective techniques, as was typical for the time. Had we isolated the artificially constructed 'perspectives' of Evelyn's garden design, from the broader discourse of optics and horology, we may have overlooked crucial interpretative clues. Only by reintegrating these other 'perspectives' into our consideration of his intentions, do we arrive at our interpretation of the Dyall garden at Sayes Court. This little garden is a theatre, in which the principal drama in play is the perspectival trace of shadows cast across the fabric of the garden by the sun and moon in their daily course. The dial conceit underpins the emblematics of the garden, it explains its primary cosmic orientation, and supports its perspectival elongation. It establishes that in Evelyn's garden, 'perspective', understood as the spectator's view, stands in dynamic balance with the more compelling dramatic perspective of the celestial lights, both intentions contributing to the primary artistic conceit of representing the harmony of the cosmos through the geometrically ordered spaces of the garden.

In the field of perspective studies, if we ignore the culturally specific connections between perspective and associated operational fields such as dialling, surveying, or indeed astronomy, and place too much emphasis on the strictly bounded field of constructional artificial perspective itself, we run the risk of misinterpreting the artistic intentions of past practitioners.

Bibliography

Manuscript and Archival Sources

British Architectural Library/RIBA, London, Drawings Collection, 'Sketch plan of parterre, attributed to John Evelyn by William Upcott but of uncertain draftsmanship'.

British Library, London, Evelyn Papers, Additional MS 15950, fols 178–88.

British Library, London, Evelyn Papers, Additional MS 15950, fols 173–74.

British Library, London, Evelyn Papers, Additional MS 78330, Cap.II: Mathematicae Disciplinae, Arithmetica, Geometria, Geographia {Astronomia'} Cosmographia, Hydrographia, Optica, Musica, fols 88–95^{v}.

British Library, London, Evelyn Papers, Additional MS 78330, fols 132^{v}–133^{r}.

British Library, London, Evelyn Papers, Additional MS 78330, fols 89^{r}–91^{v}; Joannes Hevelius, *Selenographia: sive, lunæ descriptio. Addita est, lentes expoliendi nova ratio; [...]* (Gedani: Hünefeld, 1647).

92 Jonas Moore, 'To Lovers of Ingenious Practices', in Gérard Desargues, *Mr De Sargues Universal Way of Dyaling, [...]*, ed. by Abraham Bosse, trans. by Daniel King (London: Tho. Leach, 1659), unpaginated Preface.

British Library, London, Evelyn Papers, Additional MS 78330, John Evelyn, 'Tomus tertius', fols 126^r–141^v.
British Library, London, Evelyn Papers, Additional MS 78342.
British Library, London, Evelyn Papers, Additional MS 78367.
Christie, Manson and Woods Ltd, Sale catalogue of 'The Evelyn Library', 4 vols. (1977), Items 1521, 1349 and 1316.

Primary Sources

Bosse, Abraham and Girard Desargues, *La maniere vniuerselle de mr. Desargues [...] pour poser l'essieu & placer les heures & autres choses aux cadrans au soleil* (Paris: P. Des-hayes, 1643).
Bosse, Abraham and Girard Desargues, *Maniere vniuerselle de mr. Desargues, pour pratiquer la perspectiue par petit-pied, [...]* (Paris: P. Des-Hayes, 1648).
Bosse, Abraham, *Moyen uniuersel de pratiquer la perspectiue sur les tableaux ou surfaces irregulieres. [...]* (Paris: Chez Bosse, 1653).
Boyceau, Jacques, *Traité du jardinage, selon les raisons de la nature et de l'art [...]* (Paris: M. Vanlochom, 1638).
Chambray, Roland Fréart, sieur de, *A Parallel of the Antient Architecture with the Modern; [...] With L. B. Alberti's Treatise of Statues*, trans. by John Evelyn (London: John Place, 1664).
Descartes, René, *Discours de la méthode pour bien conduire sa raison et chercher la vérité dans les sciences, plus la dioptrique, les météores et la géométrie qui sont des essais de cette méthode* (Leyde: Jan Maire, 1637).
Evelyn, John, *Sculptura: Or the History, and Art of Chalcography and Engraving in Copper [...]* (London: J. C. for G. Beedle, T. Collins and J. Crook, 1662).
Evelyn, John, *Sylva, or a Discourse of Forest-Trees, and the Propagation of Timber in His Majesties Dominions [...]* (London: Jo. Martyn and Ja. Allestry, 1664; further editions 1670 and 1679).
Evelyn, John, *The History of Religion*, ed. by R. M. Evanson, 2 vols (London: Henry Colburn, 1850).
Evelyn, John, *The Diary of John Evelyn. Now First Printed in full from the Manuscripts [...]*, ed. by E. S. de Beer (Oxford: Clarendon Press, 1955).
Evelyn, John, *Elysium Britannicum, or the Royal Gardens*, ed. by John E. Ingram (Philadelphia: University of Pennsylvania Press, 2001).
Giacomo called Il Vignola Barozzi, *[...] Le due regole della prospettiva pratica [...]* (Roma: Nella Stamparia Camerale, 1611).
Mollet, Claude, *Théâtre des plans et jardinages [...]* (Paris: Charles de Sercy, 1652).
Moore, Jonas, 'To Lovers of Ingenious Practices', in Gérard Desargues, *Mr De Sargues Universal Way of Dyaling, [...]*, ed. by Abraham Bosse, trans. by Daniel King (London: Tho. Leach, 1659).
Niceron, Jean François, *Thaumaturgus opticus, seu admiranda optices [...]* (Lutetiæ Parisiorum: typis & formis Francisci Langlois, 1646).
Schott, Gaspar, *Magia universalis naturæ et artis ... opus quadripartitum. pars. I. continet optica. ii. acoustica. iii. mathematica. iv. physica [...] cum figuris, etc* (Herbipoli, 1657).
Serlio, Sebastiano, *The First Booke of Architecture, [...]*, trans. by Robert Sir Peake (London: Simon Stafford and Thomas Snodham, 1611).

Secondary Works

Adams, William Howard, *The French Garden, 1500–1800* (London: Scolar Press, 1979).

Bennett, Jim, 'Projection and the Ubiquitous Virtue of Geometry in the Renaissance', in *Making Space for Science: Territorial Themes in the Shaping of Knowledge*, ed. by Jon Agar, Crosbie Smith, and Gerald Schmidt (Basingstoke: Macmillan, 1998), pp. 27–38.

Chambers, Douglas, '"Wild Pastoral Encounter": John Evelyn, John Beale and the Renegotiation of Pastoral in the Mid-Seventeenth Century', in *Culture and Cultivation in Early Modern England: Writing and the Land*, ed. by Michael Leslie and Timothy Raylor (Leicester: Leicester University Press, 1992), pp. 173–94.

Conan, Michel, 'The New Horizons of Baroque Garen Cultures' in *Baroque Garden Cultures: Emulation, Sublimation, Subversion*, ed. by Michel Conan (Washington, D.C.: Dumbarton Oaks Research Library and Collection, 2005), pp. 1–36.

Darley, Gillian, *John Evelyn: Living for Ingenuity* (New Haven: Yale University Press, 2006).

Elkins, James, *The Poetics of Perspective* (Ithaca and London: Cornell University Press, 1994).

Farhat, George, 'Le Nôtre and the Quarrel of the Ancients and Moderns', in *Andre Le Notre in Perspective*, ed. by Georges Farhat and Patricia Bouchenot-Dechin (New Haven: Yale University Press, 2014).

Goodchild, Peter H., '"No Phantasticall Utopia, but a Real Place": John Evelyn, John Beale and Blackbury Hill, Herefordshire', *Garden History*, 19 (1991), 105–27.

Harris, Frances, 'The Manuscripts of the "Elysium Britannicum"', in John Evelyn, *Elysium Britannicum, or the Royal Gardens*, ed. by John E. Ingram (Philadelphia: University of Pennsylvania Press, 2001), pp. 13–21.

Hazlehurst, Frank Hamilton, *Jacques Boyceau and the French Formal Garden* (Athens: University of Georgia Press, 1966).

Hazlehurst, Franklin Hamilton, *Gardens of Illusion: The Genius of Andre Le Nostre* (Nashville: Vanderbilt University Press, 1980).

Hunt, John Dixon, *Garden and Grove: The Italian Renaissance Garden and the English Imagination 1600–1750* (London: Dent, 1986).

Hunt, John Dixon, 'Evelyn's Idea of a Garden: A Theory for All Seasons', in *John Evelyn's 'Elysium Britannicum' and European Gardening*, ed. by Therese O'Malley and Joachim Wolschke-Bulmahn (Washington, DC: Dumbarton Oaks Research Library and Collection, 1998), pp. 269–87.

Hunter, Michael, 'John Evelyn in the 1650s: A Virtuoso in Quest of a Role', in *John Evelyn's 'Elysium Britannicum' and European Gardening*, ed. by Therese O'Malley and Joachim Wolschke-Bulmahn (Washington, DC: Dumbarton Oaks Research Library and Collection, 1998), pp. 79–106.

Jeffery, Sally, 'The Way of Italian Gardens', in *A Celebration of John Evelyn: Proceedings of a Conference to Mark the Tercentenary of his Death*, ed. by Mavis Batey (Surrey: The Surrey Gardens Trust, 2007).

Keynes, Geoffrey, *John Evelyn: A Study in Bibliophily with a Bibliography of his Writings* (Oxford: Clarendon Press, 1968).

Laird, Mark, 'Parterre, Grove, and Flower Garden: European Horticulture and Planting Design in John Evelyn's Time', in *John Evelyn's 'Elysium Britannicum' and European Gardening*, ed. by

Therese O'Malley and Joachim Wolschke-Bulmahn (Washington, DC: Dumbarton Oaks Research Library and Collection, 1998), pp. 171–221.

Leith-Ross, Prudence, 'A Seventeenth-Century Paris Garden', *Garden History*, 21 (1993), 50–57.

Leith-Ross, Prudence, 'The Garden of John Evelyn at Deptford', *Garden History*, 25 (1997), 138–52.

Mariage, Thierry, *The World of André Le Nôtre* (Philadelphia: University of Pennsylvania Press, 1999).

Massey, Lyle, *Picturing Space, Displacing Bodies: Anamorphosis in Early Modern Theories of Perspective* (University Park: Pennsylvania State University Press, 2007).

McTighe, Sheila, 'Abraham Bosse and the Language of Artisans: Genre and Perspective in the Academie royale de peinture et de sculpture, 1648–1670', *Oxford Art Journal*, 21 (1998), 1–26.

O'Malley, Therese, 'Introduction', in *John Evelyn's 'Elysium Britannicum' and European Gardening*, ed. by Therese O'Malley and Joachim Wolschke-Bulmahn (Washington, DC: Dumbarton Oaks Research Library and Collection, 1998), pp. 9–34.

Odgers, Juliet, 'Water in Use and Philosophy at Wotton House: John Evelyn and the History of the Trades', *arq: Architectural Research Quarterly*, 15 (2011), 237–47.

Odgers, Juliet, 'Scales of Reference: John Evelyn and Caruso St John Architects', *Made at WSA* (2013), pp. 88–97.

Odgers, Juliet, 'Resemblance and Figure in Garden and Laboratory: Gaffarel's Influence on John Evelyn', in *Jacques Gaffarel: Between Magic and Science*, ed. by Hiro Hirai (Rome: Serra, 2014), pp. 85–109.

Pérez-Gómez, Alberto, *Architecture and the Crisis of Modern Science* (Cambridge, MA: MIT Press, 1983).

Repetzki, Michael M., 'John Evelyn: Virtuoso and the Venture of Atomism', in *John Evelyn's Translation of Titus Lucretius Carus 'De rerum natura'* (Frankfurt: Peter Lang, 2000).

Taylor, F. Sherwood, 'The Chemical Studies of John Evelyn', *Annals of Science*, 8 (1952), 285–292.

Weiss, Allen S., *Mirrors of Infinity: The French Formal Garden and 17th-Century Metaphysics* (New York: Princeton Architectural Press, 1995).

Weiss, Allen S., *Unnatural Horizons: Paradox and Contradiction in Landscape Architecture* (New York: Princeton Architectural Press, 1998).

Weltman-Aron, Brigitte, *On Other Grounds: Landscape Gardening and Nationalism in Eighteenth-Century England and France* (Albany: State University of New York Press, 2001).

Woodbridge, Kenneth, *Princely Gardens: The Origins and Development of the French Formal Style* (New York: Rizzoli, 1986).

II

Writing on Perspective

ELAHEH KHEIRANDISH

Optics and Perspective in and beyond the Islamic Middle Ages

A Study of Transmission through Multidisciplinary Sources in Arabic and Persian

Introduction

The theory and practice of perspective as developed in Western Europe of the Renaissance period represents a rich and puzzling history that may be well captured through the various forms of transmission, and indeed non-transmission, of the major works and concepts behind its development. This is also the case with the study of optics, itself a major core of what is now known as 'linear perspective'.[1] To use an image from the study of optics and perspective themselves, the 'window' of transmission, revealing the circulation of ideas, has much illuminated the complex history of early optics.[2] But the same can hardly be said in the case of even the most linear forms of the much broader 'landscape' of pre-perspective developments. Most striking in both subjects is the case of Ibn al-Haytham's (d. after *c.* 1040–41) monumental *Book of Optics*, known in Arabic as *Kitāb al-Manāẓir*,

1 On the relationship between optics and perspective, see (in chronological order): Kim H. Veltman, *Optics and Perspective: A Study in the Problems of Size and Distance* (unpublished doctoral thesis, University of London, 1975), pp. 7, 15, 17; David C. Lindberg, *Theories of Vision from Al-Kindi to Kepler* (Chicago: University of Chicago Press, 1976), pp. 147–54; C. D. Brownson, 'Euclid's Optics and its Compatibility with Linear Perspective', *Archive for History of Exact Sciences*, 24 (3) (1981), 165–94 (pp. 166–85); Kim H. Veltman, in collaboration with Kenneth D. Keele, *Linear Perspective and the Visual Dimensions of Science and Art: Studies on Leonardo da Vinci I* (Munich: Deutscher Kunstverlag, 1986), pp. 48–50; Kirsti Andersen, 'Ancient Roots of Linear Perspective', in *From Ancient Omens to Statistical Mechanics: Essays on the Exact Sciences Presented to Asger Aaboe*, ed. by J. L. Berggren and B. R. Goldstein (Copenhagen: University Library, 1987), pp. 75–89 (78–79); Wilbur R. Knorr, 'On the Principle of Linear Perspective in Euclid's Optics', *Centaurus*, 34 (1991), 193–210 (pp. 193–95); and more recently, Dominique Raynaud, 'Why Did Geometrical Optics Not Lead to Perspective in Medieval Islam?' in *Raymond Boudon: A Life in Sociology*, ed. by M. Cherkaoui and P. Hamilton (Oxford: The Bardwell Press, 2009), pp. 243–66; and Dominique Raynaud, *Optics and the Rise of Perspective: A Study in Network Knowledge Diffusion* (Oxford: The Bardwell Press, 2014), Chapter 2.

2 See Elaheh Kheirandish, *The Arabic Version of Euclid's 'Optics': Kitāb Uqlīdis fī Ikhtilāf al-Manāẓir, Edited and Translated with Historical Introduction and Commentary*, 2 vols, Sources in the History of Mathematics and Physical Sciences series, ed. by G. J. Toomer (New York: Springer-Verlag, 1999), I, Preface, pp. VII–VIII; Elaheh Kheirandish, 'The Mixed Mathematical Sciences: Optics and Mechanics in the Islamic Middle Ages', in *The Cambridge History of Science*, ed. by David C. Lindberg and Michael H. Shank (Cambridge: Cambridge University Press, 2013), II, pp. 84–108 (pp. 91–103).

Elaheh Kheirandish Harvard University, kheirand@fas.harvard.edu

Perspective as Practice. Renaissance Cultures of Optics, ed. by Sven Dupré, Turnhout, 2019 (*Techne. Knowledge, Technique, and Material Culture*, 1), p. 205-239

DOI 10.1484/M.*Techne*-EB.5.117727

an increasingly important work variously referred to in the present chapter and volume for the foundational role it played in developments in optics and perspective.[3] This work was transmitted as the Latin *Perspectiva* and *De aspectibus* in the twelfth and thirteenth centuries,[4] as the Italian *Prospettiva*, in the fourteenth century,[5] and as the printed *Opticae thesaurus* in the sixteenth century.[6] The so-called 'checkered history' of Ibn al-Haytham's celebrated *Book of Optics* in terms of the 'undeserved misfortune' and 'unexpected good luck' of its respective transmissions in Islamic and European lands has long been thoroughly examined.[7] The overlapping 'fortunes' and 'misfortunes' of the transmission of the work's seven books have also seen further extensions in my recent publications and projects as applied to the earlier optical works of Euclid (*c.* 300 BCE) and Ptolemy (in the second century), and the microscopic traces of the 'footpaths' and 'footprints' of all such major works in their respective lands.[8]

Viewed primarily through the lens of transmission, questions on the development of linear perspective in early modern Europe, and explanations of the non-development of comparable forms in Islamic lands carry similar promise; much more so than those posed with reference to religion, culture, and ethnicity, given such limited forms of argumentation

3 Ibn al-Haytham, *Kitāb al-Manāẓir* ('Book of Optics'), *Books I–III: On Direct Vision*, ed. by Abdelhamid I. Sabra (Kuwait: The National Council for Culture, Arts and Letters, 1983); Ibn al-Haytham, *Kitāb al-Manāẓir* ('Book of Optics'), *Books IV–V: On Reflection, and Images Seen by Reflection*, ed. by Abdelhamid I. Sabra, 2 vols (Kuwait: The National Council for Culture, Arts and Letters, 2002); A. I. Sabra, *The Optics of Ibn al-Haytham: Books I–III on Direct Vision, Translated with Introduction and Commentary*, 2 vols, Studies of the Warburg Institute series, ed. by J. B. Trapp (London: Warburg Institute, 1989), II, pp. LXXIII–LXXIX.

4 Alhacen, *Alhacen's Theory of Visual Perception: A Critical Edition, with English Translation and Commentary, of the First Three Books of Alhacen's 'De aspectibus', the Medieval Latin Version of Ibn al-Haytham's 'Kitāb al-Manāẓir'*, ed. and trans. by A. Mark Smith, 2 vols (Philadelphia: Transactions of the American Philosophical Society, 2001); Alhacen, *Alhacen on the Principles of Reflection: A Critical Edition, with English Translation and Commentary, of Books 4 and 5 of Alhacen's 'De aspectibus'*, ed. and trans. by A. Mark Smith, 2 vols (Philadelphia: Transactions of the American Philosophical Society, 2006); Alhacen, *Alhacen on Image-Formation and Distortion in Mirrors: A Critical Edition, with English Translation and Commentary, of Book 6 of Alhacen's 'De aspectibus'*, ed. and trans. by A. Mark Smith, 2 vols (Philadelphia: Transactions of the American Philosophical Society, 2008); Alhacen, *Alhacen on Refraction: A Critical Edition, with English Translation and Commentary, of Book 7 of Alhacen's 'De aspectibus'*, ed. and trans. by A. Mark Smith, 2 vols (Philadelphia: Transactions of the American Philosophical Society, 2010).

5 Sabra, *The Optics of Ibn al-Haytham*, II, p. LXXV; A. Mark Smith, 'The Latin Source of an Italian Translation of Alhacen's "De aspectibus" (Vat. Lat. 4595)', *Arabic Sciences and Philosophy*, 11 (2001), 27–43.

6 Alhacen, *Opticae thesaurus: Alhazeni Arabis libri septem, nuncprimum editi; Eiusdem liber de crepusculis et nubium ascensionibus, item vitellonis thuringopoloni libri X*, ed. by Friedrich Risner (Basel: Bischoff, 1572; repr. New York: Johnson, 1972).

7 On 'chequered history', see Sabra, *The Optics of Ibn al-Haytham*, II, Preface, p. XI; Elaheh Kheirandish, 'Light and Dark: The "Checkered History" of Early Optics', in *God is the Light of the Heavens and the Earth: Light in Islamic Art and Culture*, ed. by Jonathan Bloom and Sheila Blair (New Haven: Yale University Press, 2015), pp. 61–85; Elaheh Kheirandish, 'Eloge A. I. Sabra (8 June 1924–18 December 2013)', *Early Science and Medicine*, 19 (3) (2014), 281–86 (pp. 282–83).

8 Elaheh Kheirandish, 'Light and Dark: The "Checkered History" of Early Optics', pp. 78–79, Plates 54–55; Elaheh Kheirandish, '"Checkered History" Recolored: The Changing Fortunes and Misfortunes of Optical Works in Islamic and European Lands': in progress, presented in the session 'The Legacy of A. I. Sabra: New Perspectives on the History of Science in Islam', Middle East Studies Association (MESA), Boston: 19 November 2016. Some features of these projects were developed as part of my contributions to the working group behind this volume through workshops in Berlin and Paris in 2012 and 2013.

and substantiation.[9] Most problematic are generalisations with 'essentialist' overtones using blanket expressions like 'religious prohibitions', 'differing mindsets' and 'iconoclastic cultures' with reference to Muslim theology, Arab ethnicity and Islamic culture respectively.[10] By contrast, the present study recalls the uniqueness of historical events and developments, and the caution called for in posing questions like why something as unique as what developed in early modern Europe as Linear Perspective in its narrow mathematical sense, did not occur before or elsewhere, in this case, Arab and Persian lands of the Islamic Middle Ages. More historically-sensitive questions, including those addressing in what forms or areas related subjects did or did not develop or overlap, offer useful frameworks. This is the case anywhere from early colour and depth manipulations, to later 'frontal and oblique projections' and other 'nonrealist' representations in Islamic sources as distinct from European linear perspective.[11] Cases devoted to examinations of transmission in the present chapter, especially those extended to conceptual and linguistic elements and shifts, provide particularly revealing directions through a multidisciplinary set of Arabic and Persian sources mostly from the Islamic Middle Ages. In these cases, alternative forms or variations of pre-perspective concepts and developments, being readily present through the verifiable forms of textual, rather than oral or visual, transmission, may be drawn directly from such primary sources, a selection of which are included here (see Appendix).

9 On some related discussions with reference to the Islamic religion, culture, and mindset, see respectively: Thomas W. Arnold, *Painting in Islam: A Study of the Place of Pictorial Art in Muslim Culture* (Oxford: Clarendon Press, 1928; repr. New York: Dover, 1965); Naṣr Allāh Pūrjavādī, 'The Concept of Perspective in "Kalīla va Dimna" and the Reasons for the Absence of Three-Dimensional Space in Islamic Paintings', *Nashr-i Dānish*, 8 (1988), 5, 18–30; Hans Belting, *Florenz und Bagdad: eine westöstliche Geschichte des Blicks* (München: C. H. Beck, 2008), translated to English as Hans Belting, *Florence and Baghdad: Renaissance Art and Arab Science*, trans. by Deborah Lucas Schneider (Cambridge, Mass.: Belknap Press of Harvard University Press, 2011). For a long list of other relevant sources and approaches, see also Raynaud, 'Why Did Geometrical Optics Not Lead to Perspective in Medieval Islam?'; Raynaud, *Optics and the Rise of Perspective*, Chapter 2.

10 Some examples of essentialist overtones, with reference to Muslim theology, Arab ethnicity and Islamic culture, and challenges to their related arguments are as follows: the expression 'prohibitions of religion' is used in the preface of Arnold's *Painting in Islam*. The more specific expression 'aniconism' is defined by Raynaud, as the 'prohibition of image-making in the Islamic tradition' in his 'Why Did Geometrical Optics Not Lead to Perspective in Medieval Islam?', 255, who challenges the argument there and also in his *Optics and the Rise of Perspective*, Chapter 2. For reactions to expressions of 'essentialisation' such as 'Arab visual theory', 'Arab culture' and 'Middle Eastern way of thinking' in Belting's *Florence and Baghdad*, see the review of David J. Roxburgh, 'Two Point Perspective: On Hans Belting's "Florence and Baghdad"', *Artforum* (2012), 61–64 (p. 62). For problems with similar expressions, such as Belting's reference to 'different mindsets' in his *Florence and Baghdad*, 33–35, see Gülru Necipoğlu, 'The Scrutinizing Gaze in the Aesthetics of Islamic Visual Cultures: Sight, Insight and Desire', in *Gazing Otherwise: Modalities of Seeing in and Beyond the Lands of Islam*, ed. by Olga Busch and Avinoam Shalem, Supplement to *Muqarnas: An Annual on the Visual Cultures of the Islamic World* (Leiden: Brill, 2015), XXXII, pp. 23 n. 50, 24–61. The expression 'mental set' occurs in Samuel Y. Edgerton, *The Renaissance Rediscovery of Linear Perspective* (New York: Basic Books, 1975), p. 32, where its opening chapter, 'The Western Window', reads: 'The other great cultures of the world — and here we may include the Chinese, the Persian, Indian, Arab, and Byzantine civilizations — [...] seem to have [not] *been interested* in this geometric-optical way of picture making' [italics mine]. The expression, 'iconoclastic cultures' is in Veltman, *Linear Perspective* with reference to an ornamental Quranic pattern (Plate 59) as 'an example of how words tend to replace pictures'. This is an isolated case very different from the essentalist expressions in Belting's *Florence and Baghdad*, from which any signs of Edgerton's *The Renaissance Rediscovery* or Veltmans extensive studies on both the Islamic and European traditions — including his *Optics and Perspective* of the same year, 1975 — is curiously absent.

11 Raynaud, 'Why Did Geometrical Optics Not Lead to Perspective in Medieval Islam?', pp. 243–66 (pp. 253–56); Raynaud, *Optics and the Rise of Perspective*, 37–57.

The key components of the comparable concepts and variants of perspective practices are presented in what follows under the four separate yet related headings of: 'angle and distance', 'direction and depth', 'outline and colour', and 'mirror and measure'. The focus is on the process and details of transmission through a selection of sources from both scientific and non-scientific literature. The sources selected range in subject, from optics and philosophy to literature and surveying, and in language from Arabic to various Persian sources from the Islamic period. These and others are discussed here with a view of looking directly at developments in optics and perspective through a close study of transmission itself, one that extends beyond the time and place of the multidisciplinary sources presented below, many here being discussed for the first time.

Angle and Distance through Arabic Optical Sources

Discussions of linear perspective as developed in early modern Europe involve at least three major components: the eye, the object, and a representational plane.[12] As such, the field of optics, with its premodern history as a mathematical study of vision, has been often identified as a natural predecessor of linear perspective in its direct involvement of two of those three components: the eye and the object, each with their own involvements of size and distance. The specific case of *The Optics of Euclid*,[13] the earliest known independent work devoted to the subject, has itself ranged from its 'compatibility with linear perspective' to its 'diametrical opposition' to it, the latter as part of the distinctions between 'natural' (angular) versus 'artificial' (linear) perspective, the respective vision versus representation of the visible world.[14] The third component, the representational plane, is another key item, itself having switched between geometrical, physical, and transparent planes over time.[15] The so-called 'picture plane', on the other hand, has been often left out of the discussion of earlier developments as an obvious off spring of a much later period. A close look at the transmission of these and other relevant items and concepts, in terms of both the *routes* of transmission, i.e. what was or was not transmitted, and the *nature* of transmission, i.e. how things were transmitted, offer new perspectives of their own.

This section is devoted to the discussion of angle and distance through the transmission of specific definitions and propositions in Euclid's *Optics* into Arabic to identify components among them that were continuous, as well as discontinuous, with later developments. These

12 Lindberg, *Theories of Vision*, 150; Brownson, 'Euclid's Optics', p. 180; Veltman, *Linear Perspective*, 38.

13 Greek text and facing Latin, see Euclid, *Euclidis opera omnia: Optica, opticorum recensio theonis, catoptrica, cum scholiis antiquis*, ed. by J. L. Heiberg and H. Menge (Leipzig: Teubner, 1895), VII. For an English translation, see H. E. Burton, 'The Optics of Euclid', *Journal of the Optical Society of America*, 35 (1945), 357–72.

14 Brownson, 'Euclid's Optics', 'established by Erwin Panofky and continued by' a few others named by Brownson (p. 165); he also cites the source of the related expressions (p. 182 n. 26), and discusses Erwin Panofsky's position (p. 183). See also Andersen, 'Ancient Roots', pp. 75–76 for the forms '*perspectiva naturalis* (or *communis*) and *perspectiva artificalis*' and Andersen, 'Ancient Roots', p. 81 for 'artificial (linear) and natural perspective (optics)'; and Knorr, 'On the Principle of Linear Perspective', p. 193, all with reference to Panosfky and questions regarding his position.

15 Lindberg, *Theories of Vision*, 150 quotes Alberti's discussion of 'veil' and 'intersection'. For references to a 'plane' cutting the visual field in the *Optics* of Euclid, see Brownson 'Euclid's Optics', pp. 172–73; Andersen, 'Ancient Roots', pp. 78–79; Knorr, 'On the Principle of Linear Perspective', pp. 193–94.

are, respectively, the second definition and eighth propositions from the Greek Euclidean text *Optika*, a text with a problematic Greek tradition significantly transmitted further through a transformed Arabic translation that is literally an Arabic 'version' of Euclidean optics itself.[16] What emerges from these cases, which involve the meanings and changing meanings of the shape of the visual cone through the second Euclidean definition, and the size — distance relation through the eighth proposition (Arabic ninth), is that textual, internal, considerations are often enough to address the most immediate issues.

To start with continuous developments, in the case of the second Euclidean definition, the author's own stated assumptions in Greek (as edited by Heiberg) are that in the case of *opseis* ('visual-rays'), an entity defined in the first Euclidean definition as extending from the eye to a distance of great magnitude with its own transformations, the figure enclosed by what is termed *shuʿāʿ* ('rays') as it becomes in its Arabic translation is a cone the apex of which is at the eye and its base at the extremity of the visible object (Appendix: 1.1.a). Here it is important to pause at the Arabic term for a concept that is as central as the visual cone to both optics and perspective. This is a term that started as *conus* in the ancient Greek original of the second Euclidean definition, as a cone with a circular base; but when translated into Arabic, it became *makhrūṭ*, a term that according to its historical meaning in both Arabic and Persian, has a much more flexible shape for its base. This is clear from the common definition of *makhrūṭ* as a 'cone whose base is a circle or any other figure' in texts as widely known as the bilingual Arabic and Persian *Kitāb al-Tafhīm* ('Book of Instruction') by the polymath Abū Rayḥān Bīrūnī (d. *c.* 1050).[17] In the Euclidean text itself, if a cone's base is 'circular', this is specifically added, indicating that other shapes are also possible. What is most important from the standpoint of the present discussion is that when the same Arabic term, *makhrūṭ*, is translated into medieval Latin, it becomes *puramis*, with all the new conceptions offered by the polygonal — rather than circular — forms of the base of the cone. The least that may be said here is that, with the translation of the second Euclidean definition from Greek into Arabic and Latin, the geometry of vision was transformed in a direction continuous with later developments in Europe; and further, that the transformation of the shape of the visual field from a circular cone to a polygonal pyramid was consistent with conceptions such as a plane cutting into that visual field, a component so critical to the development of linear perspective in Europe. Similarly, the relation of the Arabic 'visual cone' to the accuracy of vision, of its most central ray(s) to the visual field, and of the side view of the visual pyramid to the intersected cone of rays,

16 On the editions of two known Greek versions, as Euclid's *Optika* and *Recensio*, see *Euclidis opera omnia*, ed. by Heiberg and Menge, pp. 1–121, 143–247. On a re-evaluation of the problematic Greek textual tradition, and re-attribution of the two Greek texts, see Wilbur R. Knorr, 'Pseudo-Euclidean Reflections in Ancient Optics: A Re-Examination of Textual Issues Pertaining to the Euclidean "Optica" and "Catoptrica"', *Physis*, 31 (1) (1994), 1–45; Alexander Jones, 'Peripatetic and Euclidean Theories of the Visual Ray', *Physis*, 31 (1) (1994), 47–76. On the transformed translation of the Greek text into Arabic, see Elaheh Kheirandish, 'The Arabic "Version" of Euclidean Optics: Transformations as Linguistic Problems in Transmission', in *Tradition, Transmission, Transformation*, ed. by Jamil F. Ragep and Sally Ragep with Steven Livesey (Leiden: Brill, 1996), pp. 227–43; Kheirandish, *The Arabic Version of Euclid's 'Optics'*.

17 Abū Rayḥān Bīrūnī, *Kitāb al-Tafhīm li-awāʾil ṣināʿat al-tanjīm* ('The Book of Instruction in the Elements of the Art of Astrology'), facs. ed. by R. Ramsay Wright (London: Luzac & Co., 1934; repr., Frankfurt am Main: Institute for the History of Arabic-Islamic Science, 1998; Persian ed. by J. Humāʾī (Tehran: Bābak, 1362 = 1983). The death date of Bīrūnī is a subject of re-examination, more recently from 1048 CE to 1050 CE,

all have various overlaps with Euclidean propositions in ways closely comparable to later developments.

By contrast, the transmission of Euclidean propositions containing other components directly related to linear perspective represent discontinuities with later developments in Europe. A proposition in *Euclid's Optics* long noted as particularly relevant to linear perspective is one involving a line perpendicular to the ground comparable to a projection plane (Greek tenth, Arabic eleventh), a proposition posing questions about transmission like the role of a related Euclidean definition (fifth) missing in all Arabic manuscript copies.[18] But the Euclidean proposition representing more decisive discontinuities with later developments in linear perspective is Euclid's eighth proposition (Arabic ninth), one with a problematic Greco-Arabic transmission that seems to have affected the directions of both optics and perspective. This is a proposition noted with reference to Euclid's own *Optics* for being the only proposition in that text involving a relation between entities such as size of appearance, distance from the observation point, and angle of vision.[19]

But something quite critical to be added here is what occurs in the course of the translation of that Greek proposition into Arabic, the only proposition in the entire Arabic text where the expression *manāẓir* ('aspects') in the work's title appears within the text, here in the specific context of the ratio of two visual angles, as opposed to the more general sense of that term in titles Latinised as *De aspectibus* and *Perspectiva*.[20] The transmission of this proposition, especially in context, seems quite consequential for shifting the intended meaning of its most critical premise: that the appearance of an object (here, its magnitude, relative to another equal and parallel to it, as represented by their respective visual angles), is *not* in the same proportion (i.e. its ratio is not the same) as the respective distances of those magnitudes from the eye. More specifically, in the original Greek, the relative size of objects as determined by visual angles in Euclidean terms is set as not being directly related to the ratio of their distances from the same visual point, namely, the relation is not proportional. But in both the enunciation and conclusion of the proposition's Arabic version, the term used for the Greek *analogus* ('in proportion') is not the standard corresponding Arabic expression *'ala nisba* ('according to the ratio/proportion'), an expression widely used in mathematical texts as early as the ninth century, but rather, in the short form *'ala* ('according to') *aqdār* ('magnitudes') of distances without the part *nisba* ('ratio/proportion') in all its known manuscript copies (Appendix: 1.1.b). The shift to this shorter, non-technical form of such an important expression during the transmission of the text into its Arabic form, whether occurring through translation, omission, or abbreviation,

18 On the related discussions of Euclid's Proposition 10, see Brownson, 'Euclid's Optics', pp. 172–73; Andersen, 'Ancient Roots', pp. 78–79; Knorr, 'On the Principle of Linear Perspective', pp. 193–94. The latter usefully discusses the Greek variants of that proposition. On the related parts in Arabic, see Kheirandish, *The Arabic Version of Euclid's 'Optics'*, Definitions and Propositions 9 and 11 (respectively: I, pp. 2–3, 26–29, 34–37; II, pp. 6–7, 40–44, 49).

19 'The only proposition to comment on the nature of the numerical relation of the size of appearance presented at the observation point, to the distance of observation point from the object and size of the angle of vision': Brownson, 'Euclid's Optics', p. 181.

20 A. I. Sabra, 'Manāẓir, or "Ilm al-manāẓir"', in *Encyclopaedia of Islam, Second Edition, Glossary and Index of Terms*, ed. by P. J. Bearman, Th. Bianquis, C. E. Bosworth, E. van Donzel and W. P. Heinrichs (Leiden: Brill, 2012), 6 (1987), Fascicules 103–04, 376–77; Kheirandish, *The Arabic Version of Euclid's 'Optics'*, Title, II, pp. 1–7.

carries with it a different sense from the original Greek: in context, the Arabic version of the Euclidean proposition that misses the critical reference to 'ratio/proportion' teaches that relative size is 'not according to' the magnitude of distances; and along with it, a less direct conceptual relation between apparent size and distance at large. Such a relation, so central to developments in both optics and perspective, and to linear perspective through what later became known as the 'size-distance' principle, is a relation found, without its inverse component, in the *Optics* of Ptolemy (in the second century), and more fully Ibn al-Haytham's work (in the eleventh century), none of which circulated in Arabic anywhere near in Latin.[21]

Ptolemy's *Optics* was itself nowhere near as widely dispersed in Islamic lands as his *Almagest*, with the exception of Ibn al-Haytham full access to it. In *Optics*, Ptolemy had extended the extramissionist visual theory of appearances through visual angles, to incorporate the role played by distance. Ibn al-Haytham, the well known and influential Alhacen and Alhazen of medieval and early modern Europe, an appellation derived from his first name Al-Ḥasan, expanded on the work still further in his seven-book *Optics*; notably his text had a much greater impact in Europe than in Islamic regions.[22] Ibn al-Haytham introduced distinct conceptions and terms through his widely circulated *Perspectiva/ De aspectibus* in Latin, those between *bu'd/remotio* ('distance') and *masāfa/spatium* ('interval'), between *mu'tadila* ('moderate') and *mutafāwit* ('immoderate') distances, and between their *kammīya/quantitas* ('quantity') and *miqdār/mensura* ('measure'). He also specified the 'condition that the eye is at a certain *bu'd* ('distance') proportionate to the object'; and he further referred to methods such as *itsiqrā'/inductio* ('induction') and *i'tibār/experimentatio/experiential* ('experiment') (Appendix: 1.2.a); his new geometry of vision, where the outward rays of a visual cone extended the extramission mathematics of Euclidean and Ptolemaic models to *khuṭūṭ mutawwahama* ('imaginary lines'), and combined it with the *khuṭūṭ al-shu'ā'* ('inward radial lines') perpendicular to the visual surface further abstracting the intromission physics of Aristotelian and Atomist models, had also much in common with the later conceptions of geometry of perspective: these ranged from the relation of size and distances, especially distances of different measures, to the geometry of a plane as a surface collecting imaginary lines merging at a single point, all placed in methodological languages such as *maqāyīs* ('reasonings') and *barāhīn*

21 On the case of Ptolemy, see A. Mark Smith, 'Ptolemy's Theory of Visual Perception: An English Translation of the "Optics" with Introduction and Commentary', *Transactions of the American Philosophical Society*, 86 (2) (1996), p. 97 (Section 63). On the 'size-distance rule of perspective', referring to both Ptolemy and Ibn al-Haytham, see Veltman, *Linear Perspective*, 50. On the latter case, see Sabra, *The Optics of Ibn al-Haytham*, I, Chapter 2, Section 19; II, Chapter 3, Sections 67–68; III, Chapter 7, Section 14 and Appendix 1.2.a. On the circulation of these texts in Arabic and Latin, see note 22.

22 On the knowledge of Ptolemy's *Optics* in Islamic lands through an Arabic version translated into Latin, see Smith, 'Ptolemy's Theory of Visual Perception', pp. 55–57 (Arabic context). On the limited circulation of it, see Sabra, *The Optics of Ibn al-Haytham*, II, pp. LVIII–LX. On the case of Ibn al-Haytham, see Sabra, *The Optics of Ibn al-Haytham*, I, pp. LXXIII–LXXIX; A. I. Sabra, 'The Commentary that Saved the Text: The Hazardous Journey of Ibn al-Haytham's Arabic Optics', *Early Science and Medicine*, 12 (2007), 117–33. On the textual transmission of major works, see Kheirandish, *The Arabic Version of Euclid's 'Optics'*, II, pp. LXIV–LXV: Transmission Charts 1–2; Kheirandish, 'Light and Dark', pp. 78–79, Plates 54–55.

('demonstrations') (Appendix: 1.2.b).[23] The poor transmission of that seminal work across Islamic lands, as well as the similar fate of Ptolemy's *Optics*, with both texts being better known in Europe, the Arabic readers of Euclid's widely circulated *Optics* were left with an incomplete understanding of the relation between apparent size and visual angle.[24]

This is in clear contrast to developments in Europe, where besides distinctions between 'vertical', 'lateral', and most directly, 'orthogonal' distances on the one hand, and the 'extrinsic', 'median', and especially 'central', parts of the visual pyramid on the other hand, other important items were added into the equation.[25] The late medieval European author and credited inventor of linear perspective, Blasius of Parma (Biagio Pelacani da Parma, d. *c.* 1416), associated with the Perspectiva tradition as a possibly direct source for Filippo Brunelleschi (d. *c.* 1446), is cited for a 'dramatic claim' with which 'the stage was set for the discovery of Renaissance linear perspective as early as *c.* 1390.[26] A two-part statement is where the critical items of *quantitate* ('quantity'), *distantia* ('distance'), *proportio* ('proportion') and *experentia* ('experience') are explicitly used by him in two successive premises. Blasius first states that 'the proportion of visible objects follows a distinct proportion in quantity [...] if the distance be halved it will be seen as twice as large'; and secondly, that 'the proportion of objects does not follow the proportion of angles. For this is taught by experience'.[27]

Regarding the earlier Arabic equivalents of the four critical concepts and terms that form the two distinct premises above, the first two, *quantitate* or *qadr* and *distantia* or *bu'd*, are found in the Arabic version of Euclid's *Optics*, and the last two, *proportio* or *nisba* and *experentia* or *i'tibār* are found also in Ibn al-Haytham's *Optics*, works both widely transmitted through Latin translations.[28] Their most crucial components, however, namely an inverse proportional relation between size and distance (first premise), and a non-proportional

23 On the context of the discussion, see David C. Lindberg, 'The Intromission-Extramission Controversy in Islamic Visual Theory: Alkindi versus Avicenna', in *Interrelations in the History and Philosophy of Science*, ed. by Peter K. Machamer and Robert Turnbull (Columbus: Ohio State University Press, 1978, repr. in David C. Lindberg, *Studies in the History of Medieval Optics* (London: Variorum Reprints, 1983), IV, pp. 137–59; Franz Rosenthal, 'On the Knowledge of Plato's Philosophy in the Islamic World', *Islamic Culture*, 14 (1940), 412–16. For the key passages in Ibn al-Haytham, see Sabra, *The Optics of Ibn al-Haytham*, Book I. 6, [59–60] (see Appendix: 1.2.b).

24 Elaheh Kheirandish, 'What "Euclid Said" to his Arabic Readers: The Case of the Optics', in *De diversis artibus*, ed. by Gerard Simon and Suzanne Debarbat (Turnhout: Brepols, 2001), Tome 55: N. S. 18, pp. 17–28; Kheirandish 'Light and Dark', pp. 80–82, Plates 54–55.

25 For the distinctions between 'extrinsic', 'median', and 'central' rays, see Lindberg, *Theories of Vision*, 149. For the distinctions between 'vertical', 'lateral', and 'orthogonal' rays, see Veltman, *Linear Perspective*, 32.

26 On Blasius of Parma, the 1390 composition date of his *Questiones super perspectivam*, and other dates, see Lindberg, *Theories of Vision*, 130. For the Blasius quote in English and Latin, see Veltman, *Linear Perspective*, 50.

27 Veltman, *Linear Perspective*, p. 443 n. 76 has the Latin text from which I bracketed the terms: 'Proportio visibilium in quantitate insequitur proportionem distinctam ut dicam quod obiectum in certa distantia videtur sub certa quantitate, et in distantia insubdupla videtur maius in duplo. Ergo proportio objectorum non insequitar proportionem angularum. Antedecens enim docet experientia'.

28 On the Latin versions of Euclid's *Optics*, the Greco-Latin version (*Liber de visu*) and three Arabo-Latin versions (*Liber de aspectibus*, *Liber de radiis visualibus* and *Euclidis de aspectuum diversitate*), see Wilfred R. Theisen, 'The Mediaeval Tradition of Euclid's Optics' (unpublished doctoral thesis, University of Wisconsin, 1972; Facsimile, University Microfilms International, 1984). On *Liber de visu*, see also Wilfred R. Theisen, 'Liber de visu: The Greco-Latin Translation of Euclid's "Optics", Edited with Introduction and Notes', in *Mediaeval Studies*, 41 (1979), 44–105. On the Latin translations of Ibn al-Haytham's *Optics: Books I–VII*, see the publications of A. Mark Smith.

relation between size and angle (second premise), are not known through any earlier Arabic texts themselves.

The first premise and the use of quantity in a relation later known as the 'inverse size-distance' principle, could have conceivably been considered earlier, especially with the historically combined traditions of optics and surveying, and their respective treatments of both objective measures and subjective appearances beyond Euclidean optics.[29] This applies furthermore to the case of Ibn al-Haytham himself, where besides his optical works, he is credited with surveying works involving *miṣāḥa* ('measurement') and not just *miqdār/kammīya* ('measure'): these include a piece titled *Uṣūl al-misāḥa* ('On the Principles of Measurement'), claimed by the author as an early composition initially lost even to himself among other items.[30] The case of Euclid's *Optics* should also be mentioned here, the ninth-century Arabic version of which was summarised along with that of Ptolemy's *Optics* by Ibn al-Haytham early in his career.[31] The nature of textual transmission in the Euclidean optical proposition discussed above, along with a 'surveying proposition' in that text discussed below, may have further affected the different courses such developments took in Islamic and European lands, especially in combination. As for the second premise of the Blasius passage above, and its specification of *experentia/i'tibār* ('experience') for holding a non-proportional relation between objects and angles, this too seems to have been affected by transmission, here also directly related to elements in both the transformed contents and methods involved. These include the absence of advancing methods, anywhere from the elementary physical set ups of Ptolemy, and the *itsiqrā'* ('induction') and *i'tibār* ('experience/experiment') of Ibn al-Haytham with no room for *ikhtilāf* ('variation') and *tanāquḍ* ('contradiction'), all the way to the more comprehensive *i'tibar/experimentatio* ('comparative examination/experiment') of Ibn al-Haytham and his followers.[32] The limited transmission of such concepts and approaches extend the discussion of the act and role of transmission itself (precisely what was or was

29 On the relevant surveying traditions, see Veltman, *Optics and Perspective*; Veltman, *Linear Perspective*; Kheirandish, *The Arabic Version of Euclid's 'Optics'*, Propositions 19–22; Elaheh Kheirandish, 'An Early Tradition in Practical Geometry: The Telling Lines of Unique Arabic and Persian Sources', in *The Arts of Ornamental Geometry: A Persian Compendium on Similar and Complementary Interlocking Figures*, ed. by Gülru Necipoğlu, Supplement to *Muqarnas: An Annual on the Visual Cultures of the Islamic World* (Leiden: Brill, 2017), pp. 79–144.

30 On some surveying works by Ibn al-Haytham, see Kheirandish, *The Arabic Version of Euclid's 'Optics'*, I, pp. XLVI–XLVII, LVI–LVII; for those including Arabic editions and English translations, see respectively Rushdī Rāshid, *Ibn al-Haytham, théorie des coniques, constructions géométriques et géométrie pratique, les mathématiques infinitésimales du IXe au XIe siècle*, 5 vols (London: Al-Furqān Islamic Heritage Foundation, 2000), III, pp. 541–45; 546–49; and their English translation in Roshdi Rashed, *Ibn al-Haytham's Theory of Conics, Geometrical Constructions and Practical Geometry: A History of Arabic Sciences and Mathematics*, trans. by J. V. Field (London: Centre for Arab Unity Studies, Routledge, 2013), III, pp. 513–66, 567–68; on Ibn al-Haytham's lost surveying text, see his *Principles of Measurement*: Arabic edition (p. 541), and English translation (p. 513).

31 On Ibn al-Haytham's lost summary of the *Optics* of Euclid and Ptolemy, datable to the year 1027, see Sabra, *The Optics of Ibn al-Haytham*, II, p. XXXII.

32 A. I. Sabra, 'The Astronomical Origin of Ibn al-Haytham's Concept of Experiment, in *Actes XIIe congrès international d'histoire des sciences* (Paris: Albert Blanchard, 1971), III. A, pp. 133–36; Rushdī Rāshid, *Geometry and Dioptrics in Classical Islam* (London: Al-Furqān Islamic Heritage Foundation, 2005); Elaheh Kheirandish, 'Footprints of "Experiment" in Early Arabic Optics', in *Evidence and Interpretation in Studies on Early Science and Medicine: Essays in Honor of John E. Murdoch*, ed. by Edith Dudley Sylla and William R. Newman, *Early Science and Medicine*, 14 (2009), 79–104.

not passed on), to the nature and form of transmission (how exactly such concepts were passed on), the latter involving shifts in methods as well as concepts.

The Greek into Arabic transmission of the Euclidean definition and propositions presented here may also be viewed in combination with other relevant cases, to offer some internal explanation for the absence of developments in Islamic lands similar to those in Europe. From the transmission of other related works, such as the preface of a so-called Euclidean *Recension*, and the Arabo-Latin *De aspectibus* by Ya'qūb al-Kindī (d. *c.* 850), and their important inclusion of shadows, to the non-transmission of related discussions, those of distance or of physical set-ups in the works of Ptolemy and Ibn al-Haytham, all have their share in contributing to earlier pre-perspective practices in specific Islamic regions and languages.[33]

An entry devoted to *'ilm al-manāẓir* ('the science of optics') in the *Iḥṣā' al-'ulūm* ('enumeration of the sciences') of Abū Naṣr Fārābī (d. *c.* 950), well reflects the many aspects of 'appearances' in early Arabic optics, those which in addition to direct and indirect vision (*manāẓir* and *marāyā*), namely direct versus mediated appearances through reflection or refraction, included aspects of surveying (*misāḥā*), this time for determining magnitudes of objects, from the height of mountains to the width of rivers and depth of wells.[34] Within another century, the voluminous *Optics* of Ibn al-Haytham took a crucial step forward by a *tarkīb* ('combining') of *ta'līmī* ('mathematical') and *ṭabī'ī* ('natural philosophical') entities to give a final blow to the validity of the Euclidean-Ptolemaic visual-ray hypothesis and all its physical connotations. His new geometry of vision was to be consistent with later leaps in the direction of linear perspective, by not simply making the age-old dispute on the direction of the geometry of vision irrelevant, but in a language where 'imaginary lines' were validated, while physical visual-rays were made an impossibility, with no 'reason for it or an argument that supports it' (Appendix: 1.2.b). A variation stated by Leon Battista Alberti (d. *c.* 1472), a figure most directly associated with the development of linear perspective, is that the 'dispute among the ancients [in visual theories] is very difficult and useless for us'.[35] Ibn al-Haytham's new geometry of vision, and the directions and proportions involved, was itself only part of his foundational works anywhere from light and vision (Book I), to perception and visual illusions (Books II–III) and the geometry of reflection and refraction (Books IV–VII), all effectively transmitted and appropriated, through the Latin *Perspectiva / De aspectibus* and Italian *Prospettiva*, to the perspectivists, and artists, and through the printed *Opticae thesaurus* to still wider circles.[36]

---

33 For the Euclidean 'Recension', see 'Recensio' in *Euclidis opera omnia*, ed. by Heiberg and Menge, VII, pp. 143–247. For Kindī's *De aspectibus*, see Axel Anthon Björnbo and Sebastian Vogl, *Alkindi, Tideus und Pseudo-Euklid: Drei optische Werke. Abhandlungen zur Geschichte der mathematischen Wissenschaften* (Leipzig: Teubner, 1912), XXVI. 3, pp. 97–119.

34 Abū Naṣr Fārābī, *Iḥṣā' al-'ulūm* ('Enumeration of the Sciences'), ed. by 'Uthmān Amīn (Cairo: Maktabat al-Anjalū al-Miṣriyya, 1968), pp. 98–102. For the English translation of the entry on optics, see Sabra, *The Optics of Ibn al-Haytham*, II, pp. LVI–LVIII. For the divisions optics, catoptrics, and surveying, see Elaheh Kheirandish, 'The Many Aspects of "Appearances"': Arabic Optics to 950 AD.', in *The Enterprise of Science in Islam: New Perspectives*, ed. by Jan P. Hogendijk and Abdelhamid Sabra (Cambridge: MIT Press, 2003), pp. 55–83.

35 Quote from Alberti's *On Painting* is included in Lindberg, *Theories of Vision*, 149.

36 The expression 'appropriation' coined in the celebrated essay by A. I. Sabra, 'The Appropriation and Subsequent Naturalization of Greek Science in Medieval Islam: A Preliminary Statement', *History of Science*, 25 (3) (1987), 223–43, is both distinct from and simultaneous with some of the unintended transmissions discussed here. On

Nothing nearly comparable may be said about the transmission of either the *Optics* of Ptolemy or Ibn al-Haytham in the Islamic East, both largely limited to isolated references or partial uses.[37] The only case so far found of the full transmission of Ptolemy's *Optics* is an Arabic translation now known through a Latin version by Admiral Eugene of Sicily (d. *c.* 1203), as well as in Ibn al-Haytham's own *Optics*, itself fully transmitted only through an Arabic commentary by Kamāl al-Dīn al-Fārisī (d. *c.* 1318).[38] Not that there were no advancements in related areas including the outward and inward appearances on flat surfaces, or those involving artists. A unique passage by Naṣīr al-Dīn al-Ṭūsī (d. *c.* 1274), an optical author in both Arabic and Persian, and a student of Fārisī's student, Quṭb al-Dīn Shīrāzī (d. *c.* 1311), opens the subject of the next section as it contains related discussions involving geometrical directions in vision and depth perception: in the case of geometry of vision, it reveals unfamiliarity with the much earlier treatments of Ibn al-Haytham, his settlement of geometrical directions; and in the case of depth perception, it contains a rarely encountered reference to the works of artists, in quite different terms than those in Ibn al-Haytham, including the historical term for an 'artist' itself.

Direction and Depth through Arabic Philosophical Sources

This section turns to concepts such as direction and depth within the geometry of vision as they relate to questions on transmission. One of the most curious aspects of Ṭūsī's related discussions is his reference to outward and inward appearances involving *naqqāshān* ('artists') — using a verbal route historically applied to drawing, painting, and sculpting alike — especially that the discussion is not found in his optical works but in a philosophical discourse. The discussion is part of a philosophical discourse on what are later called extramission and intromission theories of vision, namely the visual-ray hypotheses of Euclid, Ptolemy, and Galen (second century) and their proponents, against their Aristotelian, Atomist, or Illuminationist opponents (Appendix: 2.1–2.2).[39]

the relevant Greek, Arabic and Latin texts and manuscripts, see David C. Lindberg, *A Catalogue of Medieval and Renaissance Optical Manuscripts* (Toronto: Pontifical Institute of Mediaeval Studies, 1975); Smith, 'The Latin Source'; Alhacen, *Alhacen's Theory of Visual Perception*, ed. and trans. by Smith, I, pp. XIX–XXIII. On the printed edition of Ibn al-Haytham's *Optics*, see Alhacen, *Opticae thesaurus*, ed. by Risner. On the differences between the Arabic and Latin versions, see Sabra, *The Optics of Ibn al-Haytham*, II, pp. LXXVI–LXXVII, where it is noted that the Latin text lacked the first three chapters of Book I, as Ptolemy's Arabo-Latin translation, lacked important introductory sections.

37 On the knowledge of the fifth book of Ptolemy's *Optics* by Abū Sa'd Ibn Sahl (d. *c.* 984), see Sabra, *The Optics of Ibn al-Haytham*, II, pp. LIX–LX. On a classification work by Fakhr al-Dīn Rāzī, referring to the twenty-two visible properties with reference to Ibn al-Haytham's *Optics*, see Sabra, 'Commentary that Saved the Text', p. 120 n. 7. See also Sabra, *The Optics of Ibn al-Haytham*, II, p. 85, Section 44, which cites seven visible properties in Ptolemy's *Optics*.

38 On the Arabic tradition of Ptolemy's *Optics*, see Smith, 'Ptolemy's Theory of Visual Perception', pp. 7–8, 55–57. On Kamāl al-Dīn's *Optics*, see Sabra, *The Optics of Ibn al-Haytham*, II, pp. LXIV–LXXIII; Sabra, 'Commentary that Saved the Text', pp. 131–33; Kheirandish, 'Light and Dark', pp. 72–77.

39 Lindberg, *Theories of Vision*, 1–17; Lindberg, 'The Intromission-Extramission Controversy', pp. 137–59; Rosenthal, 'On the Knowledge of Plato's Philosophy', pp. 412–16; Kheirandish, *The Arabic Version of Euclid's 'Optics'*, II, pp. 7–15 and notes.

The subject of vision itself is one which the optical works by or attributed to Ṭūsī are filled with, along with their various puzzles, from the chronological order of those works to the validity of their attributions and treatments, as discussed in my other publications.[40] A puzzle which must be added here is not limited to the question why Ṭūsī wrote about the subjects under discussion outside of his optical or surveying treatments, subjects readily combined in his *Recension (Taḥrīr) of Euclid's Optics*, itself one of the so-called 'Intermediary Books' (*Kutub al-Mutawassiṭāt*), studied intermediately after Euclidean geometry and in preparation for astronomy; or why Ṭūsī's other 'scientific' works did not include such subjects, works close to his various activities as the author of important astronomical works and director of the Marāgha Observatory.[41] Rather, Ṭūsī's notable reference to depth perception in reference to artists curiously appears in a philosophical work, a commentary on a text completed by his predecessor Fakhr al-Dīn al-Rāzī (d. *c.* 1210), whose own works relate to the present discussion in other ways. As part of a long discourse on vision, and the nature and direction of visual-radiation, following the standard format of a commentary, where he (meaning Rāzī) said, then I (meaning Ṭūsī himself) say (Appendix: 2.2.a–b), Ṭūsī addressed the subject of depth perception in a marginal and incomplete manner that 'since it is difficult for us to make an examination [of it], we put it away', then added, the brief yet revealing remark that 'although we see *naqqāshān* ('artists') [...] make an observer perceive the depths of bodies and the *bu'd* ('dimensions/distances') between them' (Appendix: 2.2.b). This differs strikingly from the extensive discussions of Ibn al-Haytham involving distances, where *mazūqīn* ('painters'), and *ṣuwar* ('paintings') are treated together under the discussion of *aghlāṭ* ('[visual] errors') in his great 'Optics'.[42]

Part of the puzzle related to the form and setting for Ṭūsī's discourse above may be explained by recalling that his predecessor, Fakhr al-Dīn al-Rāzī, may have been among a very few to whom the great *Optics* of Ibn al-Haytham was transmitted in some form. Rāzī's classification work in Persian with the Arabic titles of *Jāmi' al-'ulūm or Kitāb al-Sittīnī* (for the text's sixty sections), uncommonly refers to the title of Ibn al-Haytham's *Optics*, though in the close orthographic form *Kitāb al-Munāẓira* for *Kitāb al-Manāẓir* in some manuscripts; it also atypically contains the work's full list of visible properties.[43]

40 Elaheh Kheirandish, 'Mathematical Sciences through Persian Sources: The Puzzle of Ṭūsī's Optical Works', in *Sciences, techniques et instruments dans le monde iranien, xᵉ–xixᵉ siècle*, ed. by Naṣr Allāh Pūrjavādi and Živa Vesel (Tehran: Institut français de recherche en Iran, 2004), pp. 197–213; Kheirandish, *The Arabic Version of Euclid's 'Optics'*, II, pp. 55–59. On one of Ṭūsī's optical tracts without discussions of transmission, see H. J. J. Winter and W. 'Arafat, 'A Statement on Optical Reflection and "Refraction" Attributed to Nasir ud-Din at-tusi', *ISIS* 42 (2) (1951), 138–42.

41 Jamil F. Ragep, *Naṣīr al-Dīn al-Ṭūsī's Memoir on Astronomy (al-Tadhkira fī 'ilm al-hay'a)*, 2 vols (New York: Springer-Verlag, 1993); Aydin Sayili, *The Observatory in Islam and its Place in the General History of the Observatory* (Ankara: Turkish Historical Society, 1960).

42 Naṣīr al-Dīn Ṭūsī, *Talkhīṣ al-Muḥaṣṣal*, ed. by A. Nurani (Montreal Canada: McGill University Institute of Islamic Studies, 1980), pp. 173–74; Ibn al-Haytham, *Kitāb al-Manāẓir*, ed. by Sabra; Sabra, *The Optics of Ibn al-Haytham*, III, Chapter 7, Sections 39–43, pp. 295–97. For a related discussion, see also Necipoğlu, 'Scrutinizing Gaze', pp. 35–36.

43 Fakhr al-Dīn Rāzī, *Jāmiʿ al-ʿulūm = Kitāb-i Sittīnī* ('Compendium of Sciences = Book of Sixty'), ed. by Muḥammad Ḥusayn Tasbīḥī (Tehran: Kitābkhānah-i Asadī, 1346 = 1967), p. 174 (Ms. Facs.: MS. facs. *Kitāb al-Munāẓira = Kitāb al-Manāẓir*); later ed. by ʿAlī Dāvūd (Tehran: Bunyād-i Mawqūfāt-i Duktur Maḥmūd Afshār Yazdī, 1382 = 2003), p. 410.

But neither Rāzī nor his commentator, Ṭūsī reflect any knowledge of that foundational work even in their repeatedly outdated arguments for and against outward extension of *shuʿāʿ* ('rays') and inward reception of *inṭibāʿ* ('form'), both upgraded in Ibn al-Haytham *Optics* two centuries before. Closer attention to Ṭūsī's response to Rāzī on that subject further shows that even in his novel geometrical abstraction that 'early philosophers, only state [rays] issuing from the eye by metaphor (*majāz*) just as it is said [for] the Sun' (Appendix: 2.2.a), the discussion itself is not in pure optical terms, as it crosses into aspects of some neighbouring fields. Not a field such as astronomy to which Ṭūsī was particularly close through his known advancements, one associated in other contexts with developments in perspective theory through the use of astronomical instruments.[44] Ṭūsī's discussion here is, rather, one that applies comparative modes and analogies, in one case to luminous and visual radiation, and in another case, to large and remote versus small and close magnitudes, in forms quite fitting for a philosophical discourse.

More puzzling is the absence of any perspective-related discussion in Ṭūsī's optical commentary on Euclid's text, where its so-called surveying propositions are devoted to the determination of an object's height, depth, and length. There, problems in transmission take on different forms when the Arabic Euclidean proposition specifying the *kammīya* ('quantity') of an object's *ṭūl* ('length', Greek twenty first/Arabic twenty second proposition) appear in Ṭūsī's treatment under the *irtifāʿ* ('height') of an object. In this last of four surveying propositions, not only is the part 'length of the object on the plane' missing in Ṭūsī's version of that Arabic proposition, one with its own possible relation to pre-perspective developments; the proposition's variant procedure and figure represent other cases of transmission with likely impacts on such an isolated and limited discussion of depth and depth perception in Ṭūsī's philosophical discourse.[45]

A partial explanation for the passing remarks made by Ṭūsī in the passage under discussion may be that, just as in the case of the extensive discussions of Ibn al-Haytham, they reflect transmission through both textual and visual sources. For Ṭūsī, textual sources similarly started with the Arabic version of Euclid's *Optics*, a text with unintended shifts in the translation of its optical and surveying propositions, transmitted to Ṭūsī through his *Recension*, as distinct from the more conscious departures of the *Optics* of Ibn al-Haytham and Fārisī before and after him. As for visual sources, Ṭūsī's use of the expression 'we see' before the visual effects he describes (Appendix: 2.2.b) may refer to visual depictions or physical objects existent between his time and the time of Ibn al-Haytham: depictions beyond two dimensions, such as artwork that appears outward or inward on a wall; and objects representing three dimensions, such as *muqarnas* ('vaulted structure') work on a building, all instances mentioned in the contemporaneous literature.

---

44 Kim H. Veltman, 'Ptolemy and the Origin of Linear Perspective', in *Atti del convegno internazionale di studi: la prospettiva rinascimentale, Milan 1977*, ed. by Marisa Dalai-Emiliani (Florence: Centro Di, 1980), pp. 403–07 (astronomical instruments named on p. 403 are 'planisphere' and 'astrolabe').

45 Naṣīr al-Dīn Ṭūsī, *Taḥrīr al-Manāẓir* ('Recension of the Optics'), in *Rasāʾil* ('Treatises') (Hyderabad: Dāʾirat al-Maʿārif al-ʿUthmānīyah, 1358 = 1939), I, 5, pp. 2–24; Naṣīr al-Dīn Ṭūsī, *Taḥrīr al-Manāẓir* ('Recension of the Optics'), in *Revue de l'institut des manuscrits arabes*, 9, ed. by A. S. Dimirdash (1383 = 1963), pp. 251–90. The last of the four surveying Propositions 19–22, is 22 (Greek 21): see Kheirandish, *The Arabic Version of Euclid's 'Optics'*, I, pp. 66–69; II, pp. 60–61.

Outline and Colour through Persian Literary Sources

Successive descriptions of outward and inward appearances variously involving artists are further reflected through some literary sources, in ways that are comparable to both pre-perspective practices of Islamic lands, and the much more advanced developments in Europe. These are handed down from source to source and language to language, and correspond to more than one specific time and place; as indeed are the linguistic variations in the terms used for 'outward' and 'inward' appearances themselves, not to mention the various terms used in one language alone. Some of these sources have been consulted and included here from the viewpoint of the present discussion; others remain to be recorded and analysed through projects related to and beyond the present volume.[46]

The practices reflected in the literature so far noted largely concern the subject of visual illusions often involving flat surfaces and colours. These illusions are represented in terms quite different from the most primitive forms of both 'angular' and 'linear' perspectives in scientific texts, or the 'concave' and 'convex' appearances of objects with non-aligned sides and centres, in both early and late traditions including Ṭūsī's version.[47]

Certain ancient literary sources exist which may be the oldest known documents describing pre-perspective practices. Such sources were transmitted from Sanskrit and/or other ancient Indian languages, via middle Persian (Pahlavi), Syriac, and later, medieval Arabic and Persian, with some still surviving in the form of various illustrated manuscripts.[48] Persian sources from successive centuries including the Islamic Middle Ages, some with revealing verses, should also be mentioned here. The ancient Persian source, *Kalīla and Damna*, titled after the names of two jackals in a collection of stories, was transmitted from ancient sources initially through a seventh-century Middle Persian version based on Sankskirt and/or ancient Indian sources, from which a surviving eighth-century Arabic translation reads: 'Like the expert form-maker (al-*muṣawwar*) who makes forms on a wall which are seen to be outside the wall, while they [actually] are not', or are seen to be inside the wall, wile they [actually] are not'.[49] Two Persian translations from the twelfth century, based on the earlier Arabic, translated by near contemporaries Naṣr Allāh Munshī and Muḥammad Bukhārī (*c.* 1143-45) have their own terminological variations of the two visual effects (Appendix: 3.1.a–b).[50] The artists vary from the former's *ṣurat-gar* ('form-maker')

46 My project "IOTA" (Index of Optical Terms in Arabic) is a record of key terms or roots in Arabic to identify relations between texts, including related cases in Greek and Latin ("IOTA+" includes Persian).

The model used for the IOTA project to represent related texts on optics graphically was based on Project Archimedes, see Kheirandish, 'Light and Dark', p. 81, p. 85 n. 13.

47 Kheirandish, *The Arabic Version of Euclid's 'Optics'*, I, Proposition 63, pp. 218–21; II, pp. 98–99 and related notes.

48 P. Waley and Norah M. Titley, 'An Illustrated Persian Text of Kalīla and Dimna dated 707/1307–8', *The British Libray Journal*, 1 (1975), 42–61; Sofie Walzer, 'The Topkapi Saray Manuscript of the Persian "Kalīla wa-Dimna" (dated A.D. 1413)', in *Paintings from Islamic Lands*, ed. by R. Pinder-Wilson (Oxford: Cassirer, 1969), pp. 48–84; *Kalīla wa-dimna: Fables from a Fourteenth-Century Arabic Manuscript*, ed. by Esin Atıl (Washington, D.C.: Smithsonian Institution Press, 1981); Pūrjavādī, 'Concept of Perspective', p. 24, cites debates on Indian sources.

49 On the early Arabic translation of Ibn Muqaffa', see *Kalīla wa-dimna*, ed. by Louis Chikho (Bayrut: al-Maṭba'ah al-Kāthūlīkīyah, 1952), p. 67. The Arabic passage is included in Pūrjavādī, 'Concept of Perspective', p. 19 followed by successive ones in relevant languages; Arabic and Persian passages in English are all my translations.

50 On the medieval Persian translations of Munshī and Bukhārī, see respectively: Abū al-Ma'ālī Naṣr Allāh Munshī, *Kalīla wa-dimna, tarjumah-i kalīla wa-dimna*, ed. by Mujtabā Mīnuvī Ṭihrānī (Tehran: Dānishgāh-i Tihrān, 1343 = 1964); and Muḥammad Ibn, 'Abd Allāh al-Bukhārī, "Kalīla wa-dimna"', in *Dāstānhā-yi bīdpāy*, ed. by Parvīz Nātil

and *ustād* ('master') in their *pīsha* ('profession'), to the latter's *naqqāsh* ('drawer/painter/sculptor'), here being *chābuk-dast* ('[having] skilled hands') with *qalam* ('pen/brush/stick'). The effects created vary in a similar fashion, from the former's *pindār* ('thought') of being within or outside a wall (*dar* and *bīrūn* respectively), to the latter's shift of *namūd* ('appearance') *dar naẓar* ('in viewing') something [actually] *musaṭṭaḥ* ('flat') as standing out (*angīkhta*), or [actually] standing out, while appearing as flat, this time without any reference to 'inward' appearance, or a wall, surface, or representational plane.

The successive terms and forms of this passage alone suggest changes through multiple translations and linguistic transmissions.

Comparable visual effects involving both outline and colour were further transmitted in various striking variations and depictions in other sources in Arabic and Persian. An early medieval Arabic variation can be seen in the story of a competition between Greek and Chinese artists in the *Iḥyā' al-'ulūm al-Dīn* of Muḥammad al Ghazālī (d. *c.* 1111). A much later work in Arabic by the Egyptian Taqī al-Dīn al-Maqrīzī (d. *c.* 1442), an author largely known for his important history of Egypt, reports on an older competition between an artist from Egypt and another from Iraq (or non-Arab Iraq/Persia): the former paints a dancer with a white dress on a black crescent background, the latter with a red dress on a yellow crescent background, respectively appearing to be inside and outside a wall, through contrasting colours in the background and foreground creating effects of depth and thickness.[51] Persian versions of each of those variations are, in turn, depicted in successive medieval Persian poems, such as those in the *Iskandar-nāma* of Niẓāmī-yi Ganjavī (d. *c.* 1209) and *Mathnavī-yi ma'navī* of Maulānā Jalāl al-Dīn-i Rūmī (d. *c.* 1273), with their own varied representations, as well as the result of the contest: here the two competing art works are by the Greek and Chinese, but in the context of the superiority of mysticism over rational forms of knowledge. In Niẓāmī's version of the two productions, one has the form *padhīruft* ('accepted'), one, *minimūd* ('give out'), while the expertise of *naqsh bastan* ('forming impressions') is associated with the Greeks, and of *ṣayqal* ('polishing'), with the Chinese; the roles are then reversed in the later version of Rūmī, where the *nuqūsh* ('forms') shine forth, from the verb, *tāftan* ('shine'), and the *lawḥ* ('tablet') of *nishān* ('impression'), are *padhīrā* ('receptive') hearts (Appendix: 3.2.a–b).[52]

Khānlarī and Muḥammad Rawshan (Tehran: Khvārazmī, 1361 = 1982). The Persian passages (eds. pp. 66 and 77 respectively) are included in Pūrjavādī, 'Concept of Perspective', pp. 19–20, dated *c.* 538–40 H), and followed by relevant images (p. 26), including one representing combined dimensions in space (p. 28).

51 On the account of Maqrīzī, see Pūrjavādī, 'Concept of Perspective', pp. 22–24; Nasser O. Rabbat, "Ajīb and Gharīb: Artistic Perception in Medieval Arabic Sources', *The Medieval History Journal*, 91 (1) (2006), 99–113 (pp. 101–02).

52 On the Persian variations, see B. Furūzānfar, *Sources of the Stories and Depictions of the Mathnawī (Ma'ākhidh qiṣaṣ va tamthīlāt-i mathnavī)* (Tehran, 1333 = 1954), pp. 33–35. For citations, see Priscilla Soucek, 'Niẓāmī on Painters and Painting', in *Islamic Art in the Metropolitan Museum of Art*, ed. by Richard Ettinghausen (New York: Metropolitan Museum of Art, 1972), pp. 9–21 (esp. pp. 12–14), including a 1449–50 manuscript illustration of that contest (p. 13). Necipoğlu, 'Scrutinizing Gaze', pp. 46–47 includes coloured manuscripts from 1495 and 1513 Shiraz (p. 47). Belting, 'Florence and Baghdad', pp. 80–82 includes a coloured manuscript for another Niẓāmī work from 1401–11 Shiraz (p. 81): 'painted at almost the same time that linear perspective was invented in the West' where 'three dimensional space and two dimensional surface merge', in a place described itself as a 'picture' that 'is not simply "backward" [...] but rather offers an alternative to the perspective gaze' (p. 82).

Some epistemological issues are naturally associated with the visual effects of such impressions, as part of the phenomena referred to as 'beyond nature' as found in some Greek and early Arabic scientific literature.[53] The corresponding Arabic expression *'ajab* and its various cognates, meaning to evoke a sense of wonder, appear quite early on in historical sources: in mechanics, this corresponds to the Greek *para phusin* of the pseudo-Aristotelian *Mechanical Problems* and mechanical *ḥiyal* ('devices'), like a lever producing amazingly unexpected effects.[54] In optics, such an effect is expressed in the context of the similarly old problem of the circular appearance of far 'angled' objects, and an early response to a version of it in the ninth Euclidean proposition (or tenth in Arabic) through the expression *a'jab al-'ajab* ('the most amazing of amazements'), as the object's *ḥaqīqī* ('real') shape is not actually being seen.[55] In the case of the Arabic and Persian sources discussed above, the so-called 'negative philosophical value' of some false impressions given, next to the 'positive artistic values' of the expertise for the clever tricks performed in multilingual literature, may have had an impact on the limited development of perspective practices; but there are limitations to such forms of explanation.[56] Genres such as the *'ajā'ib*, and its historical counterpart, *gharā'ib*, the later wondrous and marvellous of their evolving verbal forms, are documented with positive epistemological associations in no uncertain terms, in both Arabic and Persian. This includes material from the fifteenth century, when parallel trends can be discerned in Europe in corresponding Latin expressions such as *mirabilia*, and other words with notable distinctions and increasingly positive connotations.[57] In Arabic, Maqrīzī reports the *'ajīb* ('wondrous') visual effect of a painting in a Fatimid mosque in

53 On the expression 'beyond nature' which appears in the opening lines of the pseudo-Aristotelian *Mechanical Problems* as *para phusin* as distinct from *kata phusin* ('according to nature') (847a11–15), and its context in the ancient Greek tradition, see Mark J. Schiefsky, 'Art and Nature in Ancient Mechanics', in *The Artificial and the Natural: an Evolving Polarity*, ed. by Bernadette Bensaude-Vincent and William R. Newman (Cambridge, Mass.: MIT Press, 2007), pp. 67–108.

54 For the corresponding early Arabic expressions as *yukhālif* ('contrary to') nature versus 'according to' nature, with reference to *'ajab*, or 'wonder', and its modern English translation as 'marvel', see Mohammad Abattouy, 'Nutaf min al-ḥiyal: a Partial Arabic Version of Pseudo-Aristotle's "Problemata Mechanica"', *Early Science and Medicine*, 6 (2) (2001), 96–122 (pp. 110–13). The expression *'ajab* is also in Fārābī, *Iḥṣā' al-'ulūm* ('Enumeration of the Sciences'), ed. by Amīn under 'mechanical sciences'. For English translation, see George Saliba, 'The Function of Mechanical Devices in Medieval Islamic Society', in *Science and Technology in Medieval Society*, ed. by Pamela O. Lang, Annals of the New York Academy series (New York: New York Academy of Sciences, 1985), pp. 141–51 (esp. p. 146). On other occurrences of the expression in Arabic literature, see Roy P. Mottahedeh, '"Ajā'ib" in the Thousand and One Nights', in *The Thousand and One Nights in Arabic Literature and Society*, ed. by Richard C. Hovannisian and Georges Sabagh (Cambridge: Cambridge University Press, 1997), pp. 29–39 (and below).

55 On the expression *'ajab al-ajab* ('the most amazing of amazement'), with reference to the Euclidean proposition of the circular appearance of far 'angled' objects, see my discussion of an early optical text by Aḥmad ibn 'Īsā, in Kheirandish, *The Arabic Version of Euclid's 'Optics'*, I, p. XLV; II, p. 47, and n. 182, for an excerpt.

56 Pūrjavādī, 'Concept of Perspective', pp. 25–29. The in-depth analysis of this article, which extends to the pre-Islamic period (p. 25), as well as discussions and illustrations of alternative representations, including related European distinctions (p. 28) are too extensive to be included here.

57 See Mottahedeh, '"Ajā'ib" in the Thousand and One Nights', pp. 30–31, where in addition to the historical meanings and distinctions made between *'Ajā'ib* and *gharā'ib*, modern English variants are given as 'surprising, astonishing, amazing, wonderful, fabulous, curious, marvellous' (pp. 29–30). On related European expressions, see Katherine Park, 'The Meaning of Natural Diversity: Marco Polo on the "Division" of the World', in *Texts and Contexts in Ancient and Medieval Science: Studies on the Occasion of John E. Murdoch's Seventieth Birthday*, ed. by Edith Sylla and Michael McVaugh (Leiden: Brill, 1997), pp. 134–47 (esp. p. 136) for examples from the twelfth and thirteenth centuries; and more generally, Lorraine Daston and Katherine Park, *Wonders, and the Order of Nature (1150–1750)* (New York: Zone Books, 1998).

the context of a surface appearing as a three-dimensional *muqarnas* from the centre, and as flat from the side.[58] Abū Isḥāq Kūbanānī (d. after 1483), who wrote in Arabic and Persian and completed a Persian translation of the impressive *Geometrical Constructions* of Abū al-wāfa' Būzjānī (d. *c.* 998) in Arabic, refers to the text as *kitāb-i 'ujāb* ('a wondrous book'), while the works of the initial translator of that book, were *gharā'ib* ('marvels') in the *funūn* ('arts or crafts') of *'ilm* ('science').[59] Still later, a sixteenth-century anonymous text known in both Arabic and Persian, speaks of *aṣḥāb al-isti'jāb wa al-istrighrāb* ('companions/ proponents of wonder and marvel'), offering further measures of the increasing currency of such practices, as well as the widening community of their practitioners.[60]

There are also more direct ways of explaining the different directions pre-perspective practices took in lands outside of Europe, starting with the limited access that figures with scientific and non-scientific profiles had to each other's work, even in the closest of periods and regions under discussion. Examples of these, to remain within the context of names, times, and places already discussed, include thirteenth-century northern Iran, when a scientific figure like Ṭūsī, and a literary figure like Rūmī both wrote scientifically related poems without apparent knowledge of each other's works.[61] And there is no indication as to why this would have been any different in a setting as adverse to exchanges as the post-Mongol lands of the Arab- and Persian-speaking regions, in sharp contrast to the close coordinates of European practitioners of the sciences and arts.

Mirror and Measure though Persian Surveying Sources

Developments in linear perspective involving surveying in general and the use of mirrors in particular, pose questions of transmission in quite different terms. Relevant European developments using mirrors have been dated to as far back as Greek antiquity, in respect to the ancient science of catoptrics. Early architecture and scene painting likewise demonstrate uses of mirrors.[62] A further extension of related practices involving mirrors have also been identified in later periods, for example in paintings by Giotto di Bondone (d. *c.* 1337) and in perspective demonstrations by Brunelleschi, as reported by closely dated contemporaries.[63]

58 On the uses of such expressions in fifteenth-century Arabic sources, see Nasser O. Rabbat, 'Ajīb and Gharīb', pp. 101–02.

59 On the positive uses of such expressions in fifteenth-century Persian sources, see Elaheh Kheirandish, 'An Early Tradition in Practical Geometry', Part III, Appendix III, Passage 3.

60 On the use of the same expressions in an anonymous sixteenth century optical work known through both Arabic and Persian versions, see 'Optics, History of', in *Encyclopaedia Iranica: A Comprehensive Research Tool Dedicated to the Study of Iranian Civilization in the Middle East, the Caucasus, Central Asia, and the Indian Subcontinent* (New York: Columbia University Press, 2010): http://www.iranicaonline.org/articles/optics [accessed 27-07-2017].

61 On the case of Ṭūsī and Rūmī, as well as earlier contemporary scientific and literary figures like Bīrūnī and Gurgānī, as related to astronomy, see Elaheh Kheirandish, 'Astronomical Poems from the "Four Corners" of Persia (*c.* 1000–1500 CE)' in *Essays in Islamic Philology, History, and Philosophy*, ed. by Alireza Korangy, Wheeler M. Thackston, Roy P. Mottahedeh and William Granara (Berlin: Walter de Gruyter GmbH, 2016), pp. 51–90.

62 On earlier relevant works see, Wilbur R. Knorr, 'Archimedes and Pseudo-Euclidean Catoptrics: Early Stages in the Ancient Geometric Theory of Mirrors', *Archive for History of Exact Sciences*, 35 (1985), 28–105, 114–15 (p. 31); Veltman, *Linear Perspective*, 35; Andersen, 'Ancient Roots', pp. 82–86.

63 Edgerton, *The Renaissance Rediscovery*, 5, 125, 134.

Alberti's groundbreaking *Della pittura* (*c.* 1435) dedicated to Brunelleschi, for 'judging not composing pictures', is further evidence of this tendency.[64] The elaborations of the still later Leonardo da Vinci (d. *c.* 1519), show a continuation of the 'direct link between the use of perspective and mirrors' with specific reference to fifteenth-century painters outside the context of surveying.[65]

This final section seeks to expand upon the present discussion, and address the way Arabic and Persian surveying traditions involving mirrors relate to pre-perspective developments. This is a productive question to pursue, given the specific distinctions and explicit statements of the related historical sources regarding transmission in particular.

Historical distinctions between *qadr/miqdār* ('measure'), and *misāḥā* ('measurement'), the latter being an expression through which *'ilm al-/'ilm-i misāḥa* ('the science of measure/ surveying') is itself known in Arabic and Persian, are particularly notable in this connection for cases with or without mirrors. This is a distinction made clear through methods termed as *ghayru misāḥa* ('without measurements') for the determination of an unknown quantity such as height, depth, or length from three known magnitudes through measures including the observer's position, often not involving actual measurements. Such a case is found in sources as old as a tenth-century surveying text by Sinān ibn al-Fatḥ through a proportional relation, where from three *qadr/miqdār* ('knowable measures') the fourth may become known without actual measurement or *misāḥa*.[66] This method represents a clear contrast to the slightly later treatments of Ibn al-Haytham in his *Principles of Measurement* cited above, which even includes measurement conventions.

As for surveying problems involving mirrors, this is where historical sources are even more explicit about problems specific to transmission than Ibn al-Haytham's report in his surveying text of the loss of an earlier work by him on that subject. Such reports include cases as late as a fifteenth-century testimony on the poor transmission of surveying problems and methods up until the author's own time, as discussed after a brief contextualisation.

In addition to the distinctions between theoretical treatments and actual measurements of unknown magnitudes identified here, there are numerical and geometrical traditions within surveying in cases when the proportional relations for calculations involve numerical values and geometrical magnitudes respectively, again with or without measurement. Surveying may be further identified in at least two other distinct forms, 'area surveying' and 'distance surveying', as I have called them to distinguish between aspects of the field

64 Lindberg, *Theories of Vision*, 151–52; Edgerton, *The Renaissance Rediscovery*, 134–35.

65 On the role of surveying, and the related works of Leonardo da Vinci, see Veltman: *Optics and Perspective*, 15–73 (Chapter 1: 'The Surveying Tradition and the Origin of Linear Perspective'); Veltman, *Linear Perspective*, 350–54 (Sections 10–11: 'The Mirror as Arbiter' and 'The Mirror as Creative Aid'.

66 On the tenth-century surveying text by Sinān ibn al-Fatḥ titled *al-Misāḥāt al-manāẓiriyya* ('Optical Surveying'), specifying methods such as *ghayru misāḥa* ('without measurements'), Kheirandish, *The Arabic Version of Euclid's 'Optics'*, I, p. XLVI. The facsimile reproduction of the text is included in Elaheh Kheirandish, 'The Medieval Tradition of Euclid's "Optika"', 2 vols (unpublished doctoral thesis, Harvard University, 1991), I. The author, Sinān ibn al-Fatḥ, is cited in Boris A. Rozenfel'd and Ekmeleddin İhsanoğlu, *Mathematicians, Astronomers, and Other Scholars of Islamic Civilization and Their Works, 7th–19th c.* (Istanbul: Research Center for Islamic History, Art and Culture, 2003), pp. 91–92, and the work with reference to a treatment in Bīrūnī's *Shadows*, for which see *The Exhaustive Treatise on Shadows*, ed. and trans. by E. S. Kennedy (Aleppo: University of Aleppo Press, 1976), I, p. 267; II, p. 166. On discussions of surveying with reference to this text and its non-measurement method, see also Kheirandish, 'An Early Tradition in Practical Geometry'.

involving the respective 'application of areas' after the *Elements* of Euclid, and proportional relations of magnitudes after the *Optics* by the same author.[67] Developments in 'area surveying' have had their own distinct traditions. One such case is the tenth-century *Geometrical Constructions* of Abū al-Wāfa' Būzjānī, and its rich Arabic and Persian traditions, involving methods based on the equality of areas comparable to land surveying. A further important 'anonymous compendium', this time in Persian and known only through an undated manuscript, most likely from the fifteenth century, comprises partly dependent constructions of repeat patterns or interlocking figures mostly directed at oramental geometry.[68] Aspects of 'area surveying' closely associated with ornamental geometry through constructions of various simple and complex patterns are represented in both textual and physical sources from the Islamic Middle Ages, so far without much explicit or identifiable historical links between their theoretical and practical aspects in terms of specific temporal and geographical coordinates. But whether textual constructions and demonstrations in geometrical texts, or physical patterns and ornamentations on known buildings, the methods employed are largely based on the equality of areas of the figures or patterns involved, the simplest identifiable in geometrical propositions in Euclid's *Elements* such as the proposition best known as the 'Pythagorean theorem'.[69] By contrast, aspects of 'distance surveying' directed to the determination of distances, heights, widths, and depths in surveying texts, often with specific reference to the height of mountains, width of rivers, and depth of wells, involve not the application of equal areas, but the proportion of equal ratios. Distance itself, represented as an unknown quantity calculated through a proportional relation between four entities, as in the tradition of the surveying propositions of Euclid's *Optics*, is an important part of pre-perspective developments, one further affected by a combination of non-measurement methods, and specific problems in transmission.

I have treated the details of how the principle of reflection was transmitted through a surveying proposition in Euclid's *Optics* (nineteen and twentieth in the Greek and Arabic proposition respectively) elsewhere: in short, how the Greek term *antanakeklastô* ('reflection') was translated into Arabic with a non-standard term *iniʿṭāf* ('refraction') instead of the standard form *inʿikās* ('reflection'), how the close orthography of the latter terms through transmission changed the meaning of the principle of the equality of the angles of incidence and reflection to one with four equal angles, namely, incidence, reflection, refraction, and penetration, and above all, how, the particular treatment of Ṭūsī, who added that problematic principle to the definitions of his own widely circulated *Recension* of Euclid's *Optics*, contributed to its subsequent 'puzzles' in the words of Fārisī slightly later, as reproduced further below.[70]

Turning to two unpublished surveying texts in Persian from a slightly later period (Appendix: 4.1–4.2), it may be shown how the particular transmission of the principle

---

67 Kheirandish, 'An Early Tradition in Practical Geometry'.

68 See the chapters of volume *The Arts of Ornamental Geometry*, ed. by Necipoğlu; Jan P. Hogendijk, 'Ancient and Modern Secrets of Isfahan', *Nieuw Archief voor Wiskunde*, 5 (9) (2008), p. 121 n. 2.

69 Kheirandish, 'An Early Tradition in Practical Geometry'; Thomas L. Heath, *The Thirteen Books of Euclid's Elements* (Cambridge: Cambridge University Press, 1956), I, Proposition 47. For a three-dimensional figure of the same Euclidean proposition by Leonardo Da Vinci, see Veltman, *Linear Perspective*, 188, Figure 708.

70 Kheirandish, *The Arabic Version of Euclid's 'Optics'*, I, Proposition 20, pp. 58–61; II, pp. 2, 55–59; Kheirandish, 'The Puzzle of Ṭūsī's Optical Works', pp. 197–205, 205–07 (texts); Kheirandish, 'Light and Dark', pp. 72–77.

of reflection through 'distance surveying' had continued with consequences beyond developments in optics. These two texts, authored by little-known historical figures, represent related features that capture some uncovered aspects of transmission. They belong to a period shortly after Kamāl al-Dīn Fārisī, the Persian author of the important Arabic commentary on Ibn al-Haytham, *Tanqīḥ al-Manāẓir* ('Revision of Optics'), who wrote in his preface to that work, about the *qiṣṣa* ('story') of how 'puzzles' and 'distortions' in optical principles involving the geometry of reflection and refraction led to his optical research, and eventual access to the great volumes of Ibn al-Haytham, through his teacher, Quṭb al-Dīn Shīrāzī (d. *c.* 1311), a teacher of Ṭūsī.[71] The authors of those later surveying texts, both in Persian, emerge as agents of different stories of transmission.

One of the two surveying authors has the unfamiliar name of Abū Bakr al-Khalīl al-Tājir, a figure who comes across as a pre-Renaissance 'Renaissance man', involved with many works and practices, although his exact time and place is hard to determine. The earliest point when he may have been active appears to have been the early 1300s, estimated on the basis of associated transcriptions including that of 'the Principles of Measurement', composed by none other than Ibn al-Haytham.[72] The *A'māl* ('constructions') carrying the name of Abū Bakr al-Khalīl al-Tājir al-Raṣadī, whose glosses place the latest dates for the figure named in its title in the early 1400s, are of direct interest in the context of surveying.[73] Among various subjects contained in the *Constructions of Abū Bakr*, from astronomical tables to mechanical devices, are two surveying problems, one involving two mirrors, notable for an extended application of a Euclidean surveying proposition using a plane mirror to determine height (Appendix: 4.1).

Plane mirrors have had a long and established application in both optics and perspective, in optics, as part of the ancient science of catoptrics devoted to vision through mediums such as polished surfaces, and in perspective, as an instrument for determining a point for establishing proportional relations.[74] The application of mirrors in the field of surveying, and by extension, perspective practices, is as different from the case of optics as the problems of transmission involved in each. Of the two above-mentioned surveying texts, the earlier one transmitted as *Constructions of Abū Bakr*, does not reflect the work of Ibn al-Haytham's *Principles of Measurement*, despite the name of the author linked to a copy of that work; a later surveying text datable to after the mid 1400s, provides more details and statements about the state of affairs at this later period.

---

71 See Sabra, *The Optics of Ibn al-Haytham*, II, pp. lxiv–lxxiii; Sabra, 'Commentary that Saved the Text', pp. 131–33 (translated preface); Kheirandish, 'Light and Dark', pp. 72–77, and p. 84, n. 9 (include the preface, as well as parts of it not discussed elsewhere).

72 On Abū Bakr al-Khalīl al-Tājir's transcription of one of the manuscripts of Ibn al-Haytham's *Principles of Measurement*, see Rāshid, *Ibn al-Haytham, théorie des coniques*, III, 513 and Rashed, *Ibn al-Haytham's Theory of Conics*, III, 504. Successive publications where the editions and English translations of the text are respectively included. On this figure, see also Kheirandish, 'An Early Tradition in Practical Geometry'; Gülru Necipoğlu, 'Ornamental Geometries: A Persian Compendium at the Intersection of the Visual Arts and Mathematical Sciences,' in *The Arts of Ornamental Geometry: A Persian Compendium on Similar and Complementary Interlocking Figures*, ed. by Gülru Necipoğlu, Supplement to *Muqarnas*: *An Annual on the Visual Cultures of the Islamic World* (Leiden: Brill, 2018).

73 On the textual variation of the name Abū Bakr al-Khalīl al-Tājir, see Kheirandish, 'An Early Tradition in Practical Geometry'; Necipoğlu, 'Ornamental Geometries', pp. 11–78.

74 Lindberg, *Theories of Vision*, 148.

This late Persian surveying text, containing the terms *irtifāʿ* ('height') and *mirʾāt* ('mirror') in its alternative titles, is authored by Badr al-Dīn Ṭabarī (*c.* 1482), a figure known through his only other composition, a commentary on a work by Naṣīr al-Dīn Ṭūsī.[75] To his opening lines recounting how he was asked whether or not it is possible to determine the height of upright elevations by means of a *mirʾāt* ('mirror'), he adds the key statements that 'on this matter' he has not seen anything; that from the *ahl-i ṣināʿat* ('people of this craft'), namely surveying, nothing has reached him suggesting otherwise; and that only after his *istikhrāj* ('derivation'), *ʿamal* ('construction') and *burhān* ('demonstration') it came to his *khāṭir* (mind) that it is also possible to determine the height of elevations by the method of *ʿamūd* ('rods'). His two methods which he loosely calls *wajh* ('ways') are then presented as two chapters, much in the same way as the earlier Persian surveying text by Abū Bakr al-Khalīl, without any mention of a work by him so closely related to it in provenance and language as well as subject (Appendix: 4.2).[76]

None of these cases have anything comparable to developments in linear perspective occurring around the same time in European lands. Surprisingly, there is a slightly later case involving mirrors that points to developments in Islamic lands in directions opposite to those expected. In a striking passage from an anonymous text already mentioned in both its Arabic and Persian versions, a text datable to the early 1500s through dedications to the Ottoman Sultan Selim I (*r.* 1512–20) specified as son of Bayezid II (*r.* 1481–1512), high standing *mirʾāt* ('mirrors') constructed by *ḥukamā* ('scholars') from *Farangistān* ('Europe') are reported to have been sent to the *ʿulamā* ('learned figures') in Khurāsān in northeast Iran, for the *bayān* ('explanation') of its *ḥikmat* ('rationale') through *tajriba* ('experience') and *imtiḥān* ('testing').[77] Though the striking report does not specify the time of that unexpected exchange, the descriptions of those mirrors, with features involving various visual experiences described when looking into them at close, medium, and far distances, suggest more than the supposedly more advanced state of such developments in Islamic lands of a supposedly late period. The specific use of the expression 'wondrous' in the context of practices more advanced in *Khurāsān* ('Iran') than *Farangistān* ('Europe') of the early modern period, has positive associations that were also identified above in works datable to at least the late 1400s. As such, they all challenge arguments related to the non-development of linear perspective in Islamic lands in terms of the negative values of visual illusions as a determining factor for the absence of three-dimensional space in

---

75 Kheirandish, *The Arabic Version of Euclid's 'Optics'*, I, p. L. On the text and its known manuscripts, Kheirandish, *The Arabic Version of Euclid's 'Optics'*, I, p. 1; Rozenfel'd and İhsanoğlu, *Mathematicians, Astronomers, and Other Scholars*, p. 298, no. 876, A1 ('Ṭūsī commentary'); A2 (translated as 'Altitude'). The facsimile reproduction of the text is included in Kheiranish, 'The Medieval Tradition of Euklid's "Optika"', I.

76 Badr al-Dīn Ṭabarī, *Irtifāʿ* ('Height'): The opening lines specify the author had access to works on surveying; but his treatment is comparable to those who hold that in a proportional relation between four entities, such as a relation between geometrical magnitudes (lines), if one of the magnitudes (length, depth, width) is unknown, it can be calculated from the magnitude of the other three by measurement.

77 Anonymous, *Risāla fī sabab ruʾya al-ashyāʾ wa bayān madhāhib fihi and bayān ruʾya al-ashyāʾ fī al-marāyā al-maṣnūʿa* ('Treatise on the Cause of the Visibility of Objects, Explanation of Its Traditions, and Vision of Objects in Constructed Mirrors'). On the text and passage under discussion in Arabic and Persian, see Kheirandish, 'Optics, History of'; on the Persian manuscript of that text, see Kheirandish, *The Arabic Version of Euclid's 'Optics'*, I, pp. 1, lviii.

Islamic paintings.[78] In the same way, explanations in terms of transmission discussed in detail here challenge those placed in the language of the different 'mental sets', 'mindsets', or 'iconoclastic' cultures of the Islamic East.[79]

To conclude with an imagery from optics and perspective, just as the subject of transmission acts as a 'window' to illuminate the early history of optics, the much broader 'landscape' of perspective allows for viewing developments, non-developments, and partial developments through transmission, non-transmission, and partial transmission, as the case may be. The evidence of the historical sources included in this chapter from the scientific, philosophical, literary, and practical traditions (see Appendices 1–4), all point to the consequences of the different forms and processes of transmission, both in and beyond the Islamic Middle Ages, as part of the evolving cultures of optical knowledge and perspective practices that form the focus of the present volume.

Appendix[80]

1. Angle and Distance through Arabic Optical Sources

1.1. *Kitāb Uqlīdis fī Ikhtilāf al-Manāẓir* ('The Book of Euclid on Optics/Difference of Aspects'), trans. by Ibn Sarḥūn (*c.* 827–28): Kheirandish, *The Arabic Version of Euclid's 'Optics'*, I, pp. 1–225; II (Figure 1.1).

1.1.a. A definition from Kheirandish, *The Arabic Version of Euclid's 'Optics'*, I, Definition 2, pp. 2–3; II, pp. 15–18:

> The figure enclosed by the ray is a cone (*makhrūṭ*) whose apex is next to the eye and whose base next to the extremity of the visible object.

1.1.b. A proposition from Kheirandish, *The Arabic Version of Euclid's 'Optics'*, I, Proposition 9 (8 in Greek), pp. 26–29; II, pp. 40–44:

> Enunciation: [For] equal parallel magnitudes with locations unequally distant [from the eye] their difference of aspects *(Ikhtilāf al-Manāẓir)* is not seen according to (*'ala*) the [ratio of the] magnitudes (*aqdār*) of their distances.
> [...]
> Conclusion: So it is demonstrated that [for] equal parallel magnitudes unequally distant [from the eye] their *Ikhtilaf al-Manāẓir)* is not seen according to (*'ala*) the [ratio of the] magnitudes (*aqdār*) of their distances.

78 See Pūrjavādī, 'Concept of Perspective', pp. 23–25.

79 See respectively Edgerton, *The Renaissance Rediscovery*, 5, 32; Belting, *Florence and Baghdad*, 29–31; Veltman, *Linear Perspective*, Plate 59, all cited in note 10 and discussed above.

80 Parenthetical transliterations for corresponding Arabic and Persian terms are added, as are bracketed inserts within English translations and double brackets accounting for the single brackets of published translations. All my English translations are as literal as possible to capture a sense of the transmission problems at the time.

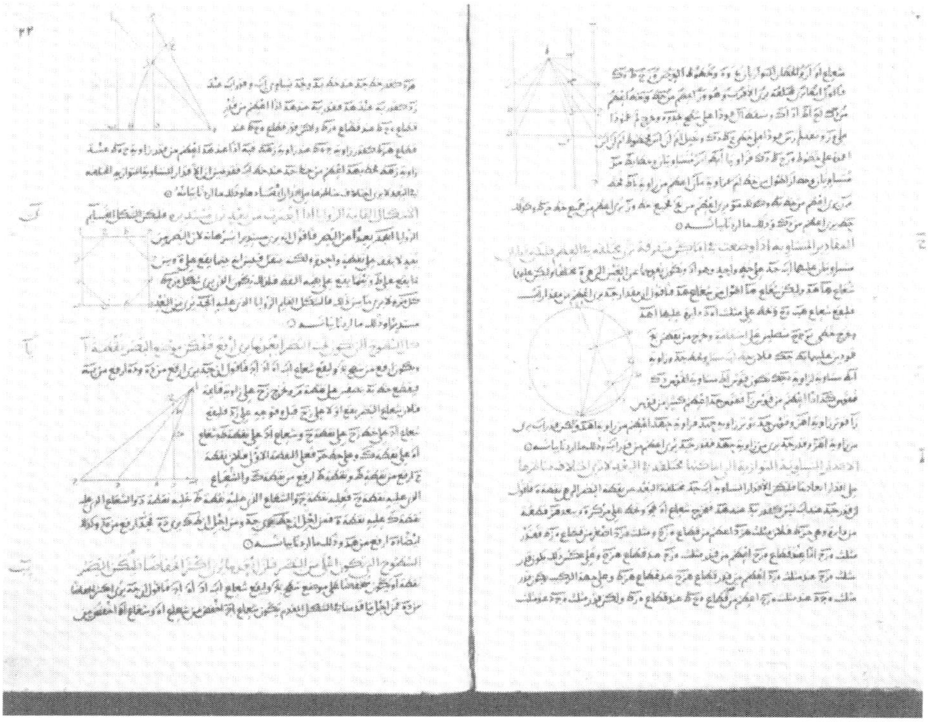

Fig. 1.1 Manuscript folios from *Kitāb Uqlīdis fī Ikhtilāf al-Manāẓir* ('The Book of Euclid on Optics/ Difference of Aspects'). Basel: Universitätsbibliothek microfilm; site of original manuscript unknown.

1.2. *Kitāb al-Manāẓir* by Ibn al-Haytham (d. after *c.* 1040–41), Selections: I, Chapter 2, Sections 20–21: Sabra (ed.), p. 69; Sabra (tr.), 1, p. 11; I, Chapter 6, Sections 59–60: Sabra (ed.), pp. 159–160; Sabra (tr.), 1, pp. 81–82; II, Chapter 3, Sections 67–68; Sabra (ed.), pp. 243–44; Sabra (tr.), 1, p. 149; III, Chapter 7, Section 14: Sabra (ed.), p. 419; Sabra (tr.), 1, pp. 283-84 (Figure 1.2).

1.2.a. Sabra, *The Optics of Ibn al-Haytham*, I, Chapter 2, Sections 20–21: 1, p. 11: [Discussions on the Forms and Perception of Distance]:

> [20] It follows from what we have stated and gathered by induction (*itsiqrā'*) regarding distances (*ab'ād*) that the distances from which an object can be perceived and those at which an object becomes invisible are according to the conditions (*aḥwāl*) and properties of the object itself, and also according to the strength or weaknesses of the sight itself that perceives it. [21] Therefore, from all that we have stated and found by induction (*itsiqrā'*) and experiment (*i'tibār*) to be uniform and subject to no variations (*ikhtilāf*) and contradictions (*tanaquḍ*), it is evident that sight does not perceive any object that exists with it in the same air and is not perceived by reflection (*in'ikās*), unless that object combines the conditions we have stated — namely: that there exists between it and the eye a certain distance (*bu'd*) proportionate [according to (*bi ḥasb*)] to that object; that it lies opposite the eye — I mean that an imaginary (*mutawahham*)

Fig. 1.2 Double manuscript folios from *Kitāb al-Manāẓir* ('Book of Optics') by Ibn al-Haytham. Courtesy of Süleymaniye Library, Istanbul Fatih 3212, book 3, fols 1v-2r, dated 476/1083.

[straight] line (*khaṭṭ*) exists between each point on its visible surface and a certain point on the surface of the eye [...] [II. 3, 67] [...] The place where a visible object is consists of three things: distance [as such] (*bu'd*), direction (*jiha*), and magnitude [or measure] (*kammīya*) of the distance [...] [68] [...] perception (*idrāk*) of distance (*bu'd*), i.e. non-contiguity, is different from perception of the measure of [spatial] interval (*masāfa*), namely the magnitude (*miqdār*) of distance (*bu'd*); nor are these properties perceived in the same manner [...] [III. 7, 14] The reason why sight perceives an object at an exceedingly great distance (*bu'd mutafāwwit*) to be smaller than its real (*ḥaqīqī*) magnitude (*miqdār*) is that the size (*'iẓam*) of an object is only perceptible by estimating (*qiyās*) the object's size by the angle of the cone (*makhrūṭ*) that surrounds it together with the magnitude of the object's distance [...] the distance at which a magnitude (*miqdār*) having a measurable (*muqtadar*) ratio (*nisba*) to the seen object does not cease to be visible is one of the moderate distances (*ab'ād mu'tadila*) at which the object is perceptible as it is [...]

1.2.b. Sabra, *The Optics of Ibn al-Haytham*, I, Chapter 6, Sections 59–60: 1, pp. 81–82, [Discussions on the Manner and Geometry of Vision]:

[59] [...] All that mathematicians who hold the doctrine of the ray (*shu'ā'*) have used in their reasonings (*maqāyīs*) and demonstrations (*barāhīn*) are imaginary (*muawahhama*) lines (*khuṭūṭ*) which they call 'lines of the ray' (*khuṭūṭ al-shu'ā'*). And we have shown that the eye cannot perceive any visible object except through these lines alone. Thus the view of those who take the radial lines to be imaginary lines is correct (*ṣaḥīḥ*), and

we have shown that vision is not affected without them. But the view of those who think that something issues from the eye other than the imaginary lines is impossible and we have shown the impossibility by the fact that it is not warranted by anything that exists, nor is there a reason (*ʿilla*) for it or an argument (*ḥujja*) that supports it. [60] It is therefore evident from all that we have shown that the eye senses the light and color that are in the surface of a visible object only through the form of that light and colour, which [form] extends from the object to the eye in the intermediate transparent body; and that the eye does not perceive any of the forms reaching it except through the straight lines which are imagined to extend between the visible object and the center of the eye and which are perpendicular (*aʿmida*) to all surfaces (*suṭūḥ*) of the coats of the eye. And that is what we wished to prove.

2. Direction and Depth through Arabic Philosophical Sources

2.1. *Kitāb Muḥaṣṣal* by Fakhr al-Dīn al-Rāzī (*c.* 1149 or 1150–1210), edition *Kitāb Muḥaṣṣal afkār al-mutaqaddimīn wa-al-mutaʾakhkhirīn min al-ʿulamāʾ wa-al-ḥukamāʾ wa-al-mutakallimīn* (Cairo: al-Ḥusaynīyah al-Miṣrīyah, 1323 = 1905).

2.2. *Talkhīṣ al-Muḥaṣṣal* by Naṣīr al-Dīn al-Ṭūsī (d. *c.* 1274), ed. by A. Nūrānī (English translation by Elaheh Kheirandish). (Figure 2)

2.2.a. Ṭūsī, *Talkhīṣ al-Muḥaṣṣal*, 173–74 [Discourses involving Direction of Radiation]

He [Fakhr al-Dīn al-Rāzī] said [re: 'extramission' arguments]
[Scholars] disagree on [the manner of] sight. Some among them say it is by rays (*shuʿāʿ*) issuing from the eye, and that is invalid, otherwise it would be necessary for sight to be disturbed by blowing winds…

And I [Naṣīr al-Dīn al-Ṭūsī] say:
Proponents of [visual-] ray (*shuʿāʿ*), who are early philosophers, only state its issuing from the eye by metaphor (*majāz*), just as it is said light issues from the Sun. And refuting that by [arguing] necessity of its disturbance near blowing winds is invalid, because the ray of the Sun and the Moon and luminous [bodies] are not disturbed by it […]

2.2.b. Ṭūsī, *Talkhīṣ al-Muḥaṣṣal*, 173–74 [Discourses Involving Depth Perception]:

And he (Fakhr al-Dīn al-Rāzī) said [re: 'intromission' arguments]:
And among them [i.e. scholars][some] say [sight] is by impression [of form] (*inṭibāʿ*), and that is [also] invalid, otherwise why would we perceive large [objects], as it is unfeasible for large [objects] to make an impression on small [objects such as the eye]; also [with sight by impression of form], why would we see the close [object] at its [own] close [distance] and the far [object] at its [own far] distance? These two sides [of the argument] require one to say: the visible [object] is an impressed form only. And [any] one who takes the impression of a small form on the [eye]ball as a condition for the perception of a large visible [object] from outside, would not go for that [position].

Fig. 2 Double manuscript folios from *Talkhīṣ al-Muḥaṣṣal* ('Abridgement [of Fakhr al-Dīn al-Rāzī's] *Kitāb al-Muḥaṣṣal*') by Naṣīr al-Dīn al-Ṭūsī. © The British Library Board Delhi Arabic 1365, fols 4ᵛ-5ʳ.

And I (Naṣīr al-Dīn al-Ṭūsī) say:

Indeed, Aristotle and his followers discussed [sight] by impression [of form], and explained the reason why vision of a large [object] from a far distance could be small. But invalidating it [sight by impression] through the unfeasibility of the impression of large [bodies] upon small [bodies such as the eye] is incorrect. Because they [intromissionists] had not, in their case, made impression of a large [body], either of itself or its size, a condition. Rather, they had spoken of the impression of the figure (*shabaḥ*) [of that body]. So perhaps the magnitude of that figure in [such] small locations requires perception of the embodiment of that figure in its large magnitude, in the same way that half of the sky and the bodies in it are impressed upon a mirror (*mir'āt*). As for vision of a close object as its close [distance], and of a far [object] at its [own] distance (*bu'd*), namely [for both] distances (*ab'ād*), what is perhaps impressed upon the eye is a configuration (*hay'a*) according to perception of depths (*a'māq*). But since it is difficult for us to make an examination of that [subject], we put it away, although we see artists (*naqqāshān*) who draw/paint/sculpt (*yanqushūn*) forms of bodies upon surfaces in a way that makes an observer perceive the depths (*a'māq*) of those bodies, and dimensions/distances (*ab'ād*) between them.

Fig. 3 Double Manuscript folios of Persian translation of *Kalīla and Dimna* by Abu al-Maʿālī, Naṣr-Allāh Munshī (twelfth century). © The British Library Board Or 13506, fols 40ᵛ-41ʳ. [See colour plate 19]

3. Outline and Colour through Persian Literary Sources

3.1. *Kalīla va Dimna,* Persian translations of an Arabic version by Ibn al-Muqaffaʿ (fl. c. 720–57) of a text in Middle Persian (not extant: Indian original(s) and Syriac version of story: extant)

3.1.a. 'A Story Involving Outward and Inward Appearances', from the Persian translation of Abu al-Maʿālī, Naṣr-Allāh Munshī (twelfth century), ed. by M. Mīnuvī (Tehran, 1343 = 1964), p. 66; a Persian transcription from the above edition and its citation is in N. Pūrjavādī, 'The Concept of Perspective', p. 20. The manuscript used here: British Library Or. 13506 (fols 4b-5a), is not included in either publication: English translation by Elaheh Kheirandish) (Figure 3).

3.1.b. 'A Story Involving Outward and Inward Appearances', from the Persian translation of Muḥammad ibn 'Abd-Allāh Bukhārī (twelfth century), ed. by P. Nātil Khānlarī and M. Rushan (Tehran: Khwārazmi, 1361 = 1982), p. 77; Persian transcription and citation of the above edition is in Pūrjavādī, 'The Concept of Perspective', p. 19 (English translation by Elaheh Kheirandish):

> The man (mard) making forms (*ṣurat-gar*), since he is a master (*ustād*) of his profession (*pīsha*),
> makes forms (*ṣurat-hā kunad*) [such] that one thinks (*pindārad*) they are in the wall (*dar dīvar*),
> while they are [actually] not in the wall;
> or that they are outside (*bīrūn*) [of the wall], while they are [actually] not outside of it.

3.2. A story about the contest of the craft (*ṣinā'a*) of people from different regions, reported in Arabic by Ghazzālī (d. *c.* 1111) in *Iḥyā' al-'ulūm al-Dīn*, III, and by Maqrīzī (*c.* 1364–1442) on the period *c.* 1050–58.

3.2.a. A poem in Persian from Niẓāmī Ganjavī (*c.* 1149–1209), *Iskandar-nāma*, ed. by B. Furūzānfar, *Ma'ākhidh qiṣaṣ*, pp. 34–35 (English translation by Elaheh Kheirandish):

> There was only one difference between them [the two competing designs]
> One was accepting (*padhīruft*), one was giving out (*minimūd*)
> The decree of that judgement was that both [forms] are assisted by sight (*baṣar yāvarī*)
> No one knows [the art of] impressed form[s] (*naqsh bast*) like the Greek[s
> While the Chin[ese] have an expert hand in polishing (*ṣayqal*)

3.2.b. A poem in Persian from *Mathnavī* of Maulānā Jalāl al-Dīn Balkhī Rūmī (Maulanā Rūmī, d. *c.* 1273), ed. by B. Furūzānfar in *Ma'ākhidh qiṣaṣ*, 213-15;; English translation of Reynold A. Nicholson, 'The story of the Contention between the Greek and the Chinese in the art of painting and picturing', *The Mathnawī of Jalāl al-Dīn Rūmī*, 189–91:

> The Chinese said "we are the better artists" (*naqqāsh*)
> The Greeks said, "The (superiority in) power and excellence belongs to us"
> "I will put you to the test in this matter", said the Sultan,
> "(and see) which of you are approved in your claim" [...]
> Ever since the form (*nuqūsh*) of the Eight Paradises have shone forth (*tāftan*),
> They have found the tablet (*lawḥ*) of their [*Ṣūfī's*] hearts receptive (*padhīrā*) of impression (*nishān*) [...]

4. Mirror and Measure though Persian Surveying Sources

4.1. *A'māl* ('Constructions'), by Abū Bakr al-Khalīl al-Tājir' (*c.* 1300s–1400s): *'Amal* (Construction) of Elevated (*murtafa'*) [Height]s [by] Reflection (*in'ikās*) through Mirrors (*mir'āt*)': Facsimile reproduction in Necipoğlu, 'Ornamental Geometries', p. 16 (English translation by Elaheh Kheirandish). (Figure 4)

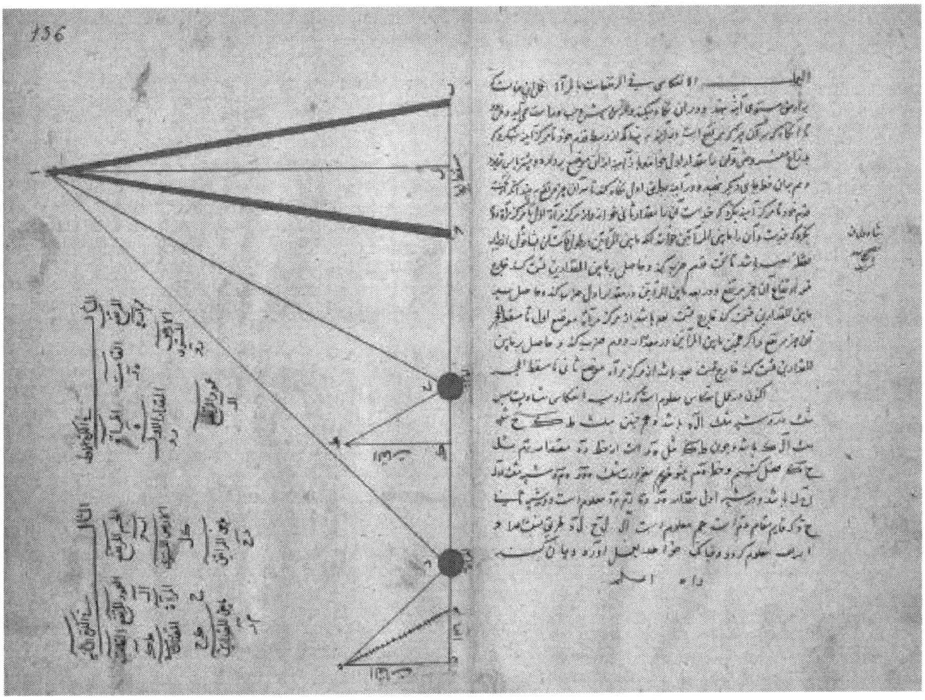

Fig. 4 Double Manuscript folio of the Persian text of *Constructions by Abū Bakr al-Khalīl al-Tājir* (ca. 1300s-1400s?): 'Construction of Elevated [Height]s [by] Reflection through Mirrors'. text (R); illustration (L). Paris, Bibliothèque nationale de France, Ms. Persan 169, fols 136^{v}-137^{r}.

The Construction is such that one places a mirror on a leveled plane and looks into it; then he keeps moving from the front and back and left and right until he sees the top of the elevated [height] in the mirror (*mir'āt*); then he looks from the middle of his steps to the center of the mirror at an assumed cubit [magnitude] and calls it the first magnitude (*miqdār*); again takes the mirror from that location whether in front or the back of it and places it on the same line in a different place. Then he looks into the mirror by the first method until he sees the top of the elevated object; then he looks to see what [magnitude] is from the bottom of his steps to the center of the mirror, and calls that the second magnitude (*miqdār thānī*). Then he looks at what is the [magnitude] from the first mirror to the second mirror, and calls it '[the magnitude] in between the two mirrors' [...] Now in the construction [by] reflection (*in'ikās*) it is known (*ma'lūm*) that the angle of reflection (*in'ikās*) is equal [to the angle of incidence] so the triangle[s are all similar] [...] and [the unknown height] becomes known (*ma'lūm*) by the way (*ṭarīq*) [i.e. method] of the ratio/proportion of four numbers (*nisbat-i a'dād-i arba'a*) [such that the unknown could be calculated from the other known ones; note that the present translation of the latter expression is conceptually and methodologically different from 'numbers for which square roots exist' in Necipoğlu, 'Ornamental Geometries', p. 17.

4.2. *Irtifāʿ* ('[Determination of] Height') by Badr al-Dīn al-Ṭabarī (*c.* 824 = 1420); Facsimile reproduction in Elaheh Kheirandish, 'The Medieval Tradition of Euclid's "Optika"' (English translation by Elaheh Kheirandish): *ʿAmal* ('Construction') and *Burhān* ('Demonstration') for Height Determination through Mirrors:

> And then, says the writer of these lines, Badr [al-Dīn] al-Ṭabarī [...] Some dear ones (*aḥibā*) asked this least [worthy subject] (*kamīna*) whether or not it is possible to [determine by] mirror (*mirʾāt*) the height of upright [figure]s (*ashkhāṣ*) from the ground. On this matter (*bāb*) a thought occurred to this impoverished [subject] (*faqīr*) without seeing [anything] anywhere, [that if] the people of this craft (*ahl-i ṣināʿat*) have opposed (*mutaʿarriḍ*), it has not reached this feeble [subject] (*ḍaʿīf*). And [only] after the construction (*ʿamal*) and demonstration (*burhān*) was derived (*istikhrāj*) then it came to [my] mind (*khāṭir*), that it is also possible to make known (*maʿlūm*) the height of [elevated] upright [figure]s by a [perpendicular] rod (*ʿamūd*) [method]. In short, [I] derived (*istikhrāj*) the construction (*ʿamal*) and demonstration (*burhān*) in both those ways (*wajh*) and brought into the two chapters of the [present] book (*kitāb*).

Bibliography

Primary Sources

Alhacen, *Opticae thesaurus: Alhazeni Arabis libri septem, nuncprimum editi; Eiusdem liber de crepusculis et nubium ascensionibus, item vitellonis thuringopoloni libri X*, ed. by Friedrich Risner (Basel: Bischoff, 1572; repr. New York: Johnson, 1972).

Alhacen, *Alhacen's Theory of Visual Perception: A Critical Edition, with English Translation and Commentary, of the First Three Books of Alhacen's 'De aspectibus', the Medieval Latin Version of Ibn al-Haytham's 'Kitāb al-Manāẓir'*, ed. and trans. by A. Mark Smith, 2 vols (Philadelphia: Transactions of the American Philosophical Society, 2001).

Alhacen, *Alhacen on the Principles of Reflection: A Critical Edition, with English Translation and Commentary, of Books 4 and 5 of Alhacen's 'De aspectibus'*, ed. and trans. by A. Mark Smith, 2 vols (Philadelphia: Transactions of the American Philosophical Society, 2006).

Alhacen, *Alhacen on Image-Formation and Distortion in Mirrors: A Critical Edition, with English Translation and Commentary, of Book 6 of Alhacen's 'De aspectibus'*, ed. and trans. by A. Mark Smith, 2 vols (Philadelphia: Transactions of the American Philosophical Society, 2008).

Alhacen, *Alhacen on Refraction: A Critical Edition, with English Translation and Commentary, of Book 7 of Alhacen's 'De aspectibus'*, ed. and trans. by A. Mark Smith, 2 vols (Philadelphia: Transactions of the American Philosophical Society, 2010).

Atıl, Esin (ed.), *Kalīla wa-dimna: Fables from a Fourteenth-Century Arabic Manuscript* (Washington, D.C.: Smithsonian Institution Press, 1981).

Bīrūnī, Abū Rayḥān, *The Exhaustive Treatise on Shadows*, ed. and trans. by E. S. Kennedy (Aleppo: University of Aleppo Press, 1976).

Bīrūnī, Abū Rayḥān, *Kitāb al-Tafhīm li-awāʾil ṣināʿat al-tanjīm*, facs. ed. by R. Ramsay Wright (London: Luzac & Co., 1934; repr., Frankfurt am Main: Institute for the History of Arabic-Islamic Science, 1998; Persian ed. by J. Humāʾī (Tehran: Bābak, 1362 = 1983).

Björnbo, Axel Anthon, and Sebastian Vogl, *Alkindi, Tideus und Pseudo-Euklid: Drei optische Werke. Abhandlungen zur Geschichte der mathematischen Wissenschaften* (Leipzig: Teubner, 1912).

Chikho, Louis (ed.), *Kalīla wa-dimna* (Bayrut: al-Maṭba'ah al-Kāthūlīkīyah, 1952).

Euclid, *Euclidis opera omnia: Optica, opticorum recensio theonis, catoptrica, cum scholiis antiquis*, ed. by J. L. Heiberg and H. Menge (Leipzig: Teubner, 1895).

Fārābī, Abū Naṣr, *Iḥṣā' al-'ulūm*, ed. by 'Uthmān Amīn (Cairo: Maktabat al-Anjalū al-Miṣriyya, 1968).

Furūzānfar, B., *Sources of the Stories and Depictions of the Mathnawī (Ma'ākhidh qiṣaṣ va tamthīlāt-i mathnavī)* (Tehran, 1333 = 1954).

Heath, Thomas L., *The Thirteen Books of Euclid's Elements* (Cambridge: Cambridge University Press, 1956).

Ibn al-Haytham, *Kitāb al-Manāẓir, Books I–III: On Direct Vision*, ed. by Abdelhamid I. Sabra (Kuwait: The National Council for Culture, Arts and Letters, 1983).

Ibn al-Haytham, *Kitāb al-Manāẓir, Books IV–V: On Reflection, and Images Seen by Reflection*, ed. by Abdelhamid I. Sabra, 2 vols (Kuwait: The National Council for Culture, Arts and Letters, 2002).

Ibn, Muḥammad, 'Abd Allāh al-Bukhārī, "Kalīla wa-dimna", in *Dāstānhā-yi bīdpāy*, ed. by Parvīz Nātil Khānlarī and Muḥammad Rawshan (Tehran: Khvārazmī, 1361 = 1982).

Kheirandish, Elaheh, *The Arabic Version of Euclid's 'Optics': Kitāb Uqlīdis fī Ikhtilāf al-Manāẓir, Edited and Translated with Historical Introduction and Commentary*, 2 vols, Sources in the History of Mathematics and Physical Sciences series, ed. by G. J. Toomer (New York: Springer-Verlag, 1999).

Munshī, Abū al-Ma'ālī Naṣr Allāh, *Kalīla wa-dimna, tarjumah-i kalīla wa-dimna*, ed. by Mujtabā Mīnuvī Ṭihrānī (Tehran: Dānishgāh-i Tihrān, 1343 = 1964).

Ragep, Jamil F., *Naṣīr al-Dīn al-Ṭūsī's Memoir on Astronomy (al-Tadhkira fī 'ilm al-hay'a)*, 2 vols (New York: Springer-Verlag, 1993).

Rashed, Roshdi, *Ibn al-Haytham's Theory of Conics, Geometrical Constructions and Practical Geometry: A History of Arabic Sciences and Mathematics*, trans. by J. V. Field (London: Centre for Arab Unity Studies, Routledge, 2013).

Rāshid, Rushdī, *Ibn al-Haytham, théorie des coniques, constructions géométriques et géométrie pratique, les mathématiques infinitésimales du IXᵉ au XIᵉ siècle*, 5 vols (London: Al-Furqān Islamic Heritage Foundation, 2000).

Rāzī, Fakhr al-Dīn, *Jāmi' al-'ulūm = Kitāb-i Sittīnī*, ed. by Muḥammad Ḥusayn Tasbīḥī (Tehran: Kitābkhānah-i Asadī, 1346 = 1967).

Rāzī, Fakhr al-Dīn, *Jāmi' al-'ulūm = Kitāb-i Sittīnī*, ed. by 'Alī Dāvūd (Tehran: Bunyād-i Mawqūfāt-i Duktur Maḥmūd Afshār Yazdī, 1382 = 2003).

Sabra, A. I., *The Optics of Ibn al-Haytham: Books I–III on Direct Vision, Translated with Introduction and Commentary*, 2 vols, Studies of the Warburg Institute series, ed. by J. B. Trapp (London: Warburg Institute, 1989).

Smith, A. Mark, 'Ptolemy's Theory of Visual Perception: An English Translation of the "Optics" with Introduction and Commentary', *Transactions of the American Philosophical Society*, 86 (2) (1996), iii–v, vii, xi, 1–61, 63–261, 263–269, 279–300.

Theisen, Wilfred R., 'Liber de visu: The Greco-Latin Translation of Euclid's "Optics", Edited with Introduction and Notes', in *Mediaeval Studies*, 41 (1979), 44–105.

Ṭūsī, Naṣīr al-Dīn, *Taḥrīr al-Manāẓir*, in *Rasā'il* (Hyderabad: Dā'irat al-Ma'ārif al-'Uthmānīyah, 1358 = 1939).
Ṭūsī, Naṣīr al-Dīn, *Taḥrīr al-Manāẓir*, in *Revue de l'institut des manuscrits arabes*, 9, ed. by A. S. Dimirdash (1383 = 1963), pp. 251–290.
Ṭūsī, Naṣīr al-Dīn, *Talkhīṣ al-Muḥaṣṣal*, ed. by A. Nurani (Montreal Canada: McGill University Institute of Islamic Studies, 1359 = 1980).

Secondary Works

Abattouy, Mohammad, 'Nutaf min al-ḥiyal: a Partial Arabic Version of Pseudo-Aristotle's "Problemata Mechanica"', *Early Science and Medicine*, 6 (2) (2001), 96–122.
Andersen, Kirsti, 'Ancient Roots of Linear Perspective', in *From Ancient Omens to Statistical Mechanics: Essays on the Exact Sciences Presented to Asger Aaboe*, ed. by J. L. Berggren and B. R. Goldstein (Copenhagen: University Library, 1987), pp. 75–89.
Arnold, Thomas W., *Painting in Islam: A Study of the Place of Pictorial Art in Muslim Culture* (Oxford: Clarendon Press, 1928; repr. New York: Dover, 1965).
Belting, Hans, *Florenz und Bagdad: eine westöstliche Geschichte des Blicks* (München: C. H. Beck, 2008).
Belting, Hans, *Florence and Baghdad: Renaissance Art and Arab Science*, trans. by Deborah Lucas Schneider (Cambridge, Mass.: Belknap Press of Harvard University Press, 2011).
Brownson, C. D., 'Euclid's Optics and its Compatibility with Linear Perspective', *Archive for History of Exact Sciences*, 24 (3) (1981), 165–94.
Burton, H. E., 'The Optics of Euclid', *Journal of the Optical Society of America*, 35 (1945), 357–72.
Edgerton, Samuel Y., *The Renaissance Rediscovery of Linear Perspective* (New York: Basic Books, 1975).
Encyclopaedia Iranica: A Comprehensive Research Tool Dedicated to the Study of Iranian Civilization in the Middle East, the Caucasus, Central Asia, and the Indian Subcontinent (New York: Columbia University Press, 2010), 'Optics, History of' (http://www.iranicaonline.org/articles/optics [accessed 27-07-2017]).
Hogendijk, Jan P., 'Ancient and Modern Secrets of Isfahan', *Nieuw Archief voor Wiskunde*, 5 (9) (2008), 121.
Jones, Alexander, 'Peripatetic and Euclidean Theories of the Visual Ray', *Physis*, 31 (1) (1994), 47–76.
Kheirandish, Elaheh, 'The Medieval Tradition of Euclid's "Optika"', 2 vols (unpublished doctoral thesis, Harvard University, 1991).
Kheirandish, Elaheh, 'The Arabic "Version" of Euclidean Optics: Transformations as Linguistic Problems in Transmission', in *Tradition, Transmission, Transformation*, ed. by Jamil F. Ragep and Sally Ragep with Steven Livesey (Leiden: Brill, 1996), pp. 227–43.
Kheirandish, Elaheh, 'What "Euclid Said" to his Arabic Readers: The Case of the Optics', in *De diversis artibus*, ed. by Gerard Simon and Suzanne Debarbat (Turnhout: Brepols, 2001), Tome 55: N. S. 18, pp. 17–28.
Kheirandish, Elaheh, 'The Many Aspects of "Appearances"': Arabic Optics to 950 AD.', in *The Enterprise of Science in Islam: New Perspectives*, ed. by Jan P. Hogendijk and Abdelhamid Sabra (Cambridge: MIT Press, 2003), pp. 55–83.

Kheirandish, Elaheh, 'Mathematical Sciences through Persian Sources: The Puzzle of Ṭūsī's Optical Works', in *Sciences, techniques et instruments dans le monde iranien, x^{e}–xixe siècle*, ed. by Naṣr Allāh Pūrjavādi and Živa Vesel (Tehran: Institut français de recherche en Iran, 2004), pp. 197–213.

Kheirandish, Elaheh, 'Footprints of "Experiment" in Early Arabic Optics', in *Evidence and Interpretation in Studies on Early Science and Medicine: Essays in Honor of John E. Murdoch*, ed. by Edith Dudley Sylla and William R. Newman, *Early Science and Medicine*, 14 (2009), 79–104.

Kheirandish, Elaheh, 'The Mixed Mathematical Sciences: Optics and Mechanics in the Islamic Middle Ages', in *The Cambridge History of Science*, ed. by David C. Lindberg and Michael H. Shank (Cambridge: Cambridge University Press, 2013), II, pp. 84–108.

Kheirandish, Elaheh., 'Eloge A. I. Sabra (8 June 1924-18 December 2013)', *Early Science and Medicine*, 19 (3) (2014), 281–86.

Kheirandish, Elaheh, 'Light and Dark: The "Checkered History" of Early Optics', in *God is the Light of the Heavens and the Earth: Light in Islamic Art and Culture*, ed. by Jonathan Bloom and Sheila Blair (New Haven: Yale University Press, 2015), pp. 61–85.

Kheirandish, Elaheh, 'Astronomical Poems from the "Four Corners" of Persia (*c.* 1000–1500 CE)' in *Essays in Islamic Philology, History, and Philosophy*, ed. by Alireza Korangy, Wheeler M. Thackston, Roy P. Mottahedeh and William Granara (Berlin: Walter de Gruyter GmbH, 2016), pp. 51–90.

Kheirandish, Elaheh, '"Checkered History" Recolored: The Changing Fortunes and Misfortunes of Optical Works in Islamic and European Lands': in progress, presented in the session 'The Legacy of A. I. Sabra: New Perspectives on the History of Science in Islam', Middle East Studies Association (MESA), Boston: 19 November 2016.

Kheirandish, Elaheh, 'An Early Tradition in Practical Geometry: The Telling Lines of Unique Arabic and Persian Sources', in *The Arts of Ornamental Geometry: A Persian Compendium on Similar and Complementary Interlocking Figures*, ed. by Gülru Necipoğlu, Supplement to *Muqarnas: An Annual on the Visual Cultures of the Islamic World* (Leiden: Brill, 2017), pp. 79–144.

Knorr, Wilbur R., 'Archimedes and Pseudo-Euclidean Catoptrics: Early Stages in the Ancient Geometric Theory of Mirrors', *Archive for History of Exact Sciences*, 35 (1985), 28–105, 114–15.

Knorr, Wilbur R., 'On the Principle of Linear Perspective in Euclid's Optics', *Centaurus*, 34 (1991), 193–210.

Knorr, Wilbur R., 'Pseudo-Euclidean Reflections in Ancient Optics: A Re-Examination of Textual Issues Pertaining to the Euclidean "Optica" and "Catoptrica"', *Physis*, 31 (1) (1994), 1–45.

Lindberg, David C., *A Catalogue of Medieval and Renaissance Optical Manuscripts* (Toronto: Pontifical Institute of Mediaeval Studies, 1975).

Lindberg, David C., 'The Intromission-Extramission Controversy in Islamic Visual Theory: Alkindi versus Avicenna', in *Interrelations in the History and Philosophy of Science*, ed. by Peter K. Machamer and Robert Turnbull (Columbus: Ohio State University Press, 1978).

Lindberg, David C., *Theories of Vision from Al-Kindi to Kepler* (Chicago: University of Chicago Press, 1976).

Lindberg, David C., *Studies in the History of Medieval Optics* (London: Variorum Reprints, 1983).

Mottahedeh, Roy P., '"Ajā'ib" in the Thousand and One Nights', in *The Thousand and One Nights in Arabic Literature and Society*, ed. by Richard C. Hovannisian and Georges Sabagh (Cambridge: Cambridge University Press, 1997), pp. 29–39.

Necipoğlu, Gülru, 'The Scrutinizing Gaze in the Aesthetics of Islamic Visual Cultures: Sight, Insight and Desire', in *Gazing Otherwise: Modalities of Seeing in and Beyond the Lands of Islam*, ed. by Olga Busch and Avinoam Shalem, Supplement to *Muqarnas: An Annual on the Visual Cultures of the Islamic World* (Leiden: Brill, 2015), XXXII, pp. 23–61.

Necipoğlu, Gülru, 'Ornamental Geometries: A Persian Compendium at the Intersection of the Visual Arts and Mathematical Sciences,' in *The Arts of Ornamental Geometry: A Persian Compendium on Similar and Complementary Interlocking Figures*, ed. by Gülru Necipoğlu, Supplement to *Muqarnas: An Annual on the Visual Cultures of the Islamic World* (Leiden: Brill, 2017).

Park, Katherine, 'The Meaning of Natural Diversity: Marco Polo on the "Division" of the World', in *Texts and Contexts in Ancient and Medieval Science: Studies on the Occasion of John E. Murdoch's Seventieth Birthday*, ed. by Edith Sylla and Michael McVaugh (Leiden: Brill, 1997), pp. 134–47.

Park, Katherine, and Lorraine Daston, *Wonders, and the Order of Nature (1150-1750)* (New York: Zone Books, 1998).

Pūrjavādī, Naṣr Allāh, 'The Concept of Perspective in "Kalīla va Dimna" and the Reasons for the Absence of Three-Dimensional Space in Islamic Paintings' (in Persian), *Nashr-i Dānish*, 8 (1988), 5, 18–30.

Rabbat, Nasser O., 'Ajīb and Gharīb: Artistic Perception in Medieval Arabic Sources', *The Medieval History Journal*, 91 (1) (2006), 99–113.

Rāshid, Rushdī, *Geometry and Dioptrics in Classical Islam* (London: Al-Furqān Islamic Heritage Foundation, 2005).

Raynaud, Dominique, 'Why Did Geometrical Optics Not Lead to Perspective in Medieval Islam?' in *Raymond Boudon: A Life in Sociology*, ed. by M. Cherkaoui and P. Hamilton (Oxford: The Bardwell Press, 2009), pp. 243–66.

Raynaud, Dominique, *Optics and the Rise of Perspective: A Study in Network Knowledge Diffusion* (Oxford: The Bardwell Press, 2014).

Rosenthal, Franz, 'On the Knowledge of Plato's Philosophy in the Islamic World', *Islamic Culture*, 14 (1940), 412–16.

Roxburgh, David J., 'Two Point Perspective: On Hans Belting's "Florence and Baghdad"', *Artforum* (2012), 61–64.

Rozenfel'd, Boris A., and İhsanoğlu, Ekmeleddin, *Mathematicians, Astronomers, and Other Scholars of Islamic Civilization and Their Works, 7th–19th c.* (Istanbul: Research Center for Islamic History, Art and Culture, 2003).

Sabra, A. I., 'The Astronomical Origin of Ibn al-Haytham's Concept of Experiment, in *Actes XIIe congrès international d'histoire des sciences* (Paris: Albert Blanchard, 1971), III. A, pp. 133–36.

Sabra, A. I., 'Manāẓir, or "Ilm al-manāẓir"', in *Encyclopaedia of Islam, Second Edition, Glossary and Index of Terms*, ed. by P. J. Bearman, Th. Bianquis, C. E. Bosworth, E. van Donzel and W. P. Heinrichs (Leiden: Brill), 6 (1987), Fascicules 103–04, 376–77.

Sabra, A. I., 'The Appropriation and Subsequent Naturalization of Greek Science in Medieval Islam: A Preliminary Statement', *History of Science*, 25 (3) (1987), 223–43.

Sabra, A. I., 'The Commentary that Saved the Text: The Hazardous Journey of Ibn al-Haytham's Arabic Optics', *Early Science and Medicine*, 12 (2007), 117–33.

Saliba, George, 'The Function of Mechanical Devices in Medieval Islamic Society', in *Science and Technology in Medieval Society*, ed. by Pamela O. Lang, Annals of the New York Academy series (New York: New York Academy of Sciences, 1985), pp. 141–51.

Sayili, Aydin, *The Observatory in Islam and its Place in the General History of the Observatory* (Ankara: Turkish Historical Society, 1960).

Schiefsky, Mark J., 'Art and Nature in Ancient Mechanics', in *The Artificial and the Natural: an Evolving Polarity*, ed. by Bernadette Bensaude-Vincent and William R. Newman (Cambridge, Mass.: MIT Press, 2007), pp. 67–108.

Smith, A. Mark, 'The Latin Source of an Italian Translation of Alhacen's "De aspectibus" (Vat. Lat. 4595)', *Arabic Sciences and Philosophy*, 11 (2001), 27–43.

Soucek, Priscilla, 'Niẓāmī on Painters and Painting', in *Islamic Art in the Metropolitan Museum of Art*, ed. by Richard Ettinghausen (New York: Metropolitan Museum of Art, 1972), pp. 9–21.

Theisen, Wilfred R., 'The Mediaeval Tradition of Euclid's Optics' (unpublished doctoral thesis, University of Wisconsin, 1972; Facsimile, University Microfilms International, 1984).

Veltman, Kim H., *Optics and Perspective: A Study in the Problems of Size and Distance* (unpublished doctoral thesis, University of London, 1975).

Veltman, Kim H., 'Ptolemy and the Origin of Linear Perspective', in *Atti del convegno internazionale di studi: la prospettiva rinascimentale, Milan 1977*, ed. by Marisa Dalai-Emiliani (Florence: Centro Di, 1980), pp. 403–07.

Veltman, Kim H., in collaboration with Kenneth D. Keele, *Linear Perspective and the Visual Dimensions of Science and Art: Studies on Leonardo da Vinci I* (Munich: Deutscher Kunstverlag, 1986).

Waley, P., and Norah M. Titley, 'An Illustrated Persian Text of Kalīla and Dimna dated 707/1307-8', *The British Libray Journal*, 1 (1975), 42–61.

Walzer, Sofie, 'The Topkapi Saray Manuscript of the Persian "Kalīla wa-Dimna" (dated A.D. 1413)', in *Paintings from Islamic Lands*, ed. by R. Pinder-Wilson (Oxford: Cassirer, 1969), pp. 48–84.

Winter, H. J. J., and W. 'Arafat, 'A Statement on Optical Reflection and "Refraction" Attributed to Nasir ud-Din at-tusi', *ISIS* 42 (2) (1951), 138–42.

A. MARK SMITH

The Roots and Routes of Optical Lore in the Later Middle Ages and Renaissance

Introduction

Today we associate the term 'perspective' (from the Latin *perspectiva*; Italian *prospettiva* or *perspettiva*) with methods of geometrizing visual space in art. But this particular association dates to no earlier than the second half of the fifteenth century. For more than two centuries before that, the term in both its Latin and Italian forms referred exclusively to the science of optics, especially geometrical optics, and it continued to do so long after it took on its twin connection with art. Precisely when it began to assume this earlier, optical denotation is unclear, but by no later than the mid-thirteenth century the connection between *perspectiva* and optics was firmly established — so firmly, in fact, that by the end of the century *perspectiva* had become the universally accepted designation for optics, understood broadly as the study of light and vision.[1]

The Core Sources

Under this designation, a set of three major works appeared between roughly 1265 and 1280, all of them taking theoretical and analytic inspiration from Alhacen's *De aspectibus*, the Latin version of Ibn al-Haytham's seven book *Kitāb al-Manāẓir*. Likely completed around 1200, if not somewhat earlier, this translated version of the treatise seems to have had little impact before 1260 or so.[2] Not long after that, though, it began to exert significant influence,

1 An early instance of the use of the word *perspectiva* to denote optics can be found in Robert Grosseteste's, *De iride*, which was probably composed sometime between 1232 and 1235. For the text, see *Die Philosophischen Werke des Robert Grosseteste, Bischofs von Lincoln*, ed. by Ludwig Baur, Beiträge zur Geschichte der Philosophie des Mittelalters, 9 (Münster: Aschendorff Verlag, 1912), pp. 72–78. On the dating of the *De iride*, see James McEvoy, 'The Chronology of Robert Grosseteste's Writings on Nature and Natural Philosophy', *Speculum*, 58 (1983), 614–55.

2 On the probable composition date and early influence of Alhacen's *De aspectibus*, see Alhacen, *Alhacen's Theory of Visual Perception: A Critical Edition, with English Translation and Commentary, of the First Three Books of Alhacen's 'De aspectibus', the Medieval Latin Version of Ibn al-Haytham's 'Kitāb al-Manāẓir'*, ed. and trans. by A. Mark Smith, 2 vols (Philadelphia: Transactions of the American Philosophical Society, 2001), I, pp. XIX–XXI, LXXXI–LXXXIII. The

A. Mark Smith University of Missouri, smitham@missouri.edu

Perspective as Practice. Renaissance Cultures of Optics, ed. by Sven DUPRÉ, Turnhout, 2019 (*Techne. Knowledge, Technique, and Material Culture*, 1), p. 241-258

DOI 10.1484/M.*Techne*-EB.5.117728

spawning its first major derivative work: Roger Bacon's *Perspectiva*. Written sometime before 1267, the year in which it reached the hands of Pope Clement IV as Part 5 of the *Opus majus*, this work enjoyed a fairly wide dissemination during the Middle Ages and Renaissance not only as an integral part of the *Opus majus* but also as an independent treatise.[3]

Within roughly a decade of the writing of Bacon's *Perspectiva*, a second major derivative work appeared under that same title. Authored by the Polish savant Witelo, the text is so faithful to its Alhacenian model that the late Renaissance thinker Giambattista della Porta accused Witelo — not entirely unfairly — of merely aping Alhacen ('ab eo eadem simia Vitellio').[4] The third major work stemming from the *De aspectibus* is John Pecham's *Perspectiva communis*, which appeared quite soon after Witelo's *Perspectiva*, most likely before February, 1279, when Pecham was installed as Archbishop of Canterbury.[5] These three works, along with Alhacen's *De aspectibus* itself, form the core of what has become known as the Perspectivist optical tradition. It is important to note, however, that the science of *perspectiva* was not limited to physical or geometrical optics, as we understand those disciplines today. It was the science of appearances and, as such, dealt with every aspect of vision, from the radiation of luminous colour into the eye to the sensitive, perceptual, and cognitive grasp of the physical objects from which the radiation originates. The science of *perspectiva*, in short, aimed to explain how we come to understand the world and its logical structure through sight. As a consequence, it extended well beyond the analysis of light and colour, into the anatomical, physiological, psychological, and epistemological grounds of vision.[6]

The dating and provenance of the extant manuscripts and manuscript fragments representing the four core Perspectivist texts listed above suggest that the science of *perspectiva* enjoyed a fairly wide diffusion between roughly 1300 and 1500. In all, those texts have come down to us in as few as one hundred and fifty manuscripts — or, depending on what is counted, as many as 210 — that are either currently available in various collections all over Europe or now lost but known to have existed in the fairly recent past.[7] This number constitutes a fraction of the manuscript copies that were in circulation between the late thirteenth century and roughly 1500. In fact, a recent study of the survival rates of medieval manuscripts suggests that it may represent as little as six percent of the actual total for that period.[8] Furthermore, if we take into account every other known medieval treatise that deals

---

authorial ascription 'Alhacen' is a Latin transliteration of Ibn al-Haytham's given name, 'al-Ḥasan'. For the earliest printed edition of the work, see *Opticae thesaurus: Alhazeni Arabis libri septem, nuncprimum editi; Eiusdem liber de crepusculis et nubium ascensionibus, item Vitellonis Thuringopoloni libri X*, ed. by Friedrich Risner (Basel: Bischoff, 1572).

3 See Roger Bacon, *Roger Bacon and the Origins of 'Perspectiva' in the Middle Ages. A Critical Edition and English Translation of Bacon's Perspectiva with Introduction and Notes*, ed. and trans. by David C. Lindberg (Oxford: Clarendon Press, 1996), pp. XVIII–XIX, c–ci.

4 Giambattista della Porta, *De refractione optices parte: libri novem* (Naples: Carlino & Pace, 1593), p. 76. An edition of all ten books of Witelo's *Perspectiva* (along with Alhacen's *De aspectibus*) can be found in *Opticae thesaurus*, ed. by Risner.

5 See John Pecham, *John Pecham and the Science of Optics: Perspectiva communis*, ed. and trans. by David C. Lindberg (Madison: The University of Wisconsin Press, 1970), pp. 17–18.

6 See e.g. A. Mark Smith, 'What is the History of Medieval Optics Really About?', *Proceedings of the American Philosophical Society*, 148 (2004), 180–94.

7 See Dominique Raynaud, *Optics and the Rise of Perspective* (Oxford: Bardwell Press, 2014), p. 64. For Raynaud's source, see David C. Lindberg, *A Catalogue of Medieval and Renaissance Optical Manuscripts* (Toronto: University of Toronto Press, 1975).

8 See Eltjo Buringh, *Medieval Manuscript Production in the Latin West* (Leiden: Brill, 2011).

more or less explicitly with the science of *perspectiva*, the number of extant manuscripts rises to slightly over three hundred, which could mean that as many as forty-five hundred were in circulation between roughly 1300 and 1500. The distribution of those numbers tells an important story. Of the one hundred fifty manuscripts containing the four core perspectivist sources, Pecham's *Perspectiva communis* is represented by sixty-four, Bacon's *Perspectiva* by thirty-nine, Witelo's *Perspectiva* by twenty-five, and Alhacen's *De aspectibus* by twenty-two. It is reasonable to assume that these numbers reflect, albeit roughly, the relative popularity of the four works.[9]

That Pecham's *Perspectiva communis* should have been the most widely read by far of the four core Perspectivist works is readily explained by his stated purpose in writing it. Prompted by 'associates', he informs us in the brief prologue, he decided to collect and edit various of his writings on optics in such a way that 'they might be of use to young students'.[10] The resulting study serves as a comprehensive, succinct, and low-level introduction to the science of optics in all its primary aspects, from direct, unimpeded vision to vision impeded by reflective and refractive interfaces. Broken into easily digestible theorematic segments, most of them quite brief and all of them introduced by enunciations, the text features clear and simple geometrical illustrations that demand no technical mathematical expertise. In short, the *Perspectiva communis* functions precisely as the basic, introductory textbook for beginners that Pecham had hoped it would be.

The same holds for Bacon's *Perspectiva*. Although much longer and more diffuse than Pecham's *Perspectiva communis*, it is nonetheless pitched at a level low enough to serve as an introductory text, which helps account for its survival in thirty-nine manuscripts. In fairly sharp contrast to these two tracts, Witelo's *Perspectiva* and Alhacen's *De aspectibus* are extremely long and highly complex, both placing considerable mathematical and technical demands on the reader.[11] Small, wonder, then, they should survive in relatively small numbers that presumably reflect the proportionately smaller readership they enjoyed in comparison to Pecham's and Bacon's mathematically simpler, more straightforward studies.

Dissemination through University Teaching

Although it is clear that Alhacen's De *aspectibus* and its Perspectivist offshoots were used in the teaching of optics at the university level during the Middle Ages, precisely how they were taught is less clear. A handful of university statutes indicates that, with the possible exception of Bacon's *Perspectiva*, the core Perspectivist works were at least occasionally

9 I have not included Bacon's *De multiplicatione specierum* in the discussion of Perspectivist texts because, although it certainly addresses optical issues at points, its primary focus is on the causal mechanisms of all forms of radiation according to the replication of species in media. See Roger Bacon, *Roger Bacon's Philosophy of Nature: A Critical Edition, with English Translation, Introduction, and Notes, of 'De multiplicatione specierum' and 'De speculis comburentibus'*, ed. and trans. by David C. Lindberg (Oxford: Clarendon Press, 1983).

10 Translation adapted from Pecham, *Perspectiva communis*, ed. and trans. by Lindberg, pp. 60–61.

11 Alhacen's *De aspectibus* is roughly 200.000 words long, whereas Witelo's *Perspectiva* extends to roughly 340.000. Furthermore, Book 1 of Witelo's *Perspectiva* consists of 137 geometrical theorems, some of them quite complex and none of them with explicit optical content; see Witelo, *Witelonis perspectivae liber primus: Book 1 of Witelo's 'Perspectiva': An English Translation, with Introduction and Commentary and Latin Edition of the Mathematical Book of Witelo's 'Perspectiva'*, ed. and trans. by Sabetai Unguru (Warsaw: Ossolineum, 1977).

mandated for instruction in the Arts during the later Middle Ages, but not always for the study of optics. A statute of 1390 at the University of Prague, for instance, prescribes Pecham's *Perspectiva communis* as the text for teaching mathematics at the Master's level. Likewise, in 1431, instructors at Oxford were directed to choose among Euclid (presumably his *Elements*), Alhacen, and Witelo for the same purpose at the Bachelor's level.[12] Unfortunately, aside from these few statutes and the texts themselves, there is precious little evidence to tell us how or at what level the discipline of optics was taught. It stands to reason, for instance, that instruction in either mathematics or optics based on Pecham's introductory text would have been fairly elementary. On the other hand, if such instruction were grounded in either Alhacen's *De aspectibus* or Witelo's *Perspectiva*, it would no doubt have been fairly advanced.

How often optics was taught at any stage in matriculation during the Middle Ages and the Renaissance is a matter of speculation. A few things seem clear enough, though. For one thing, if the number of extant manuscripts is any indication, intensive instruction in optics was not particularly common, at least not in comparison with astronomy. Whereas Pecham's *Perspectiva communis* is currently represented by some sixty-four manuscripts, its equivalent in astronomy, Sacrobosco's *De sphera*, is extant not only in hundreds of manuscripts but also in well over one hundred printed editions issued between 1472 and the early seventeenth century.[13] Moreover, the relatively high number of surviving manuscripts containing Pecham's *Perspectiva communis* — sixty-four out of one hundred fifty — suggests that it was the mainstay of optical instruction during the Middle Ages and Renaissance. Its printing history bears this suggestion out. Not only did the *Perspectiva communis* appear in nine separate Latin editions (as well as an Italian translation) between the very late fifteenth century and the end of the sixteenth, but these versions were all published in towns either containing a university or situated quite close to one.[14]

In contrast to the *Perspectiva communis*, Witelo's *Perspectiva* saw only two editions during the same period. The first was published in 1535 at Nuremberg and was reissued in 1551. The second appeared in Friedrich Risner's *Opticae thesaurus*, which saw print in 1572 at Basel and included an edition of Alhacen's *De aspectibus* (attributed to 'Alhazen' under the title *Perspectiva*). This is the first and only time Alhacen's *De aspectibus* appeared in print before modern times. Bacon's *Perspectiva* did not see print until 1614, so in that form it obviously could not have served as a teaching text before then.[15]

In addition to the core Perspectivist sources themselves, commentaries on those sources, many based either explicitly or implicitly on Pecham's *Perspectiva communis*, provided another layer of instruction in optics. Quite a few of these commentaries are anonymous and follow the format of straightforward textual explication.[16] Several, however, taking the *quaestio* approach, are only loosely connected to any particular source text and focus on specific issues that are

12 See A. Mark Smith, *From Sight to Light: The Passage from Ancient to Modern Optics* (Chicago: University of Chicago, 2015), pp. 280–81.

13 On the medieval and Renaissance fate of Sacrobosco's *De sphera*, see Olaf Pedersen, 'In Quest of Sacrobosco', *Journal for the History of Astronomy*, 16 (1985), 175–220 (pp. 183–84).

14 See Smith, *From Sight to Light*, 327–28.

15 See Smith, *From Sight to Light*, 328–29.

16 For listings of such anonymous commentaries, see, e.g., Lindberg, *A Catalogue*, 22–36.

not explicitly raised, or at least not emphasised, in the source literature.[17] As a teaching tool, this sort of narrowly focused commentary requires the reader to possess a fairly firm grasp of at least the rudiments of optical analysis. Such knowledge would presumably come from instruction at an introductory level based on either Pecham's *Perspectiva communis* or on an expository commentary tied to one of the Perspectivist texts. Gregor Reisch's cursory treatment of optics and visual perception in the tenth chapter of his *Margarita philosophica* (Freiburg, 1503) serves as an example of this sort of introductory exposition. Intended as 'an epitome of the whole of philosophy', the *Margarita philosophica* covers the full span of the liberal arts curriculum.[18] Reisch thus felt obligated to include an overview of optical theory, as propounded by 'Witelo, Alhacen, Bacon and others', the 'others' presumably referring to Pecham, who is the only clearly discernible source for Reisch's account.[19]

Optical matters were not always taught on the basis of Perspectivist texts, however. As part of the instruction in the discipline of medicine, for instance, the study of ophthalmology and visual disorders was commonly grounded at the theoretical level on works that subscribe to the extramissionist theory of sight, all of them harking back ultimately to Galen. Exemplary of such ophthalmological studies are Benvenutus Grassus's *De oculis eorumque egritudinibus et curis* (or *Ars probatissimus oculorum*) and Petrus Hispanus's *Liber de oculo*.[20] Unlike the core Perspectivist works, which assume that vision relies on luminous colour radiating into the eye from external objects, these texts posit that an illuminative, pneumatic agent disperses outwards in straight lines from the eye, to establish visual contact with the object in view. Key among the early medical works promoting this notion were Galen's *De placitis Hippocrates et Platonis* ('On the Doctrines of Hippocrates and Plato') and *De usu partium* ('On the Usefulness of the Parts [of the Body] '). Dealing explicitly with sight at certain points, both works had been translated from Greek to Arabic in the ninth century and then from Arabic to Latin in the later Middle Ages.[21] Latin translations of original Arabic works also played an important role in promoting the Galenic model of visual radiation among medieval and Renaissance physicians. Especially significant in this regard were the *Isagoge* and *De oculis* of Ḥunayn ibn ʾIsḥāq (or 'Johannitius' as he was known in the Latin West), which were translated into Latin in the very late eleventh century and which provided an explicit — and explicitly Galenic — model of ocular anatomy and visual perception. Another significant Arabic source of Galenic inspiration was ʿAlī ibn al-ʿAbbās al-Majūsī's *Kitāb al-Malikī*, which circulated variously in Latin translation as the *Pantegni* or *Liber regalis* with authorial attribution given to Haly Abbas.[22]

Nonetheless, although many physicians favored the extramissionist visual model according to the justification provided by these and other medical works, they were not unaware of

17 See Lindberg, *A Catalogue*, 43–46.

18 Gregor Reisch, *Natural Philosophy Epitomised: Books 8–11 of Gregor Reisch's Philosophical Pearl*, ed. and trans. by Andrew Cunningham and Sachiko Kusukawa (Farnham: Ashgate, 2010), pp. x, 192.

19 For additional discussion of Reisch's optical account, see Smith, *From Sight to Light*, 253–55, 284–85.

20 For some discussion of these points, see A. Mark Smith and Arnaldo Pinto Cardoso, *The Treatise on the Eyes by Pedro Hispano* (Lisbon: Alêtheia Press, 2009).

21 See Fernando Salmón, 'The Many Galens of the Medieval Commentators on Vision', *Revue d'histoire des sciences*, 50 (1997), 397–420.

22 On the translations of these Arabic sources, see the relevant articles in Charles Burnett and Danielle Jacquart, *Constantine the African and ʿAlī Ibn al-ʿAbbās al-Mağūsī: The 'Pantegni' and Related Texts* (Leiden: Brill, 1994).

its intromissionist counterpart or of the fundamental incompatibility between the two. After all, Avicenna's *Canon* was an authoritative textbook for the teaching of medicine, and Avicenna was unequivocal in supporting the Aristotelian intromissionist theory of vision in his *Sextus de naturalibus*, which deals with the psychology of perception and was widely read in the Middle Ages.[23] The same holds for Averroes, author of the *Colliget*, which became a mainstay of medical education in the later Middle Ages and Renaissance; he, too, favored strict intromissionism in his widely read long commentary on Aristotle's *De anima*.[24] Faced with such conflicting accounts on the part of Galenist and Aristotelian authorities, medical practioners reasoned that, since no one had yet provided definitive proof in favor of either theory, the extramissionist alternative was preferable because of its relative simplicity and convenience. Thus, as the fifteenth-century Parisian physician Jacques Despars put it, in his commentary on Avicenna's *Canon*, citing Avicenna himself, 'The verification, that is to say, the discussion and determination of the truth in [claims for] intromission or extramission, is of no concern to physicians'.[25] Like many other medieval and Renaissance physicians, in short, Despars took what amounts to an instrumentalist stance in regard to the physical foundations of visual theory.

The persistence of extramissionist visual theory throughout the Middle Ages and Renaissance cannot be attributed to physicians alone. Euclid's *Optics* and *Catoptrics*, Ptolemy's *Optics*, and Yaʿqūb al-Kindī's *De aspectibus*, all of them based on visual radiation rather than light radiation, continued to be studied throughout the Middle Ages and Renaissance. These treatises are in fact represented by more than 120 manuscripts, most of them dating from the period after 1300.[26] The quantity and chronology of these manuscripts suggest that, like their Perspectivist counterparts, they were used for optical instruction. This in turn suggests that the primary focus of such instruction during the Middle Ages and Renaissance was on the geometry rather than the physics of radiation.[27] The reason is fairly obvious. Although contradictory at the physical level, both the extramissionist and intromissionist theories are perfectly equivalent at the mathematical level because the two ultimately depend on a cone of radiation — whether of light or of visual flux — with its vertex at the centre of the eye. The so-called centre of sight, this point constitutes the cardinal reference point for optical analysis in both theories according to the geometry of the cone and its constituent rays, which form straight lines. The direction of radiation makes no difference whatever to the mathematical analysis of those rays. Moreover, the

23 The *Sextus de naturalibus* is the Latin version of the sixth part of Avicenna's *Kitāb al-Šifā* ('Book of Healing'), which deals with psychology. For an excellent account of Avicenna's visual theory within the broader context of this work, see Jon McGinnis, *Avicenna* (Oxford: Oxford University Press, 2010).

24 Although Averroes's medical compendium, entitled *Colliget* in Latin, was not translated until the fourteenth century, his long commentary on the *De anima* was rendered into Latin fairly early in the thirteenth, perhaps by Michael Scot. See *Averroes (Ibn Rushd) of Cordoba: Long Commentary on the 'De anima' of Aristotle*, ed. and trans. by Richard C. Taylor (New Haven: Yale University Press, 2009).

25 Quoted from Danielle Jacquart, *La médicine médiévale dans le cadre parisien, XIVᵉ–XVᵉ siècle* (Paris: Fayard, 1998), p. 413; author's translation from the French.

26 See Lindberg, *A Catalogue*, 21–22, 46–55, 74–75. See also pp. 55–56 and 76–77 for listings of works within this tradition.

27 This focus is clear in the sixteenth-century commentaries on Euclid's *Optics* by Francisco de Melo. See *Obras matemáticas: Francisco de Melo*, ed. and trans. by Bernardo Mota and Henrique Leitão (Lisbon: Biblioteca Nacional de Portugal, 2014).

fact that in some cases Perspectivist texts were explicitly mandated for the teaching of mathematics lends weight to the assumption that optical education was slanted toward the geometry rather than the physics of radiation.

Whatever the case, the conflict between extramissionism and intromissionism remained unresolved well into the late Renaissance. Indeed, if Johannes Kepler is to be credited, the conflict persisted to his day, having only recently been put to rest by Giambattista della Porta in his discussion of the eye and its function in Book 17 of the *Magiae naturalis libri XX* of 1589.[28] Despite Kepler's praise for Della Porta as the final arbiter of the extramission-intromission debate, however, the language of visual radiation remained in use well into the seventeenth century, even if only as a *façon de parler*. In the introduction to his *Siderius nuncius* of 1610, for instance, Galileo offers a description of how the telescope works in terms of visual rays.[29] And despite having accepted Kepler's theory of retinal imaging, which is firmly based on intromissionism, Galileo's contemporary and sometime adversary, Christoph Scheiner, still fell back upon the terminology of visual radiation in his discussion of sight and the eye in the *Oculus* of 1619.[30]

Although university instruction based on actual optical texts, whether Perspectivist or not, was one, obvious way to disseminate optical knowledge during the medieval and Renaissance periods, there were less obvious, more indirect avenues of such dissemination both within and outside the academy. For instance, as Katherine Tachau has shown in compelling detail, optical analysis figured prominently in fourteenth-century *Sentences* commentaries, which were composed at the very highest academic level during the Middle Ages and Renaissance.[31] Of particular significance in these commentaries was the issue of whether visual perception and the cognition arising from it were mediated by representative 'images' (or *species*, in scholastic parlance). William of Ockham, of course, was the key figure in this debate, staking out the controversial position that such mediating entities were unnecessary and, therefore, that visual perception and cognition are immediate and 'intuitive' in nature. Central to the debate about whether visual or conceptual species were necessary for cognition was the analysis of visual illusions and their epistemological implications. Consequently, the debate itself became an occasion of sorts for instruction in optics, even though at a fairly basic level.

Optical matters also figure prominently in commentaries on a fairly wide variety of Aristotelian works, including, most especially, the *Meteorology*, the *Problems*, the *Parva naturalia*, and the *De anima*. In this case the contradictions between extramissionism and intromissionism come to the fore because Aristotle seemed to support both alternatives. In his analysis of the rainbow in *Meteorology*, Book III, Chapter 4, for example, he based his explanation solely on the visual radiation that droplets in rainclouds reflect back to the sun.

Likewise, in *Problems*, Book XXXI, Problem 16, after asking why short-sighted people squint in order to see distant things more clearly, he concluded that 'they do so in order

28 See Johannes Kepler, *Ad vitellionem paralipomena* (Frankfurt: Claude de Marne and Jean Aubry Sons, 1604), p. 209.

29 See A. Mark Smith, 'Practice vs Theory: The Background to Galileo's Telescopic Work', *Atti di Giorgio Ronchi*, 54 (2001), 149–62.

30 See Smith, *From Sight to Light*, 374–75. It is worth noting, however, that in his *Dioptrique* of 1637, Descartes accepts that the eyes of nocturnal animals, such as cats, direct light toward external objects; see *La dioptrique* in *Discours de la méthode* (Leiden, 1637), p. 6.

31 Katherine Tachau, *Vision and Certitude in the Age of Ockham* (Leiden: Brill, 1988).

that the visual radiation may proceed forth in a more concentrated form' so as to increase the sharpness and clarity of the resulting view.[32] On the other hand, in both the *Parva naturalia* and *De anima* Aristotle was unequivocal in his support of intromissionism. 'It is, to state the matter generally', he summed up in *De sensu*, Chapter 2, 'an irrational notion that the eye should see in virtue of something issuing from it'.[33]

Given Aristotle's status as a leading authority in natural philosophy during the Middle Ages and Renaissance, his apparent ambivalence toward extramissionism and intromissionism no doubt encouraged the instrumentalist approach toward visual theory that so many medieval and Renaissance thinkers, physicians and otherwise, adopted.[34] Nor was Aristotle the only authoritative thinker to take such an ambivalent stance. Although heavily indebted to Alhacen and thus supposedly committed to the intromissionist camp, Roger Bacon argued that the visual act depends not just on radiation into the eye but also on radiation out from the eye.[35] Bacon, in short, lent further credence to the notion that intromissionism and extramissionism are not necessarily mutually exclusive. Moreover, Bacon's assumption that vision depends upon reciprocal radiation found its own justification in al-Kindī's *De radiis stellarum* ('On Rays from the Stars'). The central tenet of that brief treatise is that everything throughout the cosmos radiates its own influence onto everything else. Everything, in short, is connected in a web of reciprocal influences, many of them occult, according to al-Kindī's analysis.[36] Some of these influences are more palpable than others, with light and heat being obvious examples. As such, they testify to the intimate connection between astrology and optics. In an oblique way, therefore, the two sciences were seen to be mutually supportive, each validating the study of the other throughout the later Middle Ages and Renaissance.[37]

Another source of inspiration for Bacon was Robert Grosseteste, who also subscribed to the theory of reciprocal radiation in vision. Often included among the Perspectivists, Grosseteste made his mark not so much by contributing to the theoretical analysis of light as by justifying that analysis on religious grounds, both metaphysical and physical. By analogy to physical light, Grosseteste claims in *De luce*, God's creative impulse in forming the universe radiated outward in all directions as from a point, the result being the cosmic sphere and all the nested celestial spheres within it.[38] More specifically, he argues in *De lineis, angulis, et figuris*, any analysis of physical light must be based on geometry, because

32 Aristotle, *The Complete Works of Aristotle*, ed. by Jonathan Barnes (Princeton: Princeton University Press, 1984), p. 1510.

33 Aristotle, *The Complete Works*, ed. by Barnes, 696.

34 By an 'instrumentalist' approach I mean using a given theory out of convenience rather than conviction about whether it actually describes reality. Practically speaking, for instance, an explanation based on visual rays is easier to comprehend than one based on light rays; See, for example, Galileo's explanation of telescopic magnification on the basis of visual rays on fols 6^v–7^r of his *Sidereus Nuncius* (Venice: Tomaso Baglioni, 1610).

35 See Lindberg, *Roger Bacon and the Origins of 'Perspectiva'*, 101–07.

36 See Marie-Thérèse d'Alverny and Françoise Hudry, 'Al-Kindi: "De radiis", *Archives d'histoire doctrinale et littéraire du Moyen Âge*, 41 (1974), 139–260.

37 See Mary Quinlan-McGrath, *Influence: Art, Optics, and Astrology in the Italian Renaissance* (Chicago: University of Chicago Press, 2013).

38 See Cecilia Panti, 'Robert Grosseteste's "De luce": A Critical Edition', in *Robert Grosseteste and his Intellectual Milieu*, ed. by John Flood, James R. Ginther, and Joseph W. Goering, Papers in Mediaeval Studies, 24 (Toronto: Pontifical Institute of Mediaeval Studies, 2013), pp. 193–238. See also Neil Lewis, 'Robert Grosseteste's "On Light": An English Translation', in *Robert Grosseteste and his Intellectual Milieu*, ed. by Flood, Ginther, and Goering, pp. 239–47.

the natural world as a whole is organised by God on that basis in both structure and activity. Thus, 'all causes of natural effects, including those of light, 'must be expressed by means of lines, angles, and figures'.[39] In essence, then, understanding the mathematical structure of physical creation, as exemplified in optics, leads to a better understanding of God and his creative activity.

This metaphysical reasoning provided a powerful incentive for the study of optics not just on the part of Bacon, who was both a devoted follower of Grosseteste and a deeply religious thinker, but also on the part of the key Perspectivist authorities Pecham and Witelo.[40] In addition, an important source of corroboration for that justification was Saint Augustine, whose ruminations on vision, perception, and cognition are scattered throughout his writings and almost invariably illustrate theological points through analogy.[41] Indeed, as Franciscans, Bacon and Pecham shared a profoundly Augustinian world-view, as did Grosseteste who, though not a Franciscan himself, was a strong admirer and promoter of the order.

Popular Textual Conduits

As is the case today, in the Middle Ages and Renaissance, academic learning filtered down in various ways to the popular level, the result in this case being the development of what I call optical literacy. In some instances this filtering process involved the vernacularisation of Latin sources. An early example is the fourteenth-century Italian version of Alhacen's *De aspectibus*, which was no doubt commissioned at considerable expense for someone literate in Italian but without a university education and the fluency in Latin on which it was based.[42] Another example is the Italian translation of Pecham's *Perspectiva communis*. Printed at Venice toward the very end of the sixteenth century, this version of the text was clearly intended for a 'popular' audience, not educated at a university.[43] By that time, however, literacy in Latin was not necessarily a mark of university education, so the vernacular-Latin divide does not necessarily signal a distinction in intellectual achievement or between popular and elite readerships.

Aside from the dissemination of vernacular or Latin texts devoted explicitly to optics, there were other, more capacious conduits through which optical knowledge became increasingly 'public' during the Middle Ages and Renaissance. One such conduit consisted of a select group of classical and late antique sources that had survived the transition from late Roman to early medieval culture, all of them bearing in one way or another on optical matters and most of them based on the theory of visual radiation. In Calcidius's fourth-century Latin

39 Edward Grant, *A Source Book in Medieval Sciences* (Cambridge, MA: Harvard University Press, 1974), p. 385.

40 See Chapter 1 of John Pecham, *Tractatus de perspectiva*, ed. by David C. Lindberg (St Bonaventure, NY: Franciscan Institute, 1972), pp. 23–28; and the prologue to Witelo's *Perspectiva*, in *Opticae Thesaurus*, ed. by Risner, 1–2.

41 See Smith, *From Sight to Light*, 150–54.

42 See Graziella Vescovini, 'Alhazen vulgarisé: le de li aspecti d'un manuscrit du Vatican (moitié du XIVe siècle) et le troisième commentaire sur l'optique de Lorenzo Ghiberti', *Arabic Sciences and Philosophy*, 8 (1998), 67–96. See also A. Mark Smith, 'The Latin Source of the Fourteenth-Century Italian Translation of Alhacen's "De aspectibus" (Vat. Lat. 4595)', *Arabic Sciences and Philosophy*, 11 (2001), 27–43.

43 John Pecham, *I tre libri della perspettiva commune [...] annotati da Giovanni Paolo Gallucci Salodiano* (Venice: Giovanni Varisco Sons, 1593).

translation, for example, the segment on vision between sections 44d and 47e of Plato's *Timaeus* became an extraordinarily influential source of ideas about sight during the early and high Middle Ages, particularly in the period leading up to and during the twelfth century. Equally important, in his commentary on that segment, Calcidius provided a glimpse at second hand into atomist, Stoic, Peripatetic, and medical theories of sight.[44] In much the same way, Macrobius's discussion of vision in the seventh book of the *Saturnalia* and Seneca's treatment of halos and the rainbow in the first book of his *Natural Questions* offered insights into various classical theories of light, colour, and sight.[45] Within the same encyclopedic tradition as Seneca and Macrobius, finally, Isidore of Seville addressed optical issues in the *Etymologies*, a seventh-century compilation of knowledge organised somewhat randomly over a range of topics that include both natural and supernatural phenomena.[46]

In addition to highlighting particular optical issues and theoretical responses to them, these works also served as doxographical sources. In that capacity, they provided at least some insight into the optical thought of various classical and Hellenistic thinkers — Aristotle and Plato foremost among them — whose actual works were unavailable in Latin until the second half of the twelfth century or, in some cases, much later. Drawing on these doxographic sources, such twelfth-century thinkers as Bernard Sylvester and William of Conches were able to piece together a fairly comprehensive, if somewhat vague, account of visual perception that reflected classical and Hellenistic sources in a variety of sometimes surprising ways.[47] Even after the recovery of classical learning through Latin translation between roughly 1150 and 1250, encyclopedically inclined scholars drew on such doxographic sources for information about earlier theories and theorists. Perhaps best-known among these thirteenth-century encyclopedists are Albertus Magnus, Bartholomaeus Anglicus, and Vincent of Beauvais, all of whom were not only extraordinarily well and widely read but also concomitantly eclectic in their scholarship.

So far we have examined the transmission of optical knowledge through the medium of texts according to nesting or overlapping circles of influence both within and outside the medieval and Renaissance university. The core circle, as we saw, consists of Alhacen's *De aspectibus* and its three derivative Perspectivist sources. Next comes a circle of direct commentaries on these core works, followed in sequence by indirect commentaries organised according to *questiones* that address specific issues raised at certain points within the Perspectivist texts. At a further remove from the Perspectivist core is a circle of sources that bear more or less directly on an understanding of optics but whose theoretical foundations are generally, though not always, incompatible with those of the Perspectivists. These sources include several Aristotelian works, ranging from the extramission-based *Meteorology* and *Problems* to the intromission-based *Parva naturalia* and *De anima*. They also include the extramission-based optical texts of Euclid, Ptolemy, and al-Kindī, as well as the Galen-inspired ophthalmological treatises of such widely read authorities as Johannitius, Benvenutus Grassus, and Petrus Hispanus, all of whom subscribed to the theory of visual radiation. At the furthest remove from the core circle, finally, is a loose collection of classical

44 On Calcidius's translation of the Timaeus and its influence, see Smith, *From Sight to Light*, 232, 234–38.

45 On Macrobius and Seneca, see Smith, *From Sight to Light*, 232–33.

46 See Smith, *From Sight to Light*, 234.

47 See Smith, *From Sight to Light*, 238–41.

doxographic sources along with a number of late antique and medieval encyclopedic works that deal somewhat randomly and spottily with optical issues. Altogether, then, the textual sources of optical thought during the Middle Ages and Renaissance were quite disparate in content, style, emphasis, and theoretical approach.

Extra-Textual Conduits

But textual transmission was not the only conduit through which medieval and Renaissance audiences gained knowledge about optics. Extra-textual transmission played an equally significant, if not an even more significant, role in that process, particularly at the popular level. Oral communication played a vital role here, with church sermons surely being the most far-reaching means of promoting ideas in the Middle Ages and Renaissance. Churches, after all, were classrooms of Christian doctrine for both the tutored and the untutored, and the pulpit was nothing more than an elevated lectern from which the priest could dispense his lessons. Those lessons, in turn, were often larded with physical analogies and examples meant to illustrate spiritual points. That the science of optics should provide many of these analogies and examples should come as no surprise, since the Bible is rife with allusions to light and vision, as is the tradition of 'light metaphysics' exemplified in the theological thought of such figures as Robert Grosseteste and Roger Bacon.[48]

Among the analogies used to illustrate the connection between optics and theology, one of the most salient is that between physical and spiritual vision. The key Biblical warrant for this analogy comes from I Corinthians 13.12: 'For now [in this life] we see obscurely in a mirror', the Apostle counsels, 'but then [in the afterlife] we will see face to face'.[49] No doubt inspired by this passage, Bacon took the analogy even further. 'We attribute direct [spiritual] vision to God', he claims in the *Perspectiva*, '[whereas] departure from rectitude by refraction, which produces weaker vision, is suited to angelic nature; reflected vision, which is weaker [still], can be assigned to humans'.[50] Likewise, in defence of his theory of reciprocal radiation in physical vision, Bacon pointed to an equivalent reciprocity in spiritual, or moral, vision. Such vision requires 'not only that the soul should be the recipient from without of divine grace and powers, but also that it should cooperate by its own power'.[51] Just as physical light is needed for physical vision, therefore, grace provides the spiritual light that enables the spiritual eye to perceive what is morally correct.

Much like Bacon, but in a far more concerted way, Pierre de Limoges drew on optical analogies to make spiritual points in his *Tractatus moralis de oculo* ('Moral Treatise on the Eye'), a work also known as *De oculo morali*. A younger contemporary of Bacon, and like him a Franciscan, Pierre intended this work as an aid for preachers who were looking for *exempla* with which to illustrate points in their sermons. How well the *Tractatus* served

48 The tradition of 'light-metaphysics' harks back to such thinkers as Plotinus, Pseudo-Dionysius, and even St Augustine; it is no coincidence that Grosseteste was influenced by this tradition since he translated and commented on the corpus of Pseudo-Dionysius's writings.

49 This is a somewhat loose translation from the Vulgate: 'Videmus nunc per speculum in ænigmate, tunc autem facie ad faciem'.

50 Lindberg, *Roger Bacon and the Origins of 'Perspectiva'*, 329.

51 Lindberg, *Roger Bacon and the Origins of 'Perspectiva'*, 325.

this purpose is evident from its popularity. Well over 250 manuscript copies are currently known to exist, and it saw print as early as 1475 or 1476, an Italian translation appearing twenty years later.[52]

Pierre leaves no doubt about the inspiration for the analogies he will draw over the course of the *Tractatus*. 'It is clear', he says toward the beginning, 'how suitably the wisest Perspectivist spoke, the apostle Paul, when he called the vision of heaven face to face, but the vision of the journey there is mirror-like and only partial'.[53] Nonetheless, the deceptiveness of mirror-like vision in the physical world can be explained, and its artifice seen through, according to Perspectivist reflection-analysis. Likewise, when properly explained, the deceptions of spiritual vision can be seen through. For example, a rich man immersed in the liquidity of wealth might appear magnified and thus greater than he actually is to someone looking at him from the arid air of poverty. However, just as Perspectivist refraction theory teaches that things submerged in water appear larger than they actually are when seen from the air outside, so it teaches by analogy that a man submerged in wealth may appear greater than he actually is to a poor person. Thus, Pierre concludes, 'when a poor person living in the dry land of poverty sees someone overflowing with worldly riches, he will consider the rich man to be great' until proper spiritual judgement corrects the deception.[54]

Although it would be overstating things to claim that such analogies teach anything particularly useful or significant about optics, at least not at the technical level, they do draw on and reinforce a basic knowledge of optical principles and the phenomena associated with them. This is precisely the kind of knowledge — at the level of what I call optical literacy — to which the immensely popular, late fifteenth-century Florentine preacher Savonarola appealed when he counseled his audience to put on spiritual eyeglasses in order to more clearly discern what is morally right. Just as physical eyeglasses make 'small things appear bigger [...] because the species of the letters that enter the eye [...] strike on the glasses and here they are spread, widen and appear bigger', Savonarola argues in a sermon of 1494, so spiritual eyeglasses sharpen our moral discernment by an equivalent magnifying process.[55] At the conceptual level, if not the level of technical detail, Savonarola's account of magnification by eyeglasses is perfectly consonant with Perspectivist refraction theory as applied to convex lenses. And he must have expected his audience to be well enough schooled at this level not only to grasp the analogy but also to see its aptness.

Somewhat less broad in their reach than sermons were medieval and Renaissance literary works, which represented a hybrid, textual-oral form of dissemination insofar as they were frequently read aloud to audiences of several listeners. Not all medieval and Renaissance literary works were constructed around optical themes, of course, but many were. Among these, a few stand out not only for their use of such themes but also for the theoretical principles underlying them. Without doubt the most salient of such works is Dante's *Divine Comedy*. Peppered throughout with allusions to light and vision, the tale

52 See Peter of Limoges, *The Moral Treatise on the Eye*, trans. by Richard Newhauser (Toronto: Pontifical Institute of Mediaeval Studies, 2012), pp. XXIX–XXX.

53 Limoges, *The Moral Treatise*, trans. by Newhauser, 13.

54 Limoges, *The Moral Treatise*, trans. by Newhauser, 39.

55 Quotation from Vincent Ilardi, *Renaissance Vision from Spectacles to Telescopes* (Philadelphia: American Philosophical Society Press, 2007), p. 179.

told by Dante recounts his ascent from the blindness of Hell to a virtual view of God in Purgatory that is mediated by reflection and that culminates in a final, dazzling, direct gaze at the Lord himself in Paradise. The implicit reference here to the aforementioned passage in I Corinthians — 'For now [in this life] we see obscurely in a mirror' — is difficult to miss. No less difficult to miss is the general similarity between Dante's allegorical journey in the *Divine Comedy* and that described by St Bonaventure in his *Itinerarium mentis in Deum*.

More than just vague references to light and sight, however, these allusions reveal what appears to be a fairly firm grasp of optical theory on Dante's part. It is clear, for instance, that he was familiar with both extramissionism and intromissionism, since he uses both theories to describe the increasing clarity of his vision in the ascent from Hell to Paradise.[56] At one point, in Canto XV of the *Purgatorio*, moreover, he alludes to the equal angles law of reflection in order to explain how he was dazzled by the reflected light of an angel 'even as experiment and science show'.[57] Precisely where Dante gained his knowledge of optics and its principles is a subject of some controversy. Most scholars agree that Albertus Magnus is a likely source, and a handful point to Witelo's *Perspectiva* as another.[58] The key point, however, is not that Dante had any technical expertise in optics to share with his audience. It is, instead, that his use of optical allusions served to reinforce the idea that, as a science, the field of optics was a critical means for understanding our perception and intellectual grasp of both physical and metaphysical reality. Much like Savonarola, in short, Dante was merely appealing to optical literacy, not attempting to instruct his audience in the intricacies of optics.

It was presumably the same sort of optically literate audience that Chaucer had in mind a century later in the prologue to the 'Squire's Tale', when he referred to those who 'speken of Alocen and Vitulon, | And Aristotle, that writen in hir lyves | Of queynte mirours, and of perspectives'.[59] Not that he expected his readers or listeners to have actually read, much less assimilated, the writings of Alhacen or Witelo (or Aristotle, for that matter); nor in fact that he himself had actually assimilated them. It was enough that his audience get the allusion, just as we today might appreciate an allusion to Heisenberg's Uncertainty Principle without any grasp whatever of the physics of quantum theory that underlie it.

Along with religious sermons and literary works, art also served as a conduit through which optical ideas were disseminated publicly, sometimes through texts and sometimes through works of art themselves. This is especially the case after the development of Renaissance naturalism in the early fifteenth century, when many artists looked to the science of *perspectiva* for guidance in rendering their paintings visually 'accurate'. It was with this in mind, for instance, that Leon Battista Alberti designed his method of one-point perspective as a way to rationally organise visual space. No less significant than the method itself, however, was Alberti's attempt to justify and explain it on the basis of the

56 For an extended discussion of this ascent and its optical implications, see Suzanne Conklin Akbari, *Seeing Through the Veil* (Toronto: University of Toronto Press), pp. 137–77.

57 Dante Alighieri, *The Divine Comedy: Purgatorio*, trans. by Charles Singleton (Princeton: Princeton University Press, 1973), pp. 155–56.

58 For claims in favour of Witelo's influence on Dante, see Akbari, *Seeing Through the Veil*, 114–16, 165–66. For arguments against, see Simon Gilson, 'Dante and the Science of "Perspective": A Reappraisal', *Dante Studies*, 115 (1997), 185–219.

59 Geoffrey Chaucer, *The Riverside Chaucer*, ed. by Larry D. Benson (Oxford: Oxford University Press, 2008), p. 172.

Euclidean visual cone, conceived of as a right cone. Within that context, a diameter of the cone's base circle becomes Alberti's *linea centrica* ('horizon line'), the axis his *razzo centrico* or *radius centricus* ('centric ray'), the rays defining the cone's surface his *razzi estrinsici* or *radii extrinsici* ('extrinsic rays'), which delineate the boundary of what can be seen, and the remaining rays his *razzi mediani* or *radii medii* ('median rays').

Alberti's choice of the Euclidean visual model as the theoretical justification for his system of perspective did not necessarily reflect a belief in its physical reality. 'Truly', Alberti observes early in the *Della pittura*, 'there was no little debate [...] as to whether [the rays involved in sight] arise from the surface [of visible objects] or from the eyes'.[60] But, he cautioned, 'this is a really difficult controversy that [is] completely useless to us'.[61] For the sake of convenience, he concluded, we should simply conceive of the rays as filaments extending out from the centre of the eyes, a conception that conforms better and more simply to extramissionism than to intromissionism. Like so many medieval and Renaissance physicians, in other words, Alberti may be interpreted as taking an essentially instrumentalist stance in regard to the direction of radiation.

In dismissing the intromissionist alternative as unsuited to his purposes, Alberti gave no indication whether he had Perspectivist theory in mind, but it seems likely that he did. For one thing, he belonged to a Florentine circle of interest for which Perspectivist optics was 'such a hot topic', according to Samuel Edgerton, 'that artists as well as intellectuals [were] persuaded to learn something about [it]'.[62] One such artist was Lorenzo Ghiberti, who quoted extensively from the Italian version of Alhacen's *De aspectibus* in his *commentario terzo* on art.[63] Another member of that circle, Antonio Pierozzi, the Dominican friar and bishop now known as Saint Antoninus, not only had a mastery of Perspectivist optics but, like Pierre de Limoges, also drew on that mastery in finding apposite optical *exempla* for his popular sermons.[64] Dominique Raynaud's recent study bears out the conclusion that fifteenth-century Florence was a critical node in the networks through which Perspectivist optics diffused in the later Middle Ages.[65]

Exactly how, if at all, Perspectivist optics influenced artistic practice during the Renaissance is a vexed issue, especially with regard to the development of Albertian perspective theory and its various offshoots.[66] There is no question, however, that Leonardo da Vinci was one Renaissance artist who took optics seriously enough to let it guide his practice. Untutored at a formal level, Leonardo was nonetheless an avid autodidact and researcher. It is clear, for instance, that he was familiar with Perspectivist theory as well as its intromissionist alternative, and it is quite likely that he gained that familiarity not simply by word of mouth but by reading the sources themselves, Pecham and Witelo in particular. It is also clear that he recognised certain flaws in the Perspectivist analysis of radiation, especially in regard to shadow casting, and that he adjusted his technique accordingly — hence

60 Leon Battista Alberti, *On Painting*, ed. and trans. by Rocco Sinisgalli (Cambridge: Cambridge University Press, 2011), p. 26.

61 Alberti, *On Painting*, ed. and trans. by Sinisgalli, 26.

62 Samuel Y. Edgerton, *The Mirror, the Window, and the Telescope* (Ithaca: Cornell University Press, 2009), p. 31.

63 Vescovini, 'Alhazen vulgarisé'.

64 On this point, see Edgerton, *The Mirror*, 30–38.

65 See Raynaud, *Optics and the Rise of Perspective*.

66 For some discussion of this point, see Smith, *From Sight to Light*, 301–05.

his emphasis on chiaroscuro and sfumato. Nor did the direction of influence solely from theory to practice. Observation of how pinhole images are formed led him to propose a model of vision according to which the eye acts like a camera obscura, the pupil forming the small aperture through which inverted images are projected on the front of the lens.[67]

By the time Leonardo died, in 1519, the connection between optics and art was becoming tighter and more evident with the incorporation of various optical effects into paintings. These effects included increasingly exaggerated chiaroscuro, often based on unusual and unusually focused lighting effects, the use of anamorphism, frequently representing image-distortion in convex mirrors, and the inclusion of trompe l'œil effects. All of these optical strategies — and indeed the very effort to make paintings look 'real' — served as visible reminders to any onlooker not only that vision itself is easily deceived but also that visual deceptions are to some extent explicable according to optical principles. Somewhat paradoxically, then, later Renaissance painters validated perspectivist optics as a scientific account of vision while at the same time drawing on perspectivist analytic principles to show how untrustworthy vision can be.

The increasing incorporation of optical effects in later Renaissance painting reflected a broader awareness of those effects that can be attributed to certain technological developments. Crucial among these was a major step up in the quality of glass being produced, with the introduction of clear, 'crystal' glass in Venice around the middle of the fifteenth century. Along with this came a significant improvement during the sixteenth century in the backing of glass mirrors, moving from lead to a mercury-tin amalgam, which not only made glass mirrors more reflective but also contributed to the manufacture of more effective concave mirrors. Thus, by the end of the sixteenth century, glass objects in all shapes and sizes, from eyeglass lenses to vases, bottles, and even reading mirrors, were commonplace, their optical effects to be seen — and puzzled over — by a vastly expanded 'public'. Equally important, this expanded public now had the full spectrum of optical sources discussed to this point readily available in printed form, often in multiple editions. The cumulative effect of these developments was profound. Not only did it open the science of optics to a far more intensive and extensive scrutiny than ever before, but in providing new technical resources for that scrutiny, it brought many optical effects into a far clearer light. The stage was therefore set for a radical transformation of optics at both the theoretical and practical level during the first few decades of the 1600s.

67 On Leonardo as a student of optics, see Smith, *From Sight to Light*, 305–13.

Bibliography

Primary Sources

Alberti, Leon Battista, *On Painting*, ed. and trans. by Rocco Sinisgalli (Cambridge: Cambridge University Press, 2011).

Alhacen, *Opticae thesaurus: Alhazeni Arabis libri septem, nuncprimum editi; Eiusdem liber de crepusculis et nubium ascensionibus, item Vitellonis Thuringopoloni libri X*, ed. by Friedrich Risner (Basel: Bischoff, 1572).

Alhacen, *Alhacen's Theory of Visual Perception: A Critical Edition, with English Translation and Commentary, of the First Three Books of Alhacen's 'De aspectibus', the Medieval Latin Version of Ibn al-Haytham's 'Kitāb al-Manāẓir'*, ed. and trans. by A. Mark Smith, 2 vols (Philadelphia: Transactions of the American Philosophical Society, 2001).

Aristotle, *The Complete Works of Aristotle*, ed. by Jonathan Barnes (Princeton: Princeton University Press, 1984).

Averroes, *Averroes (Ibn Rushd) of Cordoba: Long Commentary on the 'De anima' of Aristotle*, ed. and trans. by Richard C. Taylor (New Haven: Yale University Press, 2009).

Bacon, Roger, *Roger Bacon's Philosophy of Nature: A Critical Edition, with English Translation, Introduction, and Notes, of 'De multiplicatione specierum' and 'De speculis comburentibus'*, ed. and trans. by David C. Lindberg (Oxford: Clarendon Press, 1983).

Bacon, Roger, *Roger Bacon and the Origins of 'Perspectiva' in the Middle Ages. A Critical Edition and English Translation of Bacon's Perspectiva with Introduction and Notes*, ed. and trans. by David C. Lindberg (Oxford: Clarendon Press, 1996).

Dante Alighieri, *The Divine Comedy: Purgatorio*, trans. by Charles Singleton (Princeton: Princeton University Press, 1973).

Descartes, René, *Discours de la méthode* (Leiden, 1637).

Galilei, Galileo, *Sidereus Nuncius* (Venice: Tomaso Baglioni, 1610).

Grosseteste, Robert, *Die Philosophischen Werke des Robert Grosseteste, Bischofs von Lincoln*, ed. by Ludwig Baur, Beiträge zur Geschichte der Philosophie des Mittelalters, 9 (Münster: Aschendorff Verlag, 1912).

Kepler, Johannes, *Ad vitellionem paralipomena* (Frankfurt: Claude de Marne and Jean Aubry Sons, 1604).

Lewis, Neil, 'Robert Grosseteste's "On Light": An English Translation', in *Robert Grosseteste and his Intellectual Milieu*, ed. by John Flood, James R. Ginther, and Joseph W. Goering, Papers in Mediaeval Studies, 24 (Toronto: Pontifical Institute of Mediaeval Studies, 2013), pp. 239–47.

Limoges, Peter of, *The Moral Treatise on the Eye*, trans. by Richard Newhauser (Toronto: Pontifical Institute of Mediaeval Studies, 2012).

Melo, Francisco de, *Obras matemáticas: Francisco de Melo*, ed. and trans. by Bernardo Mota and Henrique Leitão (Lisbon: Biblioteca Nacional de Portugal, 2014).

Panti, Cecilia, 'Robert Grosseteste's "De luce": A Critical Edition', in *Robert Grosseteste and his Intellectual Milieu*, ed. by John Flood, James R. Ginther, and Joseph W. Goering, Papers in Mediaeval Studies, 24 (Toronto: Pontifical Institute of Mediaeval Studies, 2013), pp. 193–238.

Pecham, John, *I tre libri della perspettiva commune [...] annotati da Giovanni Paolo Gallucci Salodiano* (Venice: Giovanni Varisco Sons, 1593).

Pecham, John, *John Pecham and the Science of Optics: Perspectiva communis*, ed. and trans. by David C. Lindberg (Madison: The University of Wisconsin Press, 1970).

Pecham, John, *Tractatus de perspectiva*, ed. by David C. Lindberg (St Bonaventure, NY: Franciscan Institute, 1972).

Porta, Giambattista della, *De refractione optices parte: libri novem* (Naples: Carlino & Pace, 1593).

Reisch, Gregor, *Natural Philosophy Epitomised: Books 8–11 of Gregor Reisch's Philosophical Pearl*, ed. and trans. by Andrew Cunningham and Sachiko Kusukawa (Farnham: Ashgate, 2010).

Witelo, *Witelonis perspectivae liber primus: Book 1 of Witelo's 'Perspectiva': An English Translation, with Introduction and Commentary and Latin Edition of the Mathematical Book of Witelo's 'Perspectiva'*, ed. and trans. by Sabetai Unguru (Warsaw: Ossolineum, 1977).

Secondary Works

Akbari, Suzanne Conklin, *Seeing Through the Veil* (Toronto: University of Toronto Press).

Alverny, Marie-Thérèse d', and Françoise Hudry, 'Al-Kindi: "De radiis"', *Archives d'histoire doctrinale et littéraire du Moyen Âge*, 41 (1974), 139–260.

Buringh, Eltjo, *Medieval Manuscript Production in the Latin West* (Leiden: Brill, 2011).

Burnett, Charles, and Danielle Jacquart, *Constantine the African and 'Alī Ibn al-'Abbās al-Maǧdūsī: The 'Pantegni' and Related Texts* (Leiden: Brill, 1994).

Chaucer, Geoffrey, *The Riverside Chaucer*, ed. by Larry D. Benson (Oxford: Oxford University Press, 2008).

Edgerton, Samuel Y., *The Mirror, the Window, and the Telescope* (Ithaca: Cornell University Press, 2009).

Gilson, Simon, 'Dante and the Science of "Perspective": A Reappraisal', *Dante Studies*, 115 (1997), 185–219.

Grant, Edward, *A Source Book in Medieval Sciences* (Cambridge, MA: Harvard University Press, 1974).

Ilardi, Vincent, *Renaissance Vision from Spectacles to Telescopes* (Philadelphia: American Philosophical Society Press, 2007).

Jacquart, Danielle, *La médicine médiévale dans le cadre parisien, XIVe–XVe siècle* (Paris: Fayard, 1998).

Lindberg, David C., *A Catalogue of Medieval and Renaissance Optical Manuscripts* (Toronto: University of Toronto Press, 1975).

McEvoy, James, 'The Chronology of Robert Grosseteste's Writings on Nature and Natural Philosophy', *Speculum*, 58 (1983), 614–55.

McGinnis, Jon, *Avicenna* (Oxford: Oxford University Press, 2010).

Pedersen, Olaf, 'In Quest of Sacrobosco', *Journal for the History of Astronomy*, 16 (1985), 175–220.

Quinlan-McGrath, Mary, *Influence: Art, Optics, and Astrology in the Italian Renaissance* (Chicago: University of Chicago Press, 2013).

Raynaud, Dominique, *Optics and the Rise of Perspective* (Oxford: Bardwell Press, 2014).

Salmón, Fernando, 'The Many Galens of the Medieval Commentators on Vision', *Revue d'histoire des sciences*, 50 (1997), 397–420.

Smith, A. Mark, 'Practice vs Theory: The Background to Galileo's Telescopic Work', *Atti di Giorgio Ronchi*, 54 (2001), 149–62.

Smith, A. Mark, 'The Latin Source of the Fourteenth-Century Italian Translation of Alhacen's "De aspectibus" (Vat. Lat. 4595)', *Arabic Sciences and Philosophy*, 11 (2001), 27–43.

Smith, A. Mark, 'What is the History of Medieval Optics Really About?', *Proceedings of the American Philosophical Society*, 148 (2004), 180–94.

Smith, A. Mark, and Arnaldo Pinto Cardoso, *The Treatise on the Eyes by Pedro Hispano* (Lisbon: Alêtheia Press, 2009).

Smith, A. Mark, *From Sight to Light: The Passage from Ancient to Modern Optics* (Chicago: University of Chicago, 2015).

Tachau, Katherine, *Vision and Certitude in the Age of Ockham* (Leiden: Brill, 1988).

Vescovini, Graziella, 'Alhazen vulgarisé: le de li aspecti d'un manuscrit du Vatican (moitié du XIVe siècle) et le troisième commentaire sur l'optique de Lorenzo Ghiberti', *Arabic Sciences and Philosophy*, 8 (1998), 67–96.

DOMINIQUE RAYNAUD

A Hitherto Unknown Treatise on Shadows Referred to by Leonardo da Vinci

Introduction

Advances in study sometimes lead us to revise accepted knowledge. For example, the debates around the attribution of the law of refraction to Snell or Descartes were eclipsed in 1990 by the discovery of a forgotten manuscript by Ibn Sahl, who had in fact already proposed such a concept around the year 984. Such discoveries have made the search for precursors a key question for contemporary historiography. To assess how new a historical development might actually be, knowledge of previous events is vital, although such knowledge may often be only piecemeal and scantly given. The recent discovery of several treatises of Arabic origin dating from the ninth and tenth centuries has made a valuable contribution to our expanding knowledge of optics, and has posed the question of the potential transmission of these treatises to the West in the form of Latin translations.

Shadows have long been a central topic in the study of optics. The distinction between partial shadow (or the penumbra) and full shadow has been theorised by Maurolico and Kepler.[1] Shadows are also dealt with in architecture and perspective, because they are directly involved in the rendering of volume and depth. Given his research interests, the subject of shadows was bound to capture the attention of Leonardo da Vinci, and it is known that he paved the way to the tracts on shadows that were appended to perspective treatises from the sixteenth century on.[2]

Although the ideas of Leonardo have been regarded as milestones in the identification of the *ombra composta* ('penumbra') as opposed to *ombra semplice* ('pure shadow'), there is evidence that he used previously existing sources when developing his thoughts on the subject. This chapter deals specifically with the optical sources that Leonardo may have

1 Francesco Maurolico, *Photismi de lumine & umbra* (Naples: Lanzi, 1611 [written 1521]), Theorem 18, in *Francesco Maurolico: Opera mathematica*, ed. by K. Takahashi and R. Bellè: www.maurolico.unipi.it [accessed 27-07-2017]; Johannes Kepler, *Ad Vitellionem paralipomena, quibus astronomiae pars optica traditur* (Frankfurt: Marnius, 1604), VI, 7.

2 Francesca Fiorani, 'The Theory of Shadow Projection and Aerial Perspective', in *Desargues en son temps*, ed. by J. Dhombres and J. Sakarovitch (Paris: Blanchard, 1994), pp. 267–82; Martin Kemp, *Leonardo da Vinci* (Oxford: Oxford University Press, 2006).

Dominique Raynaud Université Grenoble Alpes, dominique.raynaud@univ-grenoble-alpes.fr

Perspective as Practice. Renaissance Cultures of Optics, ed. by Sven Dupré, Turnhout, 2019 (*Techne. Knowledge, Technique, and Material Culture*, 1), p. 259-277

DOI 10.1484/M.*Techne*-EB.5.117729

used when composing his notes on shadows. Given Leonardo's way of thinking, which is to imagine many solutions without justifying them mathematically, this chapter will not ask whether Leonardo made his own contribution to shadow theory, but will focus exclusively on his knowledge of available optical sources.[3]

Leonardo's Authorship Concerning the Penumbra

'No predecessor except Leonardo da Vinci (who was without influence) dealt adequately with the shadows cast by extended luminous bodies', asserts David C. Lindberg.[4] There is a twofold reason for this assessment. Firstly, no Latin treatise offered a comparable treatment of the penumbra. Medieval authors rather relied on the classification of shadows provided by Theon of Alexandria, who had written 'A body equal in size, smaller or larger than the light source casts a cylindrical, convergent or divergent shadow'.[5] This theory, which circulated in the Latin world through the works of al-Kindī, Bacon, Witelo, and Pecham, is inaccurate because it considers that the shadow is cast by the extreme points of the light source or that the opaque body is at infinity, which is never the case, even in astronomy. Secondly, even when the question of the penumbra arose in earlier times, it remained unsolved. Leonardo's predecessor, Roger Bacon, came close to the solution in part five of *De speculis comburentibus*. But considering that the solitary rays emitted by the light source within the penumbra were too weak to be seen, he failed to explain the penumbral region.[6]

However one might still wonder whether there were no predecessors other than Bacon. Contrary to the views mentioned above, there were indeed ideas in circulation prior to Leonardo da Vinci's findings. The Arabic *'ilm al-aẓlāl* ('science of shadows'), for example, where we can find a similar dissection of light into small segments, is very much akin to Leonardo da Vinci's geometrical analysis. Granted, Lindberg did acknowledge that Ibn

3 This study has benefited from the technical resources provided by e-Leo: www.leonardodigitale.com [accessed 27-07-2017], which allows full-text searches of Leonardo's notes. Any following quotation of Leonardo da Vinci will refer to that source. Querying the e-Leo database yielded 398 occurrences of the singular *ombra*, and 418 occurrences of the plural *ombre*. Semantic cluster analysis was then used to select only the occurrences connected to optics. This core set consists of 67 folios, which are distributed among *Codex C* (32 occurrences), *Codex Atlanticus* (21 occurrences), *Codex E* (five occurrences), *Codex A* (four occurrences), *Disegni anatomici* (three occurrences), *Codex Trivulziano* (one occurrence) and *Codex D* (one occurrence). Further reference to Leonardo's texts will be abbreviated as follows: *Codex Atlanticus* (CA), *Disegni anatomici* (DA).

4 David C. Lindberg, 'Optics in Sixteenth Century Italy', in *Novità celesti e crisi del sapere*, ed. by P. Galluzzi (Florence: Istituto e Museo di Storia della Scienza, 1984), p. 133. Similar statements on Leonardo's precedence over the geometric analysis of the penumbra can be found in David C. Lindberg, 'Laying the Foundations of Geometrical Optics', in *The Discourse of Light from the Middle Ages to the Enlightenment*, ed. by David C. Lindberg and G. Cantor (Los Angeles: Clark Memorial Library, 1985), p. 32; E. Broydrick Thro, 'Leonardo's Early Studies of Shadows', *Achademia Leonardi Vinci*, 9 (1996), 42–50 (p. 45); Claire J. Farago, *Re-Reading Leonardo: The Treatise on Painting Across Europe, 1550–1900* (London: Ashgate, 2009), p. 295.

5 Euclid, *Opticorum recensio Theonis: Euclidis opera omnia*, ed. by J.L. Heiberg (Leipzig: Teubner, 1895), VII, pp. 144–47.

6 Roger Bacon, 'De speculis comburentibus', ed. by David C. Lindberg, in *Roger Bacon's Philosophy of Nature* (Oxford: Clarendon, 1983), p. 395. 'Despite his lack of success, Bacon is the first scholar in whom I have encountered any discussion *at all* of the penumbral region. His predecessors had universally lumped the penumbra with the fully illuminated region'. Lindberg, 'Laying the Foundations', p. 23.

al-Haytham 'employed the principle of punctiform analysis', but did not proceed any further by considering his study of shadows.[7] For our purposes, the key text is Ibn al-Haytham's *Maqāla fī kayfiyyat al-aẓlāl* ('Epistle on the Properties of Shadows').[8]

Ibn al-Haytham first defines general concepts such as *ẓulma* ('darkness'), which is the total absence of light, and *ẓill* ('shadow'), which is the partial lack of light. He then undertakes a geometric analysis of the varying shadow *ḤH*, which is cast by the opaque line *GD* when lit by the light source *AB*. Ibn al-Haytham cuts the light source *AB* into small segments and examines their projection on the screen *ḤṬ*. He then shows that the more segments an area of the screen receives, the lighter it becomes. As this number varies along *ḤH*, the analysis establishes the continuous variation of light within the penumbra from full light to full shade (Figure 1). As Ibn al-Haytham writes:

> The light which comes from a portion of the luminous body *AB* propagates in straight lines. Let this part be *AL*. We draw *LǦ* that cuts *ḤH* at the point *M*. All the lines extended from a point of *AL* to *G* will cut the line *HM*. Thus *HM* receives [a single] light from the portion *AL*. By contrast, any straight line produced from a point of *LB* to *HM* is stopped by *GD*. So *HM* does not receive some of the lights from the line *AB*. Consequently, there is a mix of light and shade in the line *HM*. Moreover, the line *AL* is much smaller than *AB*, so there is far more shade than light in *HM* [...] Then we make *LN* = *AL*, we draw *NG*, which meets *ṬḤ* in *G*. The light of *AL* falls on *MG* too, so that the line *MG* is lit twice by the light of the two segments *AL* and *LN* [...] Therefore light is stronger on *MG* than on *HM*, and the shadow is weaker on *MG* than on *HM*. This makes it clear that in the line *HḤ* the shadow is continuous (*ẓill muttaṣil*) and nonetheless variable (*wa-ma'a dhalika mukhtalif*): close to the point *H* it is stronger; close to the point *Ḥ* it is weaker [...] Thus in *HḤ*, the shadow is variable and this variation is gradual (*wa-ikhtilāf 'alā tadrij*), with no separation from one part to another. Similar considerations apply to *ZṬ* [...] By contrast, the line *ZH* is shrouded in darkness, because its points receive no light from *AB*. If a line is drawn from it [*AB*] to a point of *HZ*, it will be stopped by *GD*.[9]

Or as Leonardo expresses:

> By the simple proof of the lines intersected at the endpoints of the shadowy object, all the percussion of the luminous species would be of equal clarity, inasmuch AB, one fourth of the luminous [object], corresponds to GF, one fourth of the percussion, and likewise BC to GH, which is one fourth of both the luminous and the percussion, and so do the two other fourths; Experimentation does not confirm that KF is of the same luminosity [...] Since experimentation shows that the percussion of the light rays on any part of the height acquires degrees of darkness, and since this is not concluded from the first figure, the second concludes, because all the light AE sees I; and three

7 Lindberg, 'Laying the Foundations', p. 10.

8 Eilhard Wiedemann, 'Über eine Schrift von Ibn al-Haytam "Über die Beschaffenheit der Schatten"', *Sitzungsberichte der Physikalisch-medizinischen Sozietät in Erlangen*, 39 (1907), 226–48; Muṣṭafā Naẓīf, *Al-Ḥasan Ibn al-Haytham wa-buḥūthuhu wa-kushūfuhu al-baṣariyya* (Cairo: Nūrī Press, 1943), pp. 264–74.

9 Ibn al-Haytham, *Maqāla fī kayfiyyat al-aẓlāl*, Süleymaniye Kütüphanesi, Istanbul, MS Fātiḥ 3439, fol. 125ʳ; Staatsbibliothek, Berlin, or. 6919, fol. 32ʳ.

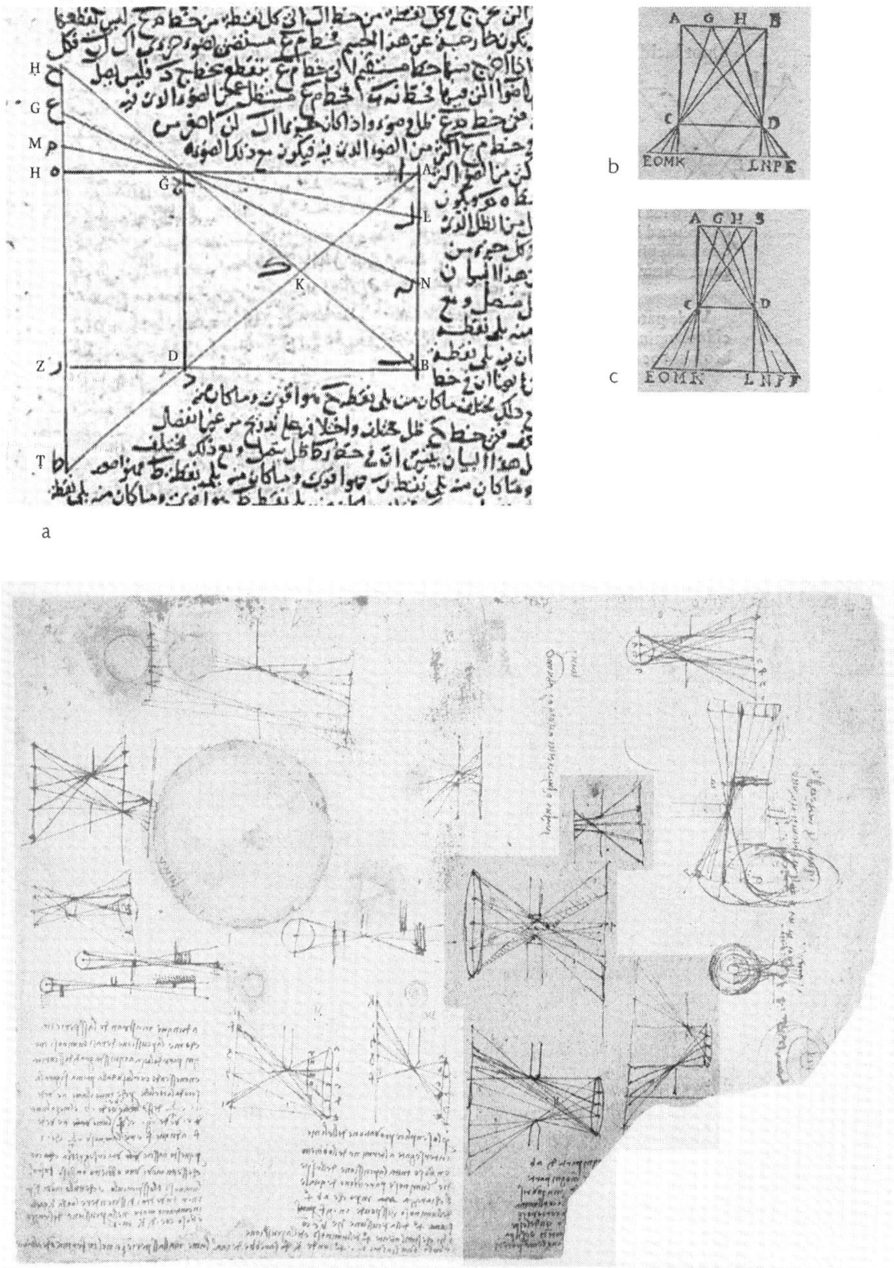

Fig. 1 (*a*) Ibn al-Haytham, *Maqāla fī kayfiyyat al-aẓlāl*, Süleymanie Kütüphanesi, Istanbul, Fātiḥ MS 3439; (*b*. and *c*.) Maurolico, Photismi de lumine et umbra (1521), Naples, 1611; (*d*) Veneranda Biblioteca Ambrosiana, Milan, Leonardo da Vinci, *Codex Atlanticus*, fol. 650^{r}.

fourths of this light sees H; and half of the light CE sees G; and one fourth of the light DE sees F. Therefore F is 3/4 as luminous as I.[10]

Maurolico states:

> Let AB be the light, CD the illuminated interval, whose shadow is cast on the plane EF. By the endpoints A C and B D we draw the rays ACK and BDL falling onto the plane. It is clear that the shadow of CD is KL, since no ray passes down from any point of the light AB to the interval KL. However K and L are not the boundaries of the shadow, terminating the shadow so that the plane outside would be fully illuminated at once. Let us take within the light AB the points G H, by whom, through the points C D, the rays GM, GP, HO and HN are drawn to the plane. Do the same with the rays AF and BE. Therefore, the interval KM is lit only by one segment AG; the interval LN, only by one segment BH; the interval MO is lit by both segments AG and GH. Likewise, the interval NP is lit by both segments BH and HG. Therefore, by the fifth supposition, the interval KM is illuminated less than MO, and LN less than NP.[11]

Ibn al-Haytham's *Maqāla fī kayfiyyat al-aẓlāl* is almost perfectly reflected in Leonardo's *Codex Atlanticus* and Maurolico's *Photismi de lumine et umbra*. They differ only in minor details. For example, in Ibn al-Haytham's analysis, the screen is lit by one third (*AL/AB*), *MG* by two thirds (*AN/AB*) and *GḤ* by the whole (*AB/AB*) of the light source *AB*. Leonardo has instead one quarter, one half, three quarters, and *tutto il lume* ('the entire light').

These striking similarities raise two main questions. Firstly, the parallels between Ibn al-Haytham, Leonardo, and Maurolico's geometric treatments of the penumbra cast doubt on the views of the historians of optics and perspective, who consider Leonardo a pioneer with no predecessor in the geometrical analysis of the penumbra. Secondly, and more fundamentally, such parallels beg the question whether developments arose due to a diffusion of Arab optics into the West or whether these advances were (re)discovered independently.[12] I will argue that Leonardo's notes on shadows result from the historical diffusion of a late scholastic optical disputation, probably connected in some way with Arabic optics.

Leonardo's Sources

Da Vinci's investigations into shadows are recorded in dozens of folios, produced between approximately 1480 and 1510. These include the *Disegni anatomici* 118^{v} (*c.* 1480),[13] *Codex A* 89^{v} (1490), *Codex C* 5^{r}, 21^{r}, 21^{v} (*c.* 1490–91), *Arundel* 112^{r}, 244^{r} (no date), *Codex Atlanticus*

10 Leonardo da Vinci, *Codex Atlanticus*, fol. 650^{r}.

11 Maurolico, *Photismi*, 13.

12 No parallelism is a proof of transmission. The splitting of the light source into small segments is not unique to Ibn al-Haytham: all authors who contributed to the *'ilm al-aẓlāl* took the same approach. The letters used in Ibn al-Haytham's and Leonardo's diagrams do not match, neither in the *abjadī* or *hijā'ī* order. Even if several Arabic works were circulated in medieval Europe, others never arrived. See Dominique Raynaud, 'Abū al-Wafā' Latinus? A Study of Method', *Historia Mathematica*, 39 (2012), 34–83. In the case at hand, there is no proof that *On the Quality of Shadows* was translated into Latin.

13 The *Disegni anatomici* are dated 1478–1516, but since folio 118^{v} is not that far from the first folio 200^{r}, it must be placed at the beginning of the period, that is, around 1480.

85^{v}, 391^{r}, 391^{v}, 988r (*c.* 1490), 104^{r}, 519^{r}, 650^{v}, 949^{r} (*c.* 1508), 625^{r}, 650^{r} (*c.* 1505–1508), 527^{v}, 817^{v} (*c.* 1508–10). As the first diagram is dated *c.* 1480 (*Disegni anatomici*, fol. 118^{v}), it is likely that Leonardo's research started in Florence, where he obtained his first commission in January 1478. However, his reflection on shadows is mainly Milanese. A first group of folios dates back to 1490, where Duke Ludovico Sforza (1482–99) requested him to work in Milan. A second group dates to 1508–10, when he was called again to Milan by Charles d'Amboise (1506–13).[14]

Leonardo's notebooks provide a first indication that he used external sources when dealing with shadows. Some notes of *Codex C* would be otherwise unintelligible. To begin with, in folio C9^{v} in *Codex C*, there is a proposition ending in the words: 'La figura destra sta bene sopra detta proposizione' ('The right figure is actually above that proposition'), although the folio has no 'right figure' at all. To solve this problem, we must recognise that the words 'figura destra' refer to the model he copied. We should read: 'In the source I used — not *Codex C* — the right diagram was situated above the proposition'. There is another mention which gives us pause in the *Disegni anatomici*. In the midst of a series of notes on shadows, Leonardo writes: 'Include in the last [section] on shadows the figures that are on the writing case of Gerardo the illuminator at San Marco in Florence'.[15] The *Disegni anatomici* contain notes for the years 1478–1516; Leonardo left Florence in 1481; Gherardo di Giovanni di Miniato died in 1497. Therefore, the meeting could only have occurred at some point between 1478 and 1480, during Leonardo's first stay in Florence. If Leonardo ever used this manuscript, we can guess that it was either intended for — or already in the possession of — the library of San Marco. A close look at the present day catalogue of the BNCF as well as the 1495 inventory of San Marco informs us that Leonardo's source was certainly not the *Tractatus de radiis et umbris*, an alternative title of al-Kindī's *Liber de causis diversitatum aspectus*, where there is no punctiform analysis of the penumbra, nor would it have been the anonymous *Liber de umbris* (Biblioteca Nazionale Centrale di Firenze, Conv. Soppr. J.V.18, fols 1^{r}–2^{r}) ascribed either to Abraham ha-Levi ibn Daud or Jordanus of Nemore.[16] This work is brief and lacks diagrams.[17] Nor was it the *Tractatus de umbris et radiis*, the title of a work in progress referred to by Dominicus of Clavasio in the course of the fifteenth proposition of his *Practica geometriae*.[18] The latter could be a candidate for identification, but no copy of it has been located to date and it is not even clear whether it was actually written.

Leonardo's notebooks also include a series of non-autograph notes on shadows which may have served as a model (Figure 2). The textual parallels between Leonardo's notes and Manus 3 (*Codex Arundel*, 100^{*v}–103^{v}) prove that Leonardo copied this source extensively. In what follows, I provide a diplomatic transcription of the text in order to further comment on its idiosyncrasies; /word/ refers to a word corrected over the line; word refers to a word deleted. I do not follow the suggestion of Pedretti and Vecce, *Il codice Arundel 263 nella*

14 Fabio Frosini and Carlo Vecce, 'Leonardo da Vinci, Il contributo italiano alla storia del pensiero', in *Filosofia*, ed. by M. Ciliberto (Rome: Istituto della Enciclopedia Italiana, 2012), pp. 140–41.

15 'Serva all'ultimo dell'ombre le figure [ch]e appariano nello scriptoio di Gera[r]do miniatore a San Marco in Firenze'. Leonardo da Vinci, *Disegni anatomici*, fol. 167^{r}.

16 Ron B. Thomson, 'Jordanus de Nemore: Opera', *Mediaeval Studies*, 38 (1976), 97–144 (p. 139).

17 José María Millás Vallicrosa, 'Una obra astronómica desconocida de Johannes Avendaut Hispanus', *Osiris*, 1 (1936), 451–75 (p. 453).

18 Hubert L. L. Busard, 'The Practica geometriae of Dominicus de Clavasio', *Archives for History of Exact Sciences*, 2 (1965), 520–75 (p. 538).

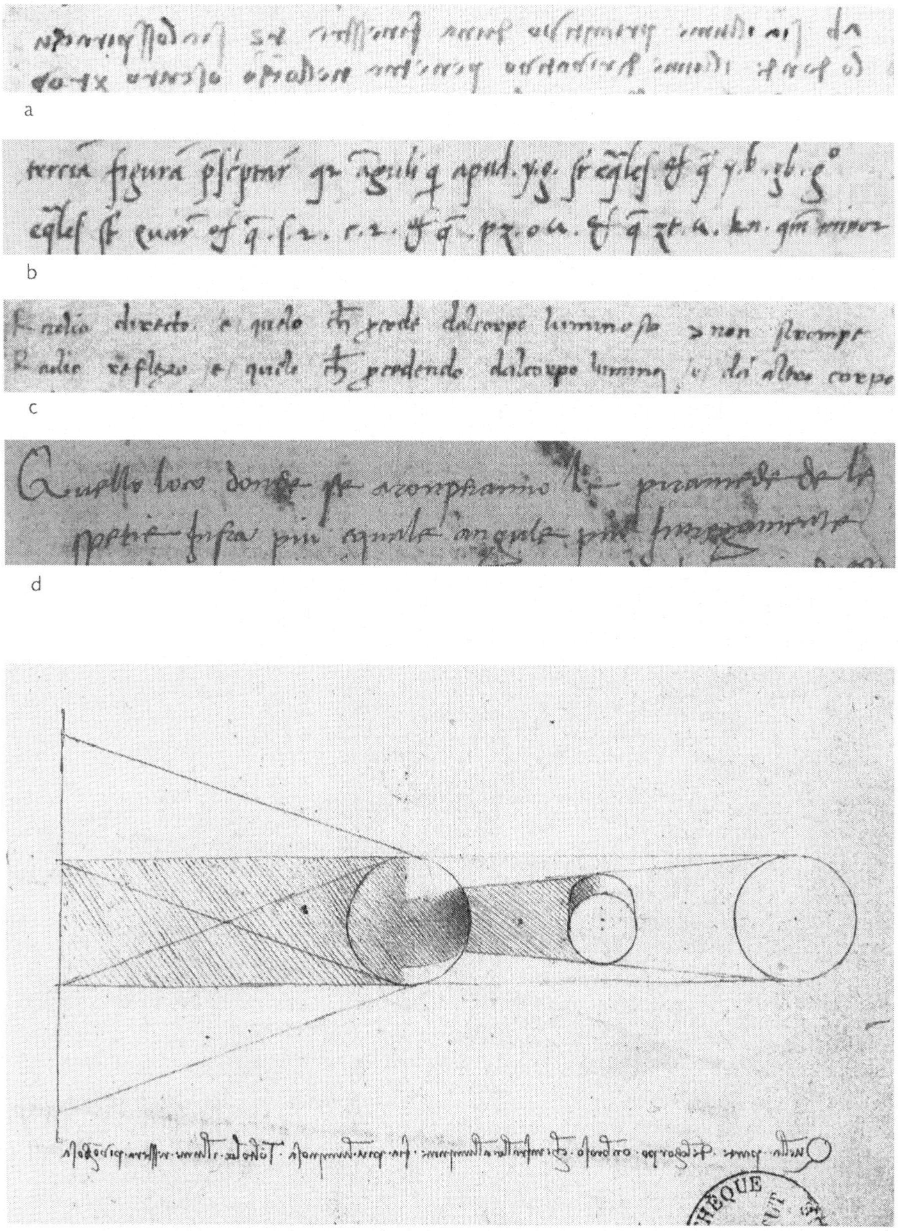

Fig. 2 (*a*) Da Vinci, CA519[r]; (*b*) Anonymous 1, CA153[r]; (*c*) Anonymous 2, CA516[r]; (*d*) Anonymous 3, Arundel 103[r]; (*e*) Da Vinci, C13[r].

British Library (Florence: Giunti, 1998), who have identified Manus 3 with the 'prospettivo milanese' (Bramante), without historical basis. The comparison of Manus 3 with a collection of autographs from the 1500s strongly suggests that the copyist of these optical notes was Caradosso Foppa (1445-1527) who, in a letter to Lucovico Sforza dated February 15, 1495, spells *avixo*, *rispoxe* and *caxone*, and has handwriting very similar to Manus 3.

Anonymous	***Leonardo***
(1) Quella /parte del/ refresso ne corpo fia piu evidente quallo terminara il loco de magiore / ob/scuritate (100^{v})	Quella parte del reflesso fia più evidente, che terminerà in loco di maggiore oscurità (C20^{r})
(2) Quanto il corpo luminoxo e circhondato da piu scura* tenebroxitate aloro parira de magiore sprandore Quello corpo luminoxo parira de magiore sprandore el quallo fia circondato da magiore tenebroxitate (100^{v})	Quel corpo luminoso parrà più splendido, il quale da più oscure* tenebre circundato fia (C5^{r})
(3) Quella parte de reflesso fia piu charia de la qualle i race de la precussione reflessione fiene piu corte (100^{v})	Quella parte del refresso fia più chiara, della quale i razzi della reflessione fien più corti (C4^{r})
(4) Quello corpo luminoxe fia de minore euidentia il quallo fia circondato da luminoxe campo (100^{v})	Quel corpo luminoso parrà di minore splendore, il quale da più luminoso campo circundato fia (C3^{r})
(5) I race sollare passate infra molti rame delle sotille ramificatione della planta a longhe andare el corpo umbroxo aluminato da quelli non fara pui che una solla umbra (100^{v})	Il corpo alluminato dai razzi solari passati in fra le sottili ramificazione delle piante, a lungo andare non farà più d'un' ombra (C2^{v})
(6) I rete termini de corpi parani roti il qualli terminarani parti in il loco tenebroxi e parti luminoxi (101^{r})	I retti termini de' corpi parranno rotti, che termineranno in loco tenebroso, rigato da percussione di luminosi razzi (C1^{v})
(7) Quello corpo umbroxo disstericha retonditate fare circulare umbra mista il qualle sara hara infra se il solle uno corpo antiposto simili ala sua quantitate qualitate (101^{r})	Quel corpo ombroso di sperica retondità farà circulare ombra mista, il quale arà in fra sé e 'l sole interposto un corpo ombroso di sua qualità (C12^{v})
(8) Infra li corpi dequalle distantie quello che piu luminoxo parra essere alocio pio propincho (101^{r})	In fra i corpi d'equal grandezza e distanzia quello che fia piu alluminato parrà all'occhio più propinquo e maggiore (C1^{r})
(9) Molti corpi umbroxi di quasi congionta vicinita essendo visti in layro luminoxi a longe distantie parrani seperati da grande intervallo (101^{r})	Se molti corpi ombrosi di quasi congiunta vicinità fieno veduti in campo luminoso, in lunga distanzia parranno separati da grande intervallo (C14^{r})
(10) Nesuno effeto e senza casone (101*^{r})	Nessuno effetto è in natura senza ragione (CA398^{v})
(11) Infra le cosse de equalle grandeze situate in layro luminoxo, quella che fia piu lontana aparira piu sciara e di minore figura (101^{v})	In fra le cose d'equal grandezza e colore quella che fia più lontana, parrà più chiara e di minor figura (C21^{v})
(12) Quante piu grosso fia languli /de/ la precussione /e risaltatione/ tante piu lontano terminara la cossa rissaltata dal suo primo moto (101^{v})	Quanto piu grosso fia l'angolo della percussione e risaltazione fatta dalla cosa movente, tanto piu distante sarà il termine del moto dal suo principio (Ar. 101^{v})
(13) Universalemente tuti e ponte dele streme ponte de le peri piramidalle spetie sono continuato et per tuto layro senza alcuno intra intervallo (103^{r})	Universalmente tutti i punti causatori delle istreme punte delle piramidali spezie delle cose son continuamente tutti per tutta l'aria insieme connessi e congiunti sanza alcuno intervallo (C20^{r})
(14) Sel lumo fia de longa figura el corpo unbroxo sia retondo lombra dirivativa sera piu larga cha alta, infine che chade in loco piano infra equale anguli (103^{r})	Il lume di lunga figura farà che l'ombra dirivativa, nata da corpo retondo, fia più larga e bassa, benché sia percossa in fra equali angoli (C18^{v})
(15) Il Quello termino del lombra dirivativa fia piu scura laquala fia circondata da piu s… chario lumo dirivativo (103^{v})	Quel termine dell'ombra dirivativa fia più oscuro, che da più chiaro lume dirivativo circundato fia (C11^{v})

This comparison proves that at least fifteen of Leonardo's propositions on shadows derive from this source in *Codex Arundel*. It is worth noting that several of the propositions contained address the variation of light within shadow. These include the descriptions: *magiore sprandore, minore chiareza, più chiara* and *più scura* among others. Furthermore, one of these propositions considers the condition responsible for the *umbra mista* (or 'penumbra'), that is, 'the shadow cast by an opaque body is smaller than the light source [when the opaque body is smaller than the light source]'.[19]

The linguistic features of these anonymous notes, which contain the words *luminoxo, umbroxo, luminoxitate, obscuritate, tenebroxitate, vicinitate,* and *sprandore*) suggest that they were written by a scribe from northern Italy. The copy is often crossed out. Some corrections are likely to be simple misreadings (the word *quantitate* is crossed out and replaced by *qualitate*). Other corrections — where for example the phrase 'Since the luminous body is surrounded' is replaced by 'This luminous body, which is surrounded by a great darkness, will appear of greater brightness'[20] — rather suggest an emendation in the very course of the redaction. (Note that using the services of a translator seems to have been common practice for Leonardo as for example in the case of DA113^{r}).

We are sure at this point that Leonardo's research on shadows and penumbrae does not consist solely of his own individual work, but rather proceeds from external, pre-existing sources, such as the diagrams held in San Marco and the anonymous notes from *Codex Arundel*. Such indications lead the present study to look now at questions of transmission and the translation of scientific texts.

The Extent of Leonardo's Source

The *Codex Atlanticus* includes a typical passage of Leonardo's notes on shadows. As the text is easily understandable and the focus is mainly on the form, I switch to Italian:

> L'ombra non si dimosterrà mai d'uniforme oscurità nel loco dove essa si taglia, se tale loco non è equidistante al corpo luminoso. *Provasi per la settima che dice*: 'quell'ombra si dimosterà più chiara o più oscura, che fia circundata da campo più oscuro o più chiaro'; *per la ottava di questo*: 'quel campo arà le sua parte tanto più scure o più chiare, quanto e' sarà più remoto o più vicino al corpo luminoso, e infra li siti d'equal distanzia al luminoso, quel si dimosterrà più alluminato, che riceve li razzi luminosi infra angoli più equali'. Sempre 'l'ombra segnata in qualunche inequalità di sito, si dimosterrà colli sua veri termini equali al corpo ombroso, se l'occhio si pon dove fu il centro del luminoso' [...] Lo spendore d'un medesimo luminoso in pari distanzia si farà di tanta maggiore potenzia, quanto fia maggiore il numero delli spiraculi, dond'esso penetra alla sua impressione. *E questo è provato nella terza* dirieto a questa faccia, eccetera. *Ma ancor si prova colla tredecesima dell'altro libro*, dove dice: 'Quella parte d'un sito sarà più alluminata, che da maggior numero di luminosi fia veduta'.[21]

19 'umbra dirivativa nata da corpo unbroxo minore di luminoxo'. Leonardo da Vinci, *Arundel*, fol. 103^{v}.

20 'Quanto il corpo luminoxo e circhondato'; 'Quello corpo luminoxo parira de magiore sprandore el quallo fia circondato da magiore tenebroxitate'. Da Vinci, *Arundel*, fol. 100^{v}.

21 Da Vinci, *Codex Atlanticus*, fol. 658^{v}.

Here Leonardo refers to numbered propositions ('Provasi per la settima [...] per la ottava di questo [...] E provato nella terza [...]'). These propositions use the same language as the anonymous notes from *Codex Arundel*. The size of the work can be estimated on that basis. The numbered propositions found throughout Leonardo's notes on shadows are first arranged according to their number (using the terms *prima, seconda, terza,* and so on), then according to their content. The symbol = refers to propositions with similar content, ≠ to propositions where content differs. This operation gives the least number of sections (books?) of the work referred to. For example, if we were to find two *proposizione quarta* with differing content this would mean that the source had (at least) two sections.

No	References	Number of sections
prima	CA642^{r} ≠ E32^{v} ≠ DA167^{r}	3
seconda	CA104^{r} ≠ DA118^{r} ≠ E32^{v}	3
terza	CA658^{v}	1
quarta	CA527^{v} = DA118^{r} = F75^{r} ≠ CA519^{r} = CA534^{v} = CA694^{v}	2
quinta	CA694^{v} ≠ CA694^{v}	2
sesta	CA694^{v}, CA705^{r}?	1 or 2
settima	CA505^{v} ≠ CA568^{v} ≠ DA118^{r}	3
ottava	CA505^{v} ≠ CA568^{v}	2
nona	CA658^{r} ≠ CA752^{r} ≠ D10^{r} ≠ D4^{v} = DA118^{r}	3 (D10^{r} is on geometry)
decima	CA505^{v}	1
undicesima	–	–
dodicesima	–	–
tredicesima[22]	CA568^{v}	2 (dell'altro libro)

As a result, the optical source accounted for in Leonardo's notes had at least three sections devoted to the study of shadows.

In the previous count I have excluded *Codex Atlanticus*, fol. 676^{r}, which consists of a book project in seven books. The first book addresses shadows in general; the second deals with *ombre primitive* ('form shadows'), the third makes a study of *ombre derivative* ('derived shadows', that is, cones of shadows), the fourth assesses cast shadows, which Leonardo called *percussioni delle ombre dirivative* ('impacts of the derived shadows').[23] The fifth book adresses the mixing of shadows, the sixth deals with the effects of reflection on the quality of a shadow, and the seventh studies distance and its effect on shadows. The reason for this exclusion is simply that the numbered propositions do not fit into Leonardo's book project. For example, the reference to Book Two does not deal with *ombre primitive* but with the intersection of shadow cones (CA658^{r}).

Between folios 1^{r}–20^{r}, *Codex C* contains forty propositions arranged in sequences of letters — for example *n, o, p ... Bi, Bl, Bm*. This lettering, which is not in Leonardo's hand, is not the source of the numbered propositions on shadows. As the three series of letters overlap, they provide a one-to-one match between letters and numbers. The letter 'a' is equivalent to the number '1', 'b' equal to '2', 'c' to '3' and so on, with 'y' equalling '21' and

22 No proposition has a number equal to or higher than fourteen.

23 Da Vinci, *Codex Atlanticus*, fol. 676^{r}.

'z' being equal to '22'. As the three series are consecutive, we deduce that Ø = I, A = II, B = III. Thus 'Bi' would mean *nona del terzo* ('ninth of the third'). Many propositions do not correspond to this numbering. For example, no proposition four (CA519^{r}, 527^{v}, 534^{v}, 694^{v}) coincides with a proposition *quarta* ('four') in *Codex C*. Similarly, no proposition nine (CA658^{r}, 752^{r}, DA118^{r}, D4^{r}, 10^{r}) coincides with any proposition *nona* in *Codex C*.

As a result, we can deduce that the numbered propositions of Leonardo's corpus are not self-referential. They refer to an external source, which is probably the same source behind the notes of *Codex Arundel*.

The Language of Leonardo's Source

Latinisms appear in many of Leonardo's propositions on shadows. Why Leonardo was using Latinisms is still an open question. Claudio Scarpati provides a list of the Latin turns of *Codex C*, written *c*. 1490–91.[24] There are instances in Leonardo's writings where the verb has been postponed. Here in C1^{r}, Leonardo writes 'Quella inferiore e superiore stremità della dirivativa ombra fie men che la laterale distinta, la quale da lume più alto che largo *causata fia*'. This is also the case in C2^{r}, 2^{v}, 5^{r}, 8^{v}, 11^{v}, and 18^{r}. Epithets are anteposed in certain of Leonardo's writings too. Even if the epithet may be indifferently placed either before or after the noun in Latin, Leonardo's antepositions sound strange in *volgare*. For example, 'Quella *inferiore* e *superiore* stremità della *dirivativa* ombra', '*solar* razzi [...] *dirivativa* ombra [...] *sperico* corpo'. Such examples can be found in C1^{r} and 12^{v}, as well as C4^{r}, 4^{v}, 7^{v}, 8^{v}, 10^{r}, 12^{v}, 18^{r} and elsewhere. There is also evidence of a certain use of Tuscan versions of Latin forms. In some places, Leonardo prefers the Latin form to the *volgare*. For example, in C1^{r}, he writes 'Parrà all'occhio più *propinquo* [*vicino*] e maggiore [...]'. Such use is also in evidence in C2^{v}, 4^{r}, 4^{v}, and 9^{v}. In C5^{r}, he writes 'quel corpo luminoso parrà più *splendido* [*splendidus: lucente*]' and in C7^{v} is written 'l'ombra sua è simile a une *retrosa* [*capovolta*] e contraria piramide'. In C10^{v} and C11^{v} we find 'impossibile è che il razzo possa *conducere* [*condurre*] [...]'. In DA118^{r} Leonardo writes 'le spezie sono in ogni parte del sito a lor *circunstante* [*circundante*]'; in DA167^{r} is written 'ombra è *alleviazion* [*allevatio: attenuazione*] di tenebre'.

These Latin turns could be explained in several ways. The first hypothesis is that all school children learned to write in Latin in fifteenth century Italy.[25] The question whether Leonardo knew Latin is misleading because knowing some vocabulary does not mean you know the entire language. The real question we should ask is at what level did Renaissance artists know Latin? The second hypothesis might be that Leonardo's use of Latin reflects his desire to appear distinguished in society. As Augusto Marinoni has written, 'The use of these latinizing devices reveals Leonardo's purpose: he wants to attain to a nobler and more dignified style'.[26] In the same sense, Scarpati claims that Leonardo's Latinisms result from the wish to confer a 'sense of grandeur' to his texts.[27]

---

24 Claudio Scarpati, *Leonardo scrittore* (Milan: Vita e Pensiero, 2001).

25 Armando F. Verde, *Lo studio fiorentino, 1473–1503* (Florence: Olschki, 1973–94).

26 Augusto Marinoni, 'Leonardo as a Writer', in *Leonardo's Legacy*, ed. by C. O'Malley (Berkeley: University of California Press, 1969), pp. 62–63.

27 Scarpati, *Leonardo scrittore*, 11.

Both hypotheses face important objections, however. There are very noticeable stylistic breaks in his writing. Latin forms disappear abruptly after the passage referred to previously. Having written 'Quel corpo luminoso [...] da più oscure tenebre circundato fia' in C5^r^, Leonardo switches to the colloquial language in the same folio: 'Se darai col martello o nel quadro q, il ferro m salterà fora' in C5^v^. Similarly, Leonardo writes: 'Lo equidistante circuito di piramidal concorso darà a li sua obietti eguale qualità d'angoli [...]' in A86^v^ and again switches to the colloquial speech: 'E leva l'uno e prova coll'altro e mai truova la verità del peso' in A88^r^. This begs the question: how should these breaks be interpreted?

The second objection might be that Leonardo mistranslated the Latin. It is plausible that *conducerre* was used instead of *condurre* ('to guide'), *o circunstante* for *circondante* ('surrounding'). It is much more difficult to accept that Leonardo had any mastery of Latin when he writes *propinquo*[28] instead of *vicino* ('close'), *retrorsa*[29] instead of *capovolta* ('upside down', C7^v^), or *splendido*[30] instead of *lucente* ('luminous', C5^r^). These seem very much to be cases of mistranslation.

Our third objection is that one can find misunderstandings elsewhere in the texts. Let us consider the sentence 'Quale parte del corpo ombroso che in fra le alluminate fia più luminosa, *toltole il lume*, resterà più ombrosa' (C13^r^, Figure 2 e).

Take the segment 'che in fra le alluminate fia più luminosa'. If one reads 'toltole il lume' as 'toglie il lume (*tollit*)', then we must translate: 'The part of the opaque body that removes the light will be darker'. This is simply nonsense, for the part that removes light is precisely the part that is lit. We must switch to the passive and read 'quale parte [...] sarà priva di lume', that is 'this part [...] will be devoid of light' (*tollitur*). The body referred to is the central sphere, the centre of which has a darker shade, once deprived of light. This allows us to translate the sentence as follows: 'When the part of the shaded body (which would be the brightest portion of the illuminated parts) is deprived of light, then it will be darker'. This translation is consistent with itself, and in accordance with the diagram. There is a certain confusion between the active *togliere (tollere)* and the passive *ser privo di (tolli)*. Leonardo's authorship over these notes on shadows is untenable, because such a misinterpretation is only possible when reading or translating.

Thusly, I conclude that Leonardo's source on shadows would have been written in Latin. Leonardo would have either translated a Latin model with the assistance of a translator, or relied on a full translation — that of *Codex Arundel*, for example. The instances where Latin appears in Leonardo's notes on shadows now make perfect sense. In the history of translation, the *ad verbum* literal method was preferred by Boethius (see *In Isagogen Porphyrii commenta*) and Burgundio of Pisa. It was criticised by the supporters of the *ad sensum* interpretive method. Its main defender, Leonardo Bruni (see *De interpretatione recta*), argued polemically with Alfonso de Cartagena (see *Declinationes super nova quadam Ethicorum Aristotelis translatione*) on this issue. The two methods were being used throughout

28 The use of the word *'propinquo'* in this sense is attested by Guinizelli (before 1276), but it is already contextualised in 'legato da vincoli di parentela' by Dante (before 1321).

29 Dante (1321) still uses *'retrorso'* for 'all'indietro', but Giamboni (before 1292) uses it only in the sense of a person 'che vuole avere poco a che fare con gli altri'.

30 *'Splendido'* already means 'magnificienza, sfarzo' in Dante (1321), 'bellissimo, stupendo, lussuoso' in Boccacio (before 1375), 'largo nello spendere e nel donare' in Filarete (before 1465).

the Middle Ages and the Renaissance, regardless of the source or target language. The literal method was a basis for translations from Greek and Arabic to Latin.[31] It was also adopted for translations from Hebrew to Spanish, from Latin to French, or from Latin to Italian.[32] Emulating the source or target language was not the translator's personal choice, rather it was conditioned by the nature of the text. *Ad sensum* translations were used for secular texts; *ad verbum* translations for theological texts. Due to their mixed status, many scientific texts in the Middle Ages were translated *ad verbum*.[33] However, some very literal translations, such as those made by Gerard of Cremona, Michael Scot and Hermann of Carinthia, were criticised by Roger Bacon (see *Opus tertium*, XXV). The choice for one or the other method was not without an impact on the process of knowledge diffusion: even though both methods can cause misunderstanding, there is always a chance to rediscover the model behind a translation *ad verbum* — which is impossible in a translation *ad sensum*. It suffices for my argument that most scientific translations made before the sixteenth century were literal translations.

The hypothesis that Leonardo's notes on shadows derived from a Latin source is supported by the fact that he frequently used calque translations of scientific texts. Below are some parallelisms between the Latin texts and Leonardo's notes *in volgare*.

> 'In eclipsi solis radius luminis a sole transiret ad nos per partes lune magis raras'.[34]
> 'Nell'eclissi di luna i razzi solari penetrebbono per alcuna parte della predetta raretà'.

> 'Inter physice considerationis studia lux iocondius afficit meditantes'.[35]
> 'Intra li studi delle naturali considerazioni la luce diletta più i contemplanti'.

---

31 'The translations of Gerard of Cremona are exceedingly literal, to the extent of retaining Arabic syntax in the Latin, and transliterating Arabic terms where no direct Latin equivalent could be found'. Charles Burnett, 'Translation Norms and Practice', in *Medieval Science, Technology, and Medicine*, ed. by T. Glick, S. J. Livesey and F. Wallis (New York: Routledge, 2005), p. 487.

32 The calque language of Hebrew, specifically, the Ladino, which differs from the Judeo-Spanish, offers a rich corpus of texts in which the translation was driven by a deliberate intention to preserve the authenticity of the source language of the Torah, without regard to the requirements of the vernacular. See Haïm Vidal Sephiha, 'Langues juives, langues calques et langues vivantes', *Linguistique*, 8 (1972), 59–68; Michèle Goyens and Pieter de Leemans, 'Traduire du grec au latin et du latin au français', in *Medieval Translation Practices*, ed. by P. Andersen (Copenhagen: Tusculanum, 2004), pp. 201–24. One must cite the case of MS Vaticano 4595, which is the Italian *ad verbum* translation of the Latin version of Alhacen's *De aspectibus*; Enrico Narducci, 'Nota intorno a una traduzione italiana fatta nel secolo decimoquarto del trattato d'ottica d'Alhazen', *Bollettino di bibliografia e di storia delle scienze matematiche e fisiche*, 4 (1871), 1–40; Graziella Federici Vescovini, 'Contributo per la storia della fortuna di Alhazen in Italia', *Rinascimento*, 5 (1965), 17–49.

33 'Boethius's model tended to prevail among medieval translators. Among the most strenuous advocates of the *ad verbum* style over the *ad sensum* style was Burgundio of Pisa, who justified this method in prefaces to his translations of theological texts from Greek. He repeats the point that, for the translation of theological and scientific works, the *ad verbum* translation is to be aimed at, even at the expense of elegant Latin style … Burgundio put this method into practice in his translations of works by Aristotle and Galen, and the *ad verbum* style was followed by his colleague James of Venice, by William of Moerbeke (who revised his own translations in the direction of literalness), and the majority of medieval Greek-Latin translators. Burnett, 'Translation Norms', p. 487. This issue is addressed in Alexander Fidora, 'Les différentes approches des traducteurs', in *Une conquête des savoirs. Les traductions dans l'Europe latine*, ed. by M. Lejbowicz (Turnhout: Brepols, 2009), pp. 46–65.

34 Albert of Saxony, *Quaestiones in libros de caelo et mundo* (Venice: Locatello, 1492), II, Quaestio 24, a work referred to in *Codex I*, cover[v]; Da Vinci, F85[r].

35 John Pecham, *John Pecham and the Science of Optics: Perspectiva communis*, ed. and trans. by David C. Lindberg (Madison: University of Wisconsin Press, 1970), proemium, p. 60; Da Vinci, *Codex Atlanticus*, fol. 543[r].

> 'Sicut cum lapis proicitur in aquam generatur circuli diversi'.[36]
> 'Si chome la pietra gittata nell'acqua si fa cientro e chausa di vari circhuli'.
>
> 'Cuiuscunque trianguli centrum gravitatis est punctum, in quo linee recte ab angulis ad dimidias bases ducte concurrunt'.[37]
> 'Ogni triangolo ha il centro della sua gravità nella intersegazione delle linie che si partano dalli angoli, e terminano nella metà delle lor opposite base'.
>
> 'Equalitas declinationis identitatem conservat ponderis'.[38]
> 'La equalità della declinatione conserva la equalità de' pesi'.
>
> 'Caput a mento ad summum verticem octavae'.[39]
> 'Dal di sotto del mento alla somita del chapo he l'octavo'.

Vitruvius, Archimedes, Jordanus of Nemore, John Pecham, Albert of Saxony, and so on: these textual parallels do not derive from any desire for 'social distinction', as mentioned earlier. All characteristics of Leonardo's notes (such as the spontaneous alternation between Latin and Tuscan, the scant evidence of a specialised knowledge of Latin, and the various misreadings and misinterpretations) are attributable to a calque translation from Latin. I thus conclude in this section that Leonardo used — beyond reasonable doubt — a Latin optical treatise for composing his notes on shadows.

The Style of Leonardo's Source

By the synthesis of sections two, three, and four, we know that this optical source, partially translated by the anonymous hand of *Codex Arundel*, was fairly extensive (at least three sections) and was originally written in Latin. Further details allow us to argue that this text was composed in the context of late scholasticism, because several features of scholastic argumentation and terminology can be found in Leonardo's notes on light and shade.

The Scholastic Argumentation

The scholastic method of argumentation follows a logical sequence, beginning with a *quesitum* ('question') and *determinatio* ('thesis'), supported by the *probatio* ('direct evidence'). The author then reviews what are termed the *sed contra* ('objections') and

36 Pecham, *Perspectiva communis*, ed. and trans. by Lindberg, I, 6 (p. 64), a work referred to in Madrid II, fol. 2v; Da Vinci, A9v.

37 Archimedes, *De planorum equilibriis*, I, 14; Marshall Clagett, *Archimedes in the Middle Ages, 2. The Translations from the Greek by William of Moerbeke* (Philadelphia: The American Philosophical Society, 1976), p. 1355, a work cited in *Codex I*, coverv; Da Vinci, *Arundel*, fol. 16v.

38 Jordanus de Nemore, 'De ponderibus', in *The Medieval Science of Weights, Scientia de ponderibus: Treatises Ascribed to Euclid, Archimedes, Thabit ibn Qurra, Jordanus de Nemore and Blasius of Parma*, ed. by E. A. Moody and M. Clagett (Madison: University of Wisconsin Press, 1952), 9, p. 188, a work referred to in Da Vinci, *Codex Atlanticus*, fol. 611r and 981r.

39 Vitruvius, *On Architecture*, ed. by F. Granger (London: Heinemann, 1934), III, 1, p. 158, a source referred to in I copv; Da Vinci, *Accademia Veneta*, 121–1A, 182.

gives a *responsum* ('response'). At the end of the investigation, he restates the solution in a clearer form, or *explicatio*.

Of course, this line of argumentation was adapted as a result of the number of objections the author faced.[40] However, regardless of the subject under discussion, the sequence is always visible. If we compare this sequence to some of Leonardo's propositions on shadows, we see that the same argumentative line is at work. Let us switch again to Italian to highlight some historical particularities:

> Dell'ombra semplice. *Perché* nelle intersegazione *a b* delle due ombre composte *e f*, *m c* si genera l'ombra semplice [...] e non si genera tale ombra semplice nelle due altre intersegazioni *c d*, fatte dalle medesime ombre composte dette di sopra? *Risposta*. Le ombre composte son miste di chiaro e di scuro, e le semplici son di semplice oscurità. Addunque delli due lumi *n o* l'uno vede l'ombra composte da un lato e l'altro vede l'ombre composte dall'altro, ma nessun lume vede le intersegazioni *a b*, e però è semplice ombra; ma nell'ombra composta vede l'uno o l'altro lume. *E qui nasce un dubbio per l'avversario, che dice*: Nelle intersegazion dell'ombra composta per necessità vede li due lumi, causa d'esse ombre [...] *Qui si risponde* esser vero il detto del *avversario* [...] *E questo è che* se vedendo li due lumi, in tale intersegazione tale ombre sarebbono annullate... *E questo si prova nell'ottava de proporrtene [i.e. De proportione], dove dice*: Tal proporzione ha semplice potenzia con semplice resistenzia, quale ha duplicata potenzia con duplicata resistenzia.[41]

Several features of scholasticism are apparent in this passage, notably the alternation of questions and answers, and the presence of a fictitious opponent, or *adversarius*. Let us have a closer look at the terminology.

The Scholastic Terminology

Probatur: There are 273 occurrences of *provasi* in Leonardo's corpus. They appear in three main clusters: *Perspectiva* (hundred occurrences), which is divided into two sub-clusters: 'Shadows' (43 occurrences) and 'Vision' (33 occurrences), *Geometria* (57 occorrences, Euclid is quoted 29 times in e-Leo) and *De ponderibus* (32 occurrences). The phrasing 'Provasi per la prima di questo [...]' ('It is proven by the first [proposition] of this [book] [...]') is the simple rendering of the scholastic wording 'Probatur per primam huius [...]'. The word *probatur* is frequently connected to a numbered proposition, which serves as a reference. Here is an example: 'E questo si prova per la nona di questo [...]' (D10^{r}). This tendency can also be found in DA167^{r}, CA505^{v}, 519^{r}, 527^{v}, 658^{v}, 694^{v}, and elsewhere.

Respondetur: The verb *respondetur* ('it is replied') also demonstrates a similar pattern. The passive *si risponde* occurs on fifty-one occasions across the three main clusters of *De caelo et mundo* (nine occurrences), *Perspectiva* (seven occurrences) and *De ponderibus* (seven occurrences) There are various examples where this verb is used that touch on the topic

40 Olga Weijers, *Queritur utrum. Recherches sur la 'disputatio' dans les universités médiévales* (Turnhout: Brepols, 2009).

41 Da Vinci, *Codex Atlanticus*, fol. 483^{r}.

of shadows. In CA310^{v}, Leonardo writes 'Qui si risponde con la quarta [...]'; CA483^{r} contains 'Qui si risponde esser vero [...]'. Folios CA437^{r}, 658^{r}, D8^{r}, and F29^{v} evidence the word further. It should be added that the Latin word *respondetur* — or *si risponde* in Italian — was of frequent use in medieval scholasticism.

Adversarius: The scholastic notion of *adversarius* ('opponent') appears ninety times in Leonardo's corpus, declined as *avversario*. The word is used in three semantic clusters, each corresponding to a special science. It appears on twenty-one occasions in *Perspectiva*, which is referred to in Madrid II 2^{v} and B58^{r}, a science divided into two sub-clusters: 'Shadows' (eleven occurrences) and 'Vision' (ten occurrences). It also appears on twenty-one occasions in *De ponderibus*[42] and eleven times in *De caelo et mundo* (a science referred to in I130^{v} and F I cover^{v}). Discussing shadows, we read 'Dice l'avversario che [...]' in CA310^{v}, or 'Come vole l'avversario [...]' in CA527^{v}. Similar features can be found in CA784^{r}, 821^{r}, D4^{r}, and E32^{r}. In Leonardo's notes, the word *avversario* always appears within the framework of a medieval science. The word has nothing to do with Leonardo's concerns of the day, such as painting, digging canals or designing war machines. Similar terminology, as well as similar sequences to those mentioned above, can be found in fourteenth century literature. Below is the beginning of John Buridan's Quaestio prima on Aristotle's *De caelo* III:

> Whether it is possible to prove from gravity and levity that the bodies are not composed of indivisible things. It is argued that it is not [...] First, it is demonstrated that every weight is divisible [...] But it seems to me that this argument is not well demonstrative against the *adversary*, because the *adversary* could at once denies it, and the greater and lesser. Let us suppose that not only the earth, but also water, and air have a weight; then the *adversary* says that is to give indivisible points from which the air is composed [...] After that, the *adversary* can say that one point of the earth is heavier that a point of water or airs.[43]

Late scholasticism made frequent use of these rhetorical devices. There are ten occurrences of the word *adversarius* in the commentary on Aristotle's *Physics* attributed to Albert of Saxony.[44] The term occurs more than thirty-four times in John Buridan's *Questiones super de caelo et mundo*.[45] This strategy seems to have been preferred by the Parisian masters

42 Leonardo knew Jordanus de Nemore's *Elementa de ponderibus* and *Liber de ratione ponderis*, Blasius de Parma's *Tractatus de ponderibus*, perhaps the Pseudo-Archimedes' *De canonio* and Thabit ibn Qurra's *Liber karastonis*. Marshall Clagett, 'Leonardo Da Vinci's Mechanics', in *Dictionary of Scientific Biography*, ed. by C. C. Gillispie (New York: Charles Scribner's Sons, 1980), VIII, p. 215.

43 'Utrum ex parte gravitatis et levitatis possit probari quod corpora non sint composita ex indivisibilibus. Et arguitur quod non [...] *Primo* ergo demonstrandum est quod omne grave est divisibile [...] Sed mihi videtur quod ista ratio non esset bene demonstrativa contra *adversarium*, quia *adversarius* statim negaret et maiorem et minorem. Ponamus enim quod non solum terra sed etiam aqua et aer habeant gravitatem; tunc *adversarius* dicet quod est dare puncta indivisibilia ex quibus componitur aer [...] Postea etiam poterit dicere *adversarius* quod unum punctum terrae est gravius uno puncto aquae vel aeris'. Johannes Buridan, *Ioannis Buridani Expositio et quaestiones in Aristotelis 'de caelo'*, ed. by B. Patar (Louvain: Peeters, 1996), pp. 513–15.

44 [Albert of Saxony], *Expositio et quaestiones in Aristotelis physicam ad Albertum de Saxonia attributae*, ed. by B. Patar (Louvain: Peeters, 1999).

45 Johannes Buridan, *Iohannis Buridani Quaestiones super libris quattuor de caelo et mundo*, ed. by E. A. Moody (Cambridge: The Medieval Academy of America, 1942).

of the late fourteenth century. Since Leonardo's notes are steeped in the same scholastic terminology and mode of argumentation, I conclude that his optical source on shadows belonged to the same tradition.

Conclusion

In this chapter, I have established that Leonardo da Vinci's study of the penumbra very much reflects an earlier Arabic text, pertaining to the *'ilm al-aẓlāl* ('science of shadows'). It should also be mentioned that Leonardo referred to a treatise on shadows held in San Marco, as well as reproduced some notes from the anonymous hand of *Codex Arundel*. We conclude therefore that most of Leonardo's writings on the topic derive from a Latin optical source, apparently a scholastic disputation, contained in three books of numbered propositions, from the late fourteenth-century Paris school. This source however has not become one of the main works on optics that have been edited or studied to date. This is as solid a conclusion that we may come to at this point, for two reasons. Firstly, David C. Lindberg's *Catalogue*, which is incomplete, lists some 105 medieval and Renaissance treatises, of which 81 have never been the subject of a critical edition or a study.[46] Secondly, there is quite a high probability that Leonardo's source has been lost. Eltjo Buringh recently estimated that twenty-five to forty percent of existing medieval manuscripts were being lost per century in the Middle Ages, with this quantity even increasing following the invention of the printing press.[47]

The features described in this chapter can be used to decide if any new candidate is indeed the source of Leonardo's research on shadows. Whatever the identity of this text may be, we know for sure that the crisscrossing translations addressed in this chapter have finally provided Leonardo with (at least) one valuable optical text on shadows through which he became acquainted with the punctiform analysis of the penumbra.

Bibliography

Manuscript and Archival Sources

Biblioteca Nazionale Centrale di Firenze, Firenze, Anonymous, *Liber de umbris*, Conv. Soppr. J.V.18, fols 1^r–2^r.

E-Leo database, www.leonardodigitale.com: Leonardo da Vinci, *Codex A*.

E-Leo database, www.leonardodigitale.com: Leonardo da Vinci, *Codex Atlanticus*.

E-Leo database, www.leonardodigitale.com: Leonardo da Vinci, *Accademia Veneta*.

E-Leo database, www.leonardodigitale.com: Leonardo da Vinci, *Arundel*.

E-Leo database, www.leonardodigitale.com: Leonardo da Vinci, *Codex C*.

E-Leo database, www.leonardodigitale.com: Leonardo da Vinci, *Codex D*.

46 David C. Lindberg, *A Catalogue of Medieval and Renaissance Optical Manuscripts* (Toronto: The Pontifical Institute of Mediaeval Studies, 1975).

47 Eltjo Buringh, *Medieval Manuscript Production in the Latin West* (Leiden: Brill, 2011).

E-Leo database, www.leonardodigitale.com: Leonardo da Vinci, *Disegni anatomici.*

E-Leo database, www.leonardodigitale.com: Leonardo da Vinci, *Codex E.*

Staatsbibliothek, Berlin, or. 6919, fol. 32r.

Süleymaniye Kütüphanesi, Istanbul, Ibn al-Haytham, *Maqāla fī kayfiyyat al-aẓlāl,* MS Fātiḥ 3439, fol. 125r.

Primary Sources

Bacon, Roger, 'De speculis comburentibus', ed. by David C. Lindberg, in *Roger Bacon's Philosophy of Nature* (Oxford: Clarendon, 1983).

Buridan, Johannes, *Iohannis Buridani Quaestiones super libris quattuor de caelo et mundo,* ed. by E. A. Moody (Cambridge: The Medieval Academy of America, 1942).

Buridan, Johannes, *Ioannis Buridani Expositio et quaestiones in Aristotelis 'de caelo',* ed. by B. Patar (Louvain: Peeters, 1996).

Euclid, *Opticorum recensio Theonis: Euclidis opera omnia,* ed. by J.L. Heiberg (Leipzig: Teubner, 1895).

Kepler, Johannes, *Ad vitellionem paralipomena, quibus astronomiae pars optica traditur* (Frankfurt: Marnius, 1604).

Maurolico, Francesco, *Photismi de lumine & umbra* (Naples: Lanzi, 1611 [written 1521]), Theorem 18, in *Francesco Maurolico: Opera mathematica,* ed. by K. Takahashi and R. Bellè: www.maurolico.unipi.it.

Nemore, Jordanus de, 'De ponderibus', in *The Medieval Science of Weights, Scientia de ponderibus: Treatises Ascribed to Euclid, Archimedes, Thabit ibn Qurra, Jordanus de Nemore and Blasius of Parma,* ed. by E. A. Moody and M. Clagett (Madison: University of Wisconsin Press, 1952).

Pecham, John, *John Pecham and the Science of Optics: Perspectiva communis,* ed. and trans. by David C. Lindberg (Madison: University of Wisconsin Press, 1970).

[Saxony, Albert of], *Expositio et quaestiones in Aristotelis physicam ad Albertum de Saxonia attributae,* ed. by B. Patar (Louvain: Peeters, 1999).

Saxony, Albert of, *Quaestiones in libros de caelo et mundo* (Venice: Locatello, 1492).

Vitruvius, *On Architecture,* ed. by F. Granger (London: Heinemann, 1934).

Secondary Works

Buringh, Eltjo, *Medieval Manuscript Production in the Latin West* (Leiden: Brill, 2011).

Burnett, Charles, 'Translation Norms and Practice', in *Medieval Science, Technology, and Medicine,* ed. by T. Glick, S. J. Livesey and F. Wallis (New York: Routledge, 2005), pp. 486–488.

Busard, Hubert L. L., 'The Practica geometriae of Dominicus de Clavasio', *Archives for History of Exact Sciences,* 2 (1965), 520–75.

Clagett, Marshall, *Archimedes in the Middle Ages, 2. The Translations from the Greek by William of Moerbeke* (Philadelphia: The American Philosophical Society, 1976).

Clagett, Marshall, 'Leonardo Da Vinci's Mechanics', in *Dictionary of Scientific Biography,* ed. by C. C. Gillispie (New York: Charles Scribner's Sons, 1980).

Farago, Claire J., *Re-Reading Leonardo: The Treatise on Painting Across Europe, 1550–1900* (London: Ashgate, 2009).

Fidora, Alexander, 'Les différentes approches des traducteurs', in *Une conquête des savoirs. Les traductions dans l'Europe latine*, ed. by M. Lejbowicz (Turnhout: Brepols, 2009), pp. 46–65.

Fiorani, Francesca, 'The Theory of Shadow Projection and Aerial Perspective', in *Desargues en son temps*, ed. by J. Dhombres and J. Sakarovitch (Paris: Blanchard, 1994), pp. 267–82.

Frosini, Fabio, and Carlo Vecce, 'Leonardo da Vinci, Il contributo italiano alla storia del pensiero', in *Filosofia*, ed. by M. Ciliberto (Rome: Istituto della Enciclopedia Italiana, 2012), pp. 140–41.

Goyens, Michèle, and Pieter de Leemans, 'Traduire du grec au latin et du latin au français', in *Medieval Translation Practices*, ed. by P. Andersen (Copenhagen: Tusculanum, 2004), pp. 201–24.

Kemp, Martin, *Leonardo da Vinci* (Oxford: Oxford University Press, 2006).

Lindberg, David C., *A Catalogue of Medieval and Renaissance Optical Manuscripts* (Toronto: The Pontifical Institute of Mediaeval Studies, 1975).

Lindberg, David C., 'Optics in Sixteenth Century Italy', in *Novità celesti e crisi del sapere*, ed. by P. Galluzzi (Florence: Istituto e Museo di Storia della Scienza, 1984), pp. 131–48.

Lindberg, David C., 'Laying the Foundations of Geometrical Optics', in *The Discourse of Light from the Middle Ages to the Enlightenment*, ed. by David C. Lindberg and G. Cantor (Los Angeles: Clark Memorial Library, 1985), pp. 3–65.

Marinoni, Augusto, 'Leonardo as a Writer', in *Leonardo's Legacy*, ed. by C. O'Malley (Berkeley: University of California Press, 1969), pp. 62–63.

Narducci, Enrico, 'Nota intorno a una traduzione italiana fatta nel secolo decimoquarto del trattato d'ottica d'Alhazen', *Bollettino di bibliografia e di storia delle scienze matematiche e fisiche*, 4 (1871), 1–40.

Naẓīf, Muṣṭafā, *Al-Ḥasan Ibn al-Haytham wa-buḥūthuhu wa-kushūfuhu al-baṣariyya* (Cairo: Nūrī Press, 1943).

Raynaud, Dominique, 'Abū al-Wafā' Latinus? A Study of Method', *Historia Mathematica*, 39 (2012), 34–83.

Scarpati, Claudio, *Leonardo scrittore* (Milan: Vita e Pensiero, 2001).

Sephiha, Haïm Vidal, 'Langues juives, langues calques et langues vivantes', *Linguistique*, 8 (1972), 59–68.

Thomson, Ron B., 'Jordanus de Nemore: Opera', *Mediaeval Studies*, 38 (1976), 97–144.

Thro, E. Broydrick, 'Leonardo's Early Studies of Shadows', *Achademia Leonardi Vinci*, 9 (1996), 42–50.

Vallicrosa, José María Millás, 'Una obra astronómica desconocida de Johannes Avendaut Hispanus', *Osiris*, 1 (1936), 451–75.

Verde, Armando F., *Lo studio fiorentino, 1473–1503* (Florence: Olschki, 1973–94).

Vescovini, Graziella Federici, 'Contributo per la storia della fortuna di Alhazen in Italia', *Rinascimento*, 5 (1965), 17–49.

Weijers, Olga, *Queritur utrum. Recherches sur la 'disputatio' dans les universités médiévales* (Turnhout: Brepols, 2009).

Wiedemann, Eilhard, 'Über eine Schrift von Ibn al-Haytam "Über die Beschaffenheit der Schatten"', *Sitzungsberichte der Physikalisch-medizinischen Sozietät in Erlangen*, 39 (1907), 226–48.

SVEN DUPRÉ

How-To Optics*

Introduction

'We shall speak of a fully perspectival view of space', Erwin Panofsky famously wrote in his *Perspective as Symbolic Form*, 'not when mere isolated objects, such as houses or furniture, are represented in foreshortening, but rather only when the entire picture has been transformed into a window, and when we are meant to believe we are looking through this window into a space'.[1] Building on Alberti's conceptualization of painting as a window, for Panofsky, perspective entailed a new geometrical conception and depiction of space. However, several scholars have posited in more recent times that perspective as defined by Panofsky is a modern construct, and Panofsky's definition of perspective still haunts present-day scholarship on the history of perspective in the histories of science and art.[2] Most recently, Hans Belting even revived Panofsky's notion of perspective as 'symbolic form' implying that it was 'expressive' of Renaissance culture.[3] The deconstruction of Panofsky's definition of perspective first and foremost entails the recognition of the polysemy of optics: the plurality of cultures, practices and meanings of perspective.

The plurality and polysemy of Renaissance perspective plays out on three levels. First, contrary to Panofsky's elevation of Albertian perspective as the *costruzione legittima*, it has been shown, on the basis of the study of the material practices of painters in imitating and representing the effects of light and space, that Renaissance artists used several, sometimes incompatible techniques to create the illusion of three dimensions on a two-dimensional surface.[4] Not only were there a variety of constructions to create the illusion of space, other

* This project has received funding from the European Research Council (ERC) under the European Union's Horizon 2020 research and innovation programme (grant agreement No 648718).

1 Erwin Panofsky, *Perspective as Symbolic Form*, trans. by Christopher S. Wood (New York: Zone Books, 1997), p. 27.

2 The most eloquent criticism of Panofsky's *Perspective as Symbolic Form* is perhaps James Elkins, *The Poetics of Perspective* (Ithaca: Cornell University Press, 1994).

3 Hans Belting, *Florence and Baghdad: Renaissance Art and Arab Science*, trans. by Deborah Lucas Schneider (Cambridge, Mass.: Belknap Press of Harvard University Press, 2011).

4 Pietro Roccasecca, 'Gentile da Fabriano, A Miracle of Saint Nicholas: A Rigorous Nonperspective Spatial Representation', *Center: Record of Activities and Research Reports, National Gallery of Art, Washington*, 21 (2001), 126–30; Pietro Roccasecca, 'Not Albertian', *Center: Record of Activities and Research Reports, National Gallery of Art, Washington*, 22 (2002), 167–69.

Sven Dupré Utrecht University and University of Amsterdam, S.G.M.Dupre@uu.nl

Perspective as Practice. Renaissance Cultures of Optics, ed. by Sven Dupré, Turnhout, 2019 (*Techne. Knowledge, Technique, and Material Culture*, 1), p. 279-300

DOI 10.1484/M.*Techne*-EB.5.117730

types of optical knowledge and experience were as important to artists as the geometry of perspective.[5] This recognition of the polysemy of perspective necessitates important corrections to David Lindberg's classic study of the history of optics from Antiquity to Johannes Kepler.[6] Following Panofsky, the categories of *perspectiva naturalis* and *perspectiva artificialis* were projected back into the period of the Renaissance, driven by the desire to find a rupture between the Middle Ages and the Renaissance. While *perspectiva naturalis* or *communis* referred to the general category of the science of optics, including questions of psychology, physiology, anatomy, physics, and mathematics, *perspectiva artificialis* was the more limited domain of the geometrical *technique* (not the science) of drawing in perspective. This presentation of *perspectiva naturalis* and *perspectiva artificialis* as two different and largely independent enterprises had serious distorting consequences for Lindberg's presentation of the role of perspective in the history of optics. Lindberg discussed perspective as an impoverished 'application' of optical theory with no development of its own and very little consequence for the route taken by the discipline of optics. As we know, the distinction between optics and perspective fails to find support in the sources. All aspects of *perspectiva*, anatomical, physiological, physical, geometrical, psychological, were in fact inseparable, and authors, such as Lorenzo Ghiberti, Leonbattista Alberti and Piero della Francesca, considered themselves to be working on *perspectiva* just as much as their thirteenth-century predecessors.

The polysemy of perspective also plays out in a second way. Panofsky's definition ties perspective to the two-dimensional picture plane and as such attributes the development of perspective to one particular group of artisans, painters. However, several historians of architecture, urban planning and gardening (such as Marvin Trachtenberg and Georges Farhat) have argued that perspective was developed as much in real sites (such as the garden and the urban piazza) as on the two-dimensional picture plane.[7] More generally, to do justice to the plurality and polysemy of Renaissance perspective, we need to treat the disciplinary histories of optics and perspective in terms of practices, a conglomerate of material, social, literary and reproductive practices, through which knowledge claims in optics were produced, promoted, legitimated and circulated in and through a variety of sites and institutions. The ways optics were used by different groups in different places (such as the university classroom, the anatomist's dissection table, the goldsmith's workshop, and the astronomer's observatory) defined the meanings of Renaissance perspective. As this period was characterized by widespread 'optical literacy', perspective was defined in different ways in different places and sites by various groups of practitioners.[8]

5 Sven Dupré, 'The Historiography of Perspective and "Reflexy-Const" in Netherlandish Art', *Nederlands Kunsthistorisch Jaarboek*, 61 (2011), 35–60; Paul Hills, *Venetian Colour: Marble, Mosaic, Painting and Glass 1250–1550* (New Haven: Yale University Press, 1999); and the contributions by Marjolijn Bol and Paul Hills in this volume.

6 David C. Lindberg, *Theories of Vision. From al-Kindi to Kepler* (Chicago: University of Chicago Press, 1976). For a development of this argument, see Dupré, 'The Historiography of Perspective and "Reflexy-Const"', pp. 35–60; with reference to Dominique Raynaud, *L'hypothèse d'Oxford. Essai sur les origins de la perspective* (Paris: Presses universitaires de France, 1998).

7 Marvin Trachtenberg, *Dominion of the Eye: Urbanism, Art and Power in Early Modern Florence* (Cambridge: Cambridge University Press, 1997); and the contributions by Marvin Trachtenberg and Georges Farhat in this volume.

8 For the term 'optical literacy', see A. Mark Smith, *From Sight to Light: The Passage from Ancient to Modern Optics* (Chicago: University of Chicago Press, 2014).

The Materiality of Perspective

Finally, the polysemy of perspective plays out on a third level, one with which I am primarily concerned in this essay: the recognition of the materiality of perspective. There are two inter-connected aspects to the materiality of perspective. First, I show how writing and reading practices in the Renaissance, that is, the materiality of texts on optics and perspective, contributed to artists' establishment as experts based on their knowledge of the secrets of perspective. Against the background of recent discussions of the artist as reader, artisanal literacy, and the role of reading, drawing and writing in Renaissance workshops, I will explore artists' readings and writings of optics. Most artist-readers, like other artisanal or vernacular readers, engaged with texts in a piecemeal fashion. Also, they were more likely to encounter optics cut and pasted as parts of recipe collections or books of secrets, which flooded the print market in the sixteenth century. Recipes and secrets were transforming vehicles for the transmission of optics. In this process of transformation, optical knowledge was separated from the context of the source text and the received conceptual apparatus of optics was left open for creative (re-)interpretations. As a consequence, readers of books of secrets and recipe collections gained a different image of perspective than that of more exceptional and scholarly readers engaging with 'complete' texts on optics.

Repackaged as secrets, optical knowledge served to establish a community of experts.[9] This is also what Federico Cesi, founder of the Academy of the ('sharp-sighted') Lynceans, had in mind when he thanked Galileo for sending him a copy of Antonio Neri's treatise *L' arte vetraria*, the first published treatise on glassmaking. Cesi wrote that the book left him 'very rich in experiences and beautiful artifices'.[10] Neri himself claimed to have been the first to reveal the 'hidden things' of this art to benefit the 'experts of this profession'.[11] He emphasized that those who wish to try out his recipes need experience, practice, a good eye and judgment.[12] The book thus served to establish a community based on shared maker's knowledge, much as envisioned by the re-packaging of optical knowledge as secrets. These secrets typically concerned visual distortions and optical illusions. Just like the pre-occupation with anamorphic images, which Stuart Clark opposed to Ivins' characterization of perspective as the 'rationalization of sight', the period's fascination with these secrets speak of the deep epistemological anxiety, widely felt in the long sixteenth century, over whether human vision could give reliable access to the real world at all.[13]

Following the work of William Eamon, in recent years it has been convincingly argued that the popularization of the tradition of secrets, peaking in the sixteenth century, laid the groundwork for the empirical culture found in seventeenth and eighteenth century

---

9 Sven Dupré, 'Trading Luxury Glass, Picturing Collections and Consuming Objects of Knowledge in Early Seventeenth-Century Antwerp', *Intellectual History Review*, 20 (1) (2010), 53–78.

10 'molto ricco d'esperienze e belli artificii'. Detlef Heikamp, *Studien zur mediceischen Glaskunst: Archivalien, Entwurfszeichnungen, Gläser und Scherben* (Florence: Kunsthistorisches Institut, 1986), p. 356.

11 Antonio Neri, *L'Arte vetraria distinta in libri sette* (Florence: de' Gunti, 1612), Address to the 'curious reader'.

12 Sven Dupré and Christine Göttler, 'Hidden Artifices', in *Knowledge and Discernment in the Early Modern Arts*, ed. by Sven Dupré and Christine Göttler (New York: Routledge, 2017), pp. 1–16.

13 Stuart Clark, *Vanities of the Eye: Vision in Early Modern European Culture* (Oxford: Oxford University Press, 2007), pp. 1–4.

scientific practice.[14] Implicitly or explicitly, this work endorses the Kuhnian distinction between a mathematical and experimental tradition in early modern science, and associates sixteenth-century books of secrets primarily with the development of 'Baconian sciences' like chemistry, metallurgy, and magnetism.

This essay is a contribution to our understanding of the impact of books of secrets on the mathematical sciences. Scrutinizing optical secrets, it explores the still little understood shared 'experimentalism' in natural magic and mathematics. Dana Jalobeanu and Cesare Pastorino have recently argued that 'this new experimentalism permeated even traditionally bookish disciplines, such as natural history, which became, in the writings of Francis Bacon, experimental, collaborative and practically oriented'.[15] They point to Bacon's posthumously published *Sylva sylvarum* (1626), which like Giambattista della Porta's *Magia naturalis* was a vast collection of miscellaneous experiments ranging from distillations to cross-breeding, and from the making of gold to 'talking heads'. In fact, Dan Garber has shown that quite a few of Bacon's experiments were taken from Della Porta's *Magia naturalis*, and placed in a different methodological and theoretical context.[16] As I have argued elsewhere, in his ground-breaking work on optics, Johannes Kepler similarly used Della Porta's *Magia naturalis* as a source book of experiments.[17] The re-packaging of optical knowledge as secrets consequentially shaped experiential knowledge in optics.

The second aspect to the materiality of perspective is that of the instruments of perspective. Precisely, the re-packaging of optical knowledge in secrets, underscored a definition of perspective as depending upon the bodily engagement with material objects and the manipulation of instruments. It is the materiality of the instruments of perspective which tends to be overlooked in Jonathan Crary's seminal *Techniques of the Observer* (1990). Convinced that a history of vision or perception 'depends on far more than an account of shifts in representational practices', Crary took as his problem, the observer. 'Vision and its effects are always inseparable from the possibilities of an observing subject who is both the historical product *and* the site of certain practices, techniques, institutions, and procedures of subjectification'.[18]

Crary's basic argument was that the early nineteenth century saw the creation of a new kind of observer. In the 1810s and 1820s he located a rupture in the *scopic regime* between a geometric model of vision (in which vision was conceived as essentially passive and independent of the subject and based on a radical distinction between interior and exterior) and a physiological model of vision (in which vision became subjective, and the product of visual experience became located in the body of the observer). Crary developed his

---

14 William Eamon, *Science and the Secrets of Nature: Books of Secrets in Medieval and Early Modern Culture* (Princeton: Princeton University Press, 1994); *Secrets and Knowledge in Medicine and Science, 1500–1800* ed. by Elaine Leong and Alisha Rankin (Aldershot: Ashgate, 2011).

15 Dana Jalobeanu and Cesare Pastorino, 'Introduction', in 'Instruments & Arts of Inquiry: Natural History, Natural Magic and the Production of Knowledge in Early Modern Europe', Special Issue of *Journal of Early Modern Studies*, 3 (2014), 9–13 (p. 10).

16 Daniel Garber, 'Merchants of Light and Mystery Men: Bacon's Last Projects in Natural History', *Journal of Early Modern Studies*, 3 (1) (2014), 91–106.

17 Sven Dupré, 'Kepler's Optics without Hypotheses', *Synthese*, 185 (2012), 501–25.

18 Jonathan Crary, *Techniques of the Observer: On Vision and Modernity in the Nineteenth Century* (Cambridge, Mass.: The MIT Press, 1990), p. 5.

argument by contrasting two instruments, the camera obscura and the stereoscope, which he considered paradigmatic for his two models of vision respectively:

> The optical devices in question, most significantly, are points of intersection where philosophical, scientific, and aesthetic discourses overlap with mechanical techniques, institutional requirements and socioeconomic forces. Each of them is understandable not simply as the material object in question, or as part of a history of technology, but for the way in which it is embedded in a much larger assemblage of events and powers.[19]

In this way, Crary kept far from any underlying assumption that artists used optical instruments to arrive at photographically realistic images, an underlying assumption developed by David Hockney's *Secret Knowledge*.[20] Crary argues that

> many accounts of the camera obscura, particularly those dealing with the eighteenth century, tend to consider it exclusively in terms of its use by artists for copying, and as an aid in the making of paintings. There is often a presumption that artists were making do with an inadequate substitute for what they really wanted, and which would soon appear, that is, a photographic camera.[21]

While this deconstruction of the history of the camera obscura as a prelude to the photographic camera is what we gain from Crary's re-focusing on the history of the observer and techniques of observation, his account also falls into the trap of declaring the camera obscura paradigmatic, with all the consequent problems of such an approach already diagnosed in Bernhard Siegert's media history.[22] Crary's sudden transposition of vision inside the body in the early nineteenth century and his very clear-cut division of the history of observation in a geometric and physiological scopic regime is one of the aspects that has repeatedly and justifiably come under attack by, among others, Erna Fiorentini in her work on the camera lucida.[23] I will argue that the observer of the camera obscura is not passive, but instead constantly tinkering with the design of the camera obscura, and that observation with this instrument depends upon a bodily engagement rather than a strict division between the subject and the object, two aspects which Crary underestimated by de-materializing the camera obscura.[24] In order to fully engage with this historiographical debate I will focus on instruments of perspective, including the camera obscura, mirrors and lenses. The tools of the draughtsman as well as the drawing techniques, occasionally codified as secrets or instructions, for example by Willem Goeree, deserve to be discussed if one adopts a focus on the materiality of perspective. However, they lie outside the scope of this chapter.

---

19 Crary, *Techniques of the Observer*, 8.

20 David Hockney, *Secret Knowledge: Rediscovering the Lost Techniques of the Old Masters* (London: Thames & Hudson, 2001); Sven Dupré, 'Optics, Instruments and Painting, 1420–1720: Reflections on the Hockney-Falco Thesis', Special Issue of *Early Science and Medicine*, 10 (2) (2005), 125–339.

21 Crary, *Techniques of the Observer*, 32.

22 Bernhard Siegert, 'Kulturtechnik', in *Einführung in die Kulturwissenschaft*, ed. by Harun Maye and Leander Scholz (Munich: Wilhelm Fink Verlag, 2011), pp. 95–118.

23 Erna Fiorentini, 'Subjective Objective. The Camera Lucida and Protomodern Observers', *Bildwelten des Wissens: Kunsthistorisches Jahrbuch für Bildkritik*, 2 (2004), 58–66.

24 For perspective and the body, specifically in relation to anamorphic images, see also Lyle Massey, *Picturing Space, Displacing Bodies: Anamorphosis in Early Modern Theories of Perspective* (University Park, Pennsylvania: The Pennsylvania State University Press, 2007).

Lorenzo Ghiberti's Note-Taking in the Third Commentary

Before I turn to the instruments of perspective, I focus on the materiality of texts on optics and perspective by scrutinizing the third commentary of Lorenzo Ghiberti. The writings of Ghiberti that have come down to us are given the title of *I commentarii*. They consist of three parts: one on ancient art, a second on modern art, and the third on vision and optics, our focus here. The nature of these writings is completely different from a treatise like, say, Alberti's *De pictura*. The third commentary is a compilation of excerpts from different authorities. Klaus Bergdolt has identified the source of most of these so that we now know that the most frequently excerpted sources are Pliny, Vitruvius' *De architectura*, and the works on *perspectiva*: the thirteenth-century optical works of Roger Bacon and John Pecham, and the work of the eleventh-century Ibn-Al Haytham, known in Latin as Alhacen, which Ghiberti studied in Italian translation.[25] Although often dismissed in the literature as a thoroughly unoriginal compilation, the third commentary becomes of interest if considered as a product of Ghiberti's practice of reading and notetaking.

Recent studies have shown that there is a much stronger continuity between medieval and early modern writing and reading practices and between manuscript and print culture than previously thought. These studies lay to rest certain misconceptions perpetrated by earlier works on print culture (as found in Elizabeth Eisenstein's seminal work) such as, for example, that books produced prior to printing were inevitably riddled with errors. A radical transformation of written culture took place connected to the establishment of universities and new religious orders. New techniques and tools for rapid consultation, and use of an ever-growing body of texts associated with the proliferating curriculum of the universities emerged, and a new type of book was invented, containing such devices as running-titles, chapter headings, tables of contents and alphabetical indexes to cater for the needs of teachers and students. Even the long medieval tradition of observing the rational order of texts came under pressure as preachers, teachers and students broke up texts, copying bits and pieces to insert them into alphabetized compendia. These techniques of reading and writing foreshadow early modern methods of commonplacing which humanists and philosophers used in response to information overload.[26] In the context of studies of artisanal literacy and the artist as reader, Heiko Damm, Michael Thimann and Claus Zittel have shown how artisans and architect-engineers adapted the humanist method of commonplacing to compile manuscripts of texts and drawings of a technical nature, carefully selected and copied.[27]

Ghiberti's third commentary is to be considered within this context of artists' adoption of humanist techniques of reading and notetaking. The study of Ghiberti's third commentary is then of interest for what it tells us about the artist's appropriation of optical knowledge.

25 Klaus Bergdolt, *Der dritte Kommentar Lorenzo Ghibertis: Naturwissenschaften und Medizin in der Kunsttheorie der Frührenaissance* (Weinheim: VCH, Acta Humaniora, 1988). Also available on [Perspectiva+]: http://perspectiva.biblhertz.it/index.html [accessed 10-08-2017], including an Italian translation of Alhacen, Vat. Lat, 4595, *Libro de li aspecti*, transcribed by Pietro Roccasecca.

26 Ann Blair, *Too Much to Know: Managing Scholarly Information before the Modern Age* (New Haven: Yale University Press, 2010).

27 Heiko Damm, Michael Thimann and Claus Zittel, *The Artist as Reader: On Education and Non-Education of Early Modern Artists* (Leiden: Brill, 2012).

Ghiberti transformed the material from his sources to different degrees. Towards the beginning of the third commentary, in particular, Ghiberti's re-writing is particularly daring and he considerably expands upon his sources.

I discuss here only one telling example of such re-writing. In the early part of the third commentary Ghiberti discusses the nature of light. He selected passages from his sources, and then expanded on his reading by linking the notes to his workshop experiences as a goldsmith, jeweller and designer of stained glass windows.[28] Immediately after distinguishing between three types of light or light-giving bodies (that is, light-giving, opaque and translucent or diaphanous bodies), a distinction, which Ghiberti took from the second book of Witelo's *Perspectiva* (as he indicated himself), he connected to examples he knew from his workshop practice:

> The first is the sun and fire and some precious stones; the second [...] is that which is the earth or other hard or dark [tenebrosa] material. The third is the translucent [diafano] body: air, water, glass, crystal, chalcedony, beryl.[29]

The connection between the nature of light and precious stones was not new. The sources Ghiberti read had already established this connection. As Alhacen wrote:

> Likewise, when transparent colored stones are in dark locations, their colors will appear dull and dark; but when intense light shines upon them, or when they are placed against a light-source so that its light shines through them, their colors will appear bright, and their transparency will be revealed by the passage of light [through them].[30]

Nevertheless, Ghiberti further expands on these notions and connections by referring to his own observations of the effects of the intensity of light on a gemstone he, as an expert, was allowed to study in a Florentine collection:

> Among the [most] remarkable things I ever saw is a wonderful engraved chalcedony which was in the collection of one of our citizens, by the name of Niccolo Landi, a very energetic researcher and investigator of many and excellent antiquities in our time, and into books of Greek and Latin writings. And among his other antiques he had this chalcedony [...] It was oval in shape, and on it was the figure of a youth holding a knife [The Rape of the Palladium]. [...] This carving was said by every expert in sculpture and painting, without exception, to be a marvellous thing [...] You could not see it well in a strong light, because when fine and polished stones are deeply cut, the strong light reflections obscure the understanding of the form. This carving could be seen best when the deeply-cut part was held against the strong light, when it could be seen perfectly.[31]

28 For Ghiberti's interest in glass and jewels, see Richard Krautheimer and Trude Krautheimer-Hess, *Lorenzo Ghiberti* (Princeton: Princeton University Press, 1982).

29 Cited from John Gage, *Colour and Meaning: Art, Science and Symbolism* (London: Thames & Hudson, 1999), p. 99.

30 Alhacen, *Alhacen's Theory of Visual Perception: A Critical Edition, with English Translation and Commentary, of the First Three Books of Alhacen's 'De aspectibus', the Medieval Latin Version of Ibn al-Haytham's 'Kitāb al-Manāẓir'*, ed. and trans. by A. Mark Smith, 2 vols (Philadelphia: Transactions of the American Philosophical Society, 2001), II, p. 347.

31 Cited from Gage, *Colour and Meaning*, 100.

Formally, Ghiberti's excerpts differ from other writings connected to the art of jewellery and glassmaking, such as a manuscript on the making of stained glass windows, compiled in the late fourteenth century, by Antonio da Pisa.[32] Antonio da Pisa's workshop was one of the most active in Florence and was involved in the making of stained glass windows for the cathedral. The book of Antonio da Pisa consists of a collection of recipes related to his art. The recipes contain knowledge of a different type than that excerpted by Ghiberti. I take one example from Antonio da Pisa. 'To make the yellow colour more intense', he advised the reader, 'to add a bit of ochre such as the painters use. But not too much because then the glass will look red'.[33] Compare this to the passage in which Ghiberti refers to stained glass windows:

> [...] when the sun's ray passes through a glass [window] or through a strongly coloured [oiled] cloth, the image of the colour appears upon the dark body [opposite].[34]

In the third commentary, this passage follows a discussion of the nature of light and species or the forms or images, which each object sends through the medium. Ghiberti excerpted this note from Roger Bacon's discussion of the multiplication of species in his *Perspectiva*, one of the key concepts of optics.[35] Compared to Antonio da Pisa's recipes, Ghiberti's commentary shows that he read optical source texts and that he appropriated optical knowledge, also of a type not found in Antonio da Pisa, for example, when Ghiberti speaks of the nature of light. Nevertheless, Ghiberti's reading is selective when excerpting and re-ordering materials from his sources. While his reading established his expertise, the theoretical framework of vision is partly lost.

Writing Optics: Recipes and Secrets

Artist-readers such as Ghiberti who engaged with 'complete' texts were exceptional. Most artist-readers, like other artisanal or vernacular readers, engaged with texts in a piece-meal fashion. They were more likely to encounter optics cut and pasted as part of collections of recipes and books of secrets. In this second part of my essay I turn to recipe collections, such as Antonio da Pisa's, with which I have compared Ghiberti's third commentary. Although there are few traces of artists' marginal annotations to collections of recipes and books of secrets to give us direct insight into their reading practices, we do have access to the collections of recipes and books of secrets themselves which artists were primarily confronted with and thus to the transformations of optical knowledge in the process of re-writing optical texts as recipes and secrets.

32 Claudine Lautier and Dany Sandron, *Antoine de Pise: L'art du vitrail vers 1400* (Paris, Comité des Travaux Historiques et Scientifiques, 2008).

33 'Si più pieno de colore volessi fare quello callo, mictine dentro um pocho de ocrea, la quale adoperano I depentori e sit u glini mittissi troppo, ritornaria el vetro rosso, ma non seria bello colore che parria uno imbratto'. Lautier and Sandron, *Antoine de Pise*, 56.

34 Cited from Gage, *Colour and Meaning*, 102.

35 Roger Bacon, *Roger Bacon's Philosophy of Nature: A Critical Edition, with English Translation, Introduction, and Notes, of 'De multiplicatione specierum' and 'De speculis comburentibus'*, ed. and trans. by David C. Lindberg (Oxford: Clarendon Press, 1983).

Recipes are probably as old as mankind's writing ability. We have clay tablets inscribed with Babylonian glass recipes, and the so-called Leiden and Stockholm Papyri contain recipes for several arts and crafts and have a long after-history reaching the early modern period.[36] By this time recipes were ubiquitous. Recipes that appeared in print in sixteenth-century books of secrets often had a pre-history in manuscript collections of recipes. Manuscript and print were and remained co-existent traditions. One point to make about these recipes and secrets is their longevity.

Avidly collected in manuscript notebooks and publicised in books of secrets flooding the print market, these recipes instructed readers how to colour glass, make gold, and brew medicine. A good part of the late medieval and early modern recipes that have come down to us concern medicine and the visual and decorative arts. In the latter category, Cennino Cennini's *Libro del' arte* is one of the more well-known collections, the fame of which is probably matched only by the recipes and workshop secrets which the physician Theodore de Mayerne compiled on the basis of conversations with Rubens, Van Dyck and their like.[37]

However, these collections are exceptions. These famous examples, connected with the name of their author or compiler, are only the tip of the iceberg. Hundreds of mostly anonymous collections of, taken all together, thousands of recipes hide beneath the water's surface. The more famous examples, like Cennini's, are also exceptional in the sense that they focus on one topic. A second point to make is that the most typical collections of recipes were miscellaneous in nature. Miscellaneous recipes also ended up in the libraries of artists. For example, Christine Sauer has identified such a manuscript, at one point, as being in the hands of Albrecht Dürer.[38] This fits the more general picture of the evolution of artisanal literacy in the early modern period; more and more artisans were able to read and write, and they increasingly possessed books and home libraries, and the number of books they possessed also increased.[39] The early modern period saw a transition in which artisans claimed authorship of knowledge specific to their craft and in which artisans also began to partly learn their trade through writings.

One of the best-known examples of a book of secrets is Giovanni Battista della Porta's *Magia naturalis*, published in 1558, and in an expanded version in 1589. Here, amidst secrets of all kinds of other productive knowledge, we find optical secrets revealing, for example, how to make a burning mirror or how to project an image in the air. This secret

36 For the Leiden and Stockholm Papyri, see Lawrence Principe, *The Secrets of Alchemy* (Chicago: University of Chicago Press, 2013), pp. 10–13; Sven Dupré, Bert de Munck and Mark Clarke, *Transmission of Artists' Knowledge* (Brussels: Royal Flemish Academy of Arts and Sciences, 2012).

37 Lara Broecke, *Cennino Cennini's Il libro dell'arte: A New English Translation and Commentary with Italian Transcription* (London: Archetype Publications, 2015). For Mayerne and his manuscript, see H. Trevor-Roper, *Europe's Physician. The Various Life of Sir Theodore de Mayerne* (New Haven: Yale University Press, 2006); Ulrike Kern, 'The Art of Conservation I: Theodore de Mayerne, the King's Black Paintings and Seventeenth-Century Methods of Restoring and Conserving Paintings', *The Burlington Magazine*, 157 (2015), 700–08.

38 Christine Sauer, 'Eine kunsttechnologische Handschrift aus dem Besitz Albrecht Dürers', in *Dürer-Forschungen* (Nürnberg: Verlag des Germanischen Nationalmuseums, 2009), II, pp. 275–96.

39 Michael Hackenberg, 'Books in Artisan Homes of Sixteenth-Century Germany', *Journal of Library History*, 21 (1986), 72–91; Bert S. Hall, 'Der Meister sol auch Kennen Schreiben und Lesen: Writings about Technology ca. 1400-ca. 1600 A.D. and their Cultural Implications', in *Early Technologies*, ed. by Denise Schmandt-Besserat (Malibu: Undena Publications, 1979), pp. 47–58; Pamela O. Long, *Openness, Secrecy, Authorship: Technical Arts and the Culture of Knowledge from Antiquity to the Renaissance* (Baltimore: Johns Hopkins University Press, 2001).

is also found in collections of recipes prior and contemporaneous to Della Porta's book of secrets. For example, the mid-sixteenth-century manuscript collection of recipes, brought together by the Antwerp apothecary Peter Van Coudenberghe, is very similar in content and organization to books of secrets published in the same period.[40] Partly written in Latin, and partly in Dutch, the recipes in Latin are diverse in nature (dealing with cooking, medicine, alchemy, and the making of glass and colours), while the longer Dutch part contains recipes which are mostly of art technological origin: the making of colours in different media, inks, glassmaking, and in this section also, optical secrets, that is, recipes for the construction of mirrors.[41] Recipe collections were vehicles for optical knowledge.

This is most tellingly the case for the circulating copies of the *Secretum philosophorum*.[42] It was originally composed in England in the late thirteenth or early fourteenth century. Devoted to the seven liberal arts, it was nevertheless more than just another university textbook. The first section on grammar consisted of recipes explaining how to construct a pen and how to make inks, and the third section, on dialectic, listed secrets on how to deceive the senses, including some on how to deceive the sense of sight. More strongly organized (according to the scheme of the liberal arts) than most recipe collections, it is nevertheless typically miscellaneous.

I want to make four points about books of secrets on the basis of the example of the *Secretum philosophorum*. The first is about ambiguity of meaning. This is one secret 'to make a mirror in which many moving images will appear in a single place':

> You can also make a mirror in which, in one glance, many moving images will appear, and this is how it is done. Take a very deep box and place in the bottom of it an ordinary mirror – that is, a convex one. Next, take six or seven other convex mirrors of the same size, and scrape off with a knife their lead which is on the concave side. But you should know that it is very difficult to scrape off all the lead cleanly, without breaking the glass. So if you want to clean the mirrors well and remove the lead, take some quicksilver and rub the lead with it, and straight away it will adhere to the lead and penetrate it, so that after a little time you can easily remove the lead completely from the mirror. Now, when they are very clean, put them in the box, but in such a way that they stand aslant on the mirror and, moreover, in different positions, which you will do thus. When the first mirror has been placed in the bottom, you will place the second mirror so that one side is attached to the first mirror and the opposite side is distant from it by one finger; and, in this way, you will put the other mirrors in the box, but in different positions. But on the top surface of the box you will put a mirror

40 Sven Dupré, 'The Value of Glass and the Translation of Artisanal Knowledge in Early Modern Antwerp', in *Trading Values in Early Modern Antwerp*, ed. by Bart Ramakers, Christine Göttler, and Joanna Woodall, Netherlands Yearbook for Art History, 64 (Leiden: Brill, 2014), pp. 138–61; E. Vandamme, 'Een 16e-eeuws Zuidnederlands receptenboek', *Jaarboek van het Koninklijk Museum voor Schone Kunsten* (1974), 101–37.

41 For example, *Om brant spiegels te ghieten* ('To cast burning mirrors'), in: Vandamme, 'Een 16e-eeuws Zuidnederlands receptenboek', 101–37 (p. 121).

42 Robert Goulding, 'Deceiving the Senses in the Thirteenth Century: Trickery and Illusion in the Secretum philosophorum', in *Magic and the Classical Tradition*, ed. by C. Burnett and W. F. Ryan (London: The Warburg Institute, 2006), pp. 135–62. See also Mark Clarke, 'Writing Recipes for Non-Specialists, c. 1300: The Anglo-Latin "Secretum philosophorum", Glasgow MS Hunterian 110', in *Sources and Serendipity: Testimonies of Artists' Practice*, ed. by Erma Hermens and Joyce H. Townsend (London: Archetype Publications, 2009), pp. 50–64.

(which has been cleaned as above) straight and not aslant and then adjust them well so that only the topmost mirror is seen. Then if you look in the mirror, you will see as many images as there are mirrors. But if you turn the mirror, you will see how one image always stays in the middle and in one position and the other images come to meet it as if they were doing a dance.[43]

The secret describes two different processes: how to make glass mirrors (starting from the then more common convex mirrors), and how to assemble the mirrors in such a way, specifying distances and positions of the respective mirrors, as to create a particular optical effect (dancing images). Depending upon our interpretation of positions and distances, there seems to be more than one way to put together this optical instrument. While it is possible to reconstruct an optical object by following this text, as I have attempted in collaboration with Carsten Wirth (Figure 1), ambiguity of meaning is nevertheless the rule in books of secrets. This ambiguity is enhanced by the typical absence of diagrams or drawings of instrument designs.[44]

A second point I want to make is about transmission. Not only did secrets travel as part of the *Secretum philosophorum*, they also travelled independently. In this process of transmission, the secrets were re-organized and appeared in different contexts. For example, Jean Fusoris was an early fifteenth-century mathematician and instrument maker, based in Paris.[45] He also showed interest in optics, in particular, in burning mirrors, and more generally, as his annotations to Witelo's *Perspectiva* and other notes show, catoptrics and image formation. His manuscript on burning mirrors (a variation on the *Libellus almukesi compositio*), now in the Bibliotheque municipale in Dijon, also contains a note on 'how to make a mirror in which many moving images of one object appear', which is taken from the *Secretum philosophorum*.[46] In short, optical secrets from the *Secretum philosophorum* travelled independently and were combined with, for example, a treatise on burning mirrors.

The Manipulation of Optical Objects

A third point: the secrets re-packaged optical knowledge aiming to create visual effects through the manipulation of objects or instruments. In the *Secretum philosophorum* these are mirrors (turned into glass spheres) and the visual effect depends upon the bodily engagement with the material objects. Only when the eye is positioned at a particular point of view with respect to the glass spheres, are the visual effects described in the secrets created. This is

43 Goulding, 'Deceiving the Senses in the Thirteenth Century', pp. 135–62 (esp. pp. 155–56).

44 On the (absence of) imagery in how-to texts, see also Sven Dupré, 'Die Sichtbarkeit und Unsichtbarkeit von Körperwissen in der Kodifikation der Künste in der frühen Neuzeit', *Paragrana: Internationale Zeitschrift für Historische Anthropologie*, 25 (1) (2016), 110–29.

45 Emmanuel Poulle, *Un constructeur d' instruments astronomiques au XVe siècle: Jean Fusoris* (Paris: Librairie Honoré Champion, 1963).

46 'speculum in quo visu uno multe apparebunt ymagines se moventes constituere'. Bibliothèque municipale, Dijon, 441 (226), fol. 206^r^. The complete passage is transcribed in Grażyna Rosińska, 'Optyka W XV wieku miedzy nauka sredniowieczna a nowozytna' ('Fifteenth Century Optics between Medieval and Modern Science'), *Studia Copernicana*, 24 (1986), 151. For the corresponding, almost identical, passage in an English manuscript, see British Library, London, Egerton 2852, fol. 19^r^ (transcribed in Rosińska, 'Optyka W XV wieku', p. 168).

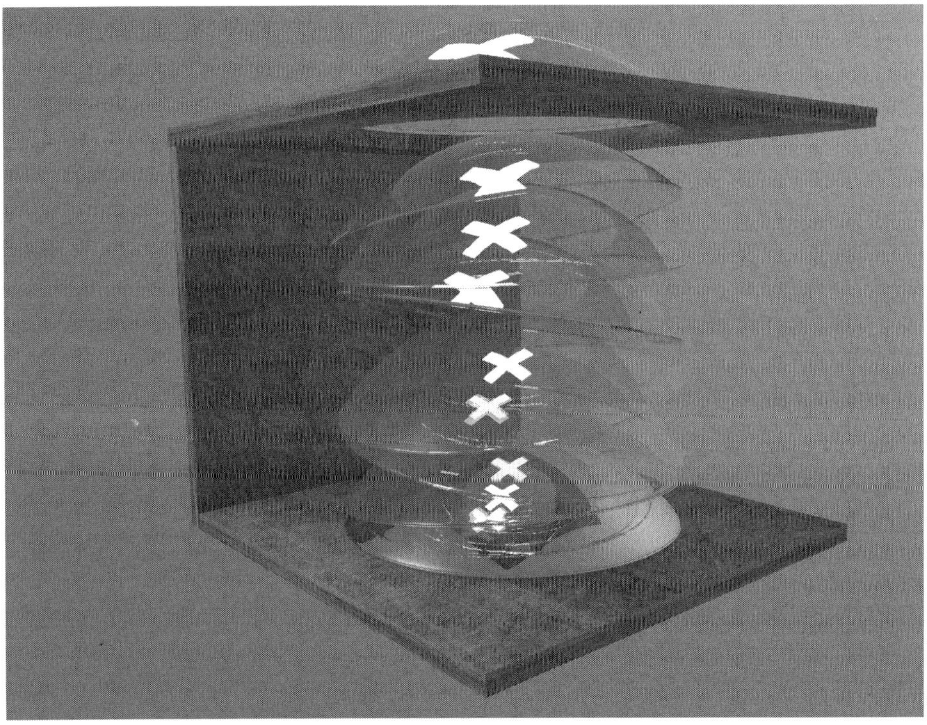

Fig. 1 Reconstruction of the secret 'To make a mirror in which many moving images will appear in a single place' in *Secretum philosophorum*, in collaboration with Carsten Wirth.

different from Crary's imagination of the camera obscura experience to which this secret is related. The emphasis on bodily engagement becomes clear in early modern descriptions of the camera obscura, such as the one by Johannes Kepler, one of the most important mathematicians and astronomers of the seventeenth century, whose book *Paralipomena*, published in 1604, laid the foundations of our modern theory of vision based on the analogy between the eye and the camera obscura. In this book, Johannes Kepler mentioned

> [...] an experimentum [...] which I saw at Dresden in the elector's theater of artifices [...] A disk thicker in the middle, or a crystalline lens, a foot in diameter, was standing at the entrance of a closed chamber against a little window, which was the only thing that was open, slanted a little to the right. Thus when the eyesight travelled through the dark emptiness, it also, fortuitously, hit upon the place of the image, nearer, in fact, than the lens. And so since the lens was weakly illuminated, it did not particularly attract the eyes. But the walls were also not particularly conspicuous through the lens, because they were in deep darkness.[47]

47 '[...] cuius experimentum vidi Dresdae in Theatro artificiali Electoris. [...] Discus in medio crassior, seu lens crystallina, pedis diametro, stabat in ingressu camerae clausae contra fenestellam, quae unica patebat, declinantem parùm ad dextram. Dum igitur oculorum acies tenebrosam capacitatem pererrant, fortuitò, et in locum imaginis incidunt, propiorem quidem quàm erat lens. Cum itaque lens malignè illustraretur, oculos non admodum erant conspicui; quia in multis tenebris'. Johannes Kepler, *Gesammelte Werke*, ed. by Max Caspar

The setting was the Dresden Kunstkammer, a place of display in which collected natural objects were juxtaposed with artificial objects, typically found at Renaissance courts.[48] In one of the rooms of the Dresden Kunstkammer, which had been turned into a room-size camera obscura, Kepler witnessed the images formed by a lens placed in the aperture of this camera obscura, which, in fact, was one of the little windows of the Kunstkammer room through which light from outside was able to enter. In this darkened room Kepler saw that 'the little window and the objects standing about it, which had the benefit of much light, lying hidden beyond the lens, set up a bright image of themselves in the air (between me and the lens)'.[49]

If we think of the camera obscura today, the object that comes readily to mind is a sort of box-type camera obscura, at least something that is fixed in terms of object and optical design, perhaps even portable. As one can clearly see from the example of Kepler's description of his camera obscura experience in Dresden, this was not the case in the early seventeenth century. The camera obscura was a darkened room, and the optics were brought and installed for the specific purpose. In Dresden a crystal ball, a gift presented to the Elector of Saxony, August I, by the Duke of Savoy in 1580 and prominently displayed in the most important room of the Dresden Kunstkammer, was occasionally moved to a dark room in the Kunstkammer to project images.[50]

This underscores the process and event character of the camera obscura experience. It also means that there was no standardized and stabilized optical design of dark room experiences. Instead, various and different optical design elements (lenses, mirrors, apertures) were brought together and assembled in diverse ways resulting in various image appearances. Nevertheless, by the seventeenth century, attempts were being made to make the camera obscura portable so that it could be used out in the field to draw landscapes. However, the 'picture box' developed by Robert Hooke, the curator of experiments at London's Royal Society, shows how difficult it must have been to create a darkened room on a smaller scale.[51] (Figure 2) How would Hooke's draughtsman have kept his balance with this picture box on his head? More than just a model of vision, the failures and successes of making the camera obscura portable highlight the practical difficulties of using the camera obscura as a drawing instrument. Rather than a model of passive, objective vision, as Crary would have wanted it, Hooke's 'picture box' was an extension of the draughtsman's body.

The ways in which optical knowledge is packaged in *Secretum philosophorum* is similar to that in the Kunstbüchlein, following in the footsteps of Dürer's *Underweysung der Messung*.[52] *Messung* was used interchangeably, as a translation of perspectiva, and it entailed

---

and Walter van Dyck, 23 vols (Munich: C. H. Beck, 1938), II, pp. 164–65. Translation in Johannes Kepler, *Optics. Paralipomena to Witelo and & Optical Part of Astronomy*, trans. by William H. Donahue (Santa Fe: Green Lion Press, 2000), p. 194.

48 Sven Dupré and Michael Korey, 'Inside the Kunstkammer: The Circulation of Optical Knowledge and Instruments at the Dresden Court', *Studies in History and Philosophy of Science*, 40 (4) (2009), 405–20.

49 'At fenestella et circumstantes res, quae multa luce fruebantur, post lentem latentes, claram sistebant in aëre (me inter et lentem) sui imaginem'. Kepler, *Gesammelte Werke*, ed. by Caspar and Van Dyck, II, p. 165. Translation in Johannes Kepler, *Optics*, trans. by Donahue, p. 194.

50 Dupré and Korey, 'Inside the Kunstkammer', pp. 405–20.

51 For the portable camera obscura, see Joachim Rees, *Die verzeichnete Fremde: Formen und Funktionen des Zeichnens im Kontext europäischer Forschungsreisen 1770–1830* (Paderborn: Wilhelm Fink, 2015), pp. 275–326.

52 Jeanne Peiffer, 'Projections Embodied in Technical Drawings: Dürer and his Followers', in *Picturing Machines 1400–1700*, ed. by Wolfgang Lefèvre, (Cambridge, Mass.: The MIT Press, 2004), pp. 245–75.

Fig. 2 Drawing instrument of Robert Hooke, *An Instrument of Use to take the Draught, or Picture of any Thing*, Communicated by Dr. Hook (sic) to the Royal Society Dec. 19, 1694, 1726, p. 295. © Max Planck Institute for the History of Science (CC-BY-SA).

a particular definition of perspective. Dürer used *Messung* to refer to constructive geometry (by ruler and compass) with an emphasis on three-dimensionality and materiality, that is on drawing artefacts placed before artists, in contrast to demonstrative geometry. Dürer's *Underweysung der Messung* is not to supply his readers with a geometrical explanation of why a construction works, but to teach them how to perform the construction. In Dürer's *Underweysung der Messung* there is an emphasis on instruments for drawing in perspective. However, while Dürer's instruments are more pedagogical embodiments of

Alberti's perspective, later German sixteenth-century writers on *Messung* emphasized the constructive use of instruments to create non-Albertian perspective. *Messung* was thus a particular embodiment of perspective, explicitly defined as based on the manipulation of instruments. It was a re-definition of perspective also pervading books of secrets.

How-To Optics

A fourth and final point: secrets packaged optical knowledge as 'how-to' guides. This engaged the reader in trying and testing the experiments. While inviting testing, recipes and secrets should not be considered experiments. More often they are the products of reading experiences, copied and pasted from other sources. For example, Della Porta's secret of how to draw a parabolic section (given the focal distance), and how to make a parabolic burning mirror, is taken from Oronce Finé's *De speculo ustorio*, a book bringing together technical and mathematical knowledge of mirrors.[53] Della Porta probably knew this book through its publication in 1587 as an appendix to Cosimo Bartoli's translation of Fine's *Protomathesis*, in an Italian translation by Ercole Bottrigaro, a Bolognese humanist who had also edited Ptolemy's *Geographia*. Della Porta's secret was taken, almost verbatim, from Finé's Propositions 8 and 9. In fact, it has been shown that Della Porta borrowed the diagrams of *De speculo ustorio* as well as Finé's mistakes.[54]

However, although Della Porta's secret is taken from a textual source, this appropriation took place in a context of making. Della Porta collaborated on the construction of a parabolic burning mirror with Jacomo Contarini, the Provveditore of the Arsenal in Venice and a collector of books, manuscripts and instruments.[55] In 1580, Della Porta's patron, the Cardinal d' Este, sent him to Venice to make or obtain a parabolic burning mirror. Looking for guidance to construct a parabolic burning mirror, he turned to Jacomo Contarini, presumably not only to provide the means, but also the skills. On 29 November 1580, Della Porta wrote to his patron that Contarini had spent a day and most of a night at the Arsenal with him supervising an attempt by one of the Arsenal craftsmen to cast a parabolic mirror. It was through Contarini that Della Porta encountered Oronce Finé's work on burning mirrors.

It is through this same network that Della Porta also came across the work of the Venetian mathematician Ettore Ausonio. When Della Porta visited Venice in 1580 to attend the construction of a parabolic mirror, he also met Paolo Sarpi, who may already have been aware of Ausonio's *Theorica speculi concavi sphaerici*, which he later copied himself.[56] What follows are a few examples of Della Porta's optical secrets which likely had their source

53 Sven Dupré, 'Printing Practical Mathematics: Oronce Fine's "De speculo ustorio" between Paper and Craft', in *The Worlds of Oronce Finé: Mathematics, Instruments and Print in Renaissance France*, ed. by Alexander Marr (Donington: Shaun Tyas, 2009), pp. 64–82.

54 Marshall Clagett, *Archimedes in the Middle Ages. Vol. 4: A Supplement on the Medieval Latin Traditions of Conic Sections* (Philadelphia: The American Philosophical Society, 1980), p. 331.

55 Paul Lawrence Rose, 'Jacomo Contarini (1536–1595), a Venetian Patron and Collector of Mathematical Instruments and Books', *Physis*, 18 (2) (1976), 117–30.

56 Sven Dupré, 'Mathematical Instruments and the "Theory of the Concave Spherical Mirror": Galileo's Optics beyond Art and Science', *Nuncius*, 15 (2) (2000), 551–88.

in his reading of Ausonio. First, Della Porta's secret of how, using a plane mirror, 'letters may be cast out and read, on a wall that is far distant':

> On the superficies of a plain Glass, make Letters with black ink, or with wax, that they may be solid to hinder the light of the Glass, and shadow it; then hold the Glass against the Sun-beams, so that the beams reflecting on the Glass, may be cast upon the opposite wall of a Chamber, it is no doubt but the light and letters will be seen in the Chamber.[57]

Ausonio's *Theorica* already proposed to project letters on a distant wall by means of a mirror and solar light. Second, Ausonio's secret of how to use candle light to read letters in an otherwise dark room was taken up by Della Porta who specified that it should be done by placing a candle in the focal point of the mirror:

> Take the Glass in your hand, and set a candle to the point of Inversion, for the parallel beams will be reflected to the place desired, and the place will be enlightened above sixty paces, and whatsoever falls between the parallels, will be clearly seen: the reason is, because the beams from the Centre to the circumference, are reflected parallel, when the parallels come to a point; and in the place thus illuminated, letters may be read, and all things done conveniently, that require great light.[58]

Third, Della Porta's secret of how to kindle fire:

> In a Concave spherical Glass the beams meeting together, kindle fire in a fourth part of the diameter under the Centre, which are directed within the side of a Hexagon from the superficies of the circle.[59]

This secret, too, and especially, the locus of the focal point of a concave spherical mirror at the fourth part of the diameter of the mirror, is taken from Ausonio.

A fourth example of appropriation concerns the secret of images in the air. In Chapter 10 Della Porta taught how 'to see an image hanging in the air' with a convex lens:

> If you put the thing to be seen behind the Lenticular, that it may pass thorow the Centre, and set your eyes in the opposite part, you shall see the Image between the Glass and your eyes; and if you set a paper against it, you shall see it clearly: so that a lighted Candle will seem to burn upon the Paper.[60]

In Chapter 13 he told of how to make an image appear in the air with a crystal ball:

> It will shew the Image in the Air, both before and behind. Let the Object be behind the Pillar, let the Pillar be between that and the eye, the Image will appear outwardly haging in the Air, above the Pillar, parted every where from the Pillar, clearly and perspicuously.[61]

---

57 Giambattista della Porta, *Natural Magick by John Baptista Porta, a Neapolitane: in Twenty Books … Wherein are Set Forth all the Riches and Delights of the Natural Sciences* (London: Printed for Thomas Young and Samuel Speed, 1658), p. 356.

58 Porta, *Natural Magick*, 362.

59 Porta, *Natural Magick*, 370.

60 Porta, *Natural Magick*, 368–69.

61 Porta, *Natural Magick*, 370.

Della Porta's secrets are variations on a secret of how to make a mirror to make an image float in the air, already found in the *Secretum Philosophorum*:

> You can also make a mirror out of a convex mirror in which an image will appear outside, and this is how it is done. Take an ordinary mirror (that is, a convex one) and scrape off the lead and put it in a box, which is not too deep, so that the convexity is towards the bottom of the box, and the concavity is outwards. Then put something dark between the bottom of the box and the mirror, such as the black cloth or some such thing, and do this so that the visual ray is better reflected. Then if you attentively gaze in the mirror, you will see your image outside the box, in the air between you and the mirror.[62]

The secret has its source in Witelo's *Perspectiva*, but travelled widely and independently, also outside the context of the *Secretum philosophorum*, in the fifteenth and sixteenth centuries.[63] In fact, the ambiguous concept of 'image in the air' is a good example of a notion which gained a life of its own without the constraints imposed by Witelo's *Perspectiva* and other perspectivist texts.[64] This ambiguity and flexibility of meaning is a consequence of the separation of the secret from the context of the source text on optics following the process of copying, breaking up and re-writing more 'rationally ordered' optical source texts as stand-alone secrets. In sixteenth-century books of secrets the context which constrained the meaning of concepts and terms disappeared, leaving the reader with a wider field of interpretation, and also room for creative misunderstandings.

Conclusion

What happened to the meaning of perspective during the process of re-packaging optical knowledge as secrets? While artisans wrote down how to go about creating their works, they were also aware of the limits of their abilities to fix their knowledge in words. It is to characterize this unspeakable property of artisanal knowledge that Dürer evoked the term *Augenmaß*. In the *Ästhetische Exkurs*, Dürer wrote:

> But if you have learned how to measure well, and added reason to practice so that you can make a thing with free assuredness, and know how to do justice to you're a thing, then it is not necessary to measure it all the time, for your accomplished art endows you with a good eye measure and your practiced hand obeys.[65]

62 Goulding, 'Deceiving the Senses in the Thirteenth Century', pp. 135–62 (esp. p. 156).

63 For its source in Witelo, see A. Mark Smith, 'Reflections on the Hockney-Falco Thesis: Optical Theory and Artistic Practice in the Fifteenth and Sixteenth Centuries', *Early Science and Medicine*, 10 (2) (2005), 163–86.

64 Sven Dupré, 'Images in the Air: Optical Games, Magic, and Imagination', in *Spirits Unseen: The Representation of Subtle Bodies in Early Modern European Culture*, ed. by Christine Göttler and Wolfgang Neuber (Leiden: Brill, 2008), pp. 71–92.

65 'Aber so du wol messen hast gelernt, und den verstandt mit sambt dem brauch uber kumen, also das du ein ding auß freyer gwißheyt kanst machen und weyst einem yetlichen ding recht zu thon, als dann ist nit alweg not, ein ydlich ding alweg zu messen, dan dein uberkumne kunst macht dir ein guten augen maß, als dann ist die geübt hand gehorsam'. Albrecht Dürer, *Schriftlicher Nachlass*, ed. by Hans Rupprich, 3 vols (Berlin: Deutscher Verlag für Kunstwissenschaft, 1969), III, p. 297.

Related to period notions of *ingenio* or *ingenium* ('ingenuity') as innate talent or natural ability without instruction, Dürer's *Augenmaß* was *Wissen* ('knowledge') partly acquired through practice. *Augenmaß* guided the hand of the skilled artisan and allowed him to avoid *yrthumb* ('error') and *falscheit* ('falseness'). Following Dürer, numerous writers on the arts used the idea of *Augenmaß* to express that essential aspects of the making of art were a matter impossible to fix in words.

Dürer's *Augenmaß* is closely related to the visual 'discernment' and 'judgment by the eye' evoked by Neri, mentioned at the beginning of my essay.[66] Neri conceptualized what the discerning eye saw as 'hidden', and thus as a 'secret' which was known only to the expert artist. As I have argued in this essay, the collections of recipes artists were primarily confronted with in their search for optical knowledge packaged optical knowledge as 'secrets'. This entailed a re-definition of perspective. First, breaking-up optical source texts in this way solicited a process of translation and appropriation of optical knowledge. Secrets repackaged optical knowledge into 'how-to' chunks, thereby separating these chunks of optical knowledge for the creation of visual effects from the context of the source text. This allowed optical knowledge to travel more easily and enabled it to reach larger audiences of readers in the sixteenth century than ever before. Given the structure of books of secrets, these readers were also invited to try and test the optical recipes, and they did; the reader's experience was not always disassociated from the worlds of making and doing. However, the structure of books of secrets also created conceptual ambiguity, because in this process the theories of light and vision disappeared into the background. Second, the emphasis of the secrets was on the manipulation of objects and instruments to create particular optical effects. In contrast to Panofsky, and acknowledging the polysemy of perspective, the materiality of the texts on optics and perspective I have discussed constructed a particular definition of perspective similar to Dürer's *Messung*. The materiality of the objects and the instruments used to draw them in perspective were crucial to this definition of perspective. In contrast to Crary's geometric scopic regime, rather than making the draughtsman into a passive observer, these instruments, including the camera obscura, relied on the bodily engagement of the observer and the artist to create visual effects and to construct perspective.

Bibliography

Manuscript and Archival Sources

Alhacen, Vat. Lat, 4595, *Libro de li aspecti*, transcribed by Pietro Roccasecca: [Perspectiva+]: http://perspectiva.biblhertz.it/index.html.

Bibliothèque municipale, Dijon, 441 (226), fol. 206r.

British Library, London, Egerton 2852, fol. 19r.

66 Dupré and Göttler, 'Hidden Artifices', pp. 1–16.

Primary Sources

Alhacen, *Alhacen's Theory of Visual Perception: A Critical Edition, with English Translation and Commentary, of the First Three Books of Alhacen's 'De aspectibus', the Medieval Latin Version of Ibn al-Haytham's 'Kitāb al-Manāẓir'*, ed. and trans. by A. Mark Smith, 2 vols (Philadelphia: Transactions of the American Philosophical Society, 2001).

Bacon, Roger, *Roger Bacon's Philosophy of Nature: A Critical Edition, with English Translation, Introduction, and Notes, of 'De multiplicatione specierum' and 'De speculis comburentibus'*, ed. and trans. by David C. Lindberg (Oxford: Clarendon Press, 1983).

Broecke, Lara, *Cennino Cennini's Il libro dell'arte: A New English Translation and Commentary with Italian Transcription* (London: Archetype Publications, 2015).

Dürer, Albrecht, *Schriftlicher Nachlass*, ed. by Hans Rupprich, 3 vols (Berlin: Deutscher Verlag für Kunstwissenschaft, 1969).

Kepler, Johannes, *Gesammelte Werke*, ed. by Max Caspar and Walter van Dyck, 23 vols (Munich: C. H. Beck, 1938).

Kepler, Johannes, *Optics. Paralipomena to Witelo and & Optical Part of Astronomy*, trans. by William H. Donahue (Santa Fe: Green Lion Press, 2000).

Neri, Antonio, *L'Arte vetraria distinta in libri sette* (Florence: de' Gunti, 1612).

Porta, Giambattista della, *Natural Magick by John Baptista Porta, a Neapolitane: in Twenty Books … Wherein are Set Forth all the Riches and Delights of the Natural Sciences* (London: Printed for Thomas Young and Samuel Speed, 1658).

Secondary Works

Belting, Hans, *Florence and Baghdad: Renaissance Art and Arab Science*, trans. by Deborah Lucas Schneider (Cambridge, Mass.: Belknap Press of Harvard University Press, 2011).

Bergdolt, Klaus, *Der dritte Kommentar Lorenzo Ghibertis: Naturwissenschaften und Medizin in der Kunsttheorie der Frührenaissance* (Weinheim: VCH, Acta Humaniora, 1988).

Blair, Ann, *Too Much to Know: Managing Scholarly Information before the Modern Age* (New Haven: Yale University Press, 2010).

Clagett, Marshall, *Archimedes in the Middle Ages. Vol. 4: A Supplement on the Medieval Latin Traditions of Conic Sections* (Philadelphia: The American Philosophical Society, 1980).

Clarke, Mark, 'Writing Recipes for Non-Specialists, *c.* 1300: The Anglo-Latin "Secretum philosophorum", Glasgow MS Hunterian 110', in *Sources and Serendipity: Testimonies of Artists' Practice*, ed. by Erma Hermens and Joyce H. Townsend (London: Archetype Publications, 2009), pp. 50–64.

Clark, Stuart, *Vanities of the Eye: Vision in Early Modern European Culture* (Oxford: Oxford University Press, 2007).

Crary, Jonathan, *Techniques of the Observer: On Vision and Modernity in the Nineteenth Century* (Cambridge, Mass.: The MIT Press, 1990).

Damm, Heiko, Michael Thimann and Claus Zittel, *The Artist as Reader: On Education and Non-Education of Early Modern Artists* (Leiden: Brill, 2012).

Dupré, Sven, 'Mathematical Instruments and the "Theory of the Concave Spherical Mirror": Galileo's Optics beyond Art and Science', *Nuncius*, 15 (2) (2000), 551–88.

Dupré, Sven, 'Optics, Instruments and Painting, 1420–1720: Reflections on the Hockney-Falco Thesis', Special Issue of *Early Science and Medicine*, 10 (2) (2005), 125–339.

Dupré, Sven, 'Images in the Air: Optical Games, Magic, and Imagination', in *Spirits Unseen: The Representation of Subtle Bodies in Early Modern European Culture*, ed. by Christine Göttler and Wolfgang Neuber (Leiden: Brill, 2008), pp. 71–92.

Dupré, Sven, 'Printing Practical Mathematics: Oronce Fine's "De speculo ustorio" between Paper and Craft', in *The Worlds of Oronce Finé: Mathematics, Instruments and Print in Renaissance France*, ed. by Alexander Marr (Donington: Shaun Tyas, 2009), pp. 64–82.

Dupré, Sven, and Michael Korey, 'Inside the Kunstkammer: The Circulation of Optical Knowledge and Instruments at the Dresden Court', *Studies in History and Philosophy of Science*, 40 (4) (2009), 405–20.

Dupré, Sven, 'Trading Luxury Glass, Picturing Collections and Consuming Objects of Knowledge in Early Seventeenth-Century Antwerp', *Intellectual History Review*, 20 (1) (2010), 53–78.

Dupré, Sven, 'The Historiography of Perspective and "Reflexy-Const" in Netherlandish Art', *Nederlands Kunsthistorisch Jaarboek*, 61 (2011), 35–60.

Dupré, Sven, 'Kepler's Optics without Hypotheses', *Synthese*, 185 (2012), 501–25.

Dupré, Sven, Bert de Munck and Mark Clarke, *Transmission of Artists' Knowledge* (Brussels: Royal Flemish Academy of Arts and Sciences, 2012).

Dupré, Sven, 'The Value of Glass and the Translation of Artisanal Knowledge in Early Modern Antwerp', in *Trading Values in Early Modern Antwerp*, ed. by Bart Ramakers, Christine Göttler, and Joanna Woodall, Netherlands Yearbook for Art History, 64 (Leiden: Brill, 2014), pp. 138–61.

Dupré, Sven, 'Die Sichtbarkeit und Unsichtbarkeit von Körperwissen in der Kodifikation der Künste in der frühen Neuzeit', *Paragrana: Internationale Zeitschrift für Historische Anthropologie*, 25 (1) (2016), 110–29.

Dupré, Sven, and Christine Göttler, 'Hidden Artifices', in *Knowledge and Discernment in the Early Modern Arts*, ed. by Sven Dupré and Christine Göttler (New York: Routledge, 2017), pp. 1–16.

Eamon, William, *Science and the Secrets of Nature: Books of Secrets in Medieval and Early Modern Culture* (Princeton: Princeton University Press, 1994).

Fiorentini, Erna, 'Subjective Objective. The Camera Lucida and Protomodern Observers', *Bildwelten des Wissens: Kunsthistorisches Jahrbuch für Bildkritik*, 2 (2004), 58–66.

Gage, John, *Colour and Meaning: Art, Science and Symbolism* (London: Thames & Hudson, 1999).

Garber, Daniel, 'Merchants of Light and Mystery Men: Bacon's Last Projects in Natural History', *Journal of Early Modern Studies*, 3 (1) (2014), 91–106.

Goulding, Robert, 'Deceiving the Senses in the Thirteenth Century: Trickery and Illusion in the Secretum philosophorum', in *Magic and the Classical Tradition*, ed. by C. Burnett and W. F. Ryan (London: The Warburg Institute, 2006), pp. 135–62.

Hackenberg, Michael, 'Books in Artisan Homes of Sixteenth-Century Germany', *Journal of Library History*, 21 (1986), 72–91.

Hall, Bert S., 'Der Meister sol auch Kennen Schreiben und Lesen: Writings about Technology ca. 1400-ca. 1600 A.D. and their Cultural Implications', in *Early Technologies*, ed. by Denise Schmandt-Besserat (Malibu: Undena Publications, 1979), pp. 47–58.

Heikamp, Detlef, *Studien zur mediceischen Glaskunst: Archivalien, Entwurfszeichnungen, Gläser und Scherben* (Florence: Kunsthistorisches Institut, 1986).

Hills, Paul, *Venetian Colour: Marble, Mosaic, Painting and Glass 1250–1550* (New Haven: Yale University Press, 1999).

Hockney, David, *Secret Knowledge: Rediscovering the Lost Techniques of the Old Masters* (London: Thames & Hudson, 2001).

Jalobeanu, Dana, and Cesare Pastorino, 'Introduction', in 'Instruments & Arts of Inquiry: Natural History, Natural Magic and the Production of Knowledge in Early Modern Europe', Special Issue of *Journal of Early Modern Studies*, 3 (2014), 9–13.

James Elkins, James, *The Poetics of Perspective* (Ithaca: Cornell University Press, 1994).

Kern, Ulrike, 'The Art of Conservation I: Theodore de Mayerne, the King's Black Paintings and Seventeenth-Century Methods of Restoring and Conserving Paintings', *The Burlington Magazine*, 157 (2015), 700–708.

Krautheimer, Richard, and Trude Krautheimer-Hess, *Lorenzo Ghiberti* (Princeton: Princeton University Press, 1982).

Lautier, Claudine, and Dany Sandron, *Antoine de Pise: L'art du vitrail vers 1400* (Paris, Comité des Travaux Historiques et Scientifiques, 2008).

Leong, Elaine, and Alisha Rankin (eds.), *Secrets and Knowledge in Medicine and Science, 1500–1800* (Aldershot: Ashgate, 2011).

Lindberg, David C., *Theories of Vision. From al-Kindi to Kepler* (Chicago: University of Chicago Press, 1976).

Long, Pamela O., *Openness, Secrecy, Authorship: Technical Arts and the Culture of Knowledge from Antiquity to the Renaissance* (Baltimore: Johns Hopkins University Press, 2001).

Massey, Lyle, *Picturing Space, Displacing Bodies: Anamorphosis in Early Modern Theories of Perspective* (University Park, Pennsylvania: The Pennsylvania State University Press, 2007).

Panofsky, Erwin, *Perspective as Symbolic Form*, trans. by Christopher S. Wood (New York: Zone Books, 1997).

Peiffer, Jeanne, 'Projections Embodied in Technical Drawings: Dürer and his Followers', in *Picturing Machines 1400-1700*, ed. by Wolfgang Lefèvre, (Cambridge, Mass.: The MIT Press, 2004), pp. 245–75.

Poulle, Emmanuel, *Un constructeur d' instruments astronomiques au XVe siècle: Jean Fusoris* (Paris: Librairie Honoré Champion, 1963).

Principe, Lawrence, *The Secrets of Alchemy* (Chicago: University of Chicago Press, 2013).

Raynaud, Dominique, *L'hypothèse d'Oxford. Essai sur les origins de la perspective* (Paris: Presses universitaires de France, 1998).

Rees, Joachim, *Die verzeichnete Fremde: Formen und Funktionen des Zeichnens im Kontext europäischer Forschungsreisen 1770-1830* (Paderborn: Wilhelm Fink, 2015).

Roccasecca, Pietro, 'Gentile da Fabriano, A Miracle of Saint Nicholas: A Rigorous Nonperspective Spatial Representation', *Center: Record of Activities and Research Reports, National Gallery of Art, Washington*, 21 (2001), 126–30.

Roccasecca, Pietro, 'Not Albertian', *Center: Record of Activities and Research Reports, National Gallery of Art, Washington*, 22 (2002), 167–69.

Rose, Paul Lawrence, 'Jacomo Contarini (1536–1595), a Venetian Patron and Collector of Mathematical Instruments and Books', *Physis*, 18 (2) (1976), 117–30.

Rosińska, Grażyna, 'Optyka W XV wieku miedzy nauka sredniowieczna a nowozytna', *Studia Copernicana*, 24 (1986).
Sauer, Christine, 'Eine kunsttechnologische Handschrift aus dem Besitz Albrecht Dürers', in *Dürer-Forschungen* (Nürnberg: Verlag des Germanischen Nationalmuseums, 2009), II, pp. 275–96.
Siegert, Bernhard, 'Kulturtechnik', in *Einführung in die Kulturwissenschaft*, ed. by Harun Maye and Leander Scholz (Munich: Wilhelm Fink Verlag, 2011), pp. 95–118.
Smith, A. Mark, 'Reflections on the Hockney-Falco Thesis: Optical Theory and Artistic Practice in the Fifteenth and Sixteenth Centuries', *Early Science and Medicine*, 10 (2) (2005), 163–86.
Smith, A. Mark, *From Sight to Light: The Passage from Ancient to Modern Optics* (Chicago: University of Chicago Press, 2014).
Trachtenberg, Marvin, *Dominion of the Eye: Urbanism, Art and Power in Early Modern Florence* (Cambridge: Cambridge University Press, 1997).
Trevor-Roper, H., *Europe's Physician. The Various Life of Sir Theodore de Mayerne* (New Haven: Yale University Press, 2006).
Vandamme, E., 'Een 16e-eeuws Zuidnederlands receptenboek', *Jaarboek van het Koninklijk Museum voor Schone Kunsten* (1974), 101–37.

JOSÉ CALVO-LÓPEZ

Teaching, Creating, and Using Perspective in Sixteenth-Century Spain

*The Architectural Notebook of Hernán Ruiz II**

Introduction

Hernán Ruiz II (*c.* 1500–69), a Spanish master mason and architect working in the mid-sixteenth century, included a number of peculiar perspective drawings and schemes in his personal sketchbook, preserved in the Biblioteca de la Escuela de Arquitectura at the Universidad Politécnica de Madrid.[1] At first sight, a number of these perspectival schemes depart from mainstream sixteenth-century theory, using angles with vertexes on the viewpoint to construct perspectives. This had led some scholars to present his work as a proof of Erwin Panofsky's theories about angular perspective and the depiction of retinal images, mentioning the publication of a translation of Euclid's *Optics* in Seville, sixteen years after Hernán Ruiz's death.[2] This chapter will analyse these problems under the light of recent research on Ruiz's career, stressing the different approaches to perspective throughout the notebook: didactics of perspective, innovation in perspective methods, and use of perspective in architectural drawings. Such approaches foster the use of different perspectival methods: principal point and diagonals in the didactic drawings, a number of idiosyncratic procedures in the experimental diagrams, and a typically Serlian mixture of orthographic projection and foreshortening in some architectural drawings. After all, as James Elkins pointed out, the notion of 'correct' perspective as a set of ideal, mutually consistent, geometrical principles,

* This study is a result of the research project 'Arquitectura renacentista y construcción pétrea' (19361/PI/14), sponsored by Fundación Séneca – Agencia Regional de Ciencia y Tecnología de la Región de Murcia under PCTIRM 2011–14. The author wishes to thank the Library of the School of Architecture of Universidad Politécnica de Madrid and Pau Natividad Vivó for their kind permissions to use images 4–16 and 3, respectively.

1 Biblioteca de la Escuela de Arquitectura de la Universidad Politécnica de Madrid, Madrid, Hernán Ruiz II, *Libro de arquitectura* (*c.* 1550). Facsimile editions include: *El libro de arquitectura de Hernán Ruiz el Joven*, ed. by Pedro Navascués Palacio (Madrid: Escuela Técnica Superior de Arquitectura, 1974); *Libro de arquitectura*, ed. by Alfonso Jiménez Martín et al. (Seville: Fundación Sevillana de Electricidad, 1998); *Libro de arquitectura*, ed. by Pedro Navascués Palacio (Madrid: Universidad Politécnica de Madrid, 2005).

2 Lino Cabezas Gelabert, *Tratadistas y tratados españoles de perspectiva desde sus orígenes hasta la geometría descriptiva de G. Monge, 1526–1803* (Barcelona: Universitat de Barcelona, 1985). See also Lino Cabezas Gelabert, 'La "perspectiva angular" y la introducción de la perspectiva artística en la España del siglo XVI', *D'art*, 15 (1989), 167–79; José María Gentil Baldrich, 'El libro de perspectiva', in Ruiz, *Libro de arquitectura*, ed. by Jiménez Martín, pp. 215–34.

José Calvo-López Technical University of Cartagena, jose.calvo@upct.es

Perspective as Practice. Renaissance Cultures of Optics, ed. by Sven Dupré, Turnhout, 2019 (*Techne. Knowledge, Technique, and Material Culture*, 1), p. 301-332
 DOI 10.1484/M.Techne-EB.5.117731

and different methods as instances of these principles, only arises in the last decades of the sixteenth century, just a few years after Hernán Ruiz's death.[3]

Hernán Ruiz II and the *Libro de arquitectura*

Hernán Ruiz II, also called *El Mozo* or *El Joven*, that is, the younger one, to differentiate him from his father, was a mason from Córdoba who attained the position of master mason of Seville cathedral in 1557.[4] He built a large number of parish churches across the huge bishophric of Seville, but he is best remembered for two exceptional works. The main Islamic mosque of Seville had been torn down in the early fifteenth century, but the minaret had been left standing. Hernán Ruiz II was commissioned with the difficult task of converting the minaret into a Christian bell tower. From this moment on, Hernán Ruiz's addition has become indivisible from the minaret, and the ensemble of both constructions is presently the main icon of Seville, the Giralda (Figure 1).

In 1558, the cathedral chapter asked him for plans for the Chapter Hall, and he designed what was possibly the first oval-plan space built in the entire European Renaissance. The church of Sant'Andrea in Via Flaminia, designed by Giacomo Barozzi da Vignola, dating from 1550–53, predates the Chapter Hall by about ten years; however, in Sant' Andrea, only the vault is oval-shaped, while the floor plan is rectangular. By contrast, in the Chapter Hall, the whole space is oval shaped, from the ground up. This is the case also in Sant'Anna dei Palafrenieri, in the Vatican, a building that dates from around 1570, which is sometimes quoted as being the first entirely oval church (Figure 2). In any case, the vault shows a remarkable mastery of stonecutting technique. When Hernán Ruiz died in 1569, the vault was still incomplete. The son of Hernán Ruiz, Hernán Ruiz III, tried to get the post of master mason, but the chapter gave this position to another man, Pedro Díaz de Palacios. Hernán Ruiz III left for Córdoba with his father's plans and templates, and Díaz de Palacios was unable to finish the vault. The chapter consulted with many architects, but the vault was not completed until 1592; this shows the complexity of the geometrical problems raised by the designs of Hernán Ruiz II.[5]

The notebook of Hernán Ruiz II is not a draft of an architectural treatise, but rather a personal sketchbook made up of many different, randomly ordered sections. These include a translation of Vitruvius's First Book, a number of architectural drawings, some

3 James Elkins, 'Renaissance Perspectives', *Journal of the History of Ideas*, 53 (2) (1992), 209–30.

4 Antonio de la Banda y Vargas, *El arquitecto andaluz Hernán Ruiz II* (Sevilla: Universidad, 1974); Pedro Navascués Palacio, 'Estudio', in Ruiz, *El libro de arquitectura de Hernán Ruiz el Joven*, ed. by Navascués Palacio; Alfredo J. Morales, *Hernán Ruiz 'El Joven'* (Madrid: Akal, 1996). Following Banda, I will call him Hernán Ruiz II where necessary, in order to differentiate him from his son, a remarkable architect who finished the cathedral built inside the Mosque of Córdoba.

5 Wolfgang Lotz, *Architecture in Italy 1500–1600* (New Haven: Yale University Press, 1995), pp. 119–20 (Previously published as Part 2 of Ludwig Heydenreich and Wolfgang Lotz, *Architecture in Italy 1500–1600* (Harmondsworth: Penguin Books, 1974)); Wolfgang Lotz, 'Die Ovalen Kirchenraume des Cinquecento', *Römische Jahrbuch für Kunstgeschichte*, 6 (1955), 7–99; Banda y Vargas, *El arquitecto andaluz Hernán Ruiz II*, 122–24; José María Gentil Baldrich, 'La traza oval y la Sala Capitular de la catedral de Sevilla. Una aproximación geométrica', in *Quatro edificios sevillanos* (Sevilla: Colegio de Arquitectos, 1996), pp. 73–147; Morales, *Hernán Ruiz 'El Joven'*, 46–51. See also Biblioteca de la Escuela de Arquitectura de la Universidad Politécnica de Madrid, Madrid, Alonso de Vandelvira, 'Libro de trazas de cortes de piedras' (*c.* 1585), fols 74^{v}–77^{r}. For a facsimile edition, see Alonso de Vandelvira, *Tratado de arquitectura*, ed. by Geneviève Barbé-Coquelin de Lisle (Albacete: Caja Provincial de Ahorros, 1977).

Fig. 1 Hernán Ruiz, 1558–68. Upper stories of the bell-tower of Seville Cathedral, known as the *Giralda*.

Fig. 2 Hernán Ruiz II and Asensio de Maeda, 1558–92. Vault of the Chapter Hall in Seville Cathedral.

stonecutting schemes, such as those for a tierceron vault and a skew arch, a number of geometrical diagrams — some quite basic, some more complex, such as ovals taken from Sebastiano Serlio — and also various perspective schemes and a few more or less finished drawings.[6] Taking into account the notebook's random composition, it is not easy to set a precise date for the whole of the manuscript, but it is safe to assume that the great majority of the drawings were prepared between the start of his work in the Hospital de la Sangre in Seville in 1558 and his death in 1569.[7]

Both Alfonso Jiménez Martín and Alfredo Morales have remarked that the notebook was probably used as a resource in the instruction of apprentices. Significantly, Hernán Ruiz signed an apprenticeship contract with Lorenzo and Juan Rodríguez in the last months of 1566. Ruiz was to give them 'three lessons each week on stonecutting tracings, architecture, and perspective'.[8] In typical medieval fashion, Ruiz's stipend as a teacher was to be subtracted from the salaries of the Rodríguez brothers as co-workers of the master in his work as contractor. Perspective lessons from an architect were not merely theoretical; in fact, Ruiz inserted perspectival compositions in such buildings as the church in Hinojosa

6 Alfonso Jiménez Martín, 'Anatomía del manuscrito', in Ruiz, *Libro de arquitectura*, ed. by Jiménez Martín, pp. 23–41; and other studies in that volume, such as Francisco Pinto Puerto, 'El libro de cantería', pp. 199–214; José Antonio Ruiz de la Rosa, 'El libro de geometría', pp. 97–141, and Gentil Baldrich, 'El libro de perspectiva', pp. 215–34.

7 For a thorough survey of the different hypotheses put forward about the date of the manuscript, see Alfonso Jiménez Martín, 'Contexto de la presente edición', in Ruiz, *Libro de arquitectura*, ed. by Jiménez Martín, pp. 16–18.

8 'tres liciones cada semana de traça e alquititura e perspetiva'. Morales, *Hernán Ruiz 'El Joven'*, 138–39.

del Duque and the House of the Inquisition in Aracena, now lost, and he prepared a number of architectural drawings in perspective, as we shall see (Figure 3).[9]

The notebook includes a fair number of perspectival drawings and schemes belonging to different categories. Some of them are introductory diagrams, probably used with didactic intent, most of them taken from Serlio. There are also a number of peculiar constructions, some using the 'angular' perspective quoted by Lino Cabezas Gelabert and Jose María Gentil Baldrich, others using more idiosyncratic methods; these I will label as 'experimental' drawings and schemes.[10] Both the constructions taken from Serlio and the experimental drawings intermingle and are grouped into two blocks running through folios 51^r to 58^v and 86^r through to 88^r.[11] It is interesting to note that the first of these drawings, a square room, carries the heading 'here begin some rules of perspective'. Since the following drawings bear no text, 'rules' should be understood rather as 'methods'. I will come back to this issue further on.

In any case, a fair number of the architectural drawings in the notebook, clearly separated from the preceding groups of diagrams, use perspective in one form or another. A great part of them present different solutions for the church of the Hospital de la Sangre in Seville, a building designed by Hernán Ruiz and built under his supervision, at least in the initial stages. Other architectural drawings could be considered mere style exercises, or perhaps solutions for other commissions carried out by Ruiz, such as a great number of churches all over the huge Seville archbishopric, which spanned parts of the present-day provinces of Seville, Cádiz, and Huelva.

In the following sections I shall analyse all three groups of drawings. Since the experimental schemes are the most complex and innovative set by far, I will deal with them in greater detail; however, neither the simpler didactic diagrams nor the architectural drawings should be left aside, since Hernán Ruiz's notebook offers an exceptional example of the uses of perspective by an architect: we can see Hernán Ruiz teaching perspective, experimenting with perspective, and using perspective in his work as a designing architect.

Didactic Diagrams

The notebook includes a number of perspectival diagrams of circles, hexagons, octagons, central plans, and a square-plan room with two lateral arches, taken more or less literally from Serlio's *Second Book*. In the semicircle and the central plans, both Serlio and Hernán Ruiz follow the same technique; first, the plan or a semicircle is drawn in a frontal, undistorted view (Figures 4 and 5).[12] Then, a vanishing point is chosen and orthogonals are traced from

9 Morales, *Hernán Ruiz 'El Joven'*, 132, 138–39; Jiménez Martín, 'Contexto de la presente edición', in Ruiz, *Libro de arquitectura*, ed. by Jiménez Martín, pp. 17–18.

10 Cabezas Gelabert, *Tratadistas y tratados españoles*; Cabezas Gelabert, 'La "perspectiva angular"'; Gentil Baldrich, 'El libro de perspectiva', in Ruiz, *Libro de arquitectura*, ed. by Jiménez Martín.

11 Folio numbers for the manuscript are taken from the 1974 facsimile edition of Navascués Palacio, 'Estudio', in Ruiz, *El libro de arquitectura de Hernán Ruiz el Joven*, ed. by Navascués Palacio.

12 Cf. Ruiz, *El libro de arquitectura*, fol. 52^v, with Sebastiano Serlio, *Il primo libro di architettura* (Paris: Iehan Barbé, 1545), fol. 34^r; Ruiz, *El libro de arquitectura*, fols 55^v and 57^v with Serlio, *Il primo libro*, fol. 33^r; Ruiz, *El libro de arquitectura*, fol. 58^v with Serlio, *Il primo libro*, fol. 35^r. This edition of Serlio, although bearing the title *Il primo libro [...]* in the frontispiece, includes the title *Il secondo libro [...]* in the internal headings.

Fig. 3 Hernán Ruiz II, *c*. 1559. Parish church of Hinojosa del Duque, window, detail. Photograph by Pau Natividad Vivó.

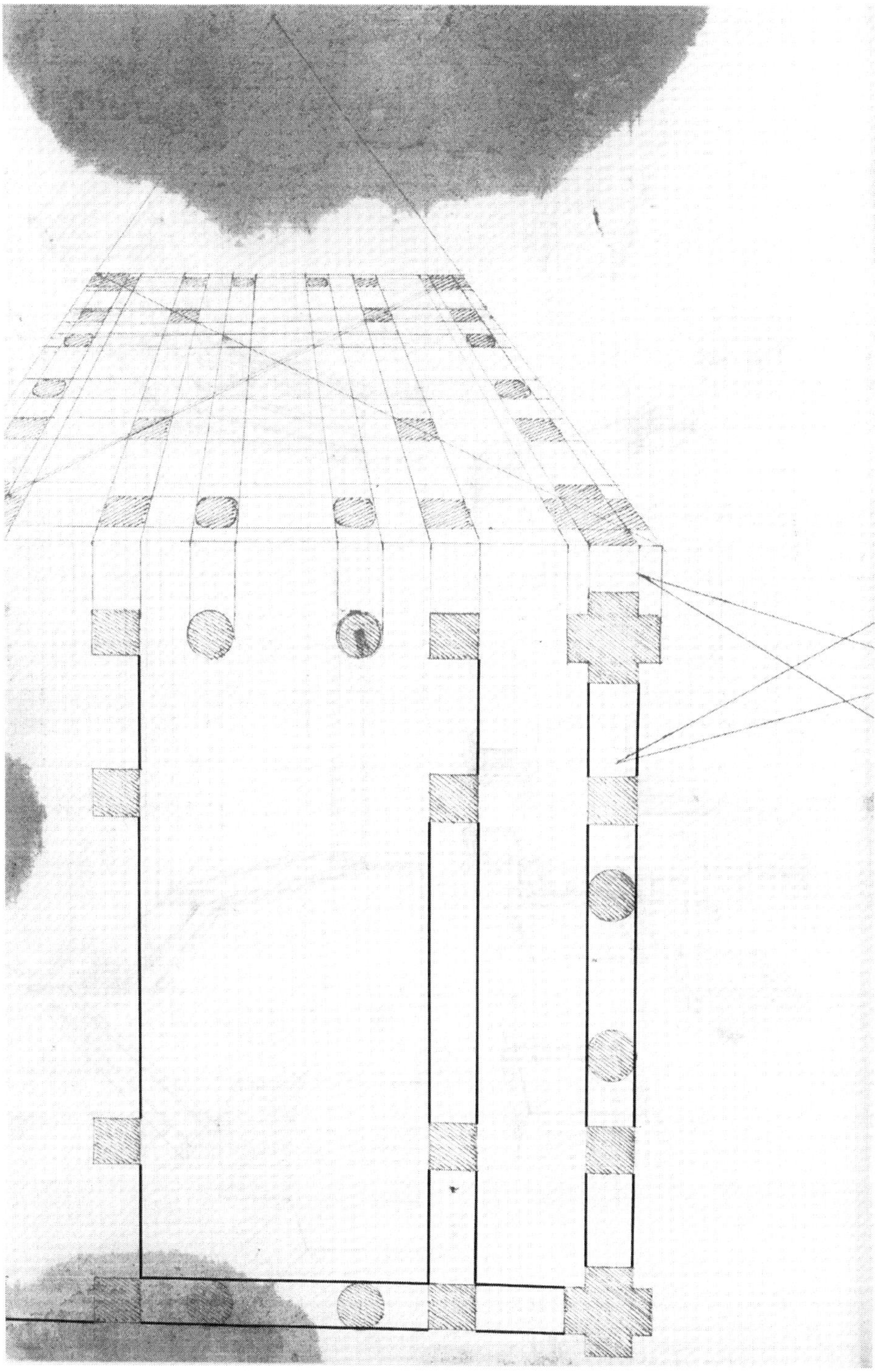

Fig. 4 Hernán Ruiz, *Libro de Arquitectura*, (*c.* 1560), fol. 52[v]. Perspective of a central plan. Hidden auxiliary lines are shown as thick lines.

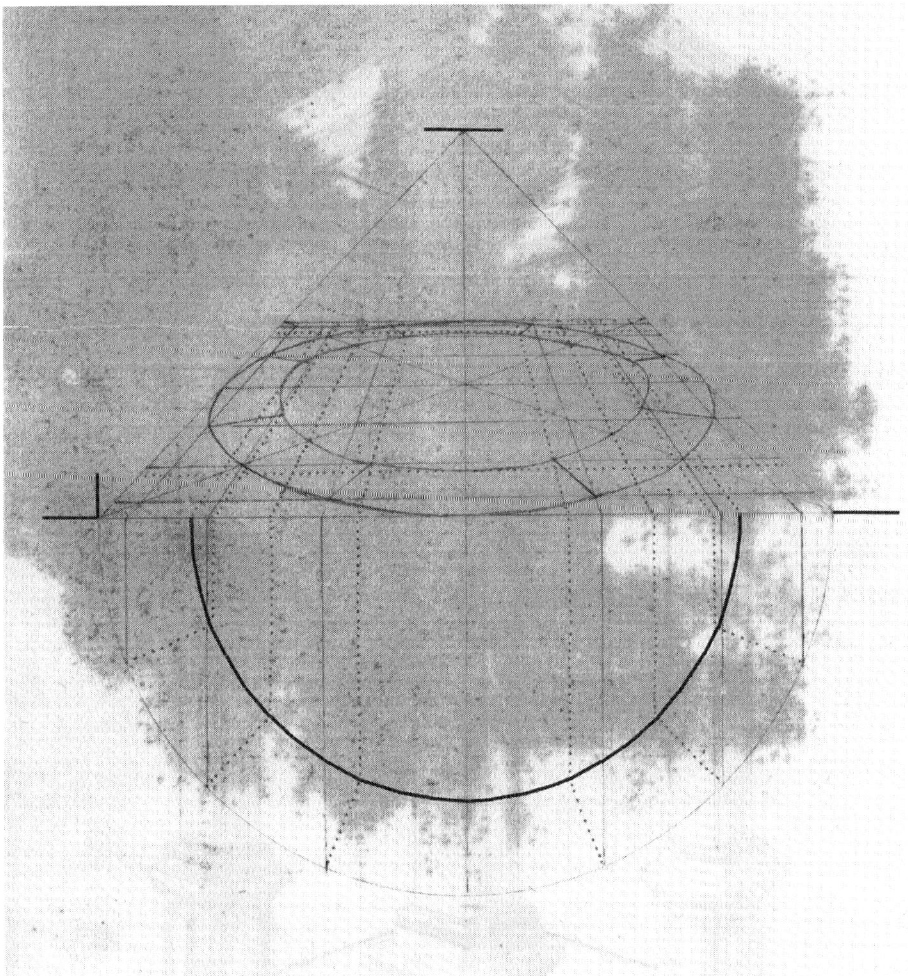

Fig. 5 Hernán Ruiz, *Libro de Arquitectura* (c. 1560), fol. 57v. Perspective of a horizontal circle. Hidden auxiliary lines are shown as thick lines.

the upper edge of the diagram, which is shown in true shape, and ending in the vanishing point. Next, the back side of the enclosing square is traced; its intersections with the side orthogonals allow Ruiz to trace one or both diagonals of the enclosing square. The back side is placed apparently at will, since the diagonals do not go further than the corner of the square. The manuscript includes a number of auxiliary, hidden lines, traced with dry point (Figures 4, 5, 7, 8, 10, 11, 13, 15).[13] In the central plans and the circles, auxiliary lines are used in order to control the general layout, or the diameter of the columns, but there

13 Unfortunately, during a restoration in the year 2000, the manuscript was pressed in order to fill the gaps in the sheets with paper paste; as a result, some of the inscribed lines are quite difficult to discern now. Thus, we cannot exclude the existence of other hidden lines in the original manuscript, in addition to the ones shown in Figures 4, 5, 7, 8, 10, 11, 13, and 15.

are no auxiliary lines extending the diagonals up to the horizon; all this suggests strongly that Ruiz did not use distance points in these constructions.

The hexagon and octagon follow a different technique, also taken literally from Serlio (Figure 6). First, the enclosing square is drawn, placing again the far side at will. No frontal view is used. In the hexagon, the near side of the polygon is traced easily, since its length equals the radius of the hexagon, that is, half the side of the enclosing square. Thus, the front and back sides of the hexagon can be traced starting from the central point of the near and far sides of the square and placing the edges of the hexagon at the points that divide both sides in four parts. Although Serlio presents this construction as somewhat simplified, or *via breve* ('short way'), up to this moment the procedure is quite precise. However, both Serlio and Hernán Ruiz are in error when duplicating the hexagon. They use the diagonals to transfer the distance between the two hexagons, measured orthogonally on one side, that is, along the apothem, to the distance between their corners, measured along the radius. As a consequence, the inner hexagon is not regular.[14]

By contrast, in the octagon, Serlio uses a rather crude simplification, mirrored by Ruiz. Once he has constructed the enclosing square in perspective, he divides the near side of the square into ten equal parts, giving four parts to the frontal side of the octagon, L in length, and three parts, that is, $0.75 \times L$, to the projection of the oblique side of the octagon over the side of the enclosing square, which should measure $1 \times \sqrt{2}/2 \times L$, that is, $0.7071 \times L$; thus, the resulting figure is the perspective applicable to an irregular octagon. However, when duplicating the octagon, both Serlio and Ruiz transfer the distance between both octagons, measured along an apothem, to another apothem; that is, both resulting octagons are irregular, yet they present the same proportions.[15] The diagram in folio 88ʳ represents a square room with lateral arches and circles in the pavement and the ceiling, also taken directly from Serlio (Figure 7).[16] In contrast with the preceding diagrams, there are several inscribed, auxiliary lines, mostly for orthogonals, transversals, and diagonals used in order to construct the circles in the pavement and the ceiling, but there is no indication of the procedure used to place the orthogonals and the transversals.

The drawing in Hernán Ruiz, *Libro de arquitectura*, folio 76ʳ, does not have a direct counterpart in Serlio's work; according to Pedro Navascués, it is a new design by Ruiz, taking elements from Donato Bramante and Serlio.[17] Moreover, there are two versions of the same plan in the same sheet (Figure 8). The upper one includes only the further half of a central, square plan; rather strikingly, it uses the diagonals of the enclosing rectangle and the side pillars are unfinished. By contrast, the other drawing, using the diagonals of the enclosing square, is almost finished, although there are *pentimenti* or alterations

14 Cf. Ruiz, *El libro de arquitectura*, fol. 54ᵛ with Serlio, *Il primo libro*, fol. 30ᵛ. About errors in Serlio's perspective, see Pietro Roccasecca, 'Sebastiano Serlio: la pratique de la perspective au service de l'architecte', in *Perspective, projections, projet: techniques de la représentation architecturale*, ed. by Fréderique Lemerle et Mario Carpo (Paris: Monum, 2005), pp. 61–70.

15 Cf. Ruiz, *El libro de arquitectura*, fol. 54ᵛ with Serlio, *Il primo libro*, fol. 31ʳ.

16 Cf. Ruiz, *El libro de arquitectura*, fol. 88ʳ with Serlio, *Il primo libro*, fol. 48ʳ.

17 Navascués Palacio, 'Estudio', in Ruiz, *El libro de arquitectura de Hernán Ruiz el Joven*, ed. by Navascués Palacio, pp. 20–21.

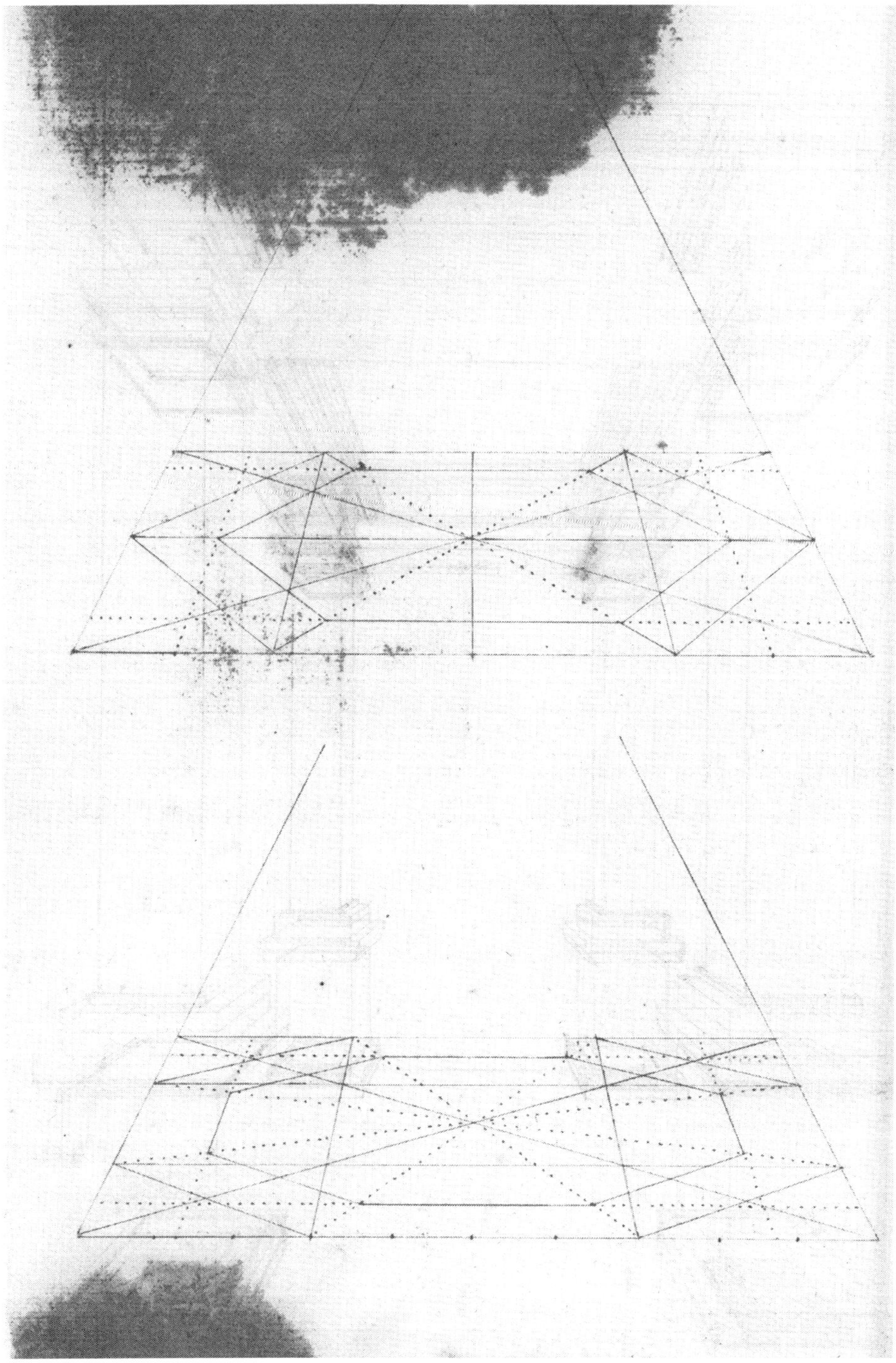

Fig. 6 Hernán Ruiz, *Libro de Arquitectura* (*c.* 1560), fol. 54v. Perspective of an hexagon and an octagon.

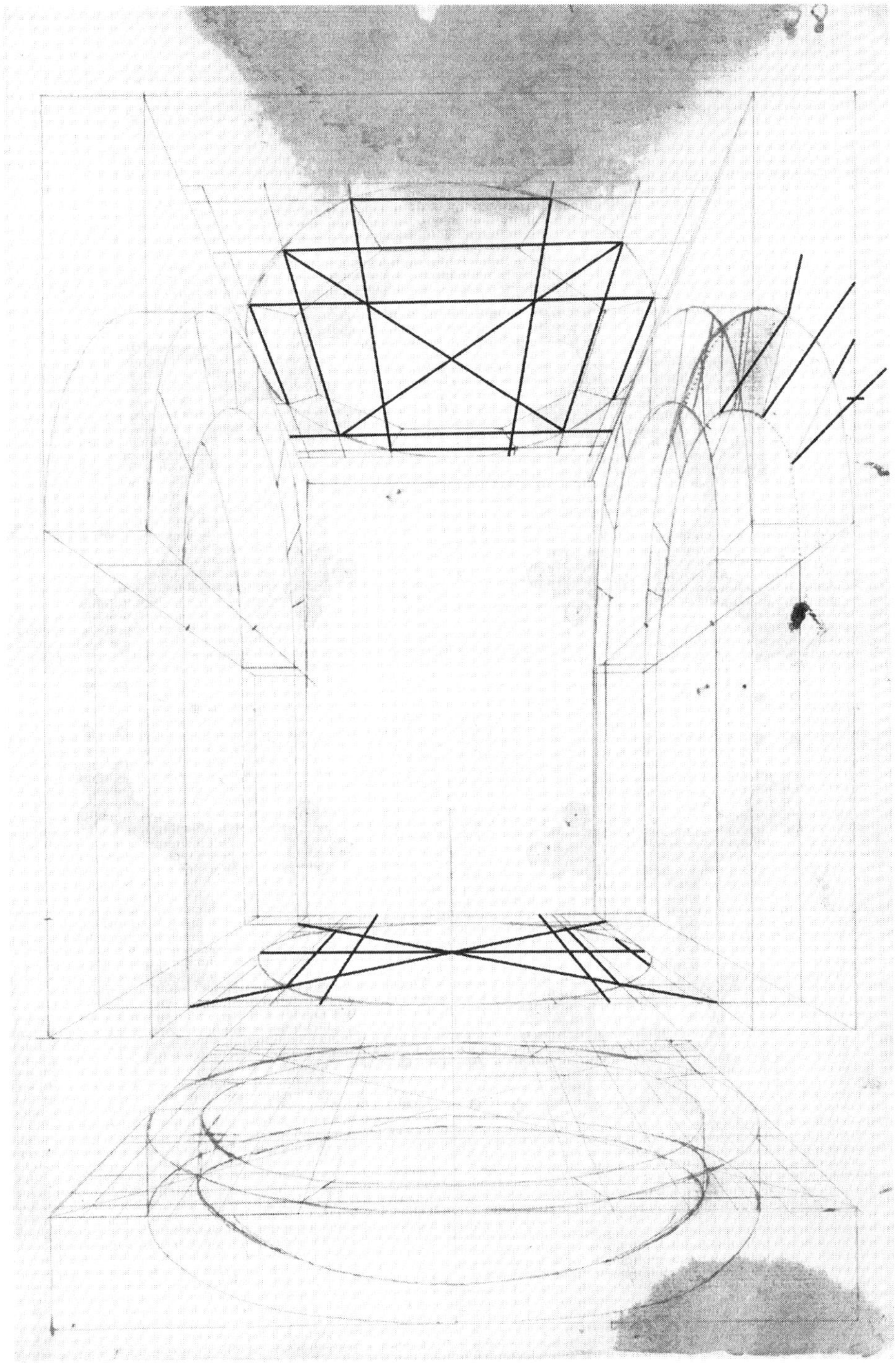

Fig. 7 Hernán Ruiz, *Libro de Arquitectura* (c. 1560), fol. 88[r]. Perspective of a square room with lateral arches and circles in the pavement and the ceiling. Hidden auxiliary lines are shown as thick lines.

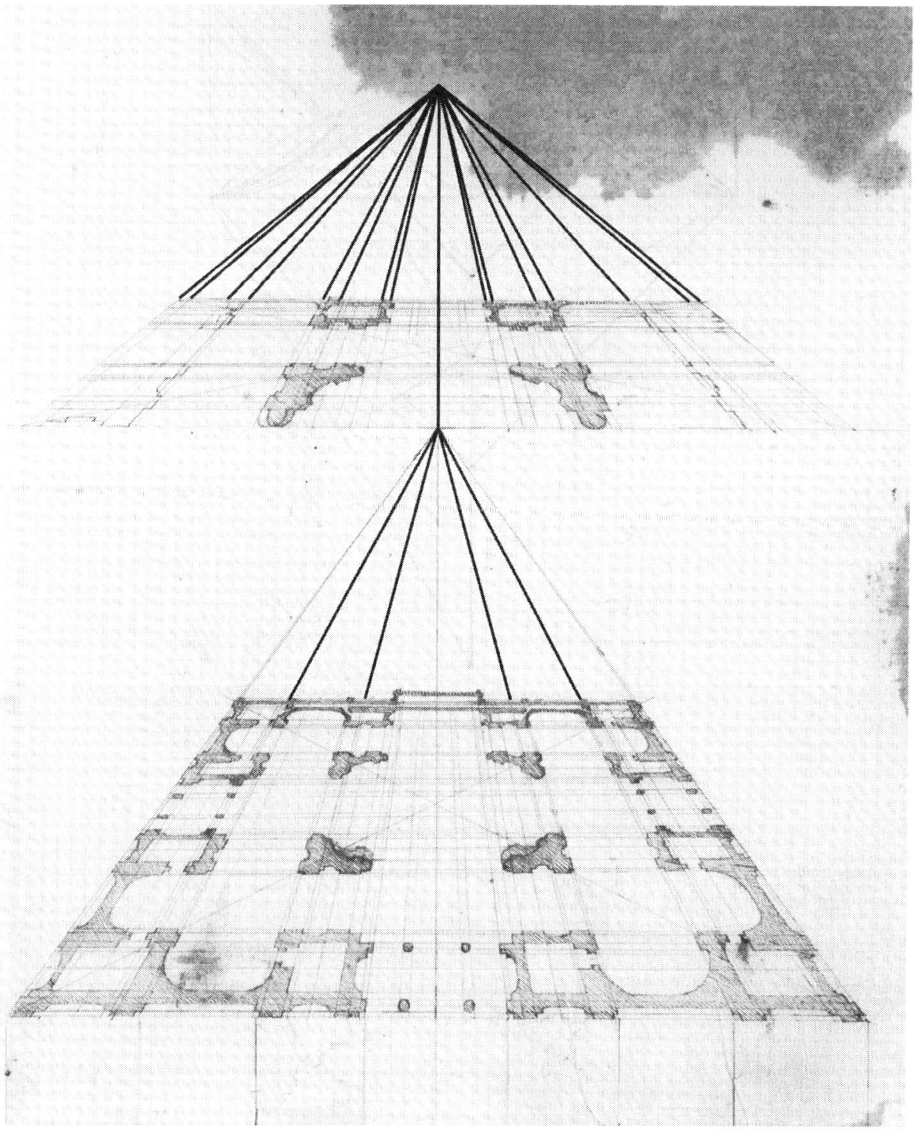

Fig. 8 Hernán Ruiz, *Libro de Arquitectura* (c. 1560), fol. 76ʳ. Two attempts at the perspective of a central plan. Hidden auxiliary lines are shown as thick lines.

in the central pillars. It seems that Ruiz learnt by trial and error that Serlio's diagonals can be applied to a square, but not to a rectangle. All this suggests that Ruiz was trying to reproduce Serlio's drawings as a way of instructing himself in Serlio's method. He was fairly successful in the task and tried to extend these techniques to a plan of his own; while this was harder, he finished the exercise with an accurate result. Of course, all this does not exclude the use of these drawings for the instruction of apprentices.

Experimental Schemes and Corresponding Drawings: The Double-Storey, Square-Plan Room

The schemes on folios 51 and 86 have attracted detailed scrutiny from Lino Cabezas Gelabert and José María Gentil Baldrich.[18] A perspectival scheme of a square room is drawn on folio 51^{r}, while folio 51^{v} includes a finished, shadowed drawing based on this scheme.[19] The scheme of a double-storey room in folio 86^{r} (Figure 9) uses similar procedures to those used in folio 51^{r}. There is a sketch based on this later scheme in folio 87^{r}, but this only includes the lower storey. Taking all this into account, it will be sufficient to analyse the scheme in folio 86^{r}, which tackles most problems with orthogonals and transversals in Hernán Ruiz's idiosyncratic perspectival method.

A number of hidden lines are present in folio 51^{r}, including two that converge, suggesting that vanishing points were not placed beforehand, but rather were located at the intersection of orthogonal lines (Figure 10). Thus, the construction of this drawing, although quite peculiar, can be followed rather easily. First, Hernán Ruiz draws the elevation of the room. Taking advantage from the square plan, this elevation functions also as a cross-section of the room. This is a typical feature of tracings made by stonecutters. Masons usually prepared large, full-scale drawings for the vaults, arches, stairways, or other constructions they were planning to build, in order to control their execution or prepare the templates for the stones of the piece.[20] Such drawings were usually inscribed in floors or walls; however, tracing lines in this way is slow and tiresome, so stonecutters only included the lines that were strictly necessary and quite frequently they reused elevations or even plans as cross-sections.

Hernán Ruiz next places a viewpoint in the cross-section and starts tracing out visual rays. Then, he repeats the operation in plan. Up to this moment, his procedure is perfectly orthodox; it is reminiscent of the way Piero della Francesca explains his methods in the Third Book of *De prospectiva pingendi*.[21] However, if we look closely at the drawing, what we see is striking. The back edge of the upper ceiling is placed much higher than it should be, while the edge line of the intermediate floor is likewise higher than its theoretical position, according to mainstream technique. Only the back edge of the floor seems to be placed at its 'correct' location.

---

18 Cabezas Gelabert, *Tratadistas y tratados españoles*; Cabezas Gelabert, 'La "perspectiva angular"'; Gentil Baldrich, 'El libro de perspectiva', in Ruiz, *Libro de arquitectura*, ed. by Jiménez Martín.

19 Navascués Palacio describes the subject of the drawings in fols 51^{r} and 51^{v} as windows opened in a wall, see: Navascués Palacio, 'Estudio', in Ruiz, *El libro de arquitectura de Hernán Ruiz el Joven*, ed. by Navascués Palacio, p. 19. Although this is tenable for both drawings as well as the sketch in fol. 87^{r}, it is not so clear for the drawing in fol. 86^{r}, which belongs to the same group and includes an intermediate layer. For the sake of clarity, I will refer to all four drawings as 'rooms', with floors, ceilings, and walls, although it would be more precise to consider them as abstract schemes.

20 Miguel Taín Guzmán, 'The Drawings on Stone in Galicia: Types, Uses and Meanings', in *Proceedings of the First International Congress on Construction History* (Madrid: Instituto Juan de Herrera, 2003), pp. 1887–98; Miguel Taín Guzmán, 'Fifteen Unedited Engraved Architectural Drawings Uncovered in Northwest Spain', in *Proceedings of the Second International Congress on Construction History: Construction History Society* (Cambridge: Construction History Society, 2006), pp. 3011–23.

21 Piero della Francesca, *De prospectiva pingendi*, ed. by G. Nicco-Fasola (Florence: Sansoni, 1984; 1st edn 1942), pp. 130–33; J. V. Field, *Piero della Francesca: A Mathematician's Art* (New Haven: Yale University Press, 2004), pp. 164–73.

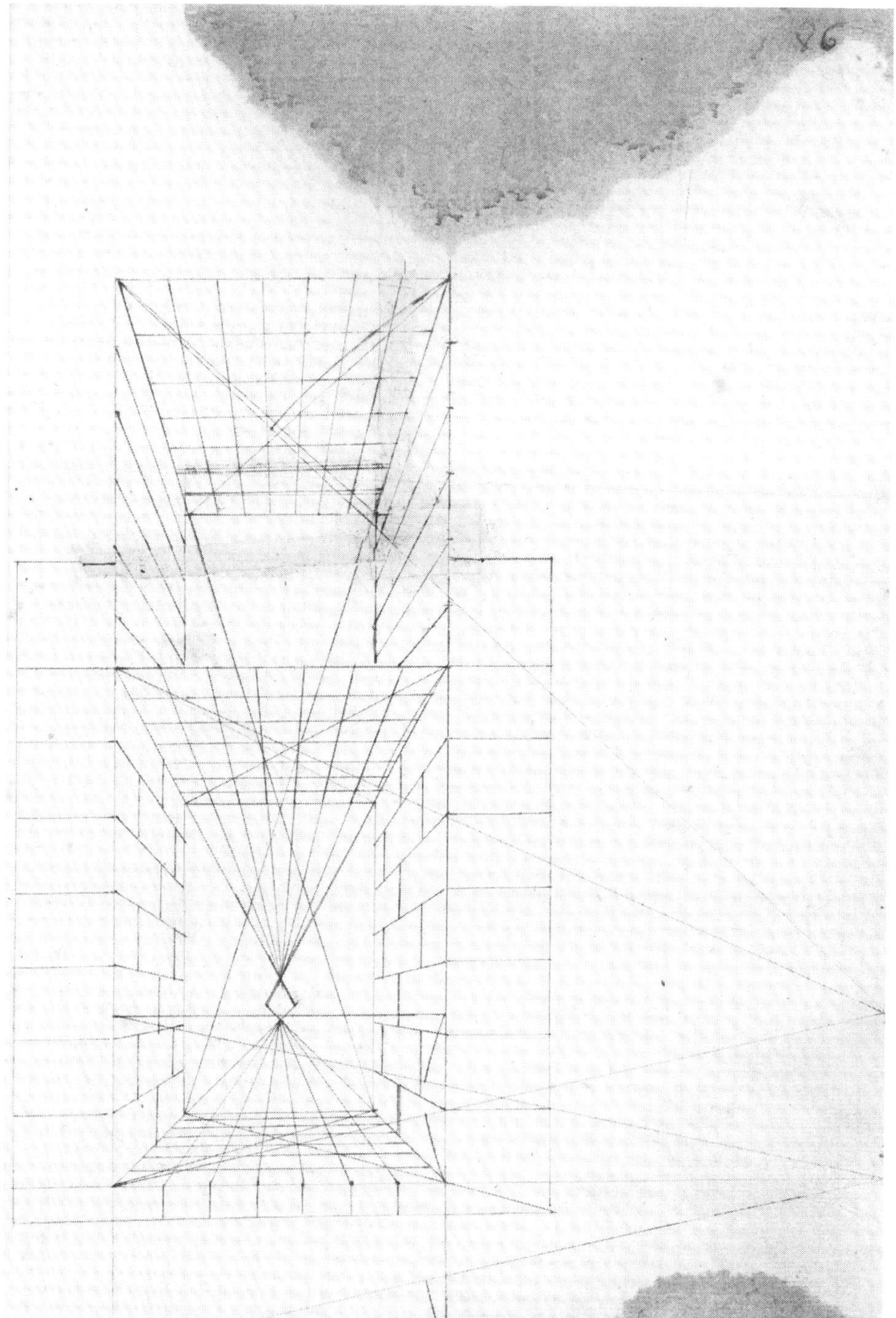

Fig. 9 Hernán Ruiz, *Libro de Arquitectura* (c. 1560), fol. 86r. Perspective scheme of a two-storey square room. [See colour plate 20]

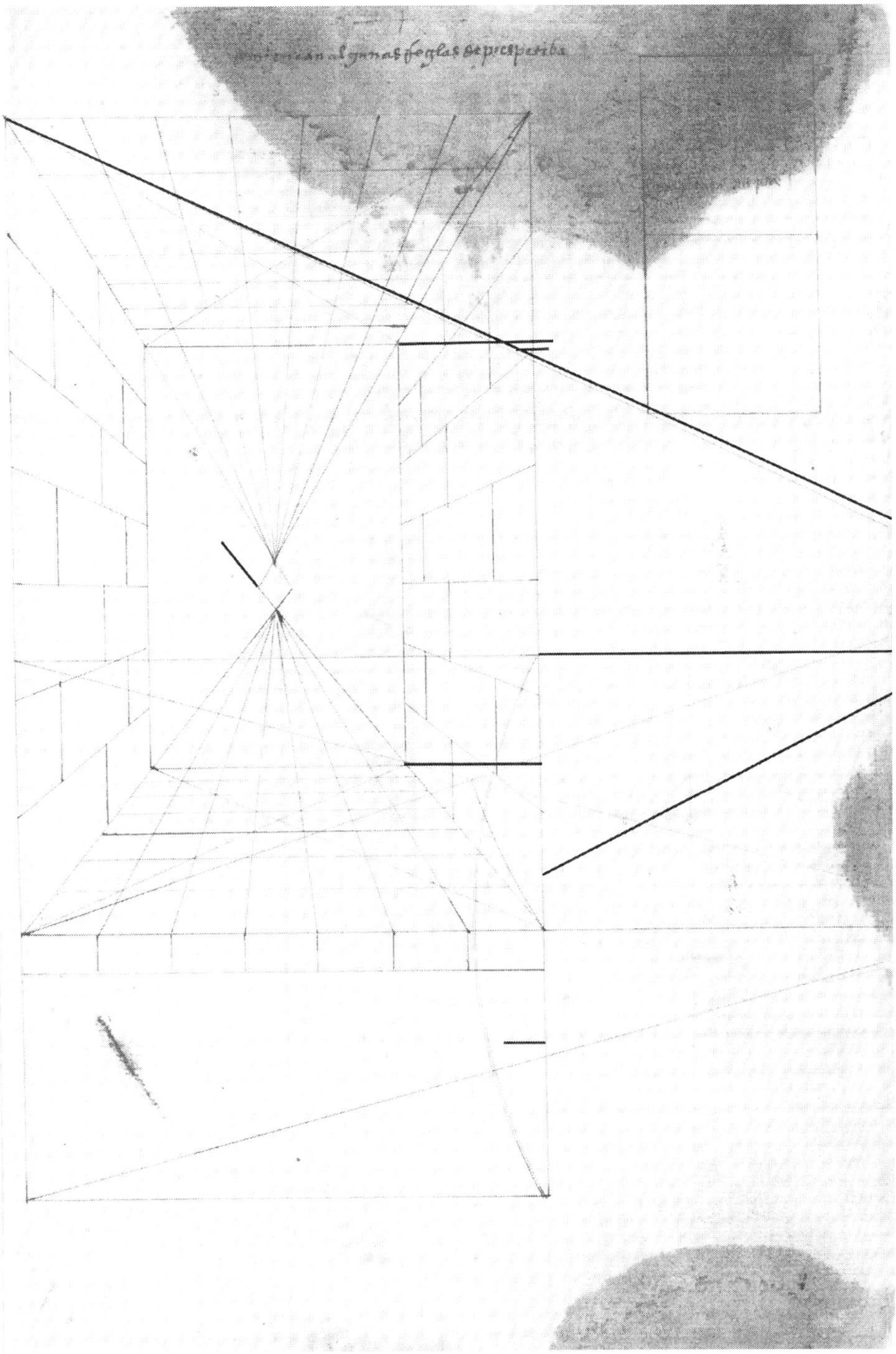

Fig. 10 Hernán Ruiz, *Libro de Arquitectura* (*c.* 1560), fol. 51^{r}. Perspective scheme of a single-storey square room. Hidden auxiliary lines are shown as thick lines.

However, the clue of this puzzling arrangement is given away by a number of small circular arcs, drawn both in cross-section and plan, with their centres placed on both viewpoints. Hernán Ruiz measures the chords of these circular arcs and transfers them to the picture plane. Cabezas and Gentil have described this peculiar method as a form of 'angular perspective', although Cabezas mentions that Ruiz is actually measuring chords, not angles.[22] This trait seems to be connected, again, with stonecutters' practices. Stonemasons did not measure angles in degrees, but rather used a typical stonecutting protractor, the *sauterelle* in French or *saltarregla* in Spanish, akin to the English bevel.[23] So, Hernán Ruiz was quite probably unable to convert angular measures into linear measures and resorted to the simpler transfer of the length of the chord to the picture plane. Another subtle detail in Hernán Ruiz's method should also be mentioned at this point. He measures the chords in orthogonal projection, either in cross-section or in plan, but not in space. We may consider the angle between the visual rays of the front and back ends of an orthogonal line. The edges of the orthogonal are at the same level, but the viewpoint is not. Thus, neither the magnitude of the angle nor the length of the chord are preserved in the horizontal projection. This small detail places the perspective method of Hernán Ruiz apart from the 'spherical' perspectives of the nineteenth and twentieth centuries and any attempt to represent the visual image on the spherical surface of the retina.[24]

In any case, once Hernán Ruiz has brought the measure of the chords to the picture plane, he is able to draw the edge lines of the ceilings; the theoretical result of the method I have described fits quite closely to the actual drawing. As for the back transversal line in the floor, both the chord method and the orthodox procedure give quite similar results; however, we may take for granted that Hernán Ruiz has also used the chord method here, since the circular arc is quite visible. The same procedure is used for the placement of the back edges of the side walls, and it fits closely with the actual result in the drawing.

At this moment, Hernán Ruiz has formulated an outline of the double-storey construction he is trying to draw in perspective, and now he must draw the orthogonals that divide floor and ceilings. Using the chord method again leads to inconsistent results, since the addition of the chords of all sections of the floor would be longer than the distance between the back edges of the side walls. Thus, in order to draw the orthogonal divisions of the floor, Hernán Ruiz goes back to more or less 'orthodox' procedures. First, he extends the side edges of the floor, until they meet at a vanishing point. It is important to stress that the lines meeting at this position are orthogonals and therefore such point is not the vanishing point of oblique lines explained by Guidobaldo del Monte, whose books were not to be published until 1600.[25] Hernán Ruiz repeats the same operation for the edge orthogonals of the intermediate ceiling. But these lines are also the upper and lower edges of the side walls, so the orthogonals for each wall meet at two additional points, giving as a result four

22 Cabezas Gelabert, *Tratadistas y tratados españoles*; Cabezas Gelabert 'La "perspectiva angular"'; Gentil Baldrich, 'El libro de perspectiva', in Ruiz, *Libro de arquitectura*, ed. by Jiménez Martín, pp. 219–20.

23 Amédée-François Frézier, *La théorie et la pratique de la coupe des pierres et des bois [...] ou traité de stéréotomie [...]* (Strasbourg and Paris: Jean Daniel Doulsseker-L. H. Guerin, 1737–39), II, Plate 28.

24 Martin Kemp, *The Science of Art* (New Haven: Yale University Press, 1990), pp. 243–49. Strictly speaking, it is impossible to draw a spherical perspective on a piece of paper, since the spherical surface is non-developable.

25 Guidobaldo del Monte, *Guidiubaldi e' marchionibus montis perspectivae libri sex* (Pesaro: Hieronymum Concordiam, 1600).

vanishing points, and not the Albertian principal point. He applies a similar method to the edge orthogonals of the upper ceiling; they meet at the centre of the entire back wall, but this seems to be a coincidence that surprised Hernán Ruiz himself.

After this, Hernán Ruiz can draw the orthogonal divisions of the floor and the intermediate and upper ceiling, meeting at their corresponding vanishing point, just as Alberti and Piero did with the principal point. Next, he does the same with the orthogonals of the side walls in the lower storey of his construction without any problem. However, things are not so easy with the orthogonals of the upper storey, which should not meet at the same point as those in the lower storey, but rather at two new vanishing points. Since this solution would have led to inconsistent results, Hernán Ruiz tampers with the vanishing point. The outcome however is not satisfactory; the lower orthogonal seems to be drawn at random. As for the transversals, Hernán Ruiz goes back to mainstream methods. Just as in an Albertian construction, he draws both diagonals of the floor and draws the transversals passing through the meeting point of each orthogonal with a diagonal.

Similar techniques seem also to have been used in the scheme for a square pavilion in folio 54r, which shows four compass marks in the otherwise blank space in the middle of the pavilion (Figure 11). Roughly, these points correspond to the vanishing points of the orthogonal cornices of the pavilion. However, here the problem is more complex. Extrapolating the construction of the square room, the orthogonals in the upper left cornice should pass through the upper vanishing point, and also through the left vanishing point; had Hernán Ruiz followed this procedure consistently, all the orthogonals in the left cornice should be parallel, so he seems to have tampered with some orthogonals that pass through both vanishing points.

Experimental Schemes and Drawings: The Octagonal Pavilion

In contrast with the scheme for a square-plan, double-storey construction, other perspectival materials in the manuscript, such as the series of schemes and drawings of an octagonal pavilion, which also feature an unorthodox perspective with multiple 'principal points', have attracted less attention.[26] Here, Ruiz included no less than four different drawings: a scheme with vanishing lines; another scheme with construction lines and some detail, but without visible vanishing lines or shadows; and two almost identical shadowed drawings, without construction lines.[27]

The scheme in folio 58r, the most abstract of the set (Figures 12 and 13), does not include a construction for the computation of arc chords, as in the square rooms. Besides, the extensions of the sides of the upper and lower octagons end at four vanishing points and do not cross at them.[28] This suggests that, in this case, the vanishing points were

26 Of course, such multiple points do not correspond to the concept of principal point in mainstream theorists. However, in order to make myself understood, I will use the term 'principal point', within quotation marks, when referring to Ruiz's method for placing orthogonals. I will also use 'tiers point' and 'distance point' in similar situations.

27 Ruiz, *El libro de arquitectura*, fols 52r, 53r, 58r, 112r.

28 For the sake of exactitude, one of the vanishing lines of the intermediate divisions of the square passes through one of the vanishing points, but this vanishing point does not seem to be placed using the vanishing line.

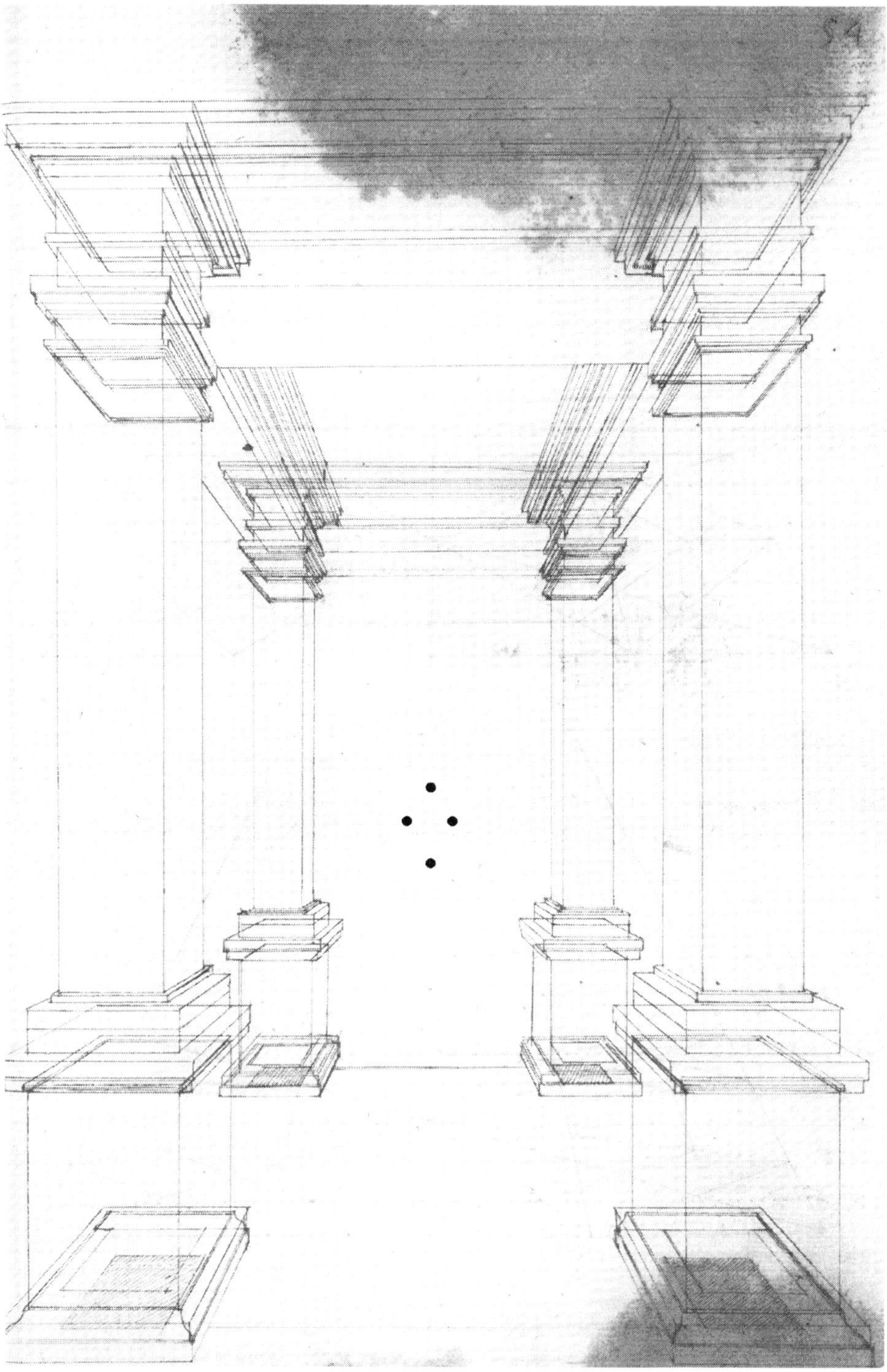

Fig. 11 Hernán Ruiz, *Libro de Arquitectura* (c. 1560), fol. 54^r. Perspective scheme of a square pavillion; compass marks are shown as large dots.

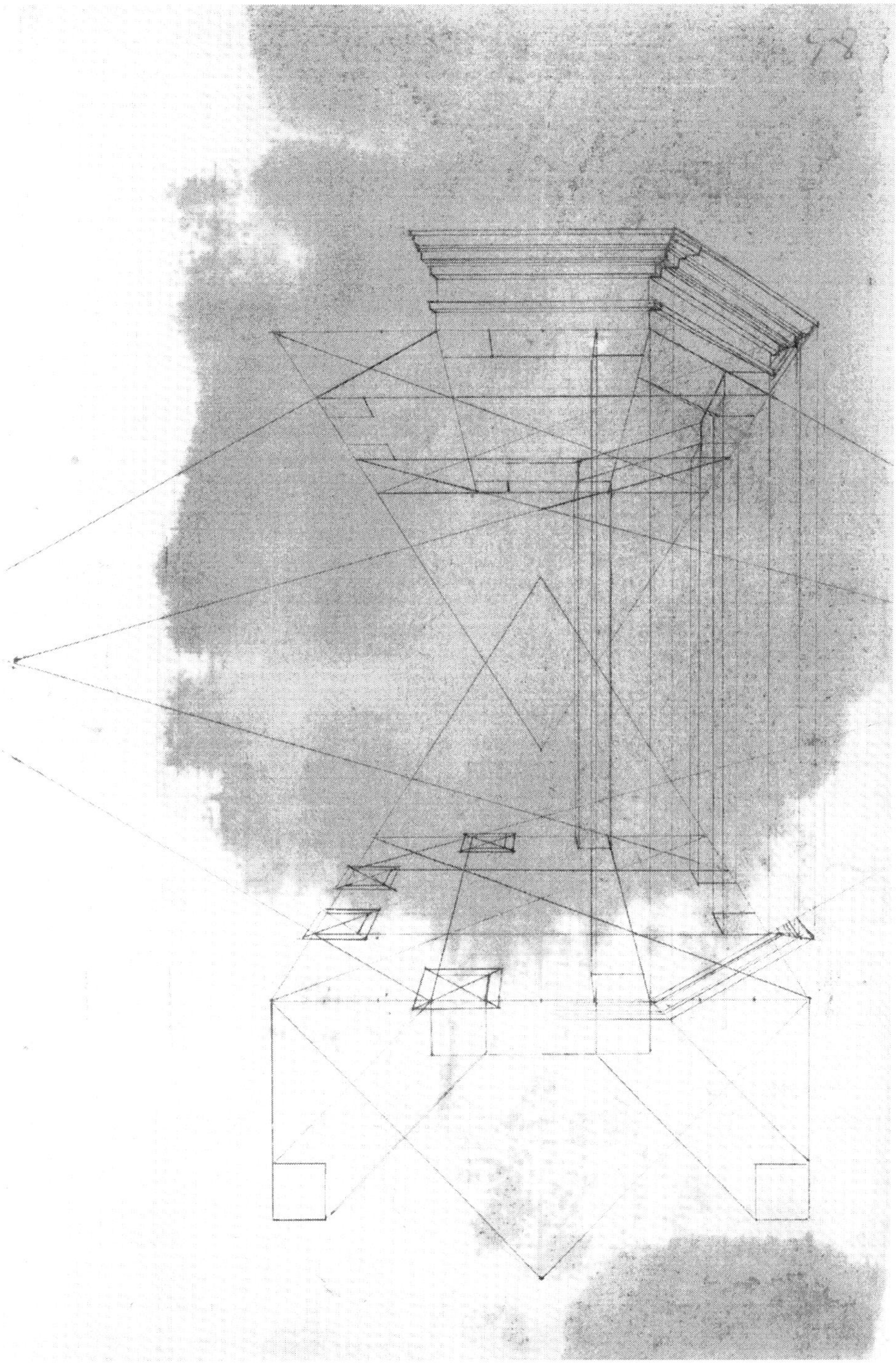

Fig. 12 Hernán Ruiz, *Libro de Arquitectura* (c. 1560), fol. 58r Perspective scheme of an octagonal pavillion.

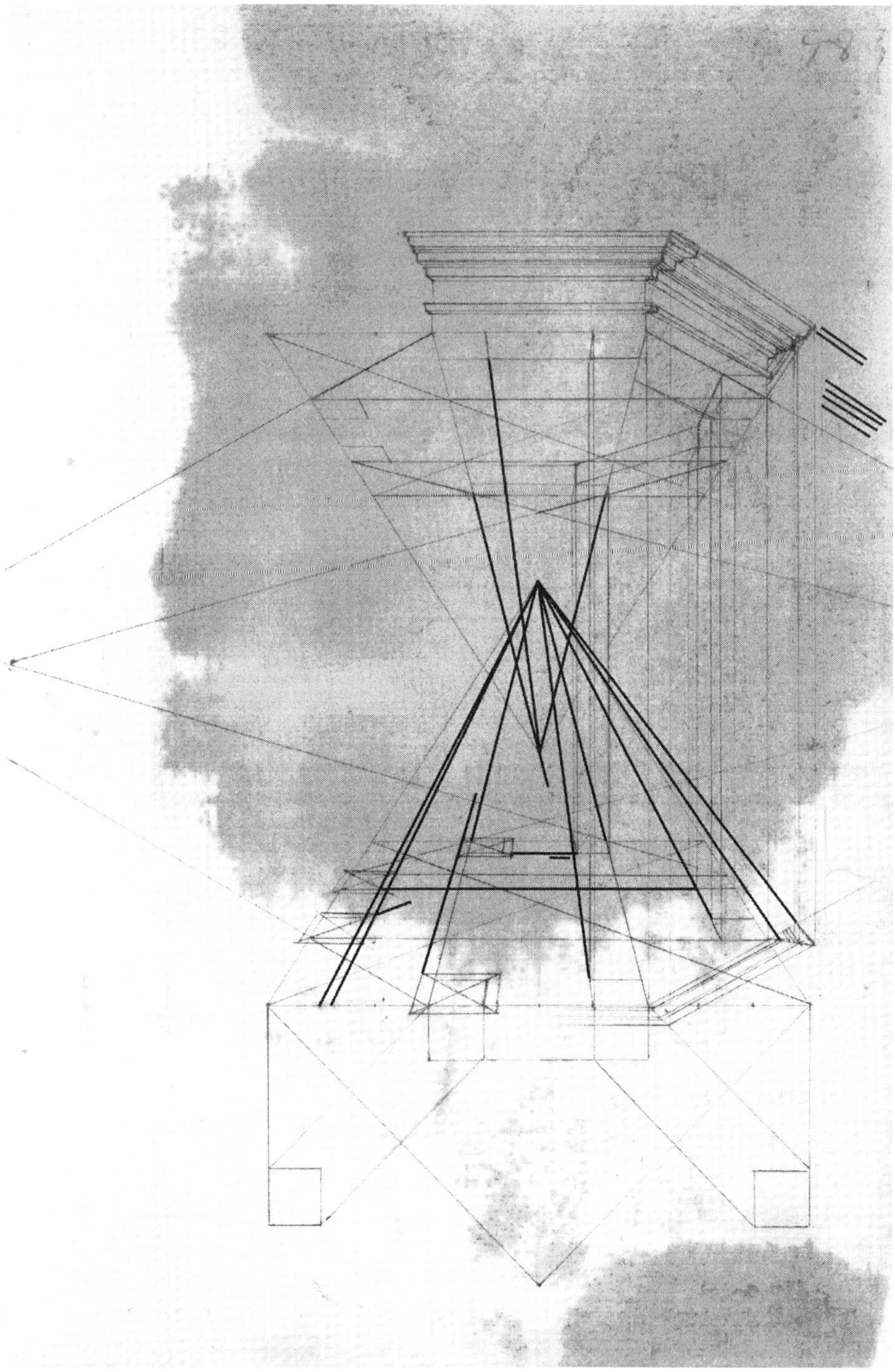

Fig. 13 Hernán Ruiz, *Libro de Arquitectura* (*c.* 1560), fol. 58[r]. Perspective scheme of an octagonal pavillion. Hidden auxiliary lines are shown as thick lines.

placed beforehand, in contrast with the square hall. We should notice also that the smaller diagonals that correspond to the sides of the octagon in the floor are extended up to the edges of the sheet; Gentil pointed out that some of these lines were drawn in other sheets, in particular folio 52.[29] Quite probably, these sheets were placed below folio 58^{r} before binding the sketchbook, so that the vanishing lines could be extended until they reached a point that resembles the 'tiers point' of other theorists. These points were also used as vanishing points of the square enclosing the octagon, but not as distance points. They were placed at roughly the same height as the 'principal point', so to speak, that corresponds to the floor. The main diagonals do not seem to be extended, neither with lines in ink nor with a dry point, but we can protract them in theory; taking into account the precision of the instruments used by Hernán Ruiz and allowing for a reasonable degree of error, they also meet the smaller diagonals at roughly the same height. The same thing happens at the cornice level; thus, we have a complete set of a 'principal point' and two 'tiers points', for the floor and another set for the cornice; all in all, four vanishing points and two principal points.

As stated, these 'principal' points and, possibly, the 'tiers' points seem to be placed at the start of the construction. Also, the other two corners of the central rhombus seem to play no particular role in this construction.[30] All this sets the method of the octagonal pavilion apart from the procedure used at the square hall. To test this, I have tried to apply the method of the square hall to the octagonal pavilion, but the results were inconsistent. In order for the chord of the arc to subtend the angle between the visual rays that pass through the front and back edges of the cornice, the viewpoint should be placed at a particular location in the elevation. But the placement of the viewpoint at the corresponding location in plan would have led to a chord that is clearly greater than this distance. In order for this chord to be equal to the distance, the viewpoint should have been placed in another different location in plan. That would lead to an inconsistency with the position of the viewpoint in elevation; in other words, the coordination between both orthogonal projections would be broken.

Thus, we must conclude that in the octagonal pavilion Hernán Ruiz used a different method for the placement of vanishing points than that used in the square rooms, although it is not easy to describe its rationale. Leaving this aside, most of the construction is more or less straightforward. Ruiz starts by drawing the plan of the pavilion in true shape, including the pillars, the sides of the octagon, and the main diagonals. Next, he draws the enclosing rectangle of the elevation and an axis of the whole composition; this allows him to place both 'principal points' along the axis, apparently at random heights. In the next step, he draws the edge orthogonals of the floor, meeting at the upper 'principal point'. After this, he may have drawn the main diagonals. This cannot be taken for granted, since

---

29 Gentil Baldrich, 'El libro de perspectiva', in Ruiz, *Libro de arquitectura*, ed. by Jiménez Martín, p. 222.

30 Gentil Baldrich, 'El libro de perspectiva', in Ruiz, *Libro de arquitectura*, ed. by Jiménez Martín, p. 222. Gentil Baldrich points out that other rhombuses can be formed between the two sets of 'tiers points' and that actually part of the rhombus corresponding to the left pair of 'tiers points' is visible in the verso of folio 52. See Gentil Baldrich, 'El libro de perspectiva', in Ruiz, *Libro de arquitectura*, ed. by Jiménez Martín, p. 222. However, the left and right corners of this rhombus play no role in the perspectival construction, and in fact the vanishing lines are not extended beyond the higher and lower corners of the rhombus.

the diagonals are not extended to the 'tiers points'. However, this is the simplest way to explain a remarkable fact: the back edge of the floor meets the side orthogonals precisely at their intersection with the main diagonals, and in turn, the intersection of the main diagonals and the side diagonals is located at the same height than the 'principal point', as we have seen. Anyhow, I should make clear that these horizontal lines are not present in the manuscript, neither in ink nor in dry point. In any case, we may assume that, in the next step, Hernán Ruiz is able to place the back edge of the floor. All this would have given him an outline of the pavilion. Drawing the orthogonal divisions, starting from the plan, would have been the next part of the process, repeating the same operation for the cornice. As a final step, he would have constructed the sides of the octagon, using side diagonals that meet the main diagonals at the 'distance points' and placed the transversals, using either the main diagonals or the sides of the octagon.

Other drawings of the octagonal pavilion are also quite interesting. For example, the intermediate scheme in folio 52^{r}, not as abstract as the scheme in folio 58^{r}, although not as finished as the drawing in folio 53^{r}, shows some hidden lines and a pair of compass marks. One of the compass marks almost coincides with the vanishing point of the lines in the floor, but these lines meet some millimetres before this point is actually reached; besides, the vanishing lines in the cornices do not meet at the other compass mark. All this suggests that Hernán Ruiz had not yet reached a finished system but was experimenting with different solutions. In fact, the finished drawings seem to have used a single vanishing point for the orthogonals, although this cannot be completely verified as I have been unable to locate hidden lines in these drawings, as in the majority of finished sketches.

Other Experimental Diagrams and Drawings

Two other drawings, representing a square pavilion in oblique perspective and a rectangular one in frontal perspective, seem to belong to the same group of the square-plan rooms and the octagonal pavilion, since they are intermingled with the square room and the octagonal pavilion and show the same graphical treatment as in this group's shadowed drawings.[31] At first sight, the rectangular pavilion uses two different vanishing points, one on top of the other, as we have seen for the octagonal pavilion (Figure 14).[32] Later on, a different draughtsman felt the need to add a scheme for the rectangular pavilion; he included a shadowed drawing, without ornamental detail, of the pavilion, in frontal one-point perspective.[33]

As for the square pavilion in oblique perspective, it seems to have been constructed using two 'tiers points' at both sides of the composition, as is the case in some diagrams by Jean Pélerin, known as Viator, or the well-known sinopia used by Paolo Uccello as a

31 Ruiz, *El libro de arquitectura*, fols 57^{r} and 56^{r}, respectively.

32 In the present state of the *El libro de arquitectura*, I have been unable to locate hidden auxiliary lines or compass marks in the drawing for the rectangular pavilion in fol. 56^{r}. This is not unusual however; hidden lines are also absent in the shadowed drawings in fols 87^{r}, 51^{v}, 55^{v}, and 53^{r}. All these drawings seem to be traced from a preparatory scheme.

33 Fol. 115^{r} in the *Libro de arquitectura* is generally considered an addition. See Navascués Palacio, 'Estudio', in Ruiz, *El libro de arquitectura de Hernán Ruiz el Joven*, ed. by Navascués Palacio, p. 54; Morales, *Hernán Ruiz el Joven*, 144.

Fig. 14 Hernán Ruiz, *Libro de Arquitectura* (*c.* 1560), fol. 56^{r}. Perspective of a rectangular pavillion.

preliminary step for the *Nativity* at the San Martino alla Scala hospital, now in the Uffizi (Figure 15).[34] However, we should take into account that the vanishing points are outside

34 Jean Pélerin (known as Viator), *De artificiali perspectiva* (Toul, 1505), fols V^{v}, VIII^{v}. Since the original book does not include folio numbers, I have taken them from the digital edition in *Museo Galileo, Bibliotheca Perspectivae*: http://bibdig.museogalileo.it/ [accessed 15-05-2019]. About the sinopia and its different interpretations, see Kemp, *The Science of Art*, pp. 37–38, and Roccasecca, 'Sebastiano Serlio', p. 66.

Fig. 15 Hernán Ruiz, *Libro de Arquitectura* (c. 1560), fol. 57[r]. Perspective drawing of a square pavillion in oblique perspective. Hidden auxiliary lines are shown as thick lines.

the sheet; as a result, the scant auxiliary hidden lines present in the sheet are rather short and the concurrence of the vanishing lines in two points is not exact.

Perspective in Architectural Drawings

Besides these experimental or didactical perspective drawings, the *Libro de arquitectura* includes a fair number of drawings of the church of the Hospital de la Sangre in Seville, designed and built by Hernán Ruiz, including no less than five different solutions, most of them including cross-sections, longitudinal sections, and elevations. Probably many of these drawings were working drawings for Hernán Ruiz's personal use. However, at least one longitudinal section seems to be prepared in order to present different alternatives to the client, since it includes a small strip of paper glued onto the main sheet; lifting the strip, the architect can show the client a different option drawn below the strip.[35] Maybe they are just style exercises or, in Navascués's words, 'speculative' solutions. Alternatively, Ruiz may have used them as 'pattern books', showing them to clients wishing to add a doorway or a crossing or facade to an existing church.[36] In any case, it is rather striking that most of these operative drawings depart completely from the techniques used in the theoretical schemes we have seen in the preceding sections. Generally speaking, facades and doorways are drawn in mainstream perspective, although on some occasions the orthogonals do not seem to converge in a single vanishing point. By contrast, in some cross-sections and longitudinal sections, Hernán Ruiz seems to have thought that mixing orthogonal projection and perspective was acceptable; he does so in a fair number of sections of the Hospital de la Sangre in Seville, which include foreshortened cornices (Figure 16). Such practices, however idiosyncratic they may seem at first sight, mirror the procedures of some drawings in Serlio's *Fifth Book*, published in 1547.[37]

The Sources of Hernán Ruiz's Perspective

As we have seen, the mainstream perspective methods used in the theoretical diagrams for central plans, polygons, and the flat ceiling room, based on a single vanishing point and the use of the diagonals of an enclosing square, were taken directly by Hernán Ruiz from Serlio. The sections for the Hospital de la Sangre, mixing orthogonal projections with foreshortening are akin to a number of Serlio's woodcuts. By contrast, Hernán Ruiz experimented with one, or maybe more than one, 'angular', or rather 'chordal' perspective

35 Ruiz, *El libro de arquitectura*, fol. 105.

36 About these drawings, see Navascués Palacio, 'Estudio', in Ruiz, *El libro de arquitectura de Hernán Ruiz el Joven*, ed. by Navascués Palacio, p. 28. Alfonso Jiménez Martín includes a thorough study of these drawings in 'El libro de las portadas', in Ruiz, *Libro de arquitectura*, ed. by Jiménez Martín, pp. 237–56.

37 Cf. Ruiz, *El libro de arquitectura*, fols 92^{r}, 94^{r}, 97^{r}, 98^{v}, 105^{r}, 107^{r}, 109^{r}, and the upper drawing in fol. 101^{r}, with Sebastiano Serlio, *Quinto libro d'architettura, di Sabastiano Serlio, [...]* (Paris: M. de Vascosan, 1547), fols 8^{v}, 10^{v}, 16^{r}, 28^{r}.

Fig. 16 Hernán Ruiz, *Libro de Arquitectura* (*c.* 1560), fol. 84v. Cross section of the church of the Hospital de la Sangre. [See colour plate 21]

methods that bear no direct resemblance to Serlio or other mainstream perspectival sources of the fifteenth or sixteenth centuries. Thus, we must ask ourselves what were the other sources influencing Hernán Ruiz's perspectival methods?

First of all, we may turn our eyes to Vitruvius, since Hernán Ruiz included a translation of his *First Book* in his manuscript. However, if we look for the well-known passage about ichnography, orthography, and scenography, we get the impression that Hernán Ruiz misunderstood Vitruvius completely. He translates Vitruvius's text with these words: 'The differences and manners of dispositions are these: ichnography, orthography, scenography. Ichnography is a moderate use of the compass and the ruler which allow taking the dimensions of dry sun forms'. The rendering of the sentence about scenography, the so-called Vitruvian perspective, is also rather awkward: 'Scenography is a shadowing and darkness of the front and the sides [...] that go back and the answer of all lines to the centre of the compass'.[38]

Thus, Vitruvius cannot be taken as a source of Hernán Ruiz's perspectival methods. Next, we can turn our eyes to Serlio and Dürer. As we have seen, Ruiz copied entire drawings from Serlio, including such details as oval vases.[39] He also took from Dürer his method for the construction of conic sections.[40] However, in contrast with the Vitruvius translation, he understood Durer's method quite clearly, so he managed to put together a fairly accurate section with symmetry around both axes, where Dürer presents an egg-like figure without symmetry around the horizontal axis.

In connection with Ruiz's perspective, Gentil has mentioned the well-known drawings by both Dürer and Serlio that have on some occasions been presented as proof of the interest

38 Ruiz, *Libro de arquitectura*, fol. 4: 'Las diferençias y maneras de las dispusiçiones son éstas: ynografia, ortografía, esçinografía. Ynografía es vn uso continente del conpás y de la regla de la qual se torna en las discusiones de las formas secas del sol. Ortografía es vna ymajen lebantada de frente, vna figura vn poco pintada en las razones de la obra que a de ser. Ecinografía es vn sonbramiento y escuridad de la frente y de los lados que se apartan y es respuesta de todas las líneas al centro del conpás'. Taken from Carmen Álvarez Márquez, 'Transcripción de los textos', in Ruiz, *Libro de arquitectura*, ed. by Jiménez Martín, p. 48. Cf. with Vitruvius, *De architectura libri decem*, Book 1, Chapter II (Ed. Auguste Choisy, *Vitruve* (Paris: Lahure, 1909), II, 18): 'Species dispositionis, quae graece dicuntur ideai, sunt haec: ichnographia; orthographia; scenographia. Ichnographia es circini regulaeque nodice continens usus equa capiuntur formarum in solis arearum descriptiones; orthographia autem est erecta frontis imago modiceque picta rationibus operis futuri figura; item scenographia est frontis et laterum abscenentium adumbratio, ad circini centrum omnium linearum responsus'. The error in the first passage seems to derive from a confusion in the interpretation of *solis* and *arearum*. Vitruvius uses *solis* as the plural ablative of *solus*, ground, but Ruiz mistakes it as the genitive of *sol*, sun. As a result, he mistranslates *arearum* the plural genitive of 'area', that is, the piece of land where a building is to stand, as 'dry'; he is probably thinking that *arearum* is a form of *aridus*. In fact, in another passage, *El libro de arquitectura*, fol. 13^r, he does not understand that the 'area' stands strictly behind a building and describes it as a plane surface prepared to make full-scale drawings, as was usual in the Middle Ages and the Early Modern period. In the passage about perspective, Ruiz translates *responsus*, a substantive derived from the verb *respondeo*, as 'is an answer'. This is the most usual meaning of *respondeo*, but Ruiz does not notice that Vitruvius is using *responsus* with the meaning of 'agreeing, according, corresponding' or, by extension, 'meeting at'.

39 Cf. Ruiz, *El libro de arquitectura*, fols 37^r–37^v with Serlio, *Il primo libro*, fols 13^v–15^v.

40 Cf. Ruiz, *El libro de arquitectura*, fols 28^r–29^r with Albrecht Dürer, *Underweysung der Messung mit dem Zirckel und Richtscheyt in Linien Ebnen unnd Gantzen Corporen [...]* (Nuremberg: s. n., 1525), fols 16^r–16^v. Since folio numbers in the original book are not systematic, I have taken them from *Museo Galileo, Bibliotheca Perspectivae*: http://bibdig.museogalileo.it/Teca/Viewer?an=000000921065 [accessed 10-01-2019].

in angular perspective in the Renaissance.[41] Hernán Ruiz must have known both, since he takes much material from Serlio's *First Book* and Dürer's *Underweyssung der Messung*. But he is in fact reversing Serlio and Dürer's methods. Both Serlio and Dürer are trying to design a column divided with uneven lines so that all sections show the same angle to the eye. By contrast, Hernán Ruiz is trying to compute the uneven angles — or chords, to be precise — subtended by a number of more or less regular figures. In any case, we must admit that both Serlio and Dürer may have fostered Hernán Ruiz's interest in these matters; and of course, this leads us to Euclid's *Optics*. Ruiz quotes Euclid on a few occasions in his manuscript, but most of these allusions seem traceable to the *Elements*.[42] He must have used either a Latin or Italian edition such as those from Campanus of Novara, Bartolomeo Zamberti or Niccolò Fontana Tartaglia or an indirect source, since the first translation of Euclid to Spanish, covering only the first six books of his *Elements*, was published in 1576, seven years after Hernán Ruiz's death. José María Gentil Baldrich has mentioned the Spanish translation of Euclid's *Optics* by Pedro Ambrosio de Ondériz in connection with Hernán Ruiz; it was published in 1585 under the title of *La perspectiva y especularia de Euclides*. Gentil has stressed also the connections of Ondériz with the *Casa de contratación* of Seville, an administrative corporation that was in charge of the supervision of maps of Spanish territories in the New World.[43]

At first sight, this line of research seems promising, since Hernán Ruiz actually worked for the *Casa de contratación*, preparing a survey of its premises.[44] However, recent studies about Spanish cartography have downplayed Ondériz's connections with Seville. In fact, he worked for the *Consejo de Indias*, another political body based in Madrid, while lecturing in the *Academia de Matemáticas* under the direction of Juan de Herrera, the architect of the Escorial, also in Madrid. It seems that Ondériz went to Seville in order to supervise the work of the mapmakers of Seville, although he died before he could fulfil his commission.[45]

---

41 Dürer, *Underweysung*, fols 43^{v}, 56^{v}. Taken from *Museo Galileo. Bibliotheca Perspectivae*; Serlio, *Il primo libro*, fol. 10. See also Gentil Baldrich, 'El libro de perspectiva', in Ruiz, *Libro de arquitectura*, ed. by Jiménez Martín, p. 229.

42 Ruiz, *El libro de arquitectura*, explains in fol. 15^{r} that solids are long, wide, and deep, following *Elements*, Book XI, Def. 1 (Cf. edition by I. L. Heiberg, *Euclidis opera omnia* (Leipzig: Teubner, 1885), IV, p. 3). In fol. 15^{v}, he mentions Euclid after explaining how to construct a square with the same area as a rectangle measuring twelve by three feet; maybe he is referring to the problems in equivalent areas tackled by the *Elements*, Book II, Propositions 1–8 (Cf. edition by I. L. Heiberg, *Euclidis opera omnia* (Leipzig: Teubner, 1883), I, pp. 119–43). In fol. 19^{r}, Ruiz presents as a conclusion from Euclid that in a figure with centre, any line passing through the centre of the figure divides it in two equal parts. Ruiz de la Rosa remarks that such a proposition is nowhere to be found in *Elements*: see Ruiz de la Rosa, 'El libro de geometría', in Ruiz, *Libro de arquitectura*, ed. by Jiménez Martín, p. 115. It is also missing in *Optics*, I should add. In fols 20^{v}, 25^{r}, and 36^{v}, Ruiz explains that the result of detracting equal figures from equal figures are equal, following *Elements*, Book 1, Axiom 3 (Heiberg, *Euclidis opera omnia*, I, p. 11). In fols 33–34, he quotes *Elements*, Book 6, Proposition 1 (I. L. Heiberg, *Euclidis opera omnia* (Leipzig: Teubner, 1884), II, pp. 73–77). In fol. 70^{v}, he defines a dodecahedron as a figure with twelve faces, mentioning the twenty-eighth definition of Euclid; he is referring to *Elements*, Book 11, Def. 28. (Cf. edition by I. L. Heiberg, *Euclidis opera omnia* (Leipzig: Teubner, 1885), IV, p. 9).

43 Euclid, *La perspectiva y especularia de Euclides [...] traduzidas en vulgar Castellano [...] por Pedro Ambrosio Ondériz [...]* (Madrid: Alonso Gómez, 1585). See also Gentil Baldrich, 'El libro de perspectiva', in Ruiz, *Libro de arquitectura*, ed. by Jiménez Martín, pp. 230–33.

44 Morales, *Hernán Ruiz 'El Joven'*, 120–21.

45 Alison D. Sandman, 'Spanish Nautical Cartography in the Renaissance', in *The History of Cartography: Cartography in the European Renaissance*, ed. by David Woodward (Chicago: University of Chicago Press, 2007), III, Part 1, pp. 1124–27, 1141.

Thus, the 1585 translation of Ondériz may not be taken as proof of Hernán Ruiz's or the Sevillian cosmographers' interest in Euclid's *Optics* in the 1550s and 1560s.[46]

Rather surprisingly, the most tangible piece of evidence that connects Ruiz with Euclid's 'angle axiom' is, again, Dürer's *Underweysung.* As Jeanne Peiffer has shown, it presents Euclid's axiom in a peculiar fashion, stating that 'Whatever is seen [in the field] enclosed by the two forked lines ab and ac and touches them, be it near or far, vertical, oblique, or curved, it appears to the eye in the same size'.[47] Moreover, Ruiz makes an extensive interpretation of this paragraph; he seems to understand that since all segments seen under the same angle appear equal to the eye, they should be represented in the perspective by the length of a chord that subtends this arc. Of course, such conception leads him quickly to unavoidable contradictions. Firstly, an angle can be subtended by different chords, as Dürer's drawing clearly shows; secondly and most importantly, when dividing an angle into three portions, as Ruiz does with the side and back faces of the square rooms, the sum of the lengths of the chords of the portions can be greater than the chord of the original angle. Probably, these contradictions led him to experiment with another different method in the octagonal pavilion and, in the end, to eschew these idiosyncratic procedures in his architectural drawings.[48]

Conclusion

All in all, Hernán Ruiz's sketchbook gives us a fairly complete picture of the perspective practices of a sixteenth-century architect in continental Europe: we have seen him copying treatises as an exercise, in particular Serlio; experimenting with one, or maybe more than

46 Other works cannot be accepted as sources of Hernán Ruiz's perspective, since they were written or published after his death. In particular, Cabezas Gelabert (in 'La "perspectiva angular"', pp. 172–75) mentions two chapters of the manuscript of Simón García, suggesting that it was written by Rodrigo Gil de Hontañón. However, Gentil Baldrich (in Gentil Baldrich, 'El libro de perspectiva', in Ruiz, *Libro de arquitectura*, ed. by Jiménez Martín, pp. 226–28) pointed out that most scholars believe that these chapters were written by Simón García in the late seventeenth century and not by Rodrigo Gil. In any case, Cabezas Gelabert's comments are interesting, since they show the persistence of 'angular' perspective methods in the seventeenth century. See also Cabezas Gelabert, *Tratadistas y tratados españoles*, who also pointed out the similarity between Ruiz, *Libro de arquitectura*, fol. 57^r and Jacques Androuet du Cerceau, *Leçons de perspective positive* (Paris: Mamert Pattison, 1576), Lesson XXVIII, implying the existence of a common source since the *Leçons* were published after Hernán Ruiz's death.

47 'Wus im gesicht zwischen den zwenen gabellinien a. b. c. beschlossen wirt unnd sie an ruret essen nahent oder ferz / ausrecht uber ort oder frum / dz schennt dem aug .a. alles in einer grosse'. Dürer, *Underweysung*, fol. 86^r. Folio number taken from *Museo Galileo. Bibliotheca Perspectivae.* See also Jeanne Peiffer, 'Projections Embodied in Technical Drawings', in *Picturing Machines*, ed. by Wolfgang Lefèvre (Cambridge, MA: The MIT Press, 2004), pp. 245–75.

48 In any case, some exchanges of knowledge between cosmography and stonecutting seem to have taken place during the sixteenth century in Spain. See José Antonio Ruiz de la Rosa and Juan Clemente Rodríguez Estévez, '"Capilla redonda en vuelta redonda" (sic): Aplicación de una propuesta teórica renacentista para la catedral de Sevilla', in *IX Congreso internacional expresión gráfica arquitectónica. Re-visión: enfoques en docencia e investigación* (La Coruña: Universidad de A Coruña, 2002), pp. 509–16; Francisco Pinto Puerto, *Las esferas de piedra. Sevilla como lugar de encuentro entre arte y ciencia en el Renacimiento* (Seville: Diputación, 2002); Francisco Pinto Puerto, 'Transformaciones. De la línea a la superficie', in *Actas del tercer congreso nacional de historia de la construcción*, ed. by Amparo Graciani García (Madrid: Instituto Juan de Herrera, 2000), pp. 815–26.

one, unorthodox perspective methods, and finally applying other, different procedures to actual architectural drawings.

He does not seem to have reached a completely consistent solution for his idiosyncratic perspective methods. One of these methods was based in chord lengths, and thus has some points of contact with Euclid's angular axiom, but in practice Ruiz used orthographic projections of the chords, rather than the actual measure of the chords or angles in space. His main sources seem to be Serlio's second book and, to a lesser extent, the fifth; Dürer's *Underweysung* and the mastery of the practical projective geometry of the stonecutting tradition; we should take into account that Dürer mentions stonemasons twice as masters of double orthographic projection.[49] Although there is no evidence of the direct use of a Latin translation of Euclid's *Optics*, it cannot be excluded. As for Vitruvius, although Hernán Ruiz includes a translation of the whole First Book, he does not seem to have understood the well-known passage about ichnography, orthography, and scenography in any way that could foster the idea of an angular perspective.

Most remarkably, Hernán Ruiz seems to have thought his systems, the more or less orthodox rules of perspective explained by Serlio and even orthographic projection were not mutually exclusive. The key to understanding such an eclectic stance lies in the heading of the main section on theoretical perspective drawings, which states: 'comiençan algunas reglas de prespetiba', that is, 'Here begin some rules [or rather, methods] of perspective'.[50] What this clearly shows is that the notion of one particular kind of perspective being absolutely correct was foreign to him; as James Elkins has pointed out, this idea does not seem to have circulated to any great degree before the end of the sixteenth century.

Bibliography

Manuscript and Archival Sources

Biblioteca de la Escuela de Arquitectura de la Universidad Politécnica de Madrid, Madrid, Hernán Ruiz II, *Libro de arquitectura* (*c.* 1550).

Biblioteca de la Escuela de Arquitectura de la Universidad Politécnica de Madrid, Madrid, Alonso de Vandelvira, 'Libro de trazas de cortes de piedras' (*c.* 1585), fols 74^{v}–77^{r}.

Primary Sources

Cerceau, Jacques Androuet du, *Leçons de perspective positive* (Paris: Mamert Pattison, 1576).

Dürer, Albrecht, *Underweysung der Messung mit dem Zirckel und Richtscheyt in Linien Ebnen unnd Gantzen Corporen [...]* (Nuremberg: s. n., 1525).

Euclid, *La perspectiva y especularia de Euclides [...] traduzidas en vulgar Castellano [...] por Pedro Ambrosio Ondériz [...]* (Madrid: Alonso Gómez, 1585).

Euclid, *Euclidis opera omnia*, ed. by I. L. Heiberg (Leipzig: Teubner, 1883–85).

49 Dürer, *Underweysung*, fols 15^{r}, 84^{v}, taken from *Museo Galileo. Bibliotheca Perspectivae*. See also Wolfgang Lefèvre, 'The Emergence of Combined Orthographic Projections', in *Picturing Machines*, ed. by Lefèvre, pp. 209–44.

50 Ruiz, *El libro de arquitectura*, fol. 51^{r}.

Francesca, Piero della, *De prospectiva pingendi*, ed. by G. Nicco-Fasola (Florence: Sansoni, 1984; 1st edn 1942).
Frézier, Amédée-François, *La théorie et la pratique de la coupe des pierres et des bois [...] ou traité de stéréotomie [...]* (Strasbourg and Paris: Jean Daniel Doulsseker-L. H. Guerin, 1737–39).
Monte, Guidobaldo del, *Guidiubaldi e' marchionibus montis perspectivae libri sex* (Pesaro: Hieronymum Concordiam, 1600).
Pélerin, Jean, (known as Viator), *De artificiali perspectiva* (Toul, 1505).
Ruiz II, Hernán, *El libro de arquitectura de Hernán Ruiz el Joven*, ed. by Pedro Navascués Palacio (Madrid: Escuela Técnica Superior de Arquitectura, 1974).
Ruiz II, Hernán, *Libro de arquitectura*, ed. by Alfonso Jiménez Martín et al. (Seville: Fundación Sevillana de Electricidad, 1998).
Serlio, Sebastiano, *Il primo libro di architettura* (Paris: Iehan Barbé, 1545).
Serlio, Sebastiano, *Quinto libro d'architettura, di Sabastiano Serlio, [...]* (Paris: M. de Vascosan, 1547).
Vandelvira, Alonso de, *Tratado de arquitectura*, ed. by Geneviève Barbé-Coquelin de Lisle (Albacete: Caja Provincial de Ahorros, 1977).
Vitruvius, *De architectura libri decem*, ed. by Auguste Choisy (Paris: Lahure, 1909).

Secondary Works

Cabezas Gelabert, Lino, 'La "perspectiva angular" y la introducción de la perspectiva artística en la España del siglo XVI', *D'art*, 15 (1989), 167–79.
Cabezas Gelabert, Lino, *Tratadistas y tratados españoles de perspectiva desde sus orígenes hasta la geometría descriptiva de G. Monge, 1526-1803* (Barcelona: Universitat de Barcelona, 1985).
Elkins, James, 'Renaissance Perspectives', *Journal of the History of Ideas*, 53 (2) (1992), 209–30.
Field, J. V., *Piero della Francesca: A Mathematician's Art* (New Haven: Yale University Press, 2004).
Gentil Baldrich, José María, 'La traza oval y la Sala Capitular de la catedral de Sevilla. Una aproximación geométrica', in *Quatro edificios sevillanos* (Sevilla: Colegio de Arquitectos, 1996), pp. 73–147.
Gentil Baldrich, José María, 'El libro de perspectiva', in Hernán Ruiz II, *Libro de arquitectura*, ed. by Alfonso Jiménez Martín et al. (Seville: Fundación Sevillana de Electricidad, 1998), pp. 215–34.
Heydenreich, Ludwig, and Wolfgang Lotz, *Architecture in Italy 1500–1600* (Harmondsworth: Penguin Books, 1974).
Jiménez Martín, Alfonso, 'Anatomía del manuscrito', in Hernán Ruiz II, *Libro de arquitectura*, ed. by Alfonso Jiménez Martín et al. (Seville: Fundación Sevillana de Electricidad, 1998), pp. 23–41.
Jiménez Martín, Alfonso, 'Contexto de la presente edición', in Hernán Ruiz II, *Libro de arquitectura*, ed. by Alfonso Jiménez Martín et al. (Seville: Fundación Sevillana de Electricidad, 1998), pp. 16–18.
Jiménez Martín, Alfonso, 'El libro de las portadas', in Hernán Ruiz II, *Libro de arquitectura*, ed. by Alfonso Jiménez Martín et al. (Seville: Fundación Sevillana de Electricidad, 1998), pp. 237–56.
Kemp, Martin, *The Science of Art* (New Haven: Yale University Press, 1990).

Lotz, Wolfgang, 'Die Ovalen Kirchenraume des Cinquecento', *Römische Jahrbuch für Kunstgeschichte*, 6 (1955), 7–99.

Lotz, Wolfgang, *Architecture in Italy 1500–1600* (New Haven: Yale University Press, 1995).

Morales, Alfredo J., *Hernán Ruiz 'El Joven'* (Madrid: Akal, 1996).

Navascués Palacio, Pedro, 'Estudio', in Hernán Ruiz el Joven, *El libro de arquitectura de Hernán Ruiz el Joven*, ed. by Navascués Palacio (Madrid: Escuela de Arquitectura de Madrid, 1974).

Peiffer, Jeanne, 'Projections Embodied in Technical Drawings. Dürer and his Followers', in *Picturing machines, 1400–1700*, ed. by Wolfgang Lefèvre (Cambridge, MA: The MIT Press, 2004), pp. 245–75.

Pinto Puerto, Francisco, 'El libro de cantería', in Hernán Ruiz II, *Libro de arquitectura*, ed. by Alfonso Jiménez Martín et al. (Seville: Fundación Sevillana de Electricidad, 1998), pp. 199–214.

Pinto Puerto, Francisco, 'Transformaciones. De la línea a la superficie', in *Actas del tercer congreso nacional de historia de la construcción*, ed. by Amparo Graciani García (Madrid: Instituto Juan de Herrera, 2000), pp. 815–26.

Pinto Puerto, Francisco, *Las esferas de piedra. Sevilla como lugar de encuentro entre arte y ciencia en el Renacimiento* (Seville: Diputación, 2002).

Roccasecca, Pietro, 'Sebastiano Serlio: la pratique de la perspective au service de l'architecte', in *Perspective, projections, projet: techniques de la représentation architecturale*, ed. by Fréderique Lemerle et Mario Carpo (Paris: Monum, 2005), pp. 61–70.

Ruiz de la Rosa, José Antonio, 'El libro de geometría', in Hernán Ruiz II, *Libro de arquitectura*, ed. by Alfonso Jiménez Martín et al. (Seville: Fundación Sevillana de Electricidad, 1998), pp. 97–141.

Ruiz de la Rosa, José Antonio and Juan Clemente Rodríguez Estévez, '"Capilla redonda en vuelta redonda" (sic): Aplicación de una propuesta teórica renacentista para la catedral de Sevilla', in *IX Congreso internacional expresión gráfica arquitectónica. Re-visión: enfoques en docencia e investigación* (La Coruña: Universidad de A Coruña, 2002), pp. 509–16.

Sandman, Alison D., 'Spanish Nautical Cartography in the Renaissance', in *The History of Cartography: Cartography in the European Renaissance*, ed. by David Woodward (Chicago: University of Chicago Press, 2007), pp. 1095–1142.

Taín Guzmán, Miguel, 'The Drawings on Stone in Galicia: Types, Uses and Meanings', in *Proceedings of the First International Congress on Construction History* (Madrid: Instituto Juan de Herrera, 2003), pp. 1887–98.

Taín Guzmán, Miguel, 'Fifteen Unedited Engraved Architectural Drawings Uncovered in Northwest Spain', in *Proceedings of the Second International Congress on Construction History: Construction History Society* (Cambridge: Construction History Society, 2006), pp. 3011–23.

Vargas, Antonio de la Banda y, *El arquitecto andaluz Hernán Ruiz II* (Sevilla: Universidad, 1974).

III

Drawing, Constructing, Painting

FILIPPO CAMEROTA

Masaccio's Elements of Painting

Geometrical Practice in the Trinity *Fresco*

Introduction

The *Trinity* of Masaccio (Figure 1a) has for a long time served as a form of training for historians involved in the study of linear perspective.[1] Not only is it the first painting in which perspective geometry appears to be applied with the evident purpose of 'breaking down the wall', challenging the viewer's ability to distinguish between appearance and reality, but it is also one of the rare works in which the constructive drawing is still visible.[2]

1 See G. Joseph von Kern, 'Das Dreifaltigkeitfresko von Santa Maria Novella. Eine Perspektivisch-Architekturgeschichtliche Studie', *Jahrbuch der Königlich preussischen Kunstsammlungen*, 34 (1913), 36–58; Piero Sanpaolesi, *Brunelleschi* (Milan: Edizioni per il Club del Libro, 1962), pp. 41–53; Horst Woldemar Janson, 'Ground Plan and Elevation in Masaccio's Trinity Fresco', in *Essays in the History of Art Presented to Rudolf Wittkower*, ed. by Douglas Fraser, Howard Hibbard, and Milton J. Lewine (London: Phaidon, 1967), pp. 83–88; Joseph Polzer, 'The Anatomy of Masaccio's Holy Trinity', *Jahrbuch der Berliner Museen*, 13 (1971), 18–59; Eugenio Battisti, *Filippo Brunelleschi* (Milan: Electa, 1976), pp. 106–07; Marshal Neal Myers, 'Observations on the Origins of Renaissance Perspective: Brunelleschi, Masaccio, Petrus Christus' (unpublished doctoral dissertation, Columbia University, 1978), pp. 26–38, 77–114; Edgar Hertlein, *Masaccios Trinität: Kunst, Geschichte und Politik der Frührenaissance in Florenz* (Florence: Olschki, 1979); J. V. Field, Roberto Lunardi, and Thomas B. Settle, 'The Perspective Scheme of Masaccio's Trinity Fresco', *Nuncius*, 4 (2) (1989), 31–118; Florian Huber, 'Das Trinitätfresko von Masaccio und Filippo Brunelleschi in Santa Maria Novella zu Florenz' (unpublished doctoral thesis, Ludwig-Maximilians-Universität, 1990), p. 64; Martin Kemp, *The Science of Art: Optical Themes in Western Art from Brunelleschi to Seurat* (New-Haven: Yale University Press, 1990), pp. 17–21; Jehane R. Kuhn, 'Measured Appearances: Documentation and Design in Early Perspective Drawings', *Journal of Warburg and Courtauld Institutes*, 53 (1990), 114–32; Jean Andrews Aiken, 'The Perspective Construction of Masaccio's Trinity Fresco and Medieval Astronomical Graphics', *Artibus et Historiae*, 16 (31) (1995), 171–87; Volker Hoffmann, 'Masaccios Trinitätfresko: die Perspectivkonstruktion und ihr Entwurfsverfahren', *Mitteilungen des Kunsthistorischen Institutes in Florenz*, 40 (1996), 43–77; J. V. Field, *The Invention of Infinity: Mathematics and Art in the Renaissance* (Oxford: Oxford University Press, 1997), pp. 43–61; Dominique Raynaud, 'Linear Perspective in Masaccio's Trinity Fresco: Demonstration or Self-Persuasion?', *Nuncius*, 18 (1) (2003), 331–44.

2 See Giorgio Vasari, *Le Vite de' più eccellenti pittori scultori e architettori* (Florence, 1568), in *Le Opere di Giorgio Vasari*, ed. by Gaetano Milanesi, 9 vols (Florence: Sansoni, 1973), II, p. 291: 'quello che vi è di bellissimo, oltre alle figure, è una volta a mezza botte, tirata in prospettiva, e spartita in quadri pieni di rosoni, che diminuiscono e scortano così bene, che pare che sia bucato quel muro' ('None the less, what is most beautiful, besides the figures, is the barrel vault drawn in perspective and divided in squares full of rosettes which are so well diminished and foreshortened that the wall appears to have holes in it'). See Giorgio Vasari, *The Lives of the Artists*, trans. and ed. by Julia Conaway Bondanella and Peter Bondanella (Oxford: Oxford University Press, 1991), p. 104.

Filippo Camerota Museo Galileo-Istituto e Museo di Storia della Scienza, Florence, f.camerota@museogalileo.it

Perspective as Practice. Renaissance Cultures of Optics, ed. by Sven Dupré, Turnhout, 2019 (*Techne. Knowledge, Technique, and Material Culture*, 1), p. 335-360

DOI 10.1484/M.*Techne*-EB.5.117732

Fig. 1 (*a*) Masaccio, *Trinity*, c. 1425, fresco, 661,44 x 303,16 cm. (*b*) Survey of the lines traced with a stylus (in yellow) and impressed on the fresh plaster by beating it with a string (in blue). Photogrammetric survey and graphic transcription by Massimo Chimenti. [See colour plate 22]

Masaccio left it impressed on the plaster, leaving us a key to the understanding of his working method. A photogrammetric survey of the lines impressed on the plaster was made during the last restoration, in 2000, and it is the basis on which the present study is carried out (Figure 1b).[3] The purpose of this study is to verify an alternative approach to the construction of the perspective composition, forgetting for a moment the long tradition which attributes to Masaccio the correct use of the so called *costruzione legittima*.[4]

3 The photogrammetric survey was made by Massimo Chimenti during the restoration carried out by the Opificio delle Pietre Dure di Florence. For a technical report of the photogrammetric survey, see Massimo Chimenti, 'Rilievo fotogrammetrico e documentazione informatica', in *La Trinità di Masaccio. Il restauro dell'anno duemila*, ed. by Cristina Danti (Florence: Edifir, 2002), pp. 89–96.

4 The tradition was inaugurated by Giorgio Vasari who first described the rule of Brunelleschi as the intersection of the visual pyramid; see *Le Opere di Giorgio Vasari*, ed. by G. Milanesi, II, p. 332. The term 'costruzione legittima' was used for the first time by critics in Heinrich Ludwig, *Leonardo da Vinci. Das Buch von der Malerei* (Wien: Wilhelm Braumüller, 1882).

I have always been convinced that the rule of Brunelleschi, whatever it was, developed out of practical geometry rather than from the reading of optical texts. Brunelleschi was an expert surveyor, as we know from his campaign of measuring ancient monuments in Rome as a young man, together with Donatello.[5] The techniques for indirect measurement of buildings were based on methodical application of the concept of intersection of visual rays, a concept that was to become fundamentally important for the rules of linear perspective. In measurement operations, the visual rays were intersected by the graduated rod of the measuring instrument or by any other instrument on which the apparent dimensions of the object sighted were marked. This method was called 'measuring by eye' or 'measuring by perspective', that is, by applying the Euclidean optical principles that established the relationships between the eye, the real object and its apparent dimensions. These principles were extensively analysed in the treatises on optics written by medieval philosophers, but were also exemplified in treatises on practical geometry. In Brunelleschi's time, for instance, a treatise entitled *De visu* ('On Vision'), composed by the renowned late-fourteenth-century abachist Grazia de' Castellani, was circulating in Florence. Although the treatise has been lost, a fragment of it has survived, transcribed by another abachist, which illustrates the application of optical principles for measuring distances indirectly.[6] Except for the title, the treatise was written in the vernacular, like all the abachist and practical geometry treatises used in teaching basic mathematics to the mercantile and artisanal classes. From the end of the twelfth century, until the sixteenth century, the abacus schools in Florence were a very important institution. Hence it is reasonable to believe that the rules of practical geometry, and most notably the indirect measurement of distances, informed the method of perspective representation known today as Brunelleschi's 'legitimate construction'.[7] My exploration of the *Trinity* has been made with this belief in mind, and pays particular attention to the kind of geometry eventually available in workshops thanks to the abacus tradition; the kind of geometry on which Leon Battista Alberti wrote in his *Elementa picturae*.[8]

---

5 See Antonio di Tuccio Manetti, *Vita di Filippo Brunelleschi, preceduta da la novella del grasso*, ed. by D. de Robertis (Milan: Il Polifilo, 1976), pp. 67–68.

6 The treatise *De visu* by Grazia de' Castellani is known to us only through an abstract concerning methods for measuring by eye, transcribed by an anonymous fifteenth-century Florentine in *Libro di praticha d'arismetrica*, Biblioteca Apostolica Vaticana, Vatican, Cod. Ottoboniano Latino 3307, fol. 172^{r}. In the final paragraph of the 'quarta distinzione che tratta del modo di misurare chol'ochio, cioè chon strumenti' ('fourth distinction which describes the method of measuring by eye, that is, with instruments'), we read: 'Se 'l tempo ci fussi, arej posto molti altrj modi a misurare choll'ochio, e' quali Maestro Gratia theologo e matematico perfetto, nel suo trattato De visu, chiaro dimostra' ('Had time allowed me to do so, I would have described many other ways of measuring by sight, which Master Gratia, perfect mathematician and theologian, clearly demonstrates in his treatise *De Visu*'), fols 407^{v}–412^{v}. See Gino Arrighi, 'Un estratto dal "De visu" di M° Grazia de' Castellani (dal Codice Ottoboniano latino 3307 della Biblioteca Apostolica Vaticana)', *Atti della Fondazione Giorgio Ronchi*, 22 (1967), 44–58; Gino Arrighi, 'La matematica in Firenze nel Rinascimento. Il Codice Ottoboniano Latino 3307 della Biblioteca Apostolica Vaticana', *Physis*, X, (1968), 70–82.

7 For these conclusions, refer to Filippo Camerota, *La prospettiva del Rinascimento. Arte architettura scienza* (Milan: Electa, 2006), pp. 22–34 ('La "perspectiva" degli abachisti'), 35–50 ('La "perspectiva" dei pittori'), 58–73 ('Brunelleschi "prespettivo"').

8 See Girolamo Mancini, *Opera inedita et pauca separatim impressa* (Florence: G.C. Sansoni, 1890); Leon Battista Alberti, *Opere volgari*, ed. by Cecil Grayson (Bari: Laterza, 1973), III, pp. 112–29; Alessandro Gambuti, 'Nuove ricerche sugli Elementa picturae', *Studi e Documenti di Architettura*, 1 (1972), 131–72.

Alberti's *Elements of Painting* is a collection of exercises apparently referring to graphic illustration of which, unfortunately, we have no documentation except in a very partial form, in the writings of Filarete.[9] In his text, Alberti intended to show painters how to improve their art by mastering the geometric tools necessary to 'guide the hand, the eye and the intellect'. After an introduction familiar to anyone who had received some instruction in an abacus school, Alberti formulated some specific concepts pertaining to geometric drawing, such as 'commensurate point', 'equal' and 'proportional' areas, and 'concentric' and 'comminuted' figures.[10] These concepts, the author believed, were crucial for understanding the processes of geometrical transformation involved in the composition of a perspective painting. Just as one learning to read must begin with the 'elements', i.e., the form of the letters, so the painter must begin with geometric figures, and the title of the text, *Elementa picturae*, inevitably evoked the well-known work of Euclid, *Elementa geometriae*. The purpose was to create a kind of painter's geometry necessary for understanding those *dirozzamenti* (instructions), which in *De pictura* 'will be easily understood by the geometer. But he who is ignorant of geometry will understand neither them nor anything else about painting'.[11]

Masaccio's *Trinity* is certainly a visual expression of Brunelleschi's supreme mastery in perspective and, probably, it was an inspirational work for Alberti as an art theorist.[12] However, it is impossible to say if Brunelleschi took part in the design of the perspective scheme, as much as to say that Masaccio made use of the legitimate construction. On the

9 Filarete (born Antonio Averlino), *Trattato di architettura* (*c.* 1461), ed. by A. M. Finoli and L. Grassi, 2 vols (Milan: Il Polifilo, 1972), pp. 639–65. For a comparison between Filarete's text and Alberti's *Elementa picturae*, see Gambuti, 'Nuove ricerche', pp. 131–72 and Alessandro Gambuti, 'I "libri del disegno": Filarete e l'educazione artistica di Galeazzo Maria Sforza', *Arte Lombarda*, 38–39 (1973), 133–43.

10 Alberti materialises Euclid's definitions with the practical language in use among painters, rephrasing the geometrical elements (point, line, surface, and body) according to their visual appearance. The profile of an object is called 'orlo' in the *De pictura* and 'lembo' in the *Elementa picturae*. The surface is called 'dorso' (*De pictura*) or 'area' (*Elementa*), and the 'area' is better defined as 'concentrica', when seen in plan or in elevation (that is, in its true geometrical shape), and 'comminuta', when seen in perspective, that is, foreshortened.

11 In the conclusion of Book I of *De pictura* (Chapter 23), Alberti declares the nature of his treatise to be more conceptual than strictly geometric: 'Qui solo raccontai i primi dirozzamenti dell'arte, e per questo così li chiamo dirozzamenti, quali ad i pittori non eruditi dieno i primi fondamenti a ben dipignere' ('Here I have related only the basic instructions of the art, and by instructions I mean that which will give the untrained painter the first fundamentals of how to paint well'). Alessandro Parronchi, 'Sul significato degli "Elementi di pittura" di L. B. Alberti', in *Cronache di archeologia e storia dell'arte* (1967), VI, pp. 107–15, maintains that *De pictura* and the *Elementa picturae* should be considered a single work. The Lucca, Biblioteca Governativa codex 1448, which presents the two works in a single volume, also contains a fragment that develops the concepts of point and line: 'De punctis et lineis apud pictores', ('On the points and lines used by painters'), see fols 53^v–54^r. The only MS in the vernacular is found at Biblioteca Capitolare, Verona, 245–CCLXXIII. Basically, Alberti associates the geometric definitions of Euclid to the more tangible geometry used by painters; this last was based on the optical appearance of points, lines, surfaces, and bodies. Alberti's text has no illustrations; a reconstruction of the operations described in the *Elementa picturae* is proposed in Alessandro Gambuti, *Nuove ricerche*, n. 8.

12 The close collaboration between Masaccio and Brunelleschi was pointed out in Vasari, *Vite*, p. 291: '[Brunelleschi] insegnò [la prospettiva] a Masaccio pittore allor giovane, molto suo amico; il quale gli fece onore in quello che gli mostrò, come appare negli edifizj dell'opere sue' ('In particular, Filippo taught perspective to Masaccio, a young painter and his very close friend at the time, who then honoured his teacher by what he displayed in the buildings he painted in his works' [Vasari, *The Lives of the Artists*, 113]). See Marcello Fagiolo, 'Eredità di Brunelleschi: la rivoluzione prospettica e il punto di vista vasariano', *Architettura e arte del Principato mediceo (1512–1737). Firenze e la Toscana, Vasari e gli Uffizi, Bollettino della Società di Studi Fiorentini*, 22, ed. by F. Canali (2013), 134–42.

other hand, we can argue that Masaccio made use of a geometrical culture rooted in the abachist tradition and which was adapted to the needs of painters.

The detailed constructive drawing impressed in the plaster of the *Trinity* fresco reveals remarkable expertise in geometry. Most of the *battiture di corda* (lines impressed on the fresh plaster by beating it with a string) and the lines directly traced with a stylus are visible under grazing light. Some markings can even be seen in normal lighting conditions, especially the orthogonal lines forming the regular squaring on the Virgin's figure, which I would like to use as the starting point of my close examination (Figure 2).

The Proportional Grid

The grid was certainly a means of enlarging the preparatory drawing, but it seems that it was used mainly as a modular pattern to size the entire architectural composition. The meshes in the grid measures 13.78 cm per side, and are equal to ¼ *braccio da terra* (55.12 cm), the measuring unit used in Tuscany by surveyors until 1781, when Grand Duke Peter Leopold of Lorraine decided to abolish it and standardise all measurements to the *braccio da panno* (58.36 cm).[13] On the Virgin's face, each mesh is further divided into four smaller meshes, each measuring 3.44 cm, that is, 1/16 *braccio da terra*. The fact that Masaccio used the surveyors' measuring unit instead of the more common *braccio da panno* indicates his familiarity with the geometric rules applied by surveyors. Although we don't have any information about his education, as the son of a notary the painter could very likely have attended an abacus school, where he would have received basic instruction in mathematics. Alternatively, his friendship with Brunelleschi could have been a secure source of mathematical learning.

According to the module adopted, the fresco measures 12 × 5.5 *braccia da terra* (661.44 × 303.16 cm). In extending the squaring over the entire painted surface, it can be seen that the painter subdivided the whole into 48 × 22 modules (Figure 3), precisely arranging the individual parts within this grid: twenty modules for the overall width from pilaster to pilaster, four modules for the pilaster to column assembly, twelve modules for the intercolumniation, twenty modules for the height of the Ionic column, two for the Ionic capital, three for the Corinthian capital, and five for the entablature. Note also that the perspective foreshortening of the barrel-vault ceiling is in part 'measured' by the same reticule that delimits the impost at both top and bottom, providing a faster method for executing the preparatory drawing. Moreover, we find that the height of the podium, midway up the painting, determined that of the slab closing off the apse at the back, and even the position of the vanishing point, suggesting that the artist employed this practical

13 Edict of 13 March 1781, see Marco Lastri, *l'Osservatore fiorentino. Sugli edifizi della sua patria* (Florence: Gaspero Ricci, 1798), V, p. 117. According to Leonardo Ximenes, in his *Del vecchio e nuovo gnomone fiorentino* (Florence: Nella Stamperia Imperiale, 1757), I, p. 3, the *braccio da terra* was subdivided into halves, thirds, quarters, and eighths. The division of the *braccio* into fourths and eighths (in our case, into sixteenths) was illustrated, at the time of Masaccio, by the Sienese engineer Mariano di Jacopo known as Taccola in his *Liber tertius de ingeneis ac edifitiis non usitatis* (1432), ed. by James H. Beck (Milan: Il Polifilo, 1969), p. 31: 'Homo [...] debet habere perticam mensuratam ["graduata"] ad bracjios divisos per quarra ["quarti"] [read "quarters"] et ottava cum punto'.

Fig. 2 The proportional grid measured with the fiorentine *braccio da terra*.

expedient not only to size the architecture but also to determine the position of some of the elements seen in perspective.

The Golden Section

The rectangle of 12 × 5.5 *braccia* is unusual in that its sides are almost exactly in the ratio of $\sqrt{5}$:1 (661.44 × 295.80 cm), and this could be a clue indicating that the rules of practical geometry were applied. A rectangle having these proportions, in fact, is described and illustrated in a fifteenth century manuscript compiled by an anonymous 'Florentine

Fig. 3 Reconstruction of the grid on the whole fresco. [See colour plate 23]

Surveyor', who proposes to extract a square from a semicircle with diameter of twelve *braccia* (Figure 4a).[14] The side of the square turns out to be 5.37 *braccia*, almost exactly the same as the rectangle of the *Trinity*. The distinctive feature of this proportional ratio, adopted later by Leonardo for his painting of the *Annunciation*, is that it contains the golden ratio that was to be hailed by Luca Pacioli, in the early years of the next century, as the 'divine proportion'.[15] The rectangle, in fact, is composed of a central square and two adjoining golden rectangles. In Luca Pacioli's subsequent commendation, the golden section would be invested with symbolic implications expressly pertaining to the Trinity.

14 See *Trattato di geometria pratica dal codice L.IV.18 (sec. XV) della Biblioteca Comunale di Siena*, ed. by Annalisa Simi (Siena: Università degli Studi di Siena, 1993), fol. 55^{v}.

15 Luca Pacioli, *De divina proporzione* (Venice: A. Paganius Paganinus, 1509). On the golden section applied to Leonardo's *Annunciation*, see also Franca Manenti Valli, *Leonardo. Il comporre armonico nella tavola dell'Annunciazione* (Milano: Silvana, 2012).

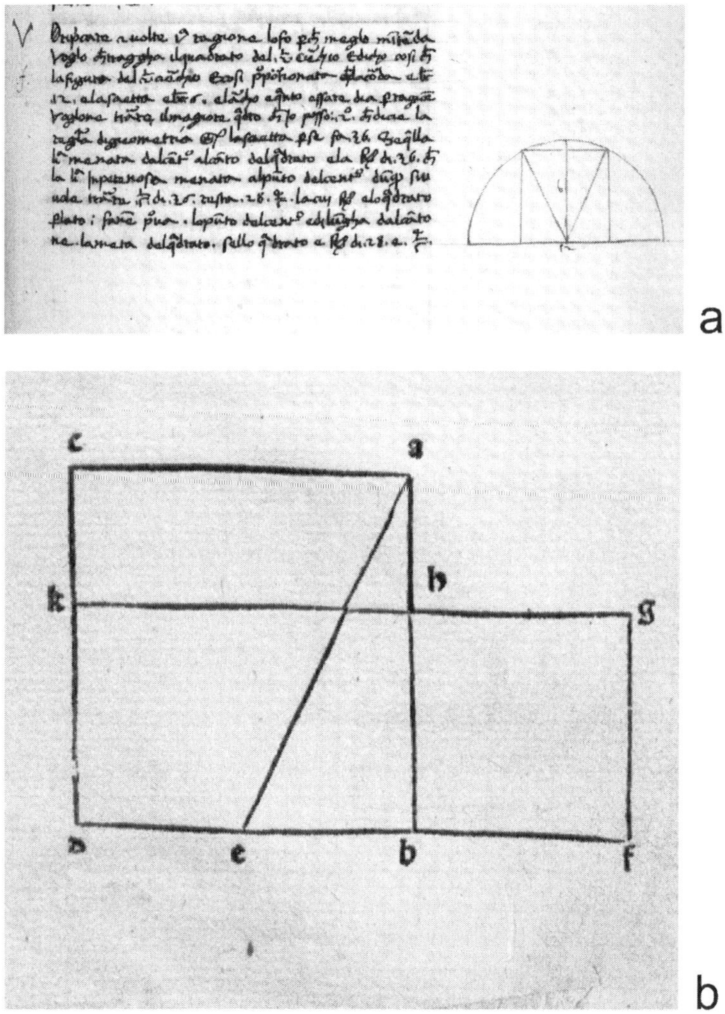

Fig. 4 (*a*) Anonymous Florentine surveyor, *Trattato di geometria pratica*, ms., *c.* 1450, Siena, Biblioteca Comunale, L.IV.18, *c.* 55^{r}. (*b*) Euclid, *Elements*, Book II, Proposition 11.

Like God, it was both *una* and *trina* (triune), being indivisible and composed of three terms (in a segment divided according to the golden section, the shorter section is to the longer as the latter is to the entire segment). But whether this concept already existed in Masaccio's intentions is impossible to say.

As an irrational proportion, 1:0.618, the best way to divide a segment in 'extreme and mean ratio', as the Abachists used to call the golden ratio, was the geometric method described in the second book of Euclid's *Elements* and commonly taught in the abacus schools (Figure 4b).[16] Having traced a square and the diagonal of its half (from the midpoint

16 Euclid, *Elements*, II, Proposition 11, in Hubert L. Busard, *Campanus of Novara and Euclid's Elements*, 2 vols (Stuttgart: Franz Steiner, 2005), I, p. 103.

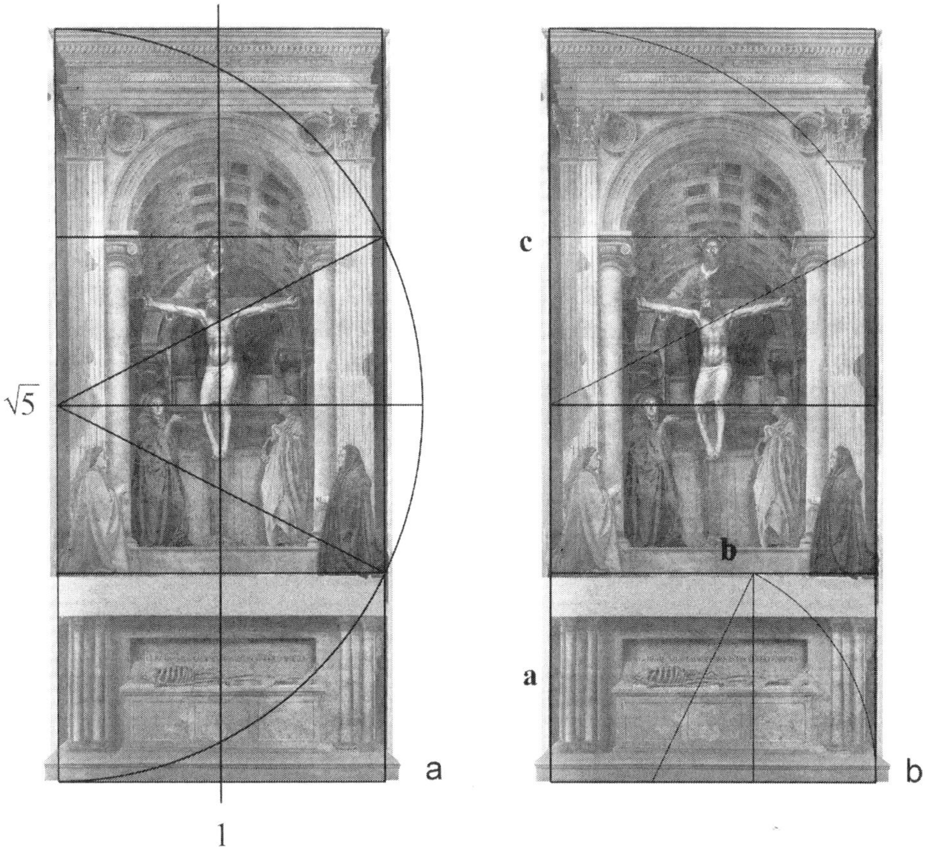

Fig. 5 (*a*) The geometrical scheme of the Florentine surveyor applied to the *Trinity* fresco. (*b*) The rectangle √5 and its decomposition in golden rectangles. [See colour plate 24]

of a side to one of the opposite angles), overturning the diagonal onto the extension of the side forms a golden segment (0.618). In this way a golden rectangle was constructed, but overturning the symmetrical diagonal as well, as shown by the 'Florentine Surveyor', produces a rectangle that is $1{:}\sqrt{5}$.

If we apply the 'Florentine Surveyor's' exercise to Masaccio's fresco, given the overall height of twelve *braccia* as diameter of a semicircle, we see that the inscribed square not only establishes the width of the fresco but also perfectly circumscribes the central group of figures, from the plane on which the donors are kneeling to the impost of the barrel vault, which is levelled with the top of God's head (Figure 5a). The horizontal centreline coincides with the top of the podium God the Father stands on, while the golden rectangles resulting from overturning the diagonals delineate the lower portion with the fresco of the *Death* and the upper part of the architecture above the impost.

In reorganizing the geometric scheme, it can be said that the entire fresco is subdivided into two golden rectangles, a smaller horizontal one (Figure 5b), with the fresco of the *Death* (b:a = 1:0.618), and a larger vertical one, with the fresco of the *Trinity* (c:b = 1:0.618). In other words, the height of the *Trinity*, nine *braccia* (three times the Trinitarian number),

Fig. 6 (*a*) The proportional grid and the tracing of the central axis. (*b*) Elevation of the chapel (left) and its perspective transformation (right). [See colour plate 25]

is subdivided according to the golden section into two segments, the greater of which establishes the width of the painting, and the lesser the height of the *Death* fresco.

Having drawn the rectangle and defined the two portions of the *Death* and the *Trinity*, the artist could have immediately traced the central axis of the whole composition and the horizontal centreline that clearly divides the celestial world of the Holy Trinity from the terrestrial one of the mourners and donors. He would then have subdivided the surface according to the grid of 48 × 22 modules (Figure 6a).

Drawing the Barrel Vault: The Longitudinal Ribs

Some features of the architectural composition suggest that Masaccio began to work as an architect, designing a detailed elevation of the chapel (Figure 6b). The two-dimensional drawing would then have taken on a corporeal shape thanks to limited but highly effective projections: brief foreshortened lines, slight curving of the abacuses and toroidal elements

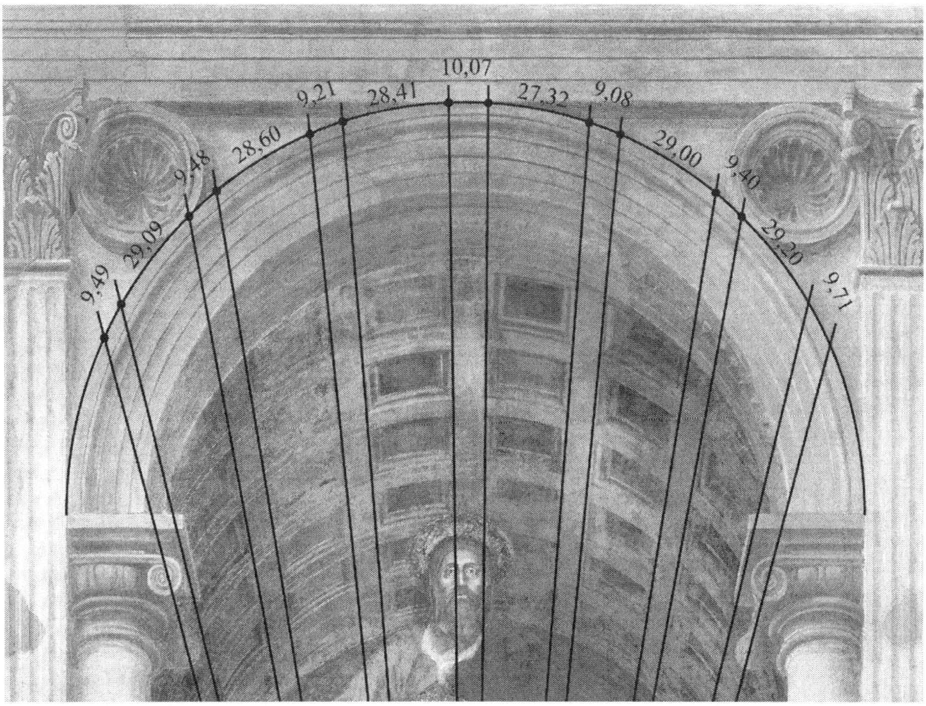

Fig. 7 Detail of the barrel vault with the partition of the front arch.

and, above all, skilful use of chiaroscuro. The frontality of the echinus on the Ionic capitals is a detail not modified in perspective, nor is the alignment of the width of the central fornix with the niche containing the sarcophagus below it, nor is the alignment of the coupled columns of the altar with the pilasters above them. Since these columns support the mensa, and thus occupy a plane more advanced than that of the Corinthian pilasters, they should not appear aligned with the pilasters. Masaccio evidently chose not to modify this part of the drawing in perspective, deeming the optical effect sufficiently convincing.

Undoubtedly, the most demanding problem was that of drawing the barrel-vault. The first step in this procedure consisted of tracing the vanishing lines of the vaulting-ribs receding in depth. By extending the lines impressed in the plaster along the vaulting-ribs (Figure 7), we find a series of reasonably regularly spaced intervals on the extrados of the arch front. The width of the coffers ranges from twenty-seven to twenty-nine cm (approximately ½ *braccio*) while that of the vaulting-ribs remains fixed at around ten cm (approximately 1/6 of a *braccio*), except for the coffers above the impost, which are notably larger than the others, but this is presumably due to a correction made later by the artist. Masaccio would have purposely omitted to paint the vaulting-rib that would be found on the impost had he continued to divide the arch according to the same modules. It would however have been a larger rib than the others, a raised band, in keeping with good architectural rules, to compensate optically for the portion concealed by the projecting cornice of the architrave.

Since the width of the vaulting-ribs is approximately one-third that of the coffers, we can imagine that the artist must have divided the semi-circumference into thirty-six equal parts, assigning one part to each of the ribs, two to the bands not appearing in the painting above the impost, and three to the coffers. The division could have been done either with a compass or, more easily, with a quadrant, marking a dot every five degrees (Figure 8).[17] Once done, the division of the two opposite arches would have allowed the tracing of the receding ribs without using the vanishing point.

However, the vanishing point was presumably established and used to control the entire composition. The lines impressed in the plaster converge with remarkable precision towards a point situated on the central axis at a height of 172.25 cm above ground level (about three *braccia*), where the nail that materially represented the vanishing point was possibly located (Figure 9). The positioning of the 'centric' point at the height of three *braccia*, codified later by Alberti in his *De pictura*, was commonly used in surveying practice, where the position of the eye was indicated by a hole or a nail on the surveyor's rod.[18] We cannot say, however, if the rope that impressed the lines in the plaster was tied to the nail, or simply stretched between the divisions of the arches.

Drawing the Barrel Vault: The Transverse Ribs

The next step would have been to draw the transverse ribs. To do this, Masaccio traced on the plaster no less than twenty-four arches: six for the centrelines of the coffers, another six for the centrelines of the ribs, and twelve for the lines of separation between coffers and ribs. Each of these arches presumes a centre, and we must imagine an entire series of nails inserted in the first coat of plaster along the central axis (Figure 9). Having tied a string to each nail and pulled it horizontally as far as the line of the impost, a line converging to the vanishing point, the artist could measure each radius and proceed to trace the arches. But how could he establish the position of the centres on the central axis? With the lack of any evidence that he applied the rule traditionally attributed to Brunelleschi, namely the intersection of the visual pyramid, we may try to think in terms of practical geometry. We can thus imagine that Masaccio performed a geometrical transformation of the type described later by Alberti in his *Elementa picturae*, obtaining the 'comminuted' form of the barrel-vault from its 'concentric' image, that is, from the plan, without recurring to perspective projection.

This hypothesis is suggested by the working drawing of the upper shaft of the Ionic column on the left, where the painter seems to have obtained the ellipse from a deformed constructive scheme undoubtedly derived from a plan (Figure 10). The constructive scheme, traced on a square with the sides divided into six parts, identifies some points of the inscribed circle, facilitating the perspective drawing of the figure. A similar procedure based on an orthogonal grid was recommended by Alberti in *De pictura* as a practical means of rapidly solving the problem of the laborious construction of the ellipse. To apply this same procedure to the perspective drawing of the barrel-vault, Masaccio would

---

17 The use of a quadrant has already been suggested in Field, Lunardi, and Settle, 'The Perspective Scheme', pp. 31–118.

18 Leon Battista Alberti, *De pictura*, ed. by Lucia Bertolini (Florence: Edizioni Polistampa, 2011), I, 19, p. 237; Arrighi, 'Un estratto dal "De visu"', pp. 44–58, Problem 1; Taccola, *Liber tertius de ingeneis*, ed. by Beck, p. 31.

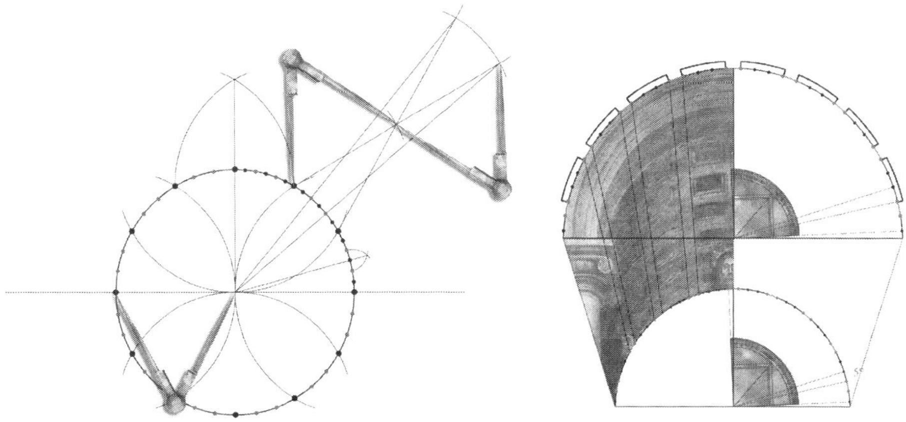

Fig. 8 Partition of the front arch, either using a compass or a quadrant. [See colour plate 26]

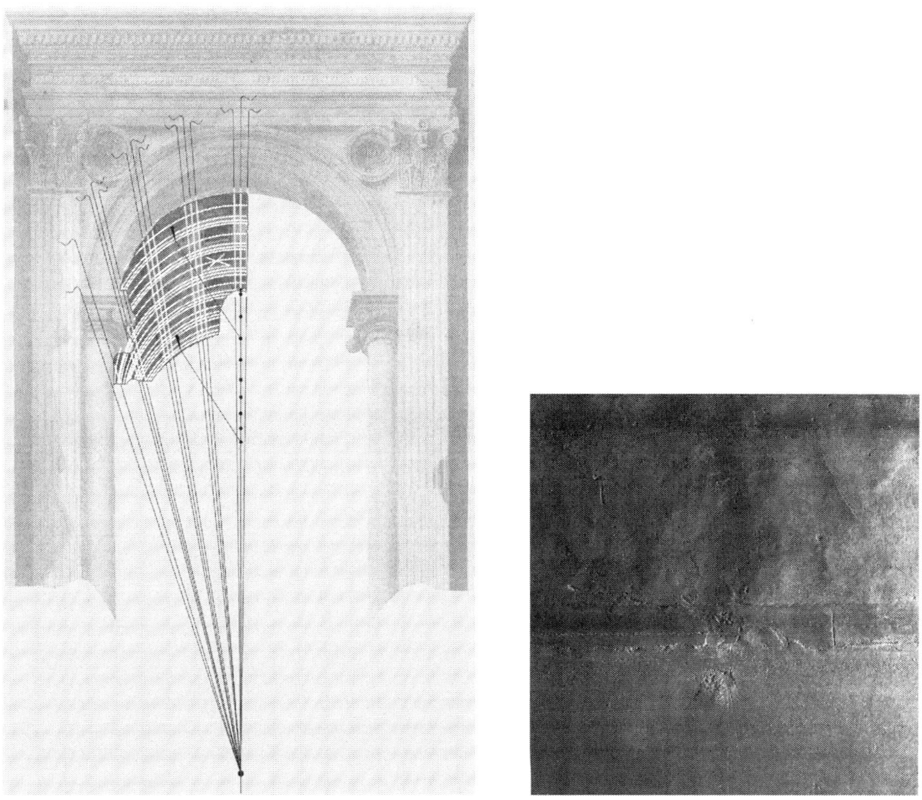

Fig. 9 Convergence of the lines impressed on the plaster toward a vanishing point still visible on the painted surface. [See colour plate 27]

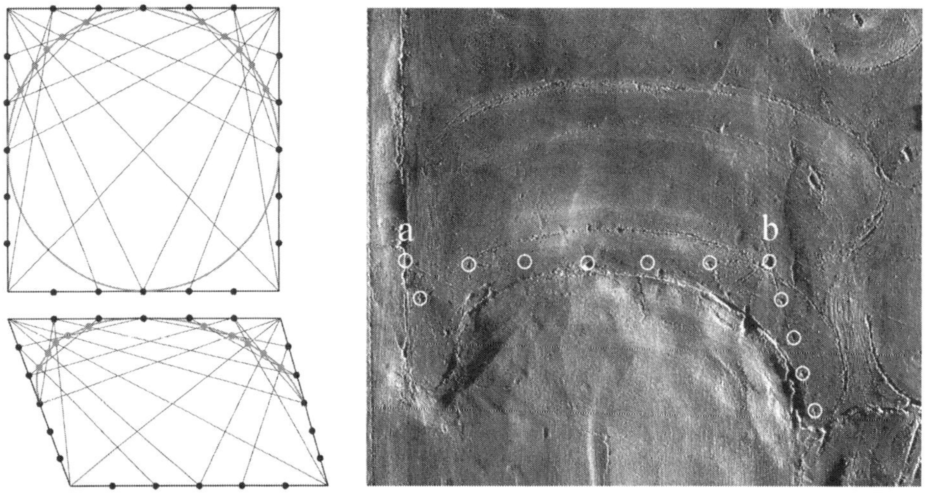

Fig. 10 Detail of the ionic capital and its geometric construction deduced from the lines traced with a stylus. [See colour plate 28]

have needed only a schematic drawing. In the painting, six rows of coffers can be seen but, due to the perspective, we must imagine that behind the front arch there is a seventh row, entirely hidden, and we can hypothesise that, for reasons of composition, the depth of the arch is equal to the width of one coffer plus one vaulting-rib. We can then imagine that the entire vault, including the front arch, is inscribed in a rectangle subdivided into eight transverse bands of the same depth (Figure 11a).

To proceed to the perspective transformation of this rectangle, it was sufficient to calculate the apparent lowering of the back. The other three sides were already determined by the diameter of the front arch and by the lines of the impost converging at the vanishing point. The lowering of the back could be easily determined by a simple proportional calculation, known to the Abachists as the 'rule of three'. The problem corresponded to Theorem 11 of Euclid's *Optics*, and could be stated as Luca Pacioli was to do later in his *Summa de arithmetica*: 'Given a horizontal table twelve *braccia* long and two wide, given a viewing distance of ten *braccia* and given the height of the eye as two *braccia*, how long will the table appear in perspective?'[19] The answer to this problem was found by applying the 'rule of three' to the known dimensions, that is, height and distance of the eye from the picture plane, and length of the 'table' or, in Masaccio's case, height and distance of the eye from the painting and length of the barrel-vault.

Having drawn on the impost a trapezoid representing the perspective view of the rectangular plan of the vault, the artist could have subdivided the figure into eight transverse bands, operating directly on the trapezoid (Figure 11b). When the central axis and the

19 See Euclid, *Euclidis Optica*, ed. by Johan L. Heiberg, in *Euclidis Opera Omnia*, ed. Johan L. Heiberg and Heinrich Menge, 9 vols (Leipzig: G. B. Teubner, 1883–1916), VII, p. 11; Luca Pacioli, *Summa de aritmetica, geometria, proportioni et proportionalità* (Venice: A. Paganius Paganinus, 1494), 'Tractatus geometriae, Distinctio octava', p. 65.

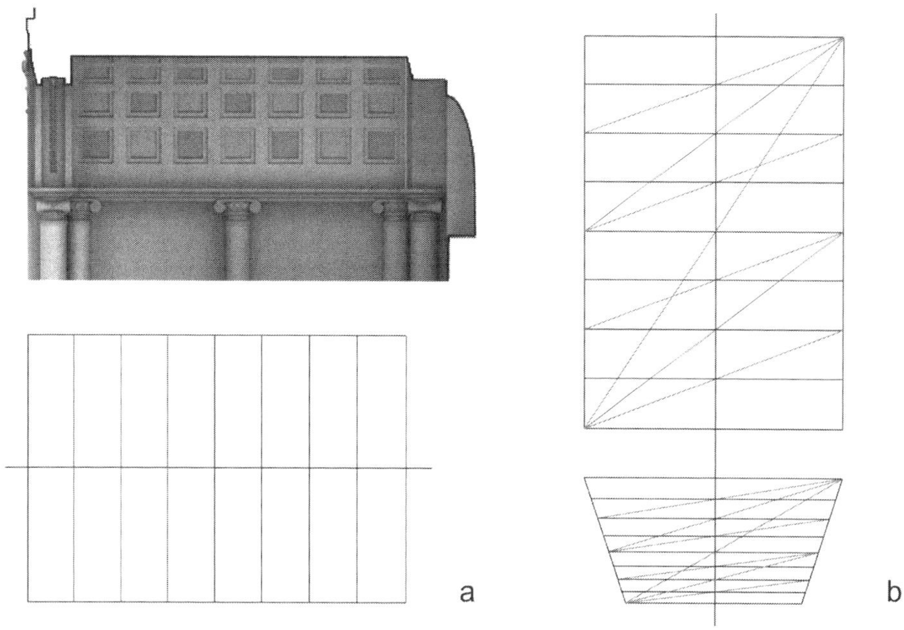

Fig. 11 (*a*) Plan and elevation of the barrel vault. (*b*) Hypothetical construction of the centres on the central axis.

diagonal had been drawn, the perspective position of the centre of the figure could be easily identified, and the figure could be divided into two smaller trapezoids. The diagonals of the latter could be used to subdivide the entire figure into four trapezoids, and then into eight, determining on the central axis the position of the centres of the arches corresponding to the middle of the crossbeams.

An example of this procedure, which has survived in the notes of Baldassarre Peruzzi, concerns a workshop practice handed down over the years that dates expressly from the time of Masaccio (Figure 12).[20] A geometric rule could be established for this practice, expressed more or less in these terms: any figure transformed 'by perspective', although changing in form, retains a biunivocal correspondence to the points in the figure it derives from (as we learn from Alberti's *Elementa picture*), or, any figure can be inscribed in a square and measured in relation to it (as was taught later by Piero della Francesca's *De prospectiva pingendi*).

Another possibility is that, having established the measurements of the coffers, and the lowering and the foreshortening of the back side, whether geometrically or arithmetically, Masaccio proceeded to draw on the impost a square in perspective, having the front side as long as the barrel vault (Figure 13). The side corresponds to the width of the fresco. Once subdivided the two parallel sides in eight equal parts and drawn the diagonal, the intersection of the latter with the lines joining the divisions gives, on the central axis, the position of the centres for the arches above.

20 Baldassarre Peruzzi's scheme is a geometrical method for the construction of the ionic volute, which closely resembles a perspective diagram (Florence, Gabinetto dei Disegni e delle Stampe degli Uffizi, 465 A).

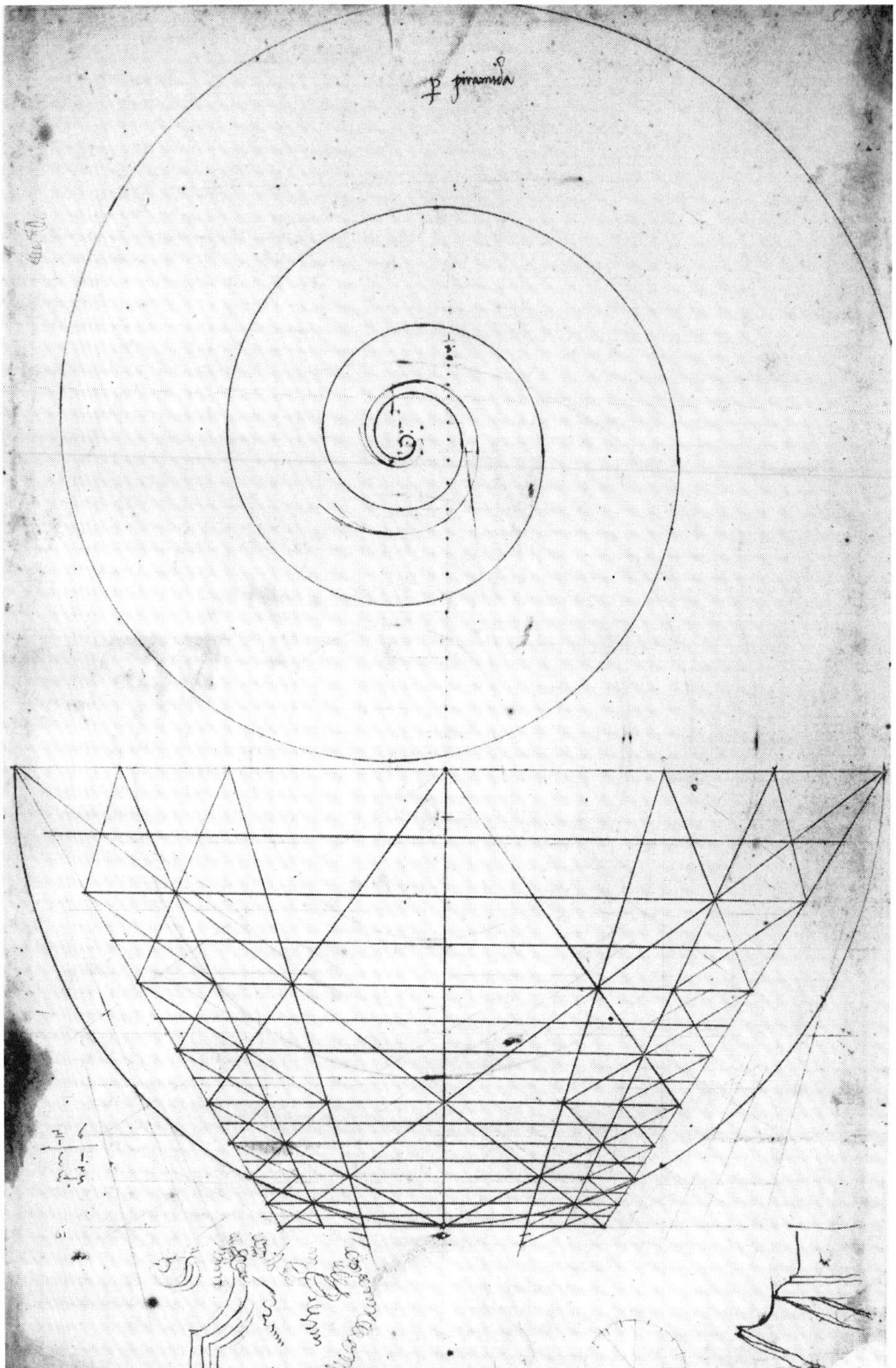

Fig. 12 B. Peruzzi, *Perspective Scheme for the Construction of the Ionic Volute*, Firenze, Gabinetto dei Disegni e delle Stampe degli Uffizi, 465 A.

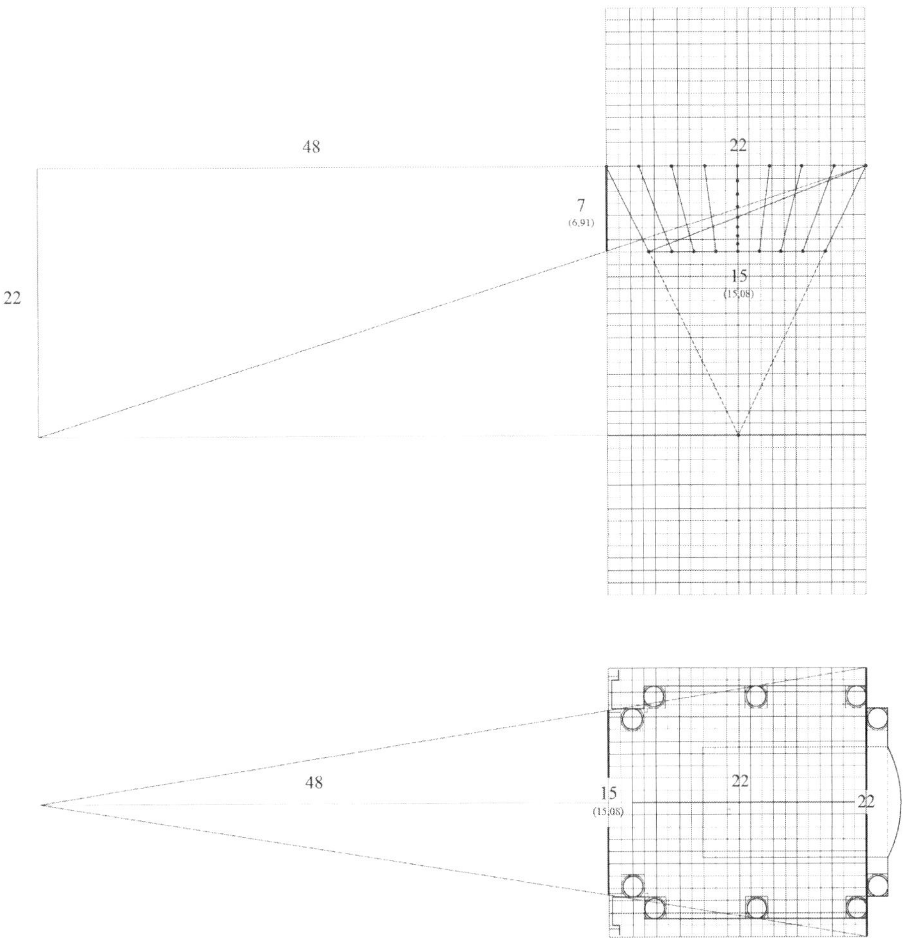

Fig. 13 Alternative hypothesis for the construction of the centres on the central axis.

The scheme, with its vertex in the vanishing point, is perfectly inscribed in the central square of the rectangle (1:√5), which circumscribes the painting. If we make just a small correction to the scheme, we find that it is perfectly inscribed in the proportional grid, allowing an easy arithmetical calculation of the perspective foreshortening. A similar perspective scheme, with the vanishing point situated on the side of a square, can be found in the MS A of the Institute de France, where Leonardo da Vinci collected simple geometrical exercises of the kind described by Alberti in the *Elementa picturae* (Figure 14).[21]

21 Institut de France, Paris, MS A, fol. 40r. In the *Codex Atlanticus* (Milan, Biblioteca Ambrosiana), fol. 683v, we find the first seven propositions of the *Elementa picturae* translated from the Latin. We read, for instance: 'Corpo chiamo quello ch'è coperto co' la superfizie sotto aspetto e con lume si può vedere' ('Body I call that which covers the surface being perceived and that can be seen with light'), or 'Lembo chiamo di ciascuna superficie veduta lo estremo circuito, del quale la terminazione sia la divisione' ('I call an 'edge' any surface seen at the extreme

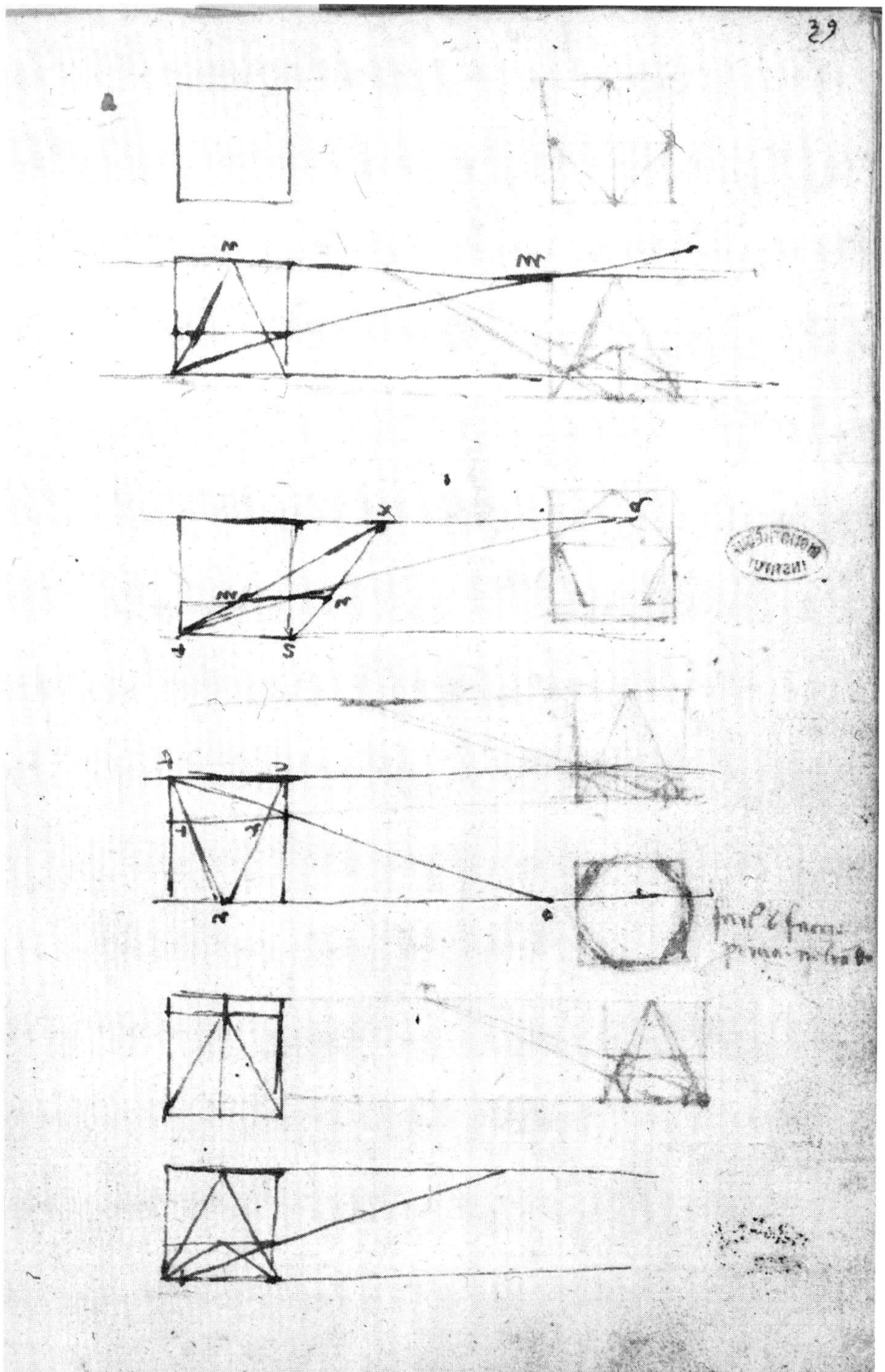

Fig. 14 Leonardo da Vinci, *Ms. A* of the Institute de France, fol. 40[r].

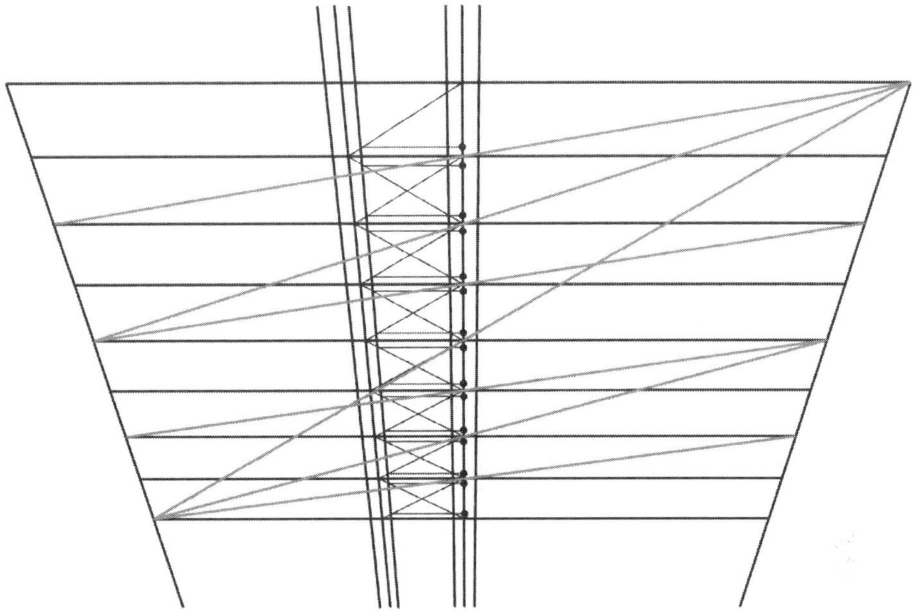

Fig. 15 Hypothetical construction of the centres of the arches circumscribing the ribs. [See colour plate 29]

To determine the centres of the arches that delimit the crossbeams, the artist could have proceeded on the grid, using the centreline of the crossbeam receding in depth nearest to the central axis (Figure 15). The resulting series of trapezoids can be considered a horizontal projection of the coffers nearest to the central axis. The diagonals of the individual trapezoids could be used to determine the position of the centres sought for on the central axis.

The Viewing Distance

Due to the precise match between the centres derived from this geometric transformation and the arches traced by Masaccio, it seems plausible to hypothesise that the vault was subdivided into eight equal parts. Concerning the real dimensions of the vault, assuming that the coffers are square, and measuring their width on the extrados of the front arch, we can estimate the width of each band as thirty-nine cm (one coffer plus one vaulting-rib), resulting in a plan of 212.16 × 312 cm. By relating the virtual depth of the ceiling (312 cm) to the lowering of the back of the impost (98.98 cm) and to the distance of the latter from the horizontal line (209.83 cm), we can determine a viewing distance of 661.4 cm (312:98.98 =

perimeter, whose end is the division'). See the corresponding Latin text in Alberti's *Elementa picturae*: 'Corpus appello id quod opertum superficie sub aspectu et lumine possit perspici', and 'Limbum appello cuiusque visae superficiei extremum ambitum, cuius terminatio sit discrimen'. Mancini, *Opera inedita*, 49.

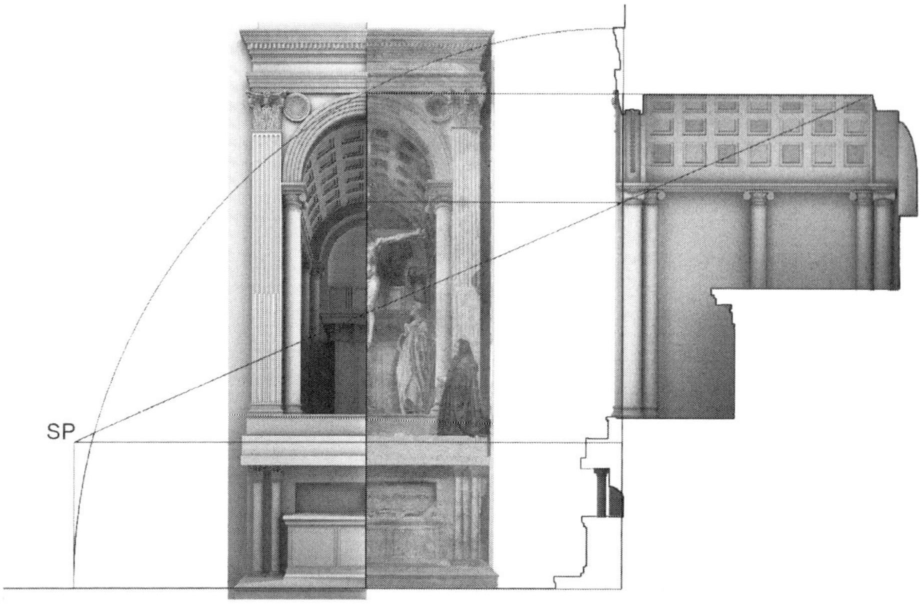

Fig. 16 Finding of the viewing distance and perspective reconstruction of the chapel. [See colour plate 30]

x:209.83, from which x = 312 · 209.83 / 98.98); the viewing distance would be equal to the height of the painting: twelve *braccia da terra* (Figure 16).[22]

Scholars usually estimate the viewing distance based on the foreshortening of the Ionic capital, taking for granted that this architectural element can be inscribed, as it should be, in a square. The diagonal of the presumed square would indicate a distance of 483.50 cm (about nine *braccia da terra*).[23] But if the ceiling were measured from this distance, its depth would be only 218.6 cm, consequently optically deforming the coffers, which would assume an atypical rectangular shape. Careful analysis shows that the Ionic capital is not, in fact, inscribed in a square but in a rectangle, so that its diagonal does not indicate the viewing distance. It seems reasonable to assume that Masaccio constructed the Ionic capital after

22 The result is midway between the 686.25 cm estimated in Field, Lunardi, and Settle, 'The Perspective Scheme' and the 642 cm estimated in Kuhn, 'Measured Appearances'. For the different results obtained by other scholars, see Kern (524 cm); Sanpaolesi (572 cm); Janson (700 cm); Polzer (210.5 cm); Battisti (through an analysis of Giovanni Degl'Innocenti, 894.2 cm); Myers (729.5 cm); Hertlein (552 cm); Huber (333 cm); Kemp (540 cm); Aiken (219.9 cm); Hoffmann (452 cm).

23 A simple verification, consisting of extending the diagonals as far as the horizon line, gives a distance of 483.50 cm, about nine *braccia da terra* (496.08 cm). Applying an arithmetical method gives the same result. The known dimensions, measurable on the fresco, are the side of the square (20 cm), the lowering of the rear side (11 cm), and the height of it above the horizon (265.93 cm). In developing the calculation, 20:11 = x: 265.93 (where x = 265.93·20/11), we obtain a viewing distance of 483.50 cm. Similar results were obtained by Field, Lunardi, and Settle in 'The Perspective Scheme'. Their analysis, however, depended on less clearly defined detail, that is, the abacus of the left capital, a figure having no direct incisions, and it was not depicted in its entirety. The difference in the result, which for them was 553.1 cm, can be attributed to the uncertainty of determining the measurement of the side.

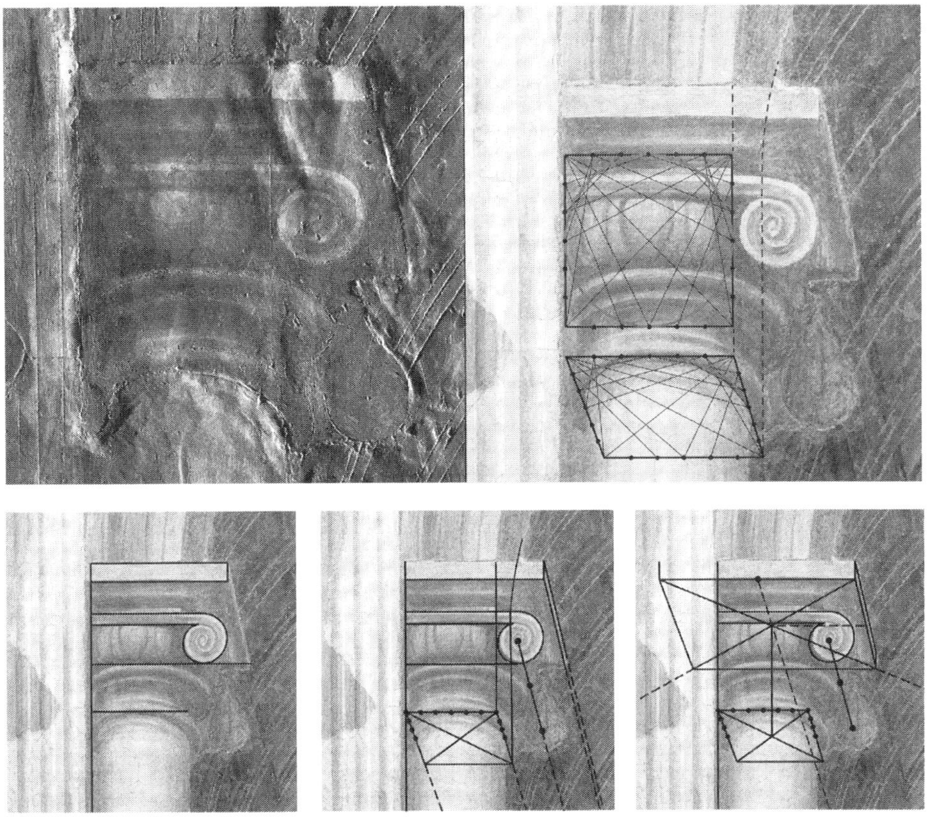

Fig. 17 Detail of the ionic capital under grazing light with an hypothesis on the geometrical construction of its foreshortening. [See colour plate 31]

having drawn the entire arch above it, that is, the front arch moulding and the intrados (Figure 17). The front of the square circumscribed with the astragal (clearly imprinted in the plaster) is equal, in fact, to the width of the arch moulding, and the depth of the foreshortened side terminates on the vertical of the foot of the semi-circle that delimits the arch in depth. Basically, the quadrangle that circumscribes the column seems to be a projection of the foot of the arch. However, since the depth of the intrados is greater than the width of the arch moulding, the foot of the arch is a rectangular figure. This explains the excessive foreshortening of the abacus and the baluster of the volute, objectively too pronounced, as has been pointed out by the critics.[24] Such a construction is in fact absurd, since it would result in an elliptical column shaft. However, the practicality of this constructive procedure, whose results are in any case good (no one would ever imagine an elliptical profile) may have led Masaccio to overlook this detail.

If Masaccio did adopt this procedure to draw the barrel-vault in perspective, we must conclude that, in what is considered to be the first perspective masterpiece of the

24 Field, Lunardi, and Settle, 'The Perspective Scheme', pp. 59–61.

Renaissance, the only calculation of diminishing perspective in the strict sense concerns the lowering of the back of the impost. All the rest would have been calculated geometrically, that is, without recurring to any projective procedure. Regardless of the procedure adopted, geometric or arithmetical, Masaccio had to calculate the lowering of the back by establishing a precise viewing distance, which as has been discussed, is equal to the height of the fresco, 661.44 cm.

In examining the modular grid, we see that the back of the impost is 7.25 modules (approx. 99 cm) lower than the front, and that its distance from the horizon is equivalent to 15.25 modules (approx. 210 cm). Considering then that the viewing distance is 48 modules (661.44 cm), we can obtain by proportion the depth of the vault as equal to 22.6 modules (approx. 312 cm). Relating this measurement to the width of the vault, 15.5 modules (212.16 cm), we find that the sides of the impost rectangle have a ratio of 1:1.46, that is, with good approximation, 1: $\sqrt{2}$. The depth would thus be equal to the diagonal of the square constructed on its width (Figure 18).

Conclusion

Examination of the signs marked on the plaster reveals refined, skilful geometric expertise, and allows us to interpret the drawing of the *Trinity* even without recurring to the perspective principles of Alberti and his followers. There are no signs traced at random. Everything appears to have been scrupulously measured. In the construction of the Ionic capitals, there is an obvious dimensional discrepancy between the depth of the intrados and the column shaft, but it is irrelevant as regards the pictorial effect. This is without doubt an extraordinary masterpiece, measured as no other work had ever been before, heralding a cultural tendency increasingly more firmly established and in rapid development.

After over a century of abacus teaching, the bases of practical geometry had begun to permeate every aspect of professional life. The adept use of irrational proportions, such as $\sqrt{2}$, the golden section, $\sqrt{5}$, indicates the increasingly widespread use of proportional standards, to which the painters recurred for any need, from sizing the paintings to composing the figures, without this necessarily implying any thorough knowledge of Euclidean problems. In similar manner, the problem of practical *perspectiva*, that is, of optics applied to the measurement of distances, was skilfully solved by geometric means, even without the theoretical support of the science of vision.

It is in this cultural context that Brunelleschi's 'rule' should be seen; it should not continue to be identified with the so-called 'legitimate construction' — which in fact, we know to have been satisfactorily illustrated only fifty years later by Piero della Francesca — but seen as a combination of geometric practice and 'perspective' calculation.[25] Knowing that the appearance of any figure changes in relation to its distance from the eye, as Euclid taught, and knowing that any figure can be inscribed in a square and measured in relation

25 Piero della Francesca, *De prospectiva pingendi*, ed. by Giusta Nicco Fasola (Florence: Sansoni 1942; repr. 1974). On Piero's perspective methods, see J. V. Field, 'Piero della Francesca's Perspective Treatise', in *The Treatise on Perspective: Published and Unpublished*, ed. by L. Massey (Washington, DC: National Gallery of Art, 2003), pp. 63–78.

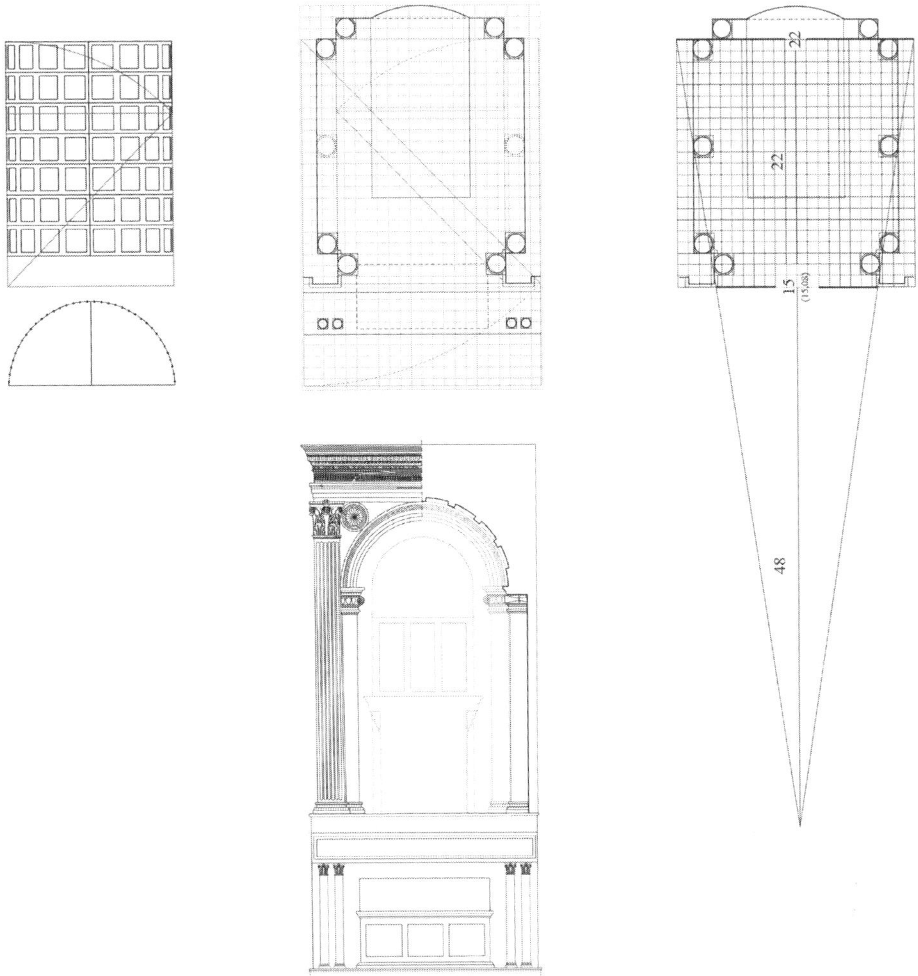

Fig. 18 Recontruction of the chapel's plan and elevation. [See colour plate 32]

to it, as the abacus masters taught, the painter had only to calculate 'by perspective' the 'comminuted' form of the square, to then obtain from it the perspective image of the inscribed figure, whether a circle or a complex capital.

Making use of 'the period eye', as defined by Baxandall, as an analytical method, is indispensable for a better understanding of the early stages of Renaissance experience with perspective.[26] The fact that the first theoretical reflection on the painters' perspective was the work of a humanist, Leon Battista Alberti, has undoubtedly influenced historiographical studies, leading scholars to interpret the phenomenon as a product of the highest mathematical culture, that of the philosophers and humanists, and often ignoring

26 See Michael Baxandal, *Painting and Experience in Fifteenth-Century Italy: A Primer in the Social History of Pictorial Style* (Oxford: Oxford University Press, 1972).

or underestimating the importance of the practical culture put to use in the art workshops. Today critics have begun to study more broadly the social implications and cultural aspects of this practical knowledge. It now seems increasingly evident that the two cultures, the practical and the speculative, were mutually enriching and coexisted without prejudice of any kind, in both painting and architecture.

Bibliography

Manuscript and Archival Sources

Biblioteca Ambrosiana, Milan, *Codex Atlanticus*, fol. 683^{v}.
Biblioteca Apostolica Vaticana, Vatican, *Libro di praticha d'arismetrica*, Cod. Ottoboniano Latino 3307, fol. 172^{r}.
Biblioteca Capitolare, Verona, 245–CCLXXIII.
Biblioteca Governativa, Lucca, Codex 1448, fols 53^{v}–54^{r}.
Institut de France, Paris, MS A, fol. 40^{r}.
Trattato di geometria pratica dal codice L.IV.18 (sec. XV) della Biblioteca Comunale di Siena, ed. by Annalisa Simi (Siena: Università degli Studi di Siena, 1993), fol. 55^{v}.

Primary Sources

Alberti, Leon Battista, *Opere volgari*, ed. by Cecil Grayson (Bari: Laterza, 1973).
Alberti, Leon Battista, *De pictura*, ed. by Lucia Bertolini (Florence: Edizioni Polistampa, 2011).
Chimenti, Massimo, 'Rilievo fotogrammetrico e documentazione informatica', in *La Trinità di Masaccio. Il restauro dell'anno duemila*, ed. by Cristina Danti (Florence: Edifir, 2002), pp. 89–96.
Euclid, *Euclidis Optica*, ed. by Johan L. Heiberg, in *Euclidis Opera Omnia*, ed. Johan L. Heiberg and Heinrich Menge, 9 vols (Leipzig: G. B. Teubner, 1883–1916).
Euclid, *Elements*, in Hubert L. Busard, *Campanus of Novara and Euclid's Elements*, 2 vols (Stuttgart: Franz Steiner, 2005).
Filarete (born Antonio Averlino), *Trattato di architettura*, ed. by A. M. Finoli and L. Grassi, 2 vols (Milan: Il Polifilo, 1972).
Francesca, Piero della, *De prospectiva pingendi*, ed. by Giusta Nicco Fasola (Florence: Sansoni 1942; repr. 1974).
Jacopo, Mariano di, known as Taccola, *Liber tertius de ingeneis ac edifitiis non usitatis*, ed. by James H. Beck (Milan: Il Polifilo, 1969).
Lastri, Marco, *l'Osservatore fiorentino. Sugli edifizi della sua patria* (Florence: Gaspero Ricci, 1798).
Ludwig, Heinrich, *Leonardo da Vinci. Das Buch von der Malerei* (Wien: Wilhelm Braumüller, 1882).
Mancini, Girolamo, *Opera inedita et pauca separatim impressa* (Florence: G.C. Sansoni, 1890).
Pacioli, Luca, *Summa de aritmetica, geometria, proportioni et proportionalità* (Venice: A. Paganius Paganinus, 1494).
Pacioli, Luca, *De divina proporzione* (Venice: A. Paganius Paganinus, 1509).

Vasari, Giorgio, *Le Vite de' più eccellenti pittori scultori e architettori* (Florence, 1568), in *Le Opere di Giorgio Vasari*, ed. by Gaetano Milanesi, 9 vols (Florence: Sansoni, 1973).
Vasari, Giorgio, *The Lives of the Artists*, trans. and ed. by Julia Conaway Bondanella and Peter Bondanella (Oxford: Oxford University Press, 1991).
Ximenes, Leonardo, *Del vecchio e nuovo gnomone fiorentino* (Florence: Nella Stamperia Imperiale, 1757).

Secondary Works

Aiken, Jean Andrews, 'The Perspective Construction of Masaccio's Trinity Fresco and Medieval Astronomical Graphics', *Artibus et Historiae*, 16 (31) (1995), 171–87.
Arrighi, Gino, 'Un estratto dal "De visu" di M° Grazia de' Castellani (dal Codice Ottoboniano latino 3307 della Biblioteca Apostolica Vaticana)', *Atti della Fondazione Giorgio Ronchi*, XXII (1967), 44–58.
Arrighi, Gino, 'La matematica in Firenze nel Rinascimento. Il Codice Ottoboniano Latino 3307 della Biblioteca Apostolica Vaticana', *Physis*, X, (1968), 70–82.
Battisti, Eugenio, *Filippo Brunelleschi* (Milan: Electa, 1976).
Baxandal, Michael, *Painting and Experience in Fifteenth-Century Italy: A Primer in the Social History of Pictorial Style* (Oxford: Oxford University Press, 1972).
Camerota, Filippo, *La prospettiva del Rinascimento. Arte architettura scienza* (Milan: Electa, 2006).
Camerota, Filippo, *Masaccio y la geometria practica: la realizacion de la Trinidad*, in Carmen Gonzales Roman, ed., *A través de la mirada. Anatomia, arquitectura y perspectiva en la tradicion artistica occidental* (Madrid: Abada Editores, 2013), pp. 127-144.
Fagiolo, Marcello, 'Eredità di Brunelleschi: la rivoluzione prospettica e il punto di vista vasariano', *Architettura e arte del Principato mediceo (1512–1737). Firenze e la Toscana, Vasari e gli Uffizi, Bollettino della Società di Studi Fiorentini*, 22, ed. by F. Canali (2013), 134–42.
Field, J. V., Roberto Lunardi, and Thomas B. Settle, 'The Perspective Scheme of Masaccio's Trinity Fresco', *Nuncius*, 4 (2) (1989), 31–118.
Field, J. V., *The Invention of Infinity: Mathematics and Art in the Renaissance* (Oxford: Oxford University Press, 1997).
Field, J. V., 'Piero della Francesca's Perspective Treatise', in *The Treatise on Perspective: Published and Unpublished*, ed. by L. Massey (Washington, DC: National Gallery of Art, 2003), pp. 63–78.
Gambuti, Alessandro, 'Nuove ricerche sugli Elementa picturae', *Studi e Documenti di Architettura*, 1 (1972), 131–72.
Gambuti, Alessandro, 'I "libri del disegno": Filarete e l'educazione artistica di Galeazzo Maria Sforza', *Arte Lombarda*, 38–39 (1973), 133–43.
Hertlein, Edgar, *Masaccios Trinität: Kunst, Geschichte und Politik der Frührenaissance in Florenz* (Florence: Olschki, 1979).
Hoffmann, Volker, 'Masaccios Trinitätfresko: die Perspectivkonstruktion und ihr Entwurfsverfahren', *Mitteilungen des Kunsthistorischen Institutes in Florenz*, 40 (1996), 43–77.
Huber, Florian, 'Das Trinitätfresko von Masaccio und Filippo Brunelleschi in Santa Maria Novella zu Florenz' (unpublished doctoral thesis, Ludwig-Maximilians-Universität, 1990).

Janson, Horst Woldemar, 'Ground Plan and Elevation in Masaccio's Trinity Fresco', in *Essays in the History of Art Presented to Rudolf Wittkower*, ed. by Douglas Fraser, Howard Hibbard, and Milton J. Lewine (London: Phaidon, 1967), pp. 83–88.

Kemp, Martin, *The Science of Art: Optical Themes in Western Art from Brunelleschi to Seurat* (New-Haven: Yale University Press, 1990).

Kern, G. Joseph von, 'Das Dreifaltigkeitfresko von Santa Maria Novella. Eine Perspektivisch-Architekturgeschichtliche Studie', *Jahrbuch der Königlich preussischen Kunstsammlungen*, 34 (1913), 36–58.

Kuhn, Jehane R., 'Measured Appearances: Documentation and Design in Early Perspective Drawings', *Journal of Warburg and Courtauld Institutes*, 53 (1990), 114–32.

Manetti, Antonio di Tuccio, *Vita di Filippo Brunelleschi, preceduta da la novella del grasso*, ed. by D. de Robertis (Milan: Il Polifilo, 1976).

Myers, Marshal Neal, 'Observations on the Origins of Renaissance Perspective: Brunelleschi, Masaccio, Petrus Christus' (unpublished doctoral dissertation, Columbia University, 1978).

Parronchi, Alessandro, 'Sul significato degli "Elementi di pittura" di L. B. Alberti', in *Cronache di archeologia e storia dell'arte* (1967), VI, pp. 107–15.

Polzer, Joseph, 'The Anatomy of Masaccio's Holy Trinity', *Jahrbuch der Berliner Museen*, 13 (1971), 18–59.

Raynaud, Dominique, 'Linear Perspective in Masaccio's Trinity Fresco: Demonstration or Self-Persuasion?', *Nuncius*, 18 (1) (2003), 331–44.

Sanpaolesi, Piero, *Brunelleschi* (Milan: Edizioni per il Club del Libro, 1962).

Valli, Franca Manenti, *Leonardo. Il comporre armonico nella tavola dell'Annunciazione* (Milano: Silvana, 2012).

PIETRO ROCCASECCA

Divided into Similar Parts

Representation of Distance and Magnitude in Leon Battista Alberti's De pictura

Introduction

While in this chapter I am interested in situating Alberti in an intertextual web of sources on optics and perspective, it is important to keep in mind that *De pictura* is a humanist text dealing with questions of knowledge and making use of various literary forms: the commentary, the university *quaestio*, the narrative, the moral precept. In general, Alberti does not make his sources clear. He never names, for example, Aristotle, even though Aristotle is the source of a number of important arguments, such as the conflict between the mathematical and physical line, the concept of place, the colours, the theory of composition. Xenophon and Galen are named, but not at the points where their writings are used. Finally, in order to introduce known or lengthily discussed topics and to cite theorems and definitions of geometry, Alberti adopts the generic terms of 'philosophers' and 'mathematicians', even if he does not always agree with them.

Moreover, the identification of a source mostly remains open to a range of possibilities, because what Alberti is quoting are ideas, thoughts and texts on natural philosophy or optics that have been transmitted from one author to another. As a young man who studied at the University of Bologna (where, besides law, he also followed courses in natural philosophy), who worked at the Vatican Curia, only a short distance from the Vatican Library run by Platina, and who wrote his brief treatise in Florence, where works by Pecham, Witelo and Alhacen himself were available, the last both in Latin and the vernacular, Alberti clearly had a range of possible opportunities to access Alhacen's work on *perspectiva*, which, I will argue, was so important to him.

For all these reasons, the comparison between *De pictura* and other texts on *perspectiva* will be of an interdiscursive nature. The various textual sources received and re-elaborated by Battista Alberti will be highlighted and the inter-textual passages — which, although rare, have already been indicated by me elsewhere — will not be shown.[1] The perspectival method in Leon Battista Alberti's *De pictura* is divided among Books One and Two. The

1 For the transmission of Alhacen's work on optics and the inter-textual quotations in *De pictura*, see: Pietro Roccasecca, *Filosofi, oratori e pittori. Una nuova lettura del De pictura di Leon Battista Alberti* (Rome: Campisano Editore, 2016).

Pietro Roccasecca Accademia di Belle Arti, Rome, p.roccasecca@gmail.com

Perspective as Practice. Renaissance Cultures of Optics, ed. by Sven Dupré, Turnhout, 2019 (*Techne. Knowledge, Technique, and Material Culture*, 1), p. 361-389

DOI 10.1484/M.*Techne*-EB.5.117733

method described in Book One has been amply studied; the one exposed in Book Two, all but ignored. The term 'perspective' is never used in *De pictura*. 'Intersection' is the word Alberti uses to define what we nowadays call perspective. The definition of painting as an 'intersection of the visual pyramid' is based on a concept fundamental to the *Scientia perspectiva*, perspectival science or optics, of the natural philosopher Ibn al-Haytam, known in Europe as Alhacen from the thirteenth century onwards.

In *De aspectibus* Alhacen illustrates how sight results from visible bodies' emission of coloured forms: visible bodies emit outgoing lines which hit the eyes' surface perpendicularly; these form a pyramid which converges at the eye's centre and vehicles coloured forms. The pyramid entering the eye is intersected by the glacialis' spheric surface. The coloured form emitted by the visible body is reproduced on the glacialis and then transmitted to the optic nerve. Vision, according to Alhacen, directly results from the intersection of the visual pyramid.[2]

Intersection and Surfaces Composition

Alberti's *De pictura* divides the painter's work into three parts: circumscription, composition, and light-reception. While intersection of the visual pyramid is operative in all three of these parts, its application is mostly in surfaces composition.[3] Composition itself is articulated in three levels: composition of surfaces, of members, and of bodies. For minor surfaces, i.e., of small dimensions, such as those which describe a face or a figure's corporeity, suffices to use an instrument Alberti claims to have invented. This tool consists of a 'veil' and is based on the 'intersection' of the visual pyramid. It facilitates the comprehension of the intersection of small surfaces for animals, man included (in the Aristotelian sense). For surfaces larger than the human figure, such as buildings, Alberti writes that it is necessary to 'find new ratios' through a graphic process which he expounds in two distinct portions of his treatise.

The 'veil' is explained in Book Two, followed by an exposition of the graphic method to obtain architecture proportionate to the human figure. One might logically expect Alberti to give, here, the method for establishing the intersection of a human figure and the ground. Instead he explains it in the conclusion of Book One. The *modo ottimo* ('optimal mode') method for establishing the magnitudes of ground and human height would, logically, be located after the description of the 'veil', just before the passage on

2 Alhacen, *Alhacen's Theory of Visual Perception: A Critical Edition, with English Translation and Commentary, of the First Three Books of Alhacen's 'De aspectibus', the Medieval Latin Version of Ibn al-Haytham's Kitāb al-Manāzir*, ed. and trans. by A. Mark Smith, 2 vols (Philadelphia: Transactions of the American Philosophical Society, 2001), I, 6. 29; I, 6. 63. See also Samuel Y. Edgerton, 'Alberti's Optics' (unpublished dissertation, University of Pennsylvania, 1965), pp. 81–83; Edgerton previously recognised the analogy between the 'intersection' of Alberti's visual pyramid and that of the radiant pyramid that intersects the glacialis.

3 Alberti wrote two different versions of the *De pictura*: the first in Italian, the second in Latin. I generally follow the vernacular edition, as edited by Lucia Bertolini: Leon Battista Alberti, *De pictura (redazione volgare)*, ed. by Lucia Bertolini (Florence: Edizioni Polistampa, 2011). Citations from the Latin edition are taken from Leon Battista Alberti, 'De pictura', in *Opere volgari*, ed. by Cecil Grayson, 3 vols (Bari: Laterza, 1973), III. The citations from the Italian version of the *De pictura* are of the translator; for the Grayson Latin version I quote Leon Battista Alberti, *On Painting and on Sculpture, the Latin Texts of 'De pictura' and 'De statua'*, ed. by Cecil Grayson (London: Phaidon, 1972).

the intersection of buildings and ground explained in Book Two. The meaning of 'optimal mode' as 'intersection' of the visual pyramid would be clearer were it treated right after the veil: the veil operatively demonstrates what Alberti means by intersection and 'why painted things never seem equal to real ones if they are not seen from a definite distance'. Ground and building intersection are structurally interdependent, one results from the other. One begins where the other finishes: the centric line — whose function was not yet clear in Book One — is essential to drawing buildings' height in Book Two.

In this essay I will study the 'intersection' of surfaces following the procedure's logical and temporal order, starting with small surfaces and ending with large ones. For this reason I will begin with the veil (Book Two) then proceed to the 'optimal mode' (Book One) and, finally, conclude with how to make the architecture proportionate to the human figures (Book Two).[4] The aim of studying this as a logical, not textual, procedure is to show the method as a whole and better to understand the theoretical, practical and aesthetic issues at stake.

As we will see, Alberti applies to painting the optical theory of the intersection of the visual pyramid, introducing the distance of observation of the intersection, *id est*, the painted surface as the ratio proportioning the representation of the distance among the bodies represented in the painting, in the operative protocol of a pre-existing method, which structured all the composition by means of purely arithmetical proportions. Even though the 'veil' and the part dedicated to proportions between 'buildings' and the human figure can be assimilated as methods of architectural survey, the novelty introduced by Alberti is not in the application of the methods of surveying to painting, but of the optics on which those methods are based. I will say more about the difference between painting and surveying later in this text.

Finally, because Alberti, in his 'optimal mode', only takes into consideration the distance of observation of the intersection (painted surface), his method is valid, but only under certain conditions. It will be Piero della Francesca who finalises a generally valid perspective method, which not only takes into consideration the distance between the eye and the intersection of the visual pyramid, but also that between the intersection and the observed object, regulating the proportions of the diminishing of the apparent magnitudes by means of what today we define as a double-ratio, which Piero called *proportione degradata* ('degraded proportion').[5] Nevertheless, *De pictura* was undoubtedly the decisive turning point that changed the theory of the visual representation of space, figures and distance, connecting it to the most up-to-date theory of vision of its time.

The Veil: Intersection of the Visual Pyramid and Drawing

The veil is an instrument that reveals and facilitates the operation of intersection (*intersecazione* or *intercisione*). Alberti's veil is not just a netted cloth but a thin cloth with an insert: a reticle

4 The veil is discussed in the vernacular edition of the *De pictura* (*redazione volgare*), ed. by Bertolini, II, 7. 11–8. 9.

5 On this subject see: Pietro Roccasecca, 'La prospettiva lineare nel Quattrocento: dalle proporzioni continuata e ordinata alla proporzione degradata', *Proportions. Science, Musique, Peinture & Architecture*, ed. by Sabine Rommevaux, Philippe Vendrix, and Vasco Zara, Études Renaissantes, 6 (Turnhout: Brepols, 2011), pp. 277–97.

of thicker and coloured threads, in between which the fabric is transparent. One does not look from a sighting device, nor does one use sheets of grid paper. The most widespread iconography of the veil is that expounded by Dürer in his 1525 treatise, *Unterweisung der Messung*. In it, there is no mention of thin cloth, only of a thickly threaded reticle; the instrument includes a sighting device, as well as the use of grid paper onto which to draw.

Alberti distinguishes the veil's ease-of-use from its two uses. His veil is convenient because it facilitates looking at the subject to be portrayed. Furthermore it makes the viewer aware of the 'power of sight' and allows him to observe how surfaces change appearance when one or several factors vary — distance, the position of the eye's centre, light. These factors' theoretical bases are treated in Book One; Book Two's 'veil' reveals how these factors play out in visual experience. The 'veil' allows discovery of the visual pyramid's true cuspid, namely, the point of the eye as projected onto the painting's surface.[6] Thanks to the 'veil' the viewer does not lose this point, even if the observed subject changes in appearance, because the reticle of threads works as an aiming device.

As for its two functions, the veil facilitates drawing a figure's outline. Alberti recommends circumscribing only the outline. Contour is understood as outlining a figure without its details. The face provides an example of how to use veil: the space between two parallel lines can make visible a forehead, another the nose or again cheeks. A panel or wall divided into parallels similar to the veil's thick reticle will make it easier for the painter to find the right place for each surface and body-part.[7]

In this case 'divided into similar parallels' means that the panel is divided into the same number of quadrangles as the veil. 'Parallels' designate the spaces between parallel lines, not the lines themselves.[8] Transfer by proportional scale requires that the two surfaces be divided into equal number of parts, whose dimensions vary according to scale. The second function of the veil is to help fill out contours with their corporeal volumes. In the Latin text Alberti clearly states that bodies are *seen* in the veil, outlined and painted in.[9] When light passes through the veil's thin cloth, it is diffracted. According to Alberti this allows the painter to see in the veil itself the 'reception of light' as though painted by nature itself. In this way the veil translates bodies' three-dimensional volumes into a two-dimensional surface, making visible how prominent and rounded forms would appear if they were painted on a flat surface.

As I said the veil consists of a very thin and rare cloth that lets through the light-colour emitted by observed bodies. The colour chosen for the veil provides the painting's half-tone and simplifies the gradients of maximal light and dark. The draughtsman must therefore mark only light and shadow onto the veil, half-tones being provided by the colour of the veil. Alberti's conception of the coloured veil recalls pen and ink drawing with white

6 Anglophone historiography calls this the vanishing point, but here it just locates the position of the eye as projected onto the surface of the artwork.

7 Alberti, *De pictura* (*redazione volgare*), ed. by Bertolini, II, 7. 17.

8 Alberti, *De pictura* (*redazione volgare*), ed. by Bertolini, I, 20. 8: 'E a questo modo mi truovo descritto tutti e' *paralelli, cioè le braccia quadrate* del pavimento nella dipintura' (italics mine) ('In this way I find described all the parallels, that is, the squared *braccia* of pavement on the painting').

9 Alberti, 'De pictura', ed. by Cecil Grayson, II, 31: 'quandoquidem rem ipsam prominentem et rotundam in istac planitie veli conscriptam et depictam videas' ('since you can see any object that is round or in relief, represented on the flat surface of the veil').

highlights on parchment or coloured paper, a technique that Cennino Cennini considered the best way to learn to paint.[10] To those who disapprove of using a veil, Alberti responds that the aim of the painter is not to endure fatigue but to achieve greatest resemblance to the subject. To those who want to test their intelligence without benefiting from the veil's usefulness, he suggests: 'note first the limits of the objects *inside* the veil's parallels' (italics mine).[11]

Given that the parallels are not the lines but the internal space delineated by parallel lines, by parallel we are given to understand the internal space of the veil's cloth. Given too that the verb 'notare' means to mark and to annotate it seems highly likely that exhorting to note 'first the limits of the objects *inside* the veil's parallels' (italics mine), Alberti is instructing to paint rounded forms and reliefs directly onto the veil's cloth, just as they are seen upon it.

Painting as Open Window: Practices and Procedures

The metaphor of painting as a window unto the world has met with the widest-spread success. It has become the very symbol of Albertian painting, of pictorial naturalism, and the immediately recognizable marker of a new age in the history of visual representation. This metaphor is based on the theoretical model of the visual pyramid: point of view, window as plane intersecting visual rays. The operation required to realise this in practice is to make the visual pyramid and the pictorial plane intersect. To achieve this, the painter must do two things. First, he must get the figures to be proportional to the floor tiles. Second, he must trace the bodies' diminishing size according to distance. This part is further subdivided into two procedures: one already known to contemporary painters, mistaken and theoretically superseded by Alberti's novel application of the theory of vision: the 'optimal mode', that is the first occurrence of perspective in painting, is a contribution that changed the Western way of seeing the world and the history of visual representation.[12] The first part of the procedure to obtain proportional figures, and the first (erroneous) part of the procedure to diminish size are two sides of the same coin. As we will see, although the method is wrong, it is based on Alhacen's optical doctrines and Euclid's theory of proportions.

The vernacular manuscripts of the *De pictura* do not offer diagrams illustrating the intersecting method. The first part of the procedure begins with determining the height of the human figures to be depicted according to the geometric shape of the picture plane (a regular quadrangle) and noting these on a side of the pictorial surface with a small dash (Figure 1). Height is then divided into three parts; one of these serves to subdivide the base of the quadrangle into as many parts as can fit. According to Alberti the average man

10 See Cennino Cennini, *Il libro dell'arte*, ed. by Franco Brunello (Vicenza: Neri Pozza, 1982), Chapter 14. For examples of drawings on tinted paper or prepared with colour, see Anna Maria Petrioli Tofani, 'I materiali e le tecniche' in *Il disegno. Forme, tecniche significati* (Monza: Amilcare Pizzi Editore, 1991), pp. 187–250.

11 'prima notino i termini de le cose drento da' paraleli del velo'. Alberti, *De pictura* (*redazione volgare*), ed. by Bertolini, II, 8. 3.

12 The first part of the method to 'intersect' is found in Alberti, *De pictura* (*redazione volgare*), ed. by Bertolini, I, 19. 1–20. 14.

Fig. 1 Determination of the height of the painted figure.

is three *braccia* high.[13] It follows that the part obtained is equivalent to one real *braccio*. At the end of this operation, the bottom line is subdivided into parts proportional to one *braccio* and the picture plane's side into three of the same parts, indicating the figures' height (Figure 2).

The next step consists in determining the ground or pavement's horizontal lines. The sentence 'ed èmmi questa linea medesima proporzionale a quella ultima quantità quale prima mi si traversò inanzi' in the vernacular is somewhat obtuse.[14] To understand what

13 The *braccio* is an antique unit of measure which corresponded to slightly different dimensions according to region. The Florentine *braccio*, for instance, measured fifty-eight centimetres. It should be mentioned that being able to start with a randomly chosen measure and dividing a randomly chosen quadrangle's base into a finite number of parts is a rather remote possibility. Nevertheless, the interpretative diagrams illustrating the *modo ottimo* all show picture planes precisely divided.

14 Lucia Bertolini, in her edited version of *De pictura* (*redazione volgare*), p. 107 ff., considers the reference to the last line to be an error introduced by the author. All the lines, however, must be proportional to each other. The Latin edition dismisses any possible doubts on this issue: 'ac mihi quidem haec ipsa iacens quadranguli linea

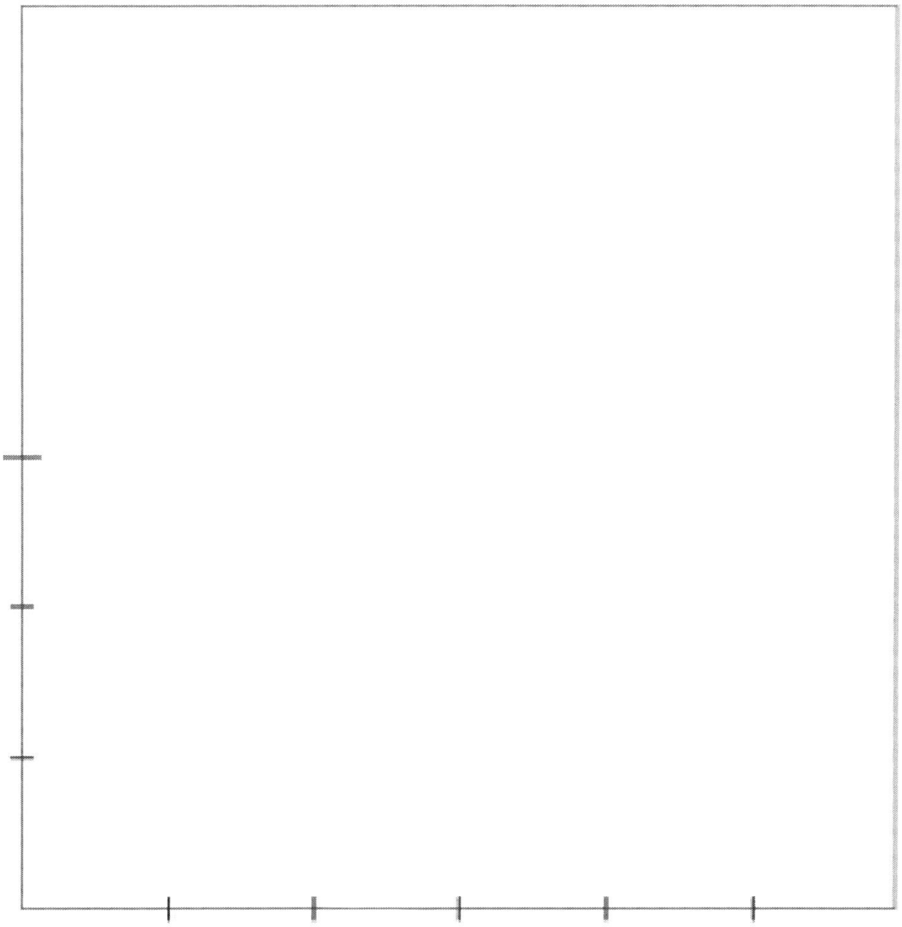

Fig. 2 Division in equal parts of the height and the base of the quadrangle to be painted.

Alberti means, recourse to the Latin version is helpful. Here Alberti specifies that the bottom line is the horizontal line (transversal line) closest to the viewer that delineates the rows of floors tiles; the horizontal lines are equal and parallel (equidistant); their magnitudes are proportional to each other; they diminish only in appearance Once this is finished, the base line is subdivided from one extremity to the other; the marked side is subdivided only for those three units that correspond to the human figures' height. Above that there are no measurement markings.

Alberti is applying what he exposed in previous chapters of the *De pictura*: man is 'mode and measure of all things'.[15] Dividing man's height into three parts and using one

est proximiori transversae et aequedistanti in pavimento visae quantitati proportionalis' ('this bottom line of the rectangle is for me proportional to the next equidistant quantity seen on the pavement'), in Alberti, 'De pictura', ed. by Grayson, I, 19.

15 'modo e misura di tutte le cose'. Alberti, *De pictura* (*redazione volgare*), ed. by Bertolini, I, 18. 14.

of these to divide the base line establishes a relation between the base of the pictorial quadrangle and the height of depicted figures, a relation which remains constant in each point of the pavement. The height of the painted human figure is in fact always equal to the width of the three tiles on which he stands and, for that reason, each and every point of the pavement's height and width is always coordinated according to one and the same proportion: everything represented maintains in each point the same ratio of height to width. Furthermore, whatever their apparent dimension, heights are always divided into three parts and always measure three *braccia* in length.

This initial phase of 'intersection' applies the first two Definitions of Book V of Euclid's *Elements*, to attune two kindred magnitudes: the 'width' of the quadrangle's base and the 'height' of the human figure to be depicted are made up of identical 'parts' or units of measure, making them multiples of the same 'part', 'magnitude' or 'size', so to speak. The *braccio* is the 'part' that quantifies both the dimension of the human figure and the width of the ground. By virtue of this operation both height and width are multiples of the same 'part' and, because the size of the 'part' is known, so is the size of the men and the things the painting depicts. The three *braccia* measure established for the human figure provides the means to compare all other heights depicted, including that of the viewer. If the average man is three *braccia* high, so is the posited viewer.

The height recorded on the quadrangle's side plays the same measuring function described in *De statua*, which Alberti calls *exempeda*.[16] A line divided into six parts, it serves to scale copies of statues. The marked side of the painting has the same function: from every point of the ground the figure shrinks, but proportionally, and always equivalent to three *braccia* in height. Alberti states that the central point is placed at will; however, the choice of its position is not utterly free. Indeed, he specifies that the central point must not be placed above the man one wants to depict. This indication leaves little room for choice: the horizontal line may be more or less thick but it must be parallel to the base of the pictorial quadrangle, at the height of the painted figure, and it indicates the point of view's maximum height.

Nowadays the vanishing point can be placed wherever, but Alberti's central point is not yet a vanishing point, solely a point of view. The central ray ends at the point where the observer's gaze is placed: 'no greater than the painted man's height'.[17] It means coordinating the observer's height to the painted figure's height, such that 'him who sees and those things that are painted appear *on* the same plane' (italics mine).[18] In other words, the painting's observer and the painted figure's feet stand on the same plane and their eyes are at the same level.

The 'intersection' expounded in the *De pictura* concerns itself with one case only: the observer and the observed stand on the same ground, as though they were in the same

16 Leon Battista Alberti, *De statua*, ed. by Marco Collareta (Pisa: Sillabe, 1998), Paragraph 6.

17 'non più che sia l'altezza dell'uomo quale ivi io abbia a dipignere'. Alberti, *De pictura (redazione volgare)*, ed. by Bertolini, I, 19. 9.

18 'chi vede e le dipinte cose vedute paiono su uno medesimo piano'. Alberti, *De pictura (redazione volgare)*, ed. by Bertolini, I, 19. 9. The vernacular edition established by Cecil Grayson which, until Bertolini's edition, was the authoritative text reads: 'paiono medesimo in *suo* uno piano' (italics mine) 'suo' (his/hers) is a misprint introduced by Cecil Grayson instead of 'su' (on). The misprint did not enable the reader to understand that the observer and the observed stand on the same ground, at the same level.

room, church or square. Alberti confirms that this is his intended meaning when, towards the chapter's conclusion, he writes: 'We see nearly all the men's heads at the same level but the feet of those remotest figures almost correspond to the knees of those nearest to us' (Figures 3–5).[19] Anyone can verify Alberti's affirmation: in a large space we see heads align and the feet furthest from us at the same level as the shins and ankles of those closest to us. Obviously in this case he who sees and he who is seen are on one and the same plane, namely, the temple floor. The metaphor of painting as open window can be misleading here. The viewer's eye is placed at the same height as the depicted figure: this means that painting is not a *piano nobile*'s open window but a door/window on the ground floor, from which we observe what is happening on the town square before us. He who observes is on the same plane as he who is depicted and, like him, the observer is three *braccia* high.

After having matched the central point with the base partitions (Figure 6), Leon Battista observes that in this way it is possible that the lines of vision follow each other 'almost to infinity'.[20] This definition has led some to believe that Alberti introduced the concept of an infinite point. The 'almost' is what disproves this and makes all the difference: *almost* to infinity means not to infinity. Moreover, what Alberti defines as 'almost to infinity' is not the distance to the point but the systematic reduction of transversal magnitudes, which, being continuous quantities, can diminish infinitely. Alberti adds 'almost' to keep transversal magnitudes within the field of human vision. Were the quantities beyond 'finite' they would not be visible to us.[21]

This last part of the procedure determines the centric point's height and, consequently, the visual angle's width. Moreover, it delimits the apparent diminution of the transversal lines. The lines traced from the central point to the divisions of the quadrangle's base line show the alterations of equidistant transversal lines, whose magnitudes diminish 'almost to infinity' and shrink in keeping with the plane delineated by the two most external lines.

The most important result of this phase of 'intersection' is the graphic schema it produces. Diminishing magnitudes succeeding each other evoke three-dimensionality. This is a crucial moment in the history of the psychology of visual perception: it establishes a convention allowing the painter to depict and the observer to recognise such visual characteristics as distance and apparent diminution of magnitudes due to distance, that is, to represent and perceive the third dimension in the flat picture plane.

---

19 'Veggiamo ne' tempî i capi degli uomini quasi tutti ad una quantità, ma i piedi de' più lontani quasi corrispondere ad i ginocchi de' più presso'. Alberti, *De pictura (redazione volgare)*, ed. by Bertolini, I, 20. 14.

20 Alberti, *De pictura (redazione volgare)*, ed. by Bertolini, I, 19. 10: 'quali segnate linee a me dimostrino in che modo, quasi persino in infinito, ciascuna traversa quantità segua alterandosi' ('These inscribed lines indicate to me in what way, almost to infinity, each transverse quantity decreases').

21 See J. I. Beare and G. R. T. Ross, 'Parva Naturalia', in *The Works of Aristotle*, ed. by W. D. Ross (Oxford: Clarendon Press, 1931), 449a20: 'That every sensible object is a magnitude, and that nothing which it is possible to perceive is indivisible, may be thus shown. The distance whence an object could not be seen is indeterminate, but that whence it is visible is determinate. [...] Now, there is, in the interval of distance, some extreme place, the last from which the object is invisible, and the first from which it is visible. This place, beyond which if the object be one cannot perceive it, while if the object be on the hither side one must perceive it, is, I presume, itself necessarily indivisible. Therefore, if any sensible object be indivisible, such object, if set in the said extreme place whence imperceptibility ends and perceptibility begins, will have to be both visible and invisible their objects, whether regarded in general or at the same time; but this is impossible'.

Fig. 3-5 The three photos taken in sequence recreate the situation described by Alberti: the viewer and the characters stand on the floor of a cathedral, their heads are at the same height, independently from their position in the church. The apparent height of the person that moves toward the altar decreases in relationship to the increasing distance from the photographer. We see her (or his?) head at the same height, while her/his feet correspond first to the knee-level, then to the elbow-level, of those in front. Photos by the author.

The schema drawn according to the *De pictura* instructions creates the illusion of a space beyond the surface of the support. A sort of double pyramid is formed: its base is the plane of intersection of vision and representation, its cusps the viewer's eye and the work's centric point, the lines that extend from the central point to the base's marked divisions mirror the outmost rays reaching from the eye to the pavement tiles' dividing lines.

The 'Superbipartienti' Method: Euclid, Boethius and Optics

Before illustrating the 'optimal mode' Alberti explains how to determine the proportions by which the side floor tiles can be foreshortened. He begins by describing a simple but erroneous method elected by 'some' painters. One traces at random a line parallel to the quadrangle's base and then divides the space between those lines into three parts. Next, a line is traced to be two-parts distance from the first one. That space is further divided into three parts, the next line always two-thirds of a distance from its predecessor, and so on, until the desired number of transversal lines have been drawn (Figure 7).

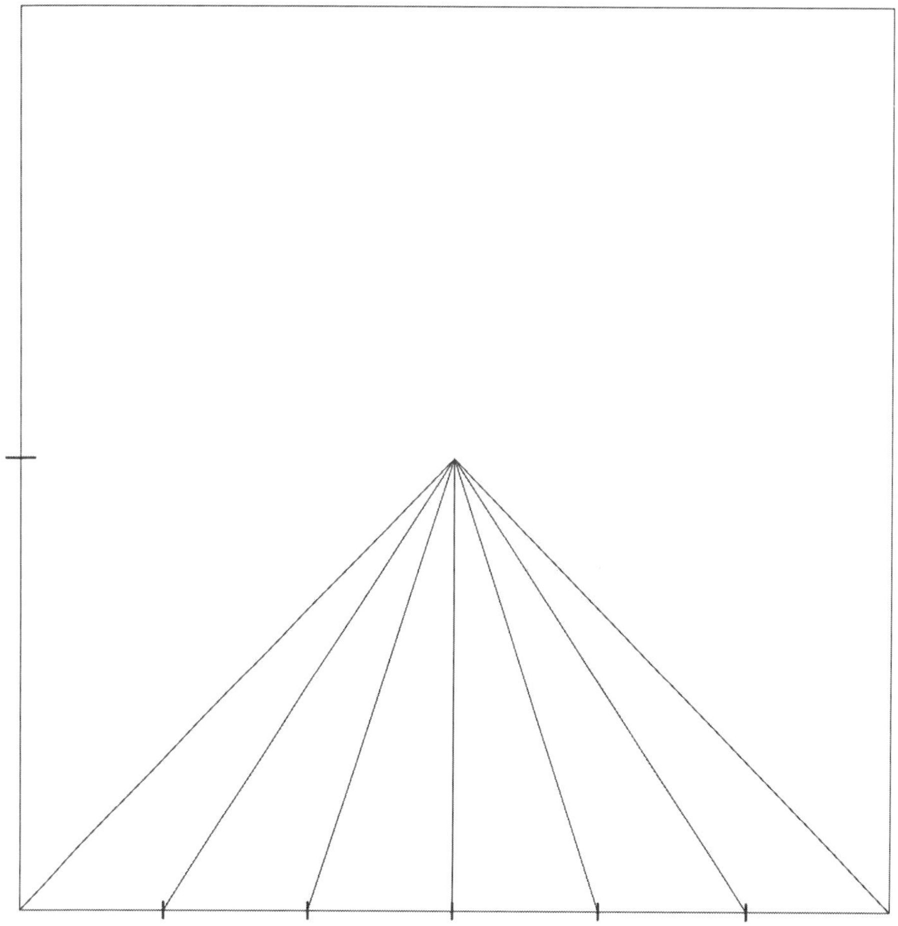

Fig. 6 Lines drawn from the central point to the base partitions.

Alberti considers this method to be mistaken because the first line is randomly positioned. Indicating spatial recession by reducing magnitudes by one-third — which to our eyes might seem an expedient solution — does not bother Alberti, however. On the contrary, it seems the only acceptable part of this method because the construction of 'successive spaces follow reason',[22] i.e., following proportions that, asserts Alberti, mathematicians call 'superbipartient'.[23] The term belongs to the nomenclature of numerical relations as defined by Boethius, but is incorrectly used by Alberti. The correct denomination

22 Alberti, *De pictura (redazione volgare)*, ed. by Bertolini, I, 19. 15: 'Però che ponendo la prima linea a caso, ben che l'altre seguano a ragione, non però sanno ove sia certo luogo alla cuspide della pirramide visiva' ('Because if the first line is placed by chance, even though the others follow in proportion, they do not know for certain the place where the point of the visual pyramid lies').

23 See Alberti, *De pictura (redazione volgare)*, ed. by Bertolini, I, 19. 13. This term has created transcription problems for copyists. In the three vernacular versions it is always written 'superbi partienti'. In some Latin manuscripts a blank space is inserted. I have personally verified the vernacular manuscripts; for the Latin manuscripts, see Leon

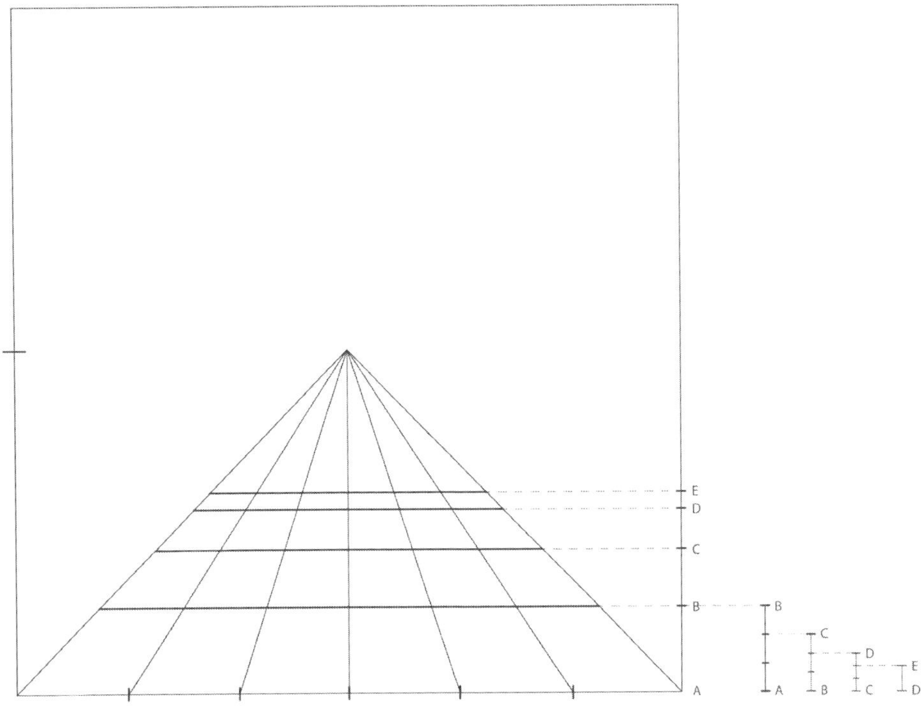

Fig. 7 The 'superbipartienti' method.

of the 3:2 relation Alberti describes is 'subsesquialter'. In any case, though the term is incorrect, his cultural context and intellectual reference remains that of Boethius' nomenclature.[24]

Recall that Alberti considers this method flawed because the first line is placed at random, instead of being deliberately positioned in relation to the 'place' of the visual pyramid's cusp. Since size does not diminish in relation to the central point, the last magnitude can very well be higher than the 'depicted man', thereby contradicting the first part of the correct method, which Alberti deems founded and mathematically justified.

Working in this way takes into account that the 'central point' is not only as 'high' as the observer but also 'distant' from him. Alberti writes: 'they do not know where a

Battista Alberti, *On Painting*, ed. by J. R. Spencer (London: Routledge and Paul, 1956), p. 110 n. 46. The Italian translation by Lodovico Domenichi correctly includes the term 'subsesquialtero'; see *La pittura di Leon Battista Alberti*, trans. by Lodovico Domenichi (Venice: Gabriel Giolito de Ferrari, 1547), p. 15, verso.

24 See Lucia Bertolini, 'Sulla precedenza della redazione Volgare del "De pictura" di Leon Battista Alberti', in *Studi per Umberto Carpi. Un saluto da allievi e colleghi pisani*, ed. by Marco Santagata and Alfredo Stussi (Pisa: Edizioni ETS, 2000), pp. 181–210 (p. 205 n. 5). The term 'superbipartienti' belongs to the nomenclature of numerical relations defined by Boethius, who follows Nicomachus of Gerasa. The latter is responsible for introducing the concept of 'denominationem partium', literally, 'denomination of parts'. This concept is foreign to modern mathematics but was crucial for the Latin tradition of the Euclidian doctrine of proportions. I owe special thanks to Sabine Rommevaux for clarification on this aspect of Medieval mathematics. See Sabine Rommevaux, 'Aperçu sur la notion de dénomination d'un rapport numérique au Moyen-Âge et à la Renaissance', *Methodos*, 1 (2001), 223–43.

given place of the visual pyramid might be'.[25] He concludes this particular paragraph by admonishing 'that no painted thing can appear equal to real ones if there is not a definite distance from which to view it'.[26] Here Alberti is *de facto* anticipating the 'optimal mode' according to which the viewer's height and his distance from the plane of intersection are the necessary parameters to establish the proportions of diminishing size non-equidistant to the pictorial plane of intersection.[27] Historiography has little interested itself in the 'superbipartient' method, considering it to be an expedient empirical practice rather than a reasoned mathematical perspectival method; a 'rule of thumb'.

The proper method, instead, has its theoretical bases in Alhacen's work on the cognition of the visual characteristics *remotio* ('separation') and *quantitas remotionis* ('magnitude of distance'); the 'superbipartient' method applies Boethius' arithmetic to Euclid's theory of proportions.[28] The visual characteristics of *remotio* and *quantitas remotionis* are minutely and prolifically treated in Alhacen's *De aspectibus*.[29] I will synthetically summarise the portions of the *De aspectibus* that explain *remotio* and the way sight measures its magnitude.

For Alhacen *remotio* is the absence of contact between two bodies. In other words: between two bodies is an interval determined by two limits. The *quantitas remotionis* is the perception of the measure of that interval. Understanding distance means perceiving the space between two bodies, which is different from understanding the magnitude of that space, i.e., measuring that space.[30]

According to Alhacen the magnitude of a distance can be certified only if between the observer and the observed visible body, or between two visible bodies, lie 'bodies arranged in successive, continuous order' whose dimensions are previously known by

---

25 'non sanno ove sia certo luogo alla cuspide della piramide visiva'. Alberti, *De pictura* (*redazione volgare*), ed. by Bertolini, I, 19. 15.

26 'che cosa niuna dipinta mai parrà pari alle vere, dove non sia certa distanza a vederle'. Alberti, *De pictura* (*redazione volgare*), ed. by Bertolini, I, 19. 17.

27 In the vernacular edition, 'certa distanza a vederle' leaves open the possibility that Alberti is referring to 'cose vere'. In the Latin text 'Tum etiam pictas res nullas veris rebus pares, nisi certa ratione distent, videri posse nemo doctus negabit' ('Besides, no learned person will deny that no objects in a painting can appear like real objects, unless they stand in a know distance'). I have slightly modified the Grayson's translation). Cf. in Alberti, 'De pictura', ed. by Grayson, I, 19; However, there is no doubt that the subject of the verb 'distent' is 'pictas res'. Thus, one must clearly read, 'certa distanza a vederle' (as 'definite distance from which to see painted things').

28 Alhacen, *Alhacen's Theory of Visual Perception*, ed. by Smith, II, 3. 67–3. 69. I have amply treated the notion of 'remotio' in my 'La prospettiva lineare nel Quattrocento: dalle proporzioni continuata e ordinata alla proporzione degradata', in *Proportions. Science, Musique, Peinture & Architecture*, ed. by Sabine Rommevaux, Philippe Vendrix, and Vasco Zara, *Études renaissantes*, 6 (Turnhout: Brepols, 2011), pp. 277–97. See also Pietro Roccasecca, 'La piramide e le intentiones: Alhacen, 'Alberti e la composizione della storia in pittura', in *La primavera del Rinascimento: la scultura e le arti a Firenze 1400–1460*, Exhibition catalogue, Florence, Palazzo Strozzi; Paris, Musée du Louvre, ed. by Beatrice Paolozzi Strozzi and Marc Bormand (Florence: Mandragora, 2013), pp. 173–79.

29 Comprehension and validation of 'remotio' are further treated by Roger Bacon, John Pecham, and Witelo; all three furnish more synthetic, accessible definitions enriched with visual examples. See Vitellonis Thvringopoloni, 'Opticae libri decem, omnes instaurati, figuris illustrati & aucti, adiectis etiam in Alhazenum commentariis', in *Opticae thesavrvs Alhazeni Arabis libri septem, nunc primum editi. Eivsdem Liber de crepusculis & nubium ascensionibus a Federico Risnero* (Basil: Episcopios, 1572), IV, p. 10; Roger Bacon, *Roger Bacon and the Origins of 'Perspectiva' in the Middle Ages, a Critical Edition, with English Translation of Bacon's 'Perspectiva' with Introduction and Notes*, ed. by David C. Lindberg (Oxford: Clarendon Press, 1996), II, 3, 3; John Pecham, *John Pecham and the Science of Optics: Perspectiva communis*, ed. and trans. by David C. Lindberg (Madison: The University of Wisconsin Press, 1970), I, 63 (66).

30 Alhacen, *Alhacen's Theory of Visual Perception*, ed. by Smith, II, 3. 68.

the observer.[31] Not everything included in a visual field is precisely measurable by sight. Only things that are at a moderate distance from the eye are precisely measurable. Bodies and spaces that are too near to or too far from the eye can be perceived by the intellect as having distance, but the intellect cannot certify their magnitude.[32]

As concerns drawing: it follows that the point of convergence, being on the horizon, exceeds moderate distance from the eye. It is thus in an area of the visual field that the intellect cannot measure, even if there are 'bodies arranged in successive, continuous order'. This might explain why in various paintings there is no trace of the point of convergence, whereas in other depictions groups of lines seem to converge in a small area instead of a single point, or far-away pavements appear thick with parallel lines.[33]

For Alhacen when the observer recognises 'bodies arranged in successive, continuous order', he recalls their dimensions and consequently measures the space that they occupy. By comparison the observer understands the measure of the distance between the ordered bodies as well as the distance that separates him from what he is observing. This, only if he is neither too near nor too far, if the distance between viewer and viewed is not at an 'inordinate distance'.

Earth is for Alhacen the ordered continuous body most commonly found between two visible bodies, or between the observer and the observed body.[34] The amount of earth between them, or between us and the visible, we 'base on the measure of the human body' using our feet and arms.[35] To illustrate how continuous ordered bodies enable the understanding of distance, Alhacen takes the example of seeing clouds above a prairie or in mountain-less landscape. The clouds seem very far away, nearly at the altitude of celestial bodies. If seen in a mountainous landscape in which the summit appears above the clouds, however, the eye can evaluate the clouds' altitude by comparison with the mountain-peaks.

The cloud example was reused by Bacon, Pecham and Witelo and is well-known.[36] Witelo adds to this example that of other known visible bodies that underscore the earth's surface: rows of trees, hills and high towers.[37] Clouds, chains of mountains, rows of trees, colonnades and high towers … these are the iconographic terms at the heart of aesthetic thinking about representing distance between bodies, from Pietro and Ambrogio Lorenzetti onwards.

The terms 'ordered' and 'continuous', for Alhacen's Latin translator, commentators and readers had a specific mathematical meaning, one no longer obvious to us today. In the Medieval Latin tradition of Euclid's *Elements*, Book V, two definitions are named 'ordered'

31 Alhacen, *Alhacen's Theory of Visual Perception*, ed. by Smith, II, 3.76.

32 Alhacen, *Alhacen's Theory of Visual Perception*, ed. by Smith, II, 3.160.

33 Examples of this case are given in my 'Evidenze materiali per una storia della prospettiva nella pittura italiana su tavola del XV secolo. Indagine su un campione di dipinti della National Gallery of Art di Washington', in Conference Proceedings of *L'artiste et l'œuvre à l'épreuve de la perspective*, ed. by Marianne Cojannot-Le Blanc, Marisa Dalai Emiliani, and Pascal Dubourg Glatigny, Collection de l'École française de Rome, 364 (Rome: École française de Rome, 2006), pp. 273–85. See also Dominique Raynaud, *L'hypothèse d'Oxford. Essai sur les origines de la perspective* (Paris: Presses Universitaires Françaises, 1998).

34 Alhacen, *Alhacen's Theory of Visual Perception*, ed. by Smith, II, 3. 150.

35 Alhacen, *Alhacen's Theory of Visual Perception*, ed. by Smith, II, 3. 151.

36 Vitellonis, 'Opticae libri decem', IV, 10; *Roger Bacon*, ed. by Lindberg, II, 3, pp. 201–15 (p. 208); Pecham, *Perspectiva communis*, ed. and trans. by Lindberg, I, 63 (66).

37 Vitellonis, 'Opticae libri decem', IV, 10.

and 'continuous'. They were handed down to us in two different Latin translations of the *Elements*, never appear in the same version of the treaty and are not found in Heiberg's nineteenth-century work on Euclid, nor in the works of Heiberg's followers. For this reason, they have been forgotten by modern editions.[38]

The notion of *ordinata proportio* is handed down only through the little influential tradition of Gerard of Cremona's translation, in which it appears as Definition XIX. The notion of *continuam proportionalitatem* is present instead as Definition V in Adelard of Bath's translation, and in those by his followers. It played a fundamental role in the mathematization of the natural sciences — which is, not incidentally, 'perspectival science' understood as physical optics. Proportionality was an essential mathematical instrument until Galileo contested its tautological structure and its inadequateness to the mathematical description of his new physics.[39]

Definition XIX of Book V in Gherardo da Cremona's translation gives ordered proportion as two series of magnitudes in which each term of one series is identically proportional to a term in the other series: given two series of magnitudes 'A. B. C'. and 'D. E. F'., A is to B as D is to E; so B is to C as E is to F. (Figure 8).[40] In editions of the *Elements* commented by Campanus of Novara, continuous proportionality consists of a series of proportions in which terms increase or decrease without interruption, reduced of a constant part of the magnitude (for example, of a third). It is not the relation between magnitudes that is at stake but their proportionality, a similarity of proportion.[41]

Ordered proportion can regulate the relationship between the relative height and width of a pictorial composition. Take for instance the height of a man and the width of the ground, or a cloister's colonnade in which the column's apparent height and the distance apparent between a pair of columns maintain the same reciprocal relation (Figure 9).

Proportional continuity can describe the apparent reduction of size in a depicted depth, such as the diminution in depth of floor-tiles or ceiling coffers. Such proportionality can

---

38 The notion of *proporzione ordinata* goes back to the Greek tradition of the *Elements* and is also found in the Arab tradition of the *Elements*, through which it reached Gerardo da Cremona. Cf. J. W. Engroff, 'The Arabic Tradition of Euclid's Elements: Book V' (unpublished dissertation, Harvard University, 1980), pp. 292–93. The notion of *proporzione continua* is instead amply known and widespread because it already appears in the work of Nicomachus of Gerasa. Leon Battista Alberti, as is well known, possessed a manuscript copy of the *Elements*, in the Campano version now in the collections the Biblioteca Marciana of Venice: Manuscript Cod. Lat. VIII. 39, coll. 3271. In it, Alberti could find the definition of *proporzione continua*.

39 On the reception of Euclid's *Elements*, see: *The First Latin Translation of Euclid's 'Elements' Commonly Ascribed to Adelard of Bath*, ed. by H. L. L. Busard (Toronto: Pont. Institute of Medieval Studies, 1983); *Johannes de Tinemue's Redaction of Euclid's 'Elements'. The So-Called Adelard III Version*, ed. by H. L. L. Busard (Franz Steiner Verlag, Stuttgart, 2001); *Campanus of Novara and Euclid's 'Elements'*, ed. by H. L. L. Busard, 2 vols (Franz Steiner Verlag, Stuttgart, 2005), I; Maurice Caveing, 'Introduction Générale', in *Euclide d'Alexandrie, 'Les Éléments', traduits du texte de Heiberg*, trans. and ed. by Bernard Vitrac, 3 vols (Paris: Presses Universitaires de France, 1990), I; Sabine Rommevaux, 'La réception des 'Élements' d'Euclide au Moyen-Âge et à la Renaissance. Introduction', *Revue d'histoire des sciences*, 56 (2003), 267–73. On Galileo Galilei and Euclid's Book V, see Enrico Giusti, *Euclides reformatus: la teoria delle proporzioni nella scuola galileiana* (Torino: Bollati Boringhieri, 1993).

40 *The Latin Translation of the Arabic Version of Euclid's Elements Commonly Ascribed to Gerard of Cremona*, ed. by H. L. L. Busard (Leiden: Brill, 1984), p. 118: 'Ordinata proportio est cum fuerit antecedens ad consequens sicut antecedens ad consequens et antecedens ad rem aliam sicut consequens ad rem aliam'.

41 *Campanus of Novara*, V, Definition V, p. 161: 'Quantitates que dicuntur continuam proportionalitatem habere, sunt quarum eque multiplicia equa sunt aut eque sibi sine interruptione addunt aut minunt'; *Campanus of Novara*, V, Definition IV, p. 161.

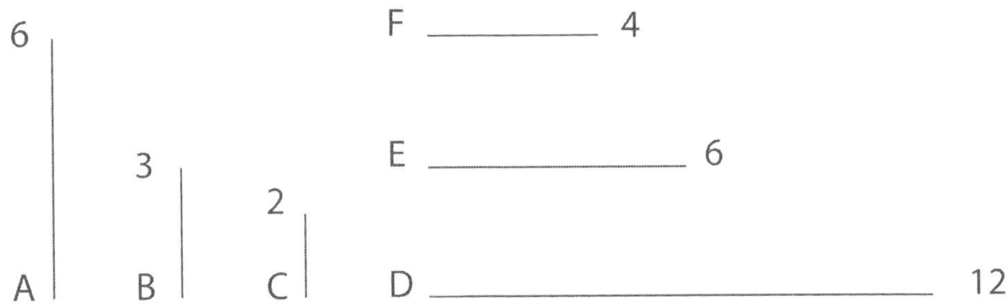

Fig. 8 Magnitudes in ordered proportion.

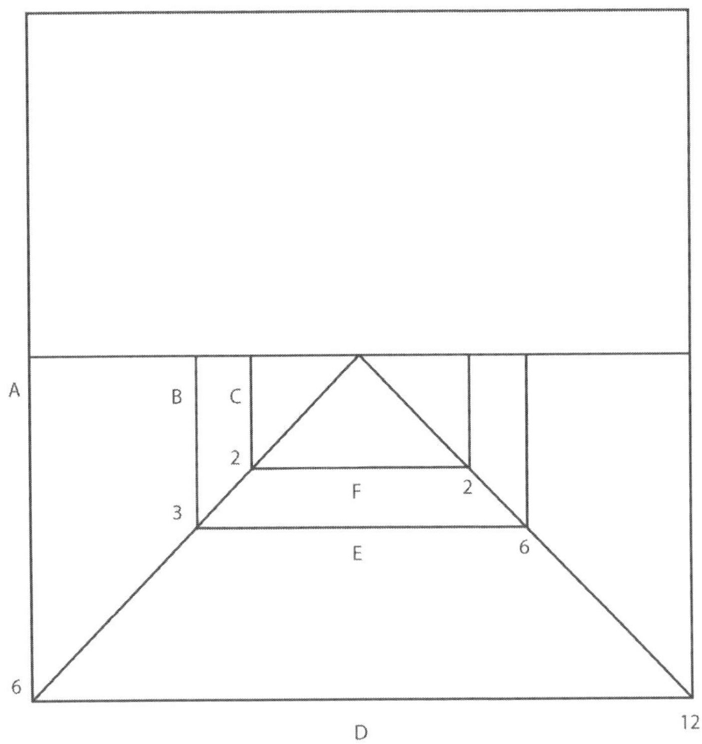

Fig. 9 Magnitudes in ordered proportion and continuous proportionality in a picture. Heights and widths A : D = B : E = C : F are in a ordered proportion; the heights A : B = B : C and the widths D : E = E : F. are continuous proportionalities.

therefore be applied to the second part of the superbipartient method. Recall that this method begins by freely choosing height and width, the first two dimensions, setting the ordered proportion. In step two one attempts to determine the third dimension by using continuous proportion, describing the continuous diminution of the magnitudes appearing in illusionistic depth. The many fourteenth- and fifteenth-century depictions of heads aligned at the same height and/or that show coffered ceilings or pavements in a 3:1 relation to the foreground figures' dimensions seem to have used the method we could call 'the ordered and continuous method'. They have the characteristic 1:3 relation between the floor-tiles' height and width. Witness, for instance, Pisanello's drawing of *Cinq hommes debout dans une architecture* in the Louvre's Départment des Arts Graphiques (inventory number 2520), Lorenzo Ghiberti's low-relief figure of *Isaac* for the *Porta del Paradiso,* or again Paolo Uccello's sinopia for the Nativity of San Martino at the Scala in Florence.[42]

The method for finding the correct diminution of intervals between transversal lines inaugurates the Renaissance in painting and in science. It is perhaps the most commented on passage of the *De pictura*. It contains the first description of an optically correct method to conceive and depict space, a method largely in continuity with previous eras, that would subsequently be mathematically improved.[43] I will now turn to the 'optimal mode' without entering into the details of the diagrams proposed by previous interpreters of the Albertian method, as it is a topic I have largely written about elsewhere.[44]

The 'modo ottimo' or Optimal Mode

As a premise Alberti warns that this mode follows the same preliminary procedures as the 'superbipartient' method. He then invites his reader to execute the 'optimal mode' in a small space. For today's reader, the logical difficulty of the 'optimal mode' lies in the instruction to divide 'into similar parts' two different magnitudes: the base of the quadrangle and the line in the small space.[45] Recall that the term 'part' refers to Euclid's lexicon of proportions. 'Similar parts' is also a concept taken from the *Elements,* specifically Book VII of Campanus' version. According to Definition XIV: are called similar parts those parts denominated by the same number.[46] What is a part's denominator? Definition XIII affirms: the denominator is the number by which a part integrates its whole.[47] Follows that any two magnitudes, if divided into the same number of parts, are 'divided into similar parts' and the number of parts into which they have been divided, the divisor, denominates them and puts them in proportion. That diverse quantities, if divided into the same number of

42 See Richard Krautheimer and Trude Krautheimer-Hess, *Lorenzo Ghiberti* (Princeton: Princeton University Press, 1982), p. 249; Pietro Roccasecca, 'Paolo Uccello, "La natività"', in *Nel segno di Masaccio: l' invenzione della prospettiva*, ed. by Filippo Camerota (Florence: Giunti, 2001), pp. 90–93; Pietro Roccasecca, 'Punti di vista non punto di fuga', *Invarianti per descrivere le trasformazioni*, 33, (1999), 41–48.

43 Discussion of the interpretations of Alberti's illustrated perspectival method has been reconstructed by and commented on by Marisa Dalai Emiliani in two different essays: Marisa Dalai Emiliani, 'La questione della prospettiva', in *La Prospettiva come 'forma simbolica'*, ed. by Erwin Panofsky (Milan: Feltrinelli, 1961), pp. 118–41 and Marisa Dalai Emiliani, 'La questione della prospettiva 1960–1968', *L'arte*, 1 (1968), 96–105. For a recent summary of the debates on perspective see Maria Mignini, 'Storia e ideologia nel dibattito critico sulla prospettiva in Italia, 1957–1977', in the Conference Proceedings of *L'artiste et l'œuvre à l'épreuve de la perspective* (2006), pp. 449–68; Kim H. Veltman, *Linear Perspective and the Visual Dimension of Science* (Munich: Dt. Kunstverl., 1986), pp. 391–92; Pietro Roccasecca, 'Il "modo optimo" di Leon Battista Alberti', *Studi di storia dell'arte*, 4 (1993), 245–62.

44 For an in-depth discussion of the 'modo ottimo', see Roccasecca, 'Il "modo optimo" di Leon Battista Alberti', pp. 245–62 and its detailed bibliography.

45 In the Latin version Alberti uses the phrase 'divido per eas partes' (Cf. in Alberti, 'De pictura', ed. by Grayson, I, 20); reference to Definition XIV of Book VII of the *Elements* is less explicit but no less cogent. Regarding the various hypotheses on this topic, see Roccasecca, 'Il "modo optimo" di Leon Battista Alberti'.

46 *Campanus of Novara*, VII, Definition XIV.

47 *Campanus of Novara*, VII, Definition XIII; Sabine Rommevaux, 'La Proportionnalité numérique dans le livre VII des *Eléménts* de Campanus', *Revue d'histoire des mathématiques*, 5 (1999), 83–126.

parts, can be proportional, is an idea already encountered in the method's first step. The depicted human figures have apparently different heights but are all divided into three parts, making the closest and furthest men divided into similar parts whose denominator is three. The same is true for the row of floor-tiles between them.

Thus, to divide the small construction's line 'into parts similar to the line laying down the quadrangle' means dividing it into the same number of parts as the large construction.[48] Alberti is applying the theory of numerical proportion here, and establishing the similarity of relations between the two base lines and their respective parts. In Definition XXI of Book VII of the *Elements*, in fact, two base lines divided into the same number of parts are proportional in their parts, having the same denominator, and are therefore divided into similar parts.[49] In other words Alberti puts two constructions to scale: the one in the future painting and the one in the 'little space'. At this point it should be clear that Alberti is instructing to draw in the small space a line divided into the same number of parts as the base of the quadrangle (Figure 10).

The next step is to mark a point at the same height as the 'central point'. It follows that one of the small space's dimensions must be at least as big as the height initially set in the quadrangle. In the vernacular edition Alberti makes no recommendations about how to position the point, which can be freely placed along the line at the height of the 'central point'. A restrictive variant appears in a group of six Latin codices: the small construction's point must be placed above the line's extremity.[50] This optimises the procedure, better to use the distance between the point and the perpendicular, but it does not substantially change Alberti's method. For this reason we have adopted it for our drawing.

From the point situated above one of the extremities are traced the lines connecting the base partitions. In the diagram is then found the distance from which the painting is to be observed, and from that point is drawn a perpendicular line. The only way to verify that this procedure yields the same results, if applied to a smaller space, is to draw two diagrams with base lines of different lengths but divided into the same number of parts, then in both diagrams to place the point at the same height before drawing their connecting lines. The distance at which to draw the perpendiculars is in both diagrams chosen by counting the same number of the subdivided line's parts. Following these instructions, the connecting lines intersect the two perpendiculars at identical 'magnitudes' (Figure 11).

To summarise the 'optimal mode' procedure: first, take a sheet of paper, trace a horizontal line, even a short one. Second, place a point above one of its extremities. The point must be as high above the horizontal line as the centric point is high above the picture plane's base line. Next, draw lines connecting this point to the horizontal line's partitions. The distance between the viewer's eye and the painting is then established by drawing a perpendicular to the horizontal line. The points at which intersect perpendicular and connecting lines graduate all subsequent transversal lines: this determines each and every perceived reduction of tridimensional magnitudes (Figure 12).

48 'in simile parte in quale divisi la linea che giace nel quadrangolo'. Alberti, *De pictura* (*redazione volgare*), ed. by Bertolini, II, 20. 3.

49 *Campanus of Novara*, VII, Definition XXI: 'Similes sive una alii eadem dicuntur proportiones, quae eandem denominationem recipiunt. Maiore, vero qua maiorem. Minor autem, quam minorem'.

50 See Alberti, 'De pictura', ed. by Grayson, III, p. 322 .

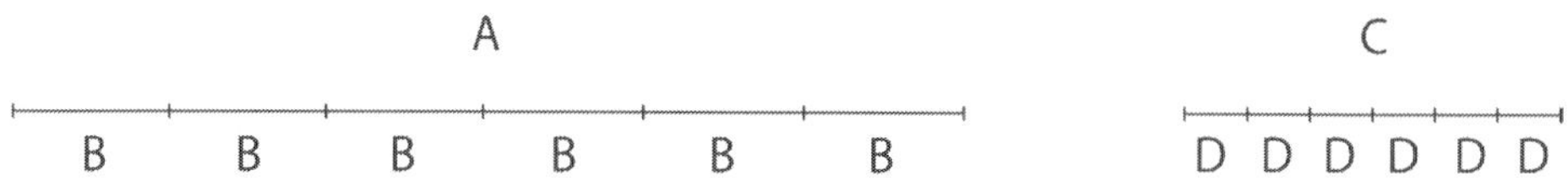

Fig. 10 Lines of different magnitudes divided in similar parts are in proportion. The line A., that represents the base of the quadrangle to be painted, is divided into 6 parts B. The line C., that represents the base line of the diagram drawn in the small space, is divided into the same number of parts D. as the base of the quadrangle to be painted. The lines and their parts are in the following proportion: A : B = C : D.

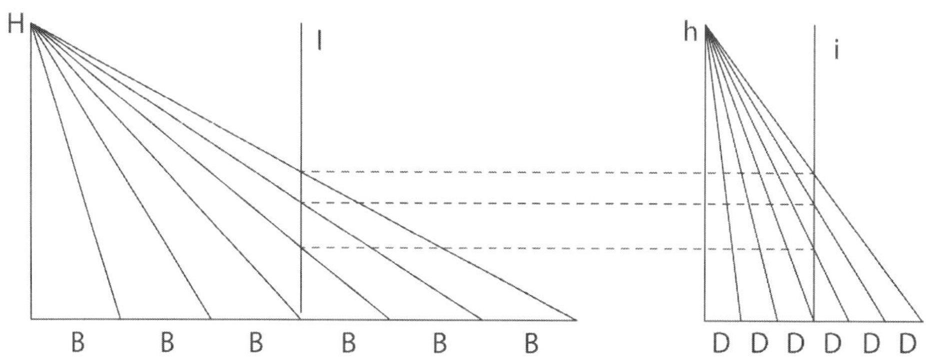

Fig. 11 Two diagrams with base lines of different lengths are divided into the same number of parts, in both diagrams a perpendicular is dropped on the same part (in this case the third one) of the subdivided lines: the connecting lines intersect the two perpendiculars defining identical 'magnitudes'.

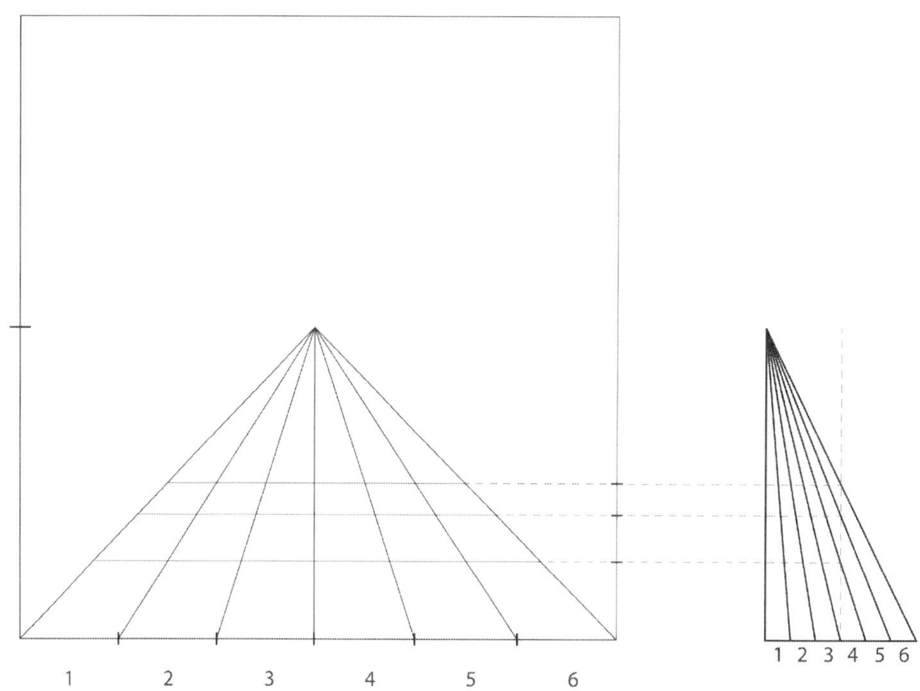

Fig. 12 The *modo ottimo* ('optimal mode').

We might wonder why Alberti determines the distance between the viewer's eye and the painting in a small space on a separate sheet of paper, and not directly onto the painting's support. Finding the proportion of diminution of magnitudes is mere 'calculation' and Alberti prefers not to leave any trace of calculations on the surface to be painted. Further, because diminution is 'calculated' on a separate sheet the result can be kept as a study and serve as an 'instrument' for the painter's future works.

Alberti warns the reader: 'painted things will never seem equal to real things if viewing distance is not taken into account'.[51] Making viewing distance a parameter to illusionistic depiction constitutes a paradigm shift in the history of visual representation. The proportion that describes the diminution of apparent size is to be found in the geometry of the visual pyramid. Because sight is predicated on a triangle that is a section of the visual pyramid, optical illusion must take into account the height of the viewer's eye and its distance from the picture plane.[52] The diminution of represented size is thus determined by the theorem of proportional triangles.[53] The procedure for making illusionistic representation has definitively changed here: continuous proportion enables the certification of diminished magnitudes. Reduction is no longer a merely numerical function, as it was according to 'subsesquialter' or 'superbipartient' operations, but a geometrical function determined by the intersection of the visual pyramid.

Note however that Alberti does not establish the distance from which visible bodies need be observed. He specifies only the distance at which sight intersects the picture plane; that is, the distance from which to observe the 'intersection', the painting. Besides, the distance from which to observe the picture plane is by definition smaller than the painting's base. This is not without consequence for mathematically correct representation. In hindsight and from the vantage point of linear perspective we can assert that Alberti made a mistake that limits the validity of the 'optimal mode'.

The *De pictura* does not explain how to transfer onto the painting's quadrangle the intersections of the small space's perpendiculars. We can hypothesise that with the help of a ruler, perhaps the ruler used to trace the perpendicular lines, the intersections Alberti calls 'parallels' and 'squared *braccia*' can be transferred onto and individuated in the ground plane. If the parallels are drawn correctly, specifies Alberti, one line will be diagonal to 'several squared *braccia*'.[54] This has led some to suppose that Alberti knew the so-called 'distance point' method whose first (surviving) literary description is found in

51 'E sappi che cosa niuna dipinta mai parrà pari alle vere, dove non sia certa distanzia a vederle'. Alberti, *De pictura (redazione volgare)*, ed. by Bertolini, I, 19. 17.

52 Alberti, *De pictura (redazione volgare)*, ed. by Bertolini, I, 6. 7: 'Onde si suole dire che al vedere si fa triangolo' ('It is said that vision makes a triangle').

53 See *Euclid's Elements. All Thirteen Books in One Volume*, ed. by Dana Densmore and trans. by T. L. Heath (Santa Fe: Green Lion Press, 2002), VI, Proposition 2.

54 Alberti attributes to *parallelo* a meaning different than we do nowadays. For today's reader it means the lines by which cartography subdivides the terrestrial globe; for Alberti a parallel is the space comprised between two transversal lines: 'e' paraleli, cioè le braccia quadrate del pavimento nella dipintura' ('parallels, that is the squared *braccia* of pavement on the painting'), see Alberti, *De pictura (redazione volgare)*, ed. by Bertolini, I, 20. 8). The lines that delimitate *paralleli* are called *equidistanti*.

the writings of Piero della Francesca, a method later codified by Jacopo Barozzi as the second rule of perspective.[55]

Alberti bases his method on the distance between the eye and the painting, instead of on the distance between the eye and the thing seen, which relies on the 'distance point'. Indeed, the diagonal intersects *several*, not *all*, quadrangles — as would be the case if Alberti were illustrating the 'distance point method'. The Humanist chooses his words carefully and if he writes several quadrangles this means that in the construction derived from the 'optimal mode' the diagonal line does not intersect all the quadrangles; the more the better the illusion, however. The diagonal line verifies the correct implementation of the method. This means that Alberti is aware that the property of the diagonal when intersecting homothetic squares (squares from an orthogonal grid inserted in a quadrangular frame, such as floor-tiles) is not lost if the apparent reduction in size is proportional.

The ground's foreshortened diagonal signals progress in the geometrical correctness of perspectival methods. Continuous proportionality produces pavements in which each quadrangle has its own diagonal and the series of diagonals trace a concave or convex curve. Alberti's 'optimal mode' enables 'several' floor-tiles to be intersected. With Piero della Francesca the diagonal intersects all squares shown in perspective and also becomes an expedient way to depict foreshortening.[56]

To return once more to the 'optimal mode': the distance of observation is always inferior to the picture plane's base line and the resulting foreshortening will be as abhorrent as is short the distance at which is traced the point's vertical line.[57] In sum: the 'optimal mode' produces the quadrangle's foreshortening, with credible and acceptable results, only for a distance from eye to painting that is at least half of the available base line; the diagonal line cannot run through all the squares from one corner to another of the foreshortened pavement because the distance from the observer is by definition always shorter than the base of the quadrangle.

Alberti ends this phase of the 'intersection' by asking the reader to draw a line, equidistant to the lines below', passing through the central point and extending to both sides of the picture plane. Alberti affirms: 'this line represents for me the limit above which no visible magnitude that is not higher than the observer's eye can be assigned, subject to correct judgement'.[58] This means that the centric line is the limit above which the observer's eye

55 See Piero della Francesca, *De prospectiva pingendi*, ed. by Giusta Nicco Fasola (Florence, Le Lettere, 1984); *Le due regole della prospettiva pratica di M. Iacomo Barozzi da Vignola, con i commentarij del R. P. M., Egnatio Danti dell'ordine dei predicatori, matematico dello studio di Bologna, in Roma per Francesco Zannetti, 1583*, ed. by Maria Walcher Casotti (Vignola: Cassa di Risparmio di Vignola, 1974).

56 Gino Casara, 'Piero della Francesca, e i fondamenti geometrici della prospettiva pittorica', in *Atti e memorie della regia accademia petrarca*, 32–33 (1944), 93–135 (p. 109).

57 See Roccasecca, 'Il "modo optimo" di Leon Battista Alberti'.

58 In the vernacular edition this sentence is not clear and consequently, neither is this line's function. Cf. Alberti, *De pictura* (*redazione volgare*), ed. by Bertolini, I, 20. 11: 'Questa linea a me tiene uno termine quale niuna veduta quantità, non più alta che l'occhio che vede, può sopr'agiudicare'. Bertolini's commentary of the vernacular edition of *De pictura* proposes a paraphrase in modern Italian that clarifies Alberti's meaning; see Alberti, *De pictura* (*redazione volgare*), ed. by Bertolini, p. 188: 'Questa linea rappresenta per me il limite sopra il quale nessuna quantità vista, che non sia più alta dell'occhio dell'osservatore, può essere assegnata, destinata per corretto giudizio' ('This line represents for me the limit above which any seen quantity, that is not higher than the eye of the observer, can be assigned, designed for correct judgment'). The corresponding sentence in the Latin

can judge and measure magnitudes. In other words, the line drawn from one side to the other, passing by the central point, is a 'height' limit of the area in which the *ordinata* and *continuata* proportions rule. This line divides the picture plane quadrangle into two parts, and it is in the lower area that the observer can exercise correct judgment of magnitudes.

In fact, adds Alberti, we see the figures' heads line up under a single line and the most distant of their feet at the level of the 'knees' in the foreground. This optical effect, as we have seen, is real (provided we observe people in a large space and that all stand on the same ground) and this demonstrates that for Alberti the observer and the observed stand 'on the same plane' (Figures 3–5).[59]

Although no higher than the viewer's eye, this new line is not a horizon line; it does not limit distance but the depicted figures' height and, again, divides the painting's quadrangle into two parts. The lower part commensurates all magnitudes no higher than the observer; the upper part, all those magnitudes higher of the viewer's eye. At this point Alberti interrupts his instructions on the 'optimal mode' but the procedure does not end here. In Book Two he continues with magnitudes placed above the viewer's eye.

Buildings: Proportions Between Architecture and Human Figures

House foundations are drawn in the floor's parallel lines. Alberti specifies that in nature 'of no square body with right angles can we immediately see the surroundings beyond two adjacent facets'.[60] This confirms that the painter/observer stands on the same plane as the architectural foundations he is about to draw and, therefore, the draughtsman sees only two facets of the buildings and palaces which have a quadrangular base. From a dominant position, from an upper story window or terrace, one might see the roofs of the houses opposite and two walls at an angle.[61]

To draw buildings one traces latitudinal and longitudinal lines using the pavement's parallels, thereby marking the building's foundations. One then determines the building's height by using a compass to find the correct proportion. One draws first the 'latitudes and longitudes' of the foreground buildings' foundations, preferably starting with those buildings parallel to the intersection/picture plane. Alberti's text is not specific here, but it seems evident that only one of the two sides of the foundations can be equidistant to the 'intersection', the other is placed at an angle and follows the ground lines.

The wall's *braccia*-length is equal to the number of 'parallels' it occupies. Walls can even begin at the diagonal's intersection, at the centre of the floor-tiles. Building heights are

edition confirms her interpretation. Cf. Alberti, 'De pictura', ed. by Grayson, I, 20: 'Haec mihi quidem linea est terminus atque limes, quem nulla non plus alta quam sit visentis oculus quantitas excedat' ('This line is for me a limit or boundary, which no quantity exceeds that is not higher than the eye of the spectator').

59 See Alberti, *De pictura* (*redazione volgare*), ed. by Bertolini, I, 19. 9: 'però che così e chi vede e le dipinte cose vedute paiono medesimo in su uno piano' ('thus both the observer and the painted things he sees appear to be on the same plane').

60 'di niuno quadrato corpo, quale abbia retti angoli, a un tratto posso vedere d'intorno più che due facce congiunte'. Alberti, *De pictura* (*redazione volgare*), ed. by Bertolini, II, 9. 10.

61 The building with a polygonal base in Perugino's *Sposalizio della vergine*, and even more so Rafael's painting of the same subject, perhaps not accidentally, belie this proposition.

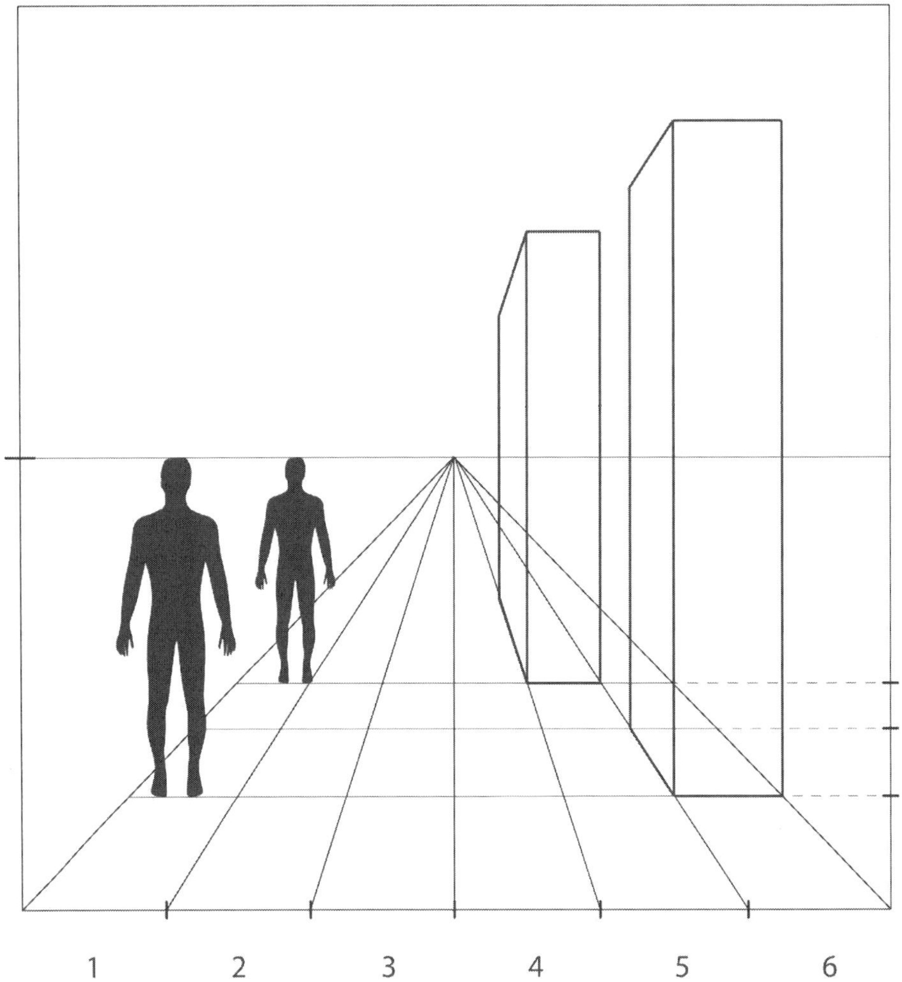

Fig. 13 Building heights are calculated using the height of the depicted figures as constant term.

calculated using the height of the depicted figures as their common, constant term. As we have already seen, from any point in the ground the distance from the centric line always measures three *braccia* in length. The ordered proportion active under the centric line is also valid for the area above the centric line. Thus to draw a building twelve *braccia* high, writes Alberti: 'three times as much will you go above the centric line, as high from the centric line as that is from the ground'.[62] A measure under the centric line and three above it make four measures of three *braccia* in length — for a total amount of twelve *braccia* (Figure 13). In fact, to determine the proportional height of the buildings one need only place a compass point on the centric line and mark as many times as desired above the centric line.

62 'tre volte tanto andrai su in alto quanto sia dalla centrica linea persino a quel luogo del pavimento'. Alberti, *De pictura* (*redazione volgare*), ed. by Bertolini, II, 9. 17.

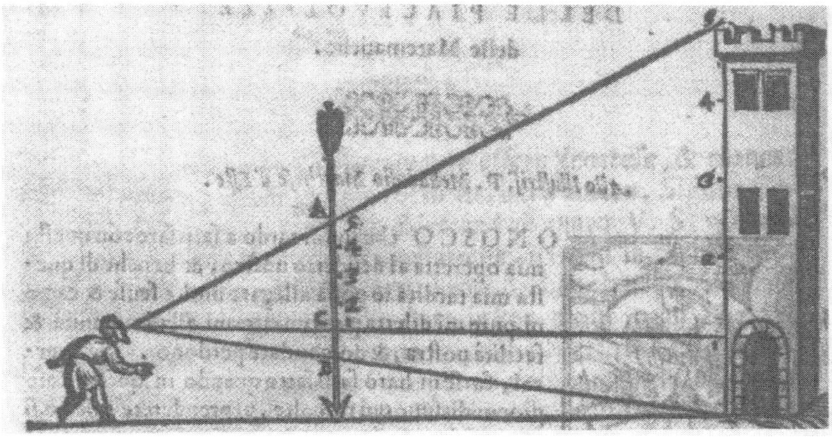

Fig. 14 How to measure the height of a tower using sight alone, from Cosimo Bartoli, *Opuscoli morali di Leon Batista Alberti gentil'huomo firentino [!] ne'quali si contengono molti ammaestramenti, necessarij al viuer de' l'huomo, così posto in dignità, come priuato. Tradotti, & parte corretti da m. Cosimo Bartoli*, in Venetia appresso Francesco Franceschi, sanese, 1568.

Conclusion

Constructing building heights poses the same mathematical problem as a well-known 'Libro d'Abaco' problem that Alberti presents in *Ludi rerum mathematicarum*, a collection of mathematical games dedicated to Meliaduse d'Este, written about fifteen years after the *De pictura*. In the first, Alberti teaches his readers how to measure the height, using sight alone, of a tower at the back of a city square. On the square's opposite side, plant a dart through which one sees the tower at a known distance. Looking at the cusp, the base and a known and measurable point of the tower, one marks in wax the point where the gaze hits the dart. In this particular case the known measure is given by the height of tower's door. The three wax markings are called, respectively, A, B and C (Figure 14).[63]

Given that the observer, the tower and the buildings are on the same ground, that the tower and the buildings are the visual pyramid's base, and that the dart and the painting are their intersection, measuring the tower and drawing the building proportionally to the human figure are two applications of the same theorem of proportional triangles.[64] The obvious correspondence between this part of the intersection and methods for measuring distance seem to support the idea that the intersection of Alberti's visual pyramid is an application of surveying methods to painting.

In a certain sense, the portrait of a face seen through a veil can be considered to be an architectural or topographical survey technique applied to pictorial representation, and the method to find the proportional height of the buildings, a solution to an Abacus

63 Leon Battista Alberti, 'Ludi rerum mathematicarum', in *Opere volgari*, ed. by Cecil Grayson, 3 vols (Bari: Laterza, 1973), III, p. 135. I have elsewhere indicated the correspondence between the method for representing buildings and measuring the tower in the *Ludi*: Cf. Roccasecca, 'Il "modo optimo" di Leon Battista Alberti'.

64 See *Euclid's Elements*, ed. by Densmore and trans. by Heath, VI, Proposition 2.

problem. That said, it is not the culture or community of measurers that produced either the 'optimal mode' or the theory of visual representation explained in the *De pictura*. If anything, portraying and surveying have a common origin in the theory of the visual pyramid's intersection. It is not given that the measure of real magnitudes should enable the pictorial representation of diminishing apparent magnitudes. In other words, it is far from obvious that a single measuring experiment can formulate a theory of pictorial representation.

The difference between measuring real objects and the graphic or pictorial representation of fictional bodies and distances is an important one: topographical survey is inferred and recorded only with adequate tools such as the one described by Grazia de' Castellani, which gives apparent measures to things that really exist.[65] In artistic invention apparent measures must be 'invented' by drawing and the draughtsman must contrive a mathematical relation between magnitudes that renders the representation credible.[66] Measuring magnitudes by sight and rendering them visually are applications of the same part of Alhacen's theory of vision. However, where the practice of measuring at a distance seeks certainty of measure, painting strives for illusionistic proportions and to that end, above and beyond *remotio*, painting requires larger parts of visual theory and such characteristics of vision as light, location, figure, corporeity, movement, to cite only the most relevant for painting of Alhacen's twenty-one visual characteristics. If the portrait of a face and surveying an antique monument have much in common, nonetheless measuring distance and pictorial illusionism are independent from each other and, despite their commonalities in Alhacen's theory, they follow diverse paths, to different ends.

Bibliography

Manuscript and Archival Sources

Biblioteca Marciana, Venice: Manuscript Cod. Lat. VIII. 39, coll. 3271.

Primary Sources

Alberti, Leon Battista, 'De pictura', in *Opere volgari*, ed. by Cecil Grayson, 3 vols (Bari: Laterza, 1973).

65 Grazia de' Castellani, theologian and mathematician active in Florence in the first decades of the 1400s, describes a surveying instrument composed of three perches: a base and two perpendiculars 'three *braccia* long, that is, of a man's stature'. The perches serve as sighting devices and to that end bear an eye-hole. Of variable dimensions, the base is divided into the same number of 'parts' as the two perpendiculars. See Gino Arrighi, 'Un estratto del "De visu" di Ma. Grazia de' Castellani (dal codice Ottoboniano Latino 3307 della Biblioteca Apostolica Vaticana)', *Atti della fondazione Giorgio Ronchi e contributi dell'Istituto Nazionale di Ottica*, 22 (1967), 44–58 (p. 47): 'di lunghezza di tre braccia, cioè alla statura di un uomo' (of length of three *braccia*, that is the height of a man).

66 On the topic of measuring from a distance, see Giorgio Vasari il Giovane, *Raccolto fatto dal Cav.re Giorgio Vasari, di varii instrumenti per misurare con la vista*, ed. by Filippo Camerota (Florence: Giunti, 1996); with regards to the possible relations between these methods and pictorial representation, see Filippo Camerota, *La prospettiva del Rinascimento: arte, architettura, scienza* (Milan: Electa, 2006).

Alberti, Leon Battista, *La pittura di Leon Battista Alberti*, trans. by Lodovico Domenichi (Venice: Gabriel Giolito de Ferrari, 1547).

Alberti, Leon Battista, *On Painting*, ed. by J. R. Spencer (London: Routledge and Paul, 1956).

Alberti, Leon Battista, *On Painting and on Sculpture, the Latin Texts of 'De pictura' and 'De statua'*, ed. by Cecil Grayson (London: Phaidon, 1972).

Alberti, Leon Battista, 'Ludi rerum mathematicarum', in *Opere volgari*, ed. by Cecil Grayson, 3 vols (Bari: Laterza, 1973).

Alberti, Leon Battista, *De statua*, ed. by Marco Collareta (Pisa: Sillabe, 1998).

Alberti, Leon Battista, *De pictura (redazione volgare)*, ed. by Lucia Bertolini (Florence: Edizioni Polistampa, 2011).

Alhacen, *Alhacen's Theory of Visual Perception: A Critical Edition, with English Translation and Commentary, of the First Three Books of Alhacen's 'De aspectibus', the Medieval Latin Version of Ibn al-Haytham's 'Kitāb al-Manāẓir'*, ed. and trans. by A. Mark Smith, 2 vols (Philadelphia: Transactions of the American Philosophical Society, 2001).

Bacon, Roger, *Roger Bacon and the Origins of 'Perspectiva' in the Middle Ages, a Critical Edition, with English Translation of Bacon's 'Perspectiva' with Introduction and Notes*, ed. by David C. Lindberg (Oxford: Clarendon Press, 1996).

Beare, J. I., and G. R. T. Ross, 'Parva Naturalia', in *The Works of Aristotle*, ed. by W. D. Ross (Oxford: Clarendon Press, 1931).

Busard, H. L. L. (ed.), *The First Latin Translation of Euclid's 'Elements' Commonly Ascribed to Adelard of Bath* (Toronto: Pont. Institute of Medieval Studies, 1983).

Busard, H. L. L. (ed.), *The Latin Translation of the Arabic Version of Euclid's Elements Commonly Ascribed to Gerard of Cremona* (Leiden: Brill, 1984).

Busard, H. L. L. (ed.), *Johannes de Tinemue's Redaction of Euclid's 'Elements'. The So-Called Adelard III Version* (Franz Steiner Verlag, Stuttgart, 2001).

Busard, H. L. L. (ed.), *Campanus of Novara and Euclid's 'Elements'*, 2 vols (Franz Steiner Verlag, Stuttgart, 2005).

Cennini, Cennino, *Il libro dell'arte*, ed. by Franco Brunello (Vicenza: Neri Pozza, 1982).

Euclid, *Euclid's Elements. All Thirteen Books in One Volume*, ed. by Dana Densmore and trans. by T. L. Heath (Santa Fe: Green Lion Press, 2002).

Francesca, Piero della, *Le due regole della prospettiva pratica di M. Iacomo Barozzi da Vignola, con i commentarij del R. P. M., Egnatio Danti dell'ordine dei predicatori, matematico dello studio di Bologna, in Roma per Francesco Zannetti, 1583*, ed. by Maria Walcher Casotti (Vignola: Cassa di Risparmio di Vignola, 1974).

Francesca, Piero della, *De prospectiva pingendi*, ed. by Giusta Nicco Fasola (Florence, Le Lettere, 1984).

Pecham, John, *John Pecham and the Science of Optics: Perspectiva communis*, ed. and trans. by David C. Lindberg (Madison: The University of Wisconsin Press, 1970).

Vasari il Giovane, Giorgio, *Raccolto fatto dal Cav.re Giorgio Vasari, di varii instrumenti per misurare con la vista*, ed. by Filippo Camerota (Florence: Giunti, 1996).

Vitellonis Thvringopoloni, 'Opticae libri decem, omnes instaurati, figuris illustrati & aucti, adiectis etiam in Alhazenum commentariis', in *Opticae thesavrvs Alhazeni Arabis libri septem, nunc primum editi. Eivsdem Liber de crepusculis & nubium ascensionibus a Federico Risnero* (Basil: Episcopios, 1572).

Secondary Works

Arrighi, Gino, 'Un estratto del "De visu" di Ma. Grazia de' Castellani (dal codice Ottoboniano Latino 3307 della Biblioteca Apostolica Vaticana)', *Atti della fondazione Giorgio Ronchi e contributi dell'Istituto Nazionale di Ottica*, 22 (1967), 44–58.

Bertolini, Lucia, 'Sulla precedenza della redazione Volgare del "De pictura" di Leon Battista Alberti', in *Studi per Umberto Carpi. Un saluto da allievi e colleghi pisani*, ed. by Marco Santagata and Alfredo Stussi (Pisa: Edizioni ETS, 2000).

Camerota, Filippo, *La prospettiva del Rinascimento: arte, architetura, scienza* (Milan: Electa, 2006).

Casara, Gino, 'Piero della Francesca, e i fondamenti geometrici della prospettiva pittorica', in *Atti e memorie della regia accademia petrarca*, 32–33 (1944), 93–135.

Caveing, Maurice, 'Introduction Générale', in *Euclide d'Alexandrie, 'Les Éléments', traduits du texte de Heiberg*, trans. and ed. by Bernard Vitrac, 3 vols (Paris: Presses Universitaires de France, 1990).

Edgerton, Samuel Y., 'Alberti's Optics' (unpublished dissertation, University of Pennsylvania, 1965).

Emiliani, Marisa Dalai, 'La questione della prospettiva', in *La Prospettiva come 'forma simbolica'*, ed. by Erwin Panofsky (Milan: Feltrinelli, 1961), pp. 118–41.

Emiliani, Marisa Dalai, 'La questione della prospettiva 1960–1968', *L'arte*, 1 (1968), 96–105.

Engroff, J. W., 'The Arabic Tradition of Euclid's Elements: Book V' (unpublished dissertation, Harvard University, 1980).

Giusti, Enrico, *Euclides reformatus: la teoria delle proporzioni nella scuola galileiana* (Torino: Bollati Boringhieri, 1993).

Krautheimer, Richard, and Trude Krautheimer-Hess, *Lorenzo Ghiberti* (Princeton: Princeton University Press, 1982).

Mignini, Maria, 'Storia e ideologia nel dibattito critico sulla prospettiva in Italia, 1957–1977', in Conference Proceedings of *L'artiste et l'œuvre à l'épreuve de la perspective*, ed. by Marianne Cojannot-Le Blanc, Marisa Dalai Emiliani, and Pascal Dubourg Glatigny, Collection de l'École française de Rome, 364 (Rome: École française de Rome, 2006), pp. 449–68.

Raynaud, Dominique, *L'hypothèse d'Oxford. Essai sur les origines de la perspective* (Paris: Presses Universitaires Françaises, 1998).

Roccasecca, Pietro, 'Il "modo optimo" di Leon Battista Alberti', *Studi di storia dell'arte*, 4 (1993), 245–62.

Roccasecca, Pietro, 'Punti di vista non punto di fuga', *Invarianti per descrivere le trasformazioni*, 33, (1999), 41–48.

Roccasecca, Pietro, 'Paolo Uccello, "La natività"', in *Nel segno di Masaccio: l' invenzione della prospettiva*, ed. by Filippo Camerota (Florence: Giunti, 2001), pp. 90–93.

Roccasecca, Pietro, 'Evidenze materiali per una storia della prospettiva nella pittura italiana su tavola del XV secolo. Indagine su un campione di dipinti della National Gallery of Art di Washington', in Conference Proceedings of *L'artiste et l'œuvre à l'épreuve de la perspective*, ed. by Marianne Cojannot-Le Blanc, Marisa Dalai Emiliani, and Pascal Dubourg Glatigny, Collection de l'École française de Rome, 364 (Rome: École française de Rome, 2006), pp. 273–85.

Roccasecca, Pietro, 'La prospettiva lineare nel Quattrocento: dalle proporzioni continuata e ordinata alla proporzione degradata', in *Proportions. Science, Musique, Peinture & Architecture*, ed. by Sabine Rommevaux, Philippe Vendrix, and Vasco Zara, Études Renaissantes, 6 (Turnhout: Brepols, 2011), pp. 277–97.

Roccasecca, Pietro, 'La piramide e le intentiones: Alhacen, 'Alberti e la composizione della storia in pittura', in *La primavera del Rinascimento: la scultura e le arti a Firenze 1400–1460*, Exhibition catalogue, Florence, Palazzo Strozzi; Paris, Musée du Louvre, ed. by Beatrice Paolozzi Strozzi and Marc Bormand (Florence: Mandragora, 2013), pp. 173–79.

Roccasecca, Pietro, *Filosofi, oratori e pittori. Una nuova lettura del De pictura di Leon Battista Alberti* (Rome: Campisano Editore, 2016).

Rommevaux, Sabine, 'La Proportionnalité numérique dans le livre VII des *Éleménts* de Campanus', *Revue d'histoire des mathématiques*, 5 (1999), 83–126.

Rommevaux, Sabine, 'Aperçu sur la notion de dénomination d'un rapport numérique au Moyen-Âge et à la Renaissance', *Methodos*, 1 (2001), 223–43.

Rommevaux, Sabine, 'La réception des 'Élements' d'Euclide au Moyen-Âge et à la Renaissance. Introduction', *Revue d'histoire des sciences*, 56 (2003), 267–73.

Tofani, Anna Maria Petrioli, 'I materiali e le tecniche' in *Il disegno. Forme, tecniche significati* (Monza: Amilcare Pizzi Editore, 1991), pp. 187–250.

Veltman, Kim H., *Linear Perspective and the Visual Dimension of Science* (Munich: Deutscher Kunstverlag, 1986).

J. V. FIELD

The Use of Perspective in the Art of Piero della Francesca

Introduction

In his lifetime, and for some time thereafter, Piero della Francesca (*c.* 1412–1492) was well known as a competent mathematician. His reputation as a painter was less robust. Styles of painting changed very rapidly in the period concerned and his work soon looked outmoded. Nevertheless, the existence of copies suggests that his works still had their admirers even in the seventeenth century. The exquisite detail of the finish — still visible today in better preserved passages of many of his works — must surely have been regarded as a proof of his competence as a craftsman even by viewers who found the overall effect too formal or too stilted. With the recovery of his reputation as a painter, from the 1920s onwards, Piero has been recognised not only for the formal qualities of his work but also for the delicacy of his handling of colour. Indeed, when looking at the originals, the colour and the sense of solid form tend to distract the eye from noticing that Piero is also an excellent draftsman, though skill as a draftsman is what is principally required in applying the geometrical rules of perspective. Infrared reflectograms of parts of Piero's fresco cycle *The Story of the True Cross* (chancel, San Francesco, Arezzo) have revealed underdrawings that not only allow us to see the careful transfer of preliminary drawings but also show freehand work of great elegance and fluidity (Figures 3 to 6). These roughly life-size underdrawings are of historical importance in providing very strong evidence that Piero did indeed apply the sometimes rather cumbersome mathematical techniques for making drawings in correct perspective that are described in his treatise *On Perspective for Painting* (*De prospectiva pingendi*), which was probably composed, or at least finished, in the later 1460s, after the completion of the fresco cycle in Arezzo.

However, Piero's use of the techniques of geometrical perspective is not a simple matter of applying the necessary mathematics. When we come to the specific task of paining, there is also a question about the applicability of what is set out in the treatise. The clearest case is that of frescos, because frescos are — in today's phrase — site-specific. That is, the exact viewing conditions are known, and they are rather unlikely to be susceptible of much change, though we shall look at a possible exception below. Moreover, in almost all cases, including that of the pictures in *The Story of the True Cross*, the viewing conditions do not

J. V. Field Birkbeck, University of London, jv.field@hist-art.bbk.ac.uk

Perspective as Practice. Renaissance Cultures of Optics, ed. by Sven Dupré, Turnhout, 2019 (*Techne. Knowledge, Technique, and Material Culture*, 1), p. 391-408

DOI 10.1484/M.*Techne*-EB.5.117734

correspond at all closely with those described in Piero's treatise (and in most subsequent treatises dating from earlier than the seventeenth century).

Piero's mathematical works, including his treatise on perspective, show due respect for logical rigour and for proving results before using them. His paintings are, however, more ambiguous. Like other paintings of his time, when one examines them closely they show a tension between naturalistic representation of appearances, that is getting the geometrical optics right, and the requirement that the picture be readable and attractive. Piero's paintings show many compromises with mathematical rigour. We can be sure they are indeed compromises, rather than mere errors, since his writings make it absolutely clear that Piero was a master of the theory. It is not clear how far his patrons or his fellow painters were aware of the compromises, but Piero himself was certainly aware of them.

Such compromises were inevitable and, as I have remarked elsewhere, the matter can best be summed up by saying that, as a mathematician, Piero was a good mathematician, but as a painter he was a good painter.[1] I have no wish to retract that opinion. This essay is about Piero the painter, but we still need to bear constantly in mind that Piero was also a good mathematician.

In the present context it is, moreover, as well to be a little more careful than art historians sometimes are about the use of period terminology. We shall keep to the norm for the history of science in adopting 'actors' categories' as far as possible. Thus the term 'artist' will be avoided because to his contemporaries Piero was a painter, a craftsman, that is someone who worked with his hands and had been trained by apprenticeship in a workshop. The term 'mathematician' is perhaps also a little misleading, since Piero's mathematics was largely, though not exclusively, of the kind associated with the practical 'abacus' tradition rather than that of the four mathematical sciences of the *quadrivium*. Abacus mathematics was written in the local vernacular (in Piero's case the Tuscan *volgare*) while learned mathematical texts were in Latin. There was undoubtedly some social and intellectual division between these two traditions in fifteenth-century Italy, but its extent is not very clear, and Piero's own work, some of which appeared in Latin translation in his lifetime, shows that the dividing line could sometimes be crossed.[2] Moreover, it seems that in the 1470s Piero's knowledge of Latin was sufficient for reading a mathematical text since, working from a manuscript in the Vatican library, he made a copy of a Latin translation of the works of Archimedes.[3] We may note that the Vatican library, of which a cousin of Piero's was a librarian, also contained an Italian translation of the optical work of Alhacen (Ibn al-Haytham, *c.* AD 965–1040), to which Piero may have had access.[4] This brings us to 'perspective'.

1 J. V. Field, 'Theory and Practice of Perspective in the Works of Piero della Francesca', *1492, Rivista della Fondazione Piero della Francesca, Anno 1* (1–2) (2008), 35–54; J. V. Field, 'Lo sguardo matematico nell' arte rinascimentale Italiana', in *La Matematica, Vol. 3, Suoni, forme, parole*, ed. by Claudio Bartocci and Piergiorgio Odifreddi (Turin: Einaudi, 2011), pp. 319–43.

2 See J. V. Field, *Piero della Francesca: A Mathematician's Art* (New Haven: Yale University Press, 2005), esp. Chapter 7.

3 James R. Banker, 'A Manuscript of the Works of Archimedes in the Hand of Piero della Francesca', *The Burlington Magazine*, 147 (2005), 165–69.

4 On the Alhacen translation see Graziella Federici Vescovini, 'Contributo per la storia della fortuna di Alhazen in Italia: il volgarizzamento del MS Vat. 4595 e il commentario terzo del Ghiberti', *Rinascimento*, 2nd Series, 5 (1965), 17–49.

Perspectiva

The period term, obviously cognate to the modern one, is the Latin *perspectiva* (occasionally *prospectiva*), which appears in Tuscan as either *perspettiva* or *prospettiva*. The variant spellings may reflect the slightly different vernacular terms in different dialects — we are in the world of phonetic spelling — but they show no divergence in meaning. However, the meaning of the term is itself slightly slippery. Latin *perspeciva* is the complete science of sight, including the functioning of the human eye as well as the properties of light. There was, of course, a geometrical component to this, and this was the part that was extended to provide rules for painters (or sculptors) wishing to use mathematical constructions to give a sense of depth (or a sense of greater depth) in their works. As the use of such techniques became fashionable in Italy from the 1430s onwards, a distinction came to be made between *perspectiva communis* (*perspettiva commune*) which dealt with appearances in the everyday world and *perspectiva artificialis* (*perspettiva artifiziale*) or *perspectiva pingendi* (*perspettiva dei pittori*) which was the part of special interest to painters and sculptors. Thus when Piero wrote a treatise on what we should now call 'linear perspective', a term introduced, in English, in the eighteenth century,[5] he gave it the Latin title *De prospectiva pingendi*. And at the very beginning hetells us explicitly what the book is about:

> Painting has three principal parts, which we say are drawing, proportion and colouring (*colorare*). Drawing we understand as meaning outlines and contours contained in things. Proportion we say is these outlines and contours positioned in proportion in their places. Colouring we mean as giving the colours as they are shown in the things, light and dark according as the light makes them vary. Of the three parts I intend to deal only with proportion, which we call perspective, mixing in with it some part of drawing, because without this perspective cannot be shown in action; colouring we shall leave out, and we shall deal with that part which can be shown by means of lines, angles and proportions, speaking of points, lines, surfaces and bodies […].[6]

This seems to give 'perspective' a rather narrowly mathematical sense, but it becomes clear elsewhere in the treatise that Piero regards this mathematical perspective as an extension of *perspectiva* proper. For example, in the last proposition of Book 1 he calls upon a standard model of the working of the human eye, namely that the eyebeams cannot spread out through more than a right angle, in order to fix limits on the possible width of a perspective picture. Thus, in today's terms, the 'soft' (anatomical cum physiological) end of the subject

5 Brook Taylor, *Linear Perspective: Or, a New Method of Representing Justly All Manner of Objects as They Appear to the Eye in All Situations* (London: R. Knaplock, 1715).

6 'La pittura contiene in sè tre parti principali, quali diciamo essere disegno, commensuratio e colorare. Disegno intendiamo essere profili e contorni che nella cosa se contene. Commensuratio diciamo essere essi profili e contorni proportionalmente posti nei luoghi loro. Colorare intendiamo dare i colori commo nelle cose si dimostrano, chiari e uscuri secondo che i lumi le devariano. Delle quali tre parti intendo tractare solo della commensuratione, quale diciamo prospectiva, mescolandoci qualche parte de desegno. perchiò che senza non se po dimostrare in opera essa prospectiva; il colorare lasciaremo stare e tractaremo de quelle parte che con line angoli e proportioni se po dimostrare, dicendo de puncti, linee, superficie et de corpi […]'. Piero della Francesca, *De prospectiva pingendi*, ed. by G. Nicco Fasola (Florence: Sansoni, 1942; repr. Florence: Casa Editrice Le Lettere, 1984), p. 63 (translation J. V. Field).

is called upon to deal with a problem arising in the 'hard' (mathematical) part.[7] In any case, Piero is taking it for granted that it is the painter's business to portray the world as it is seen by the eye, and his approach to how this can be achieved implies an understanding that in today's terms contains elements that would be considered 'scientific' — indeed some of them would have been considered 'scientific' in Piero's time too, since they were mathematical. On the other hand, there is no evidence that Piero, unlike his younger contemporary Leonardo da Vinci (1452–1519), ever saw himself as a natural philosopher. Though no less observant than Leonardo, Piero seems never to focus his attention on understanding any phenomenon in its own right but instead looks for what is required to represent its effect on the eye. We have, for instance, no evidence that Piero ever wrote about the phenomena of reflection, but his paintings show he had a very good grasp of what reflections looked like.

We may note, also, that Piero is using the term 'in proportion' (*proportionalmente* in his spelling) in a sense that applies specifically to the correct representation of the sizes of objects as shown in a painting. Also, here and in the introductory paragraph to the third book of his treatise, Piero associates his 'proportion' with commensurability. Both these terms obviously go back to Euclid's *Elements*, but in the present context the source of the relevant mathematics is Euclid's work on optics, in the fifteenth century usually known as *De aspectibus varietate* ('On the Variety of Appearances') in which there is much use of similar triangles as lines of various lengths are shown as they appear to the eye, by subtending various angles, when they are at various distances from it. Unfortunately, the word 'proportion' is used in many different contexts, in different senses, in the fifteenth and sixteenth centuries.[8] Thus Piero's technical use of 'in proportion' has sometimes been misunderstood. Since Piero gives mathematical proofs, it is absolutely certain that in this context 'in proportion' does not mean we have a specified ratio, or a series of ratios, between sizes. This matter of objects being shown the 'right size' is important because in its simplest form it provides one of Piero's characteristic means of constructing a sense of the third dimension, and the employment of this particular technique has been taken as an indicator of when particular pictures were painted.

Stylistic Dating of Piero's Paintings

Much has sometimes been made of the difficulty of dating Piero's paintings by their style. There has, however, long been general agreement on a rough dating scheme that depends upon the method Piero used to indicate depth. Three styles can be distinguished here. A few definite dates have been available for a long time, so the styles can be put in chronological order. The earliest is that of indicating depth by showing objects of known real size in varying sizes in the picture. A good example of this is the painting of *St Jerome as a Penitent*

7 See J. V. Field, 'Piero della Francesca's Treatment of Edge Distortion', *Journal of the Warburg and Courtauld Institutes*, 49 (1986), 66–99 (Plate 21c).

8 On the multiple senses and uses of 'proportion', see Sabine Rommevaux, Philippe Vendrix and Vasco Zara, *Proportions. Science, musique, peinture et architecture: actes du LIe colloque international d'études humanistes, CESR, Tours, June–July 2008* (Turnhout: Brepols, 2011).

(Figure 1), which is signed and dated 1450 on the *cartellino* at the lower right. The trees of the grove behind the saint are clearly at different distances from the eye and the house provides another approximate way of gauging distance. A similar use of trees, human figures and a distant city (a portrait of Piero's native Sansepolcro) is found in the *Baptism of Christ* (National Gallery, London). The sense of depth provided is not precise, that is, one could probably not use the depth clues to generate a precise ground plan of the scene, but the pictures convey a convincing impression that the figures are positioned in a landscape.

Later paintings show a use of mathematical techniques like those described in Piero's perspective treatise, that is they use straight lines, and specifically images of orthogonals that converge to a point (usually close to the central axis), a point to which Piero gives no name, though he consistently calls it 'A'; Leon Battista Alberti (1404–72) uses the term 'centric point', which is the one adopted here. The arch example of converging images of orthogonals is the *Flagellation of Christ* (Figure 2), signed but not dated, which is the only fifteenth-century painting for which it has been possible to reconstruct a coherent scheme of geometrical perspective, that is to find a viewing distance and to draw a ground plan and sections of the scene in the painting.[9] Pictures of the third period look perfectly coherent but do not show lines that allow one to construct a ground plan. Examples of this style are provided by the *Madonna di Senigallia* (Galleria Nazionale delle Marche, Urbino), the *Montefeltro Altarpiece* (Figure 7) and *The Nativity* (Figure 8). Piero did not, however, abandon each style successively and move on to the next, so this scheme is not a complete guide to putting his works in roughly chronological order. Some confirmation of the ordering was, however, derived from the examination of the materials Piero used. In Piero's lifetime, the use of oil as a binding medium was being introduced into Italy — apparently largely through Netherlandish painters — so it is reasonable to suppose that the use of egg tempera rather than oil indicates a relatively early date for a picture. But this is not an infallible method of dating; Piero seems sometimes to have used whatever was to hand.

Over the last twenty-five years or so, painstaking work in local archives in the regions by Italy in which Piero lived has brought to light numerous documents that provide information about his life and occasionally suggest fairly precise datings for some of his works. The principal contributor to this accumulation of evidence, James R. Banker, eventually put all 261 documents together as a book, and went on to publish a biography of Piero based in large part on this new evidence.[10]

This new biography of Piero — the first state-of-the-art biography of Piero since that of Giorgio Vasari (1511–74) — seems unlikely to put an end to scholarly wrangling over the precise dating of particular paintings, but it does confirm the overall correctness of the schema outlined above.[11] Since there is now documentary evidence that this stylistic evolution is real, rather than merely an artefact of the methods adopted by art historians,

---

9 See M. A. Lavin, *Piero della Francesca: The Flagellation of Christ* (New York: Viking Press, 1972; repr. (with additional bibliography) Chicago: University of Chicago Press, 1990) and Field, *Piero della Francesca*.

10 James R. Banker, *Documenti fondamentali per la conoscenza dellla vita e dell'arte di Piero della Francesca* (Selci-Larma: Editrice Pliniana, 2013); James R. Banker, *Piero della Francesca: Artist and Man* (Oxford: Oxford University Press, 2014).

11 Giorgio Vasari, *Le vite dei piu eccellenti architetti, pittori e scultori italiani da Cimabue a tempi nostril*, 2nd edn. (Florence, 1568; 1st edn. 1550).

Fig. 1 Piero della Francesca, *St Jerome as a Penitent*, tempera on panel, 51.5 x 38 cm, 1450, signed and dated on *cartellino* at lower right. Berlin, Gemäldegalerie der Staatlichen Museen, Stiftung Preußischer Kulturbesitz. [See colour plate 33]

Fig. 2 Piero della Francesca, *The Flagellation of Christ*, tempera on panel, 59 x 81.5 cm, probably painted between 1465 and 1475. Urbino, Galleria Nazionale delle Marche. [See colour plate 34]

Fig. 3 Piero della Francesca, underdrawing of head of a bystander in the scene of King Solomon receiving the Queen of Sheba, *The Story of the True Cross*, San Francesco, Arezzo. Infrared reflectogram courtesy of the Opificio delle Pietre Dure, Florence.

Fig. 4 Piero della Francesca, underdrawing of head of an attendant in the scene of the Queen of Sheba worshipping the wood, *The Story of the True Cross*, San Francesco, Arezzo. Infrared reflectogram courtesy of the Opificio delle Pietre Dure, Florence.

Fig. 5 Piero della Francesca, underdrawing of head of a kneeling woman in the scene of the Proving of the True Cross, *The Story of the True Cross*, San Francesco, Arezzo. Infrared reflectogram courtesy of the Opificio delle Pietre Dure, Florence.

Fig. 6 Piero della Francesca, underdrawing of head of King Chosroes in the scene of his execution, *The Story of the True Cross*, San Francesco, Arezzo. Infrared reflectogram courtesy of the Opificio delle Pietre Dure, Florence. The same underdrawing was used for the head of God the Father in the Annunciation scene to the right of this one.

we may look at it a little more closely. Piero is changing the means he uses to show the spatial relationships between the objects represented in his painting. That is, he is paying attention to the very aspect of fifteenth-century Italian art that is always singled out as its distinctive contribution to the development of Western Art as a whole, namely perspective. But here 'perspective' has its period sense, because what Piero is using is not, initially, the new geometrical techniques we find in his treatise (which was, as far as is known, the first treatise to be written on artificial perspective) but rather the whole array of methods inherited from the Euclidean tradition. Some of these show little or no change over the range of Piero's work. One such element was the direct portrayal of light.

Shadows

Shadows an object casts on itself, for instance those in the folds of drapery, had long been used to indicate modelling. On the whole, the shadows are indicated by using a darker version of the same colour, thus a pale yellow cloak will have deep yellow shadows. Piero tends to abide by this convention, but he is unusually careful, even by the standards of fifteenth-century Tuscan art, in showing highlights and cast shadows as well as reflections and refraction. All these effects belong to *perspectiva*, and are of interest to painters in Piero's time. That no doubt explains why, in the introductory paragraph of his perspective treatise that we quoted above, he states explicitly that the work will not concern itself with the variation of colours with lighting. Piero may have chosen to mention that particular matter because the skill concerned was one that all painters would learn from their master during their apprenticeship. Piero's treatise is in many respects a kind of workshop manual to take a pupil through the steps of learning to draw in artificial perspective.

In his section on artificial perspective in his *Underweysung der Messung* (1525, Latin version 1536) Albrecht Dürer (1471–1528) discusses the shadow an object casts on the ground when illuminated by sunlight.[12] Some of Dürer's discussion is closely similar to what we find in *De prospectiva pingendi* and it seems highly probable that Dürer had access to Piero's writings in the 1490s, in Venice, where he met Luca Pacioli (*c.* 1445–1517), who seems to have come into possession of Piero's mathematical writings after his death.[13] It is, perhaps, a little strange that Piero does not mention this type of problem in his treatise, since it is a purely geometrical one that can be solved by what Piero calls 'the power of the lines and angles which are obtained from it' [*sc.* perspective].[14] Nevertheless, Piero's paintings contain many cast shadows, some of compositional importance, as in the apse shown in the *Montefeltro Altarpiece* (Figure 7) and the shadow cast by the roof of the stable in *The Nativity* (Figure 8). In both these pictures, the shadows also serve to establish three-dimensional relationships. We may note, however, that the barrel vault on which the cast shadow falls in the former picture seems to have been the subject of a detailed perspective construction such as those described in Piero's perspective treatise — which accords well with its being an obvious tribute to the barrel vault in the *Trinity* fresco by Masaccio (born Tommaso di Ser Giovanni, 1401–28) in Santa Maria Novella, Florence (painted *c.* 1426), which, when Piero saw it, still had the rosettes in the coffers of the vault that are mentioned by Vasari in his *Life* of Masaccio.[15]

Cast shadows are also prominent in *The Flagellation of Christ* (Figure 2). The relief of the coffered ceiling is brought out by strong cast shadows. Light from below also shines on the golden statue (presumably a pagan god) that stands on top of the column. The light also catches the face of Christ. If it is intended to be natural, the light seems to be too bright to be anything but sunlight, but it comes from a hidden source at the lower right, the exact opposite of the usual direction for natural light, which is shown above Pilate's throne on the left. The cast shadows from the light on the right are so precise, and the positions of the ribs that define the coffers are known so exactly from the perspective, that the location of the light source can be found precisely. But, disconcertingly, it has no obvious significance. My theory, for what it is worth, is that Piero is being literal: showing the scene lit by reflected sunlight as it might have been in a pageant. Since sunlight is bright, one does not need a very good mirror. A sheet of polished metal, perhaps a very thin sheet attached to a wooden panel, will do. This may sound unacceptably naïve, but

12 Albrecht Dürer, *Underweysung der Messung mit dem Zirkel und Richtscheyt* (Nuremberg 1525, Latin version tr. Camerarius, 1536); French translation Albrecht Dürer, *La géométrie*, translated from the German and with an introduction by J. Peiffer (Paris: Éditions du Seuil, 1995).

13 Field, *Piero della Francesca*.

14 'la forza delle linee e degli angoli, che da essa se producano'. Piero della Francesca, *De prospectiva pingendi*, III, Introduction.

15 Field, *Piero della Francesca*; J. V. Field, R. Lunardi and T. B. Settle, 'The Perspective Scheme of Masaccio's "Trinity" Fresco', *Nuncius*, 4 (2) (1988), 31–118.

Fig. 7 Piero della Francesca, *The Virgin and Child Enthroned with Saints and Angels* (known as *The Montefeltro Altarpiece*), tempera and oil on panel, 251 x 172.5 (±0.5) cm, probably 1475–77, Milan, Pinacoteca di Brera. It is unlikely that this picture was intended to be an altarpiece. [See colour plate 35]

Fig. 8 Piero della Francesca, *The Nativity* (properly an *Adoration*), oil on panel, 124.3 x 123 cm, 1486-92. London, The National Gallery. [See colour plate 36]

Piero made other acknowledgements to the pageant tradition, for instance by showing the disc haloes that were used in such representations — a fairly common practice among painters — and on occasion, for example in the saints in the Sant' Antonio altarpieces (1467–69), showing the heads of the halo-wearers reflected in the polished haloes, though for the Christ-child we only see a red cross. So it seems conceivable, to me at any rate, that the apparently non-natural light from the lower right in the meticulously mathematical *Flagellation* is in fact a salute to theatrical practice.

Reflections and Refraction

Another element of *perspectiva* in painting that was in principle susceptible of geometrical construction was reflection. However, when there was specular reflection from a curved surface, as for armour or the shiny cylindrical ointment pot held by Mary Magdalene in Piero's fresco in Arezzo cathedral, this would have presented a rather hard problem (though help would have been available from Alhacen) and it seems more likely that Piero simply made some detailed studies from real objects. In any case, his success was striking. In his *Life* of Piero, Vasari, who was born in Arezzo and spent his youth there, praises the realistic glitter of the armour of the soldier immediately behind the Emperor Constantine in the scene of the Battle of the Milvian Bridge in *The Story of the True Cross*. Vasari, who had painted a good many frescos himself, remarks that it was particularly difficult to achieve such an effect in fresco. The soldier in question is wearing full fifteenth-century steel armour, though elsewhere in the fresco cycle there are elements that show an engagement with archaeological correctness.

No such ambiguity arises for the armour worn by Federigo da Montefeltro (b. 1422, ruled Urbino 1444–82), who kneels in adoration in front of the Virgin and Child with saints and angels shown in the large picture that is usually called *The Montefeltro Altarpiece* but was probably designed to be placed above his grave slab (Figure 7).[16] Piero would no doubt have had access to Federigo's armour, so here he certainly could have made a detailed study.

The glittering reflections of the armour, and the reflections on Mary Magdalene's ointment pot and its gilded mounts (which are daringly placed close to a belt that seems to have been finished with gold leaf) had a compositional purpose in catching the eye: in the case of Mary Magdalene because the pot of ointment was her identifying symbol and in the case of the soldier's armour because his figure establishes the position of the picture plane. Piero is particularly careful to do this in the scenes of the fresco cycle, where he was obviously constantly aware, as the reader of an illustrated art book often is not, that these pictures are parts of a wall.

Some years before painting the large panel for Federigo, Piero made a pair of portraits of Federigo and his wife Battista Sforza (1446–72), showing them in profile against an aerial view of a landscape. This use of distant landscape is no doubt influenced by Netherlandish models but, as any visitor to Urbino can testify, it is also naturalistic: the sitters have been shown as if against an open window of their palace, and through the window we can see far out over their domain. The landscape is articulated by small hills, their shapes seen against wisps of early-morning mist, several buildings and a winding river. Piero's blue pigments have tended to fade with age (as have Leonardo's), but it seems that in this landscape, which runs continuously behind both figures, he has made some use of colour changes due to distance. In the foreground, what catches the eye is the bright red of Federigo's large sculptural hat and the glitter of Battista's jewels. Here there can be little doubt that Piero has been shown the jewels and has made a portrait of them with the same care that he drew the profile of their owner.

Battista's jewels help her hold her own in visual terms in the pair of portraits. Jewels are used to mark rank, so that even the humble Virgin in the *Nativity* (Figure 8) wears a few beautifully observed jewels, including some pearls in her hair. Jewels again mark rank, and

16 Field, *Piero della Francesca*; Guido Ugolini, *La pala dei Montefeltro: una porta per il mausoleo dinastico di Federico* (Pesaro: Nobili, 1985).

serve to attract the eye, in the picture of Federigo da Montefeltro and the enthroned Virgin and Child (Figure 7). Here the hem of the Virgin's robe is embroidered with jewels, each tilted to the light in a slightly different way, and the Virgin originally also wore a jewel on her forehead, which would have provided an additional central accent for the composition, helping to balance the distant illuminated ostrich egg, which is now more prominent than Piero intended. The jewels form a kind of frame within a frame for the most important figures. Further reflections and refractions appear in the rock-crystal cross held by one of the saints. The rendering of the complicated interaction of the transparent colourless crystal with the incident light must also presumably be the result of observation rather than geometrical construction.

Light in the Picture and Light on the Picture

The above examples of Piero's careful handling of the effects of light all refer to local phenomena within the picture. Instances could be multiplied considerably and almost all of them are closely similar to what is found in contemporary works from North of the Alps.[17] It is sometimes asserted that the use of oil paint made it easier to convey subtle gradations of colour, but Piero shows no sign of preferring oil for that reason. Some of his most subtle effects are achieved in tempera, for instance the rendering of the sheen of the pale grey silk robe of the man standing with his back to us in the *Flagellation* (Figure 2). Piero can also produce very delicate gradations of colour in fresco. What we seem to be seeing is a painter who is clear in his own mind what effect he wants to produce and is prepared to make the available medium do it. We can see something of the same in his attitude to pigments. Following the usual practice, the restorers working on the Arezzo frescos carried out extensive chemical tests on the painted plaster before they started their work. They found, to their astonishment, that in several places colours that looked exactly the same were chemically distinct. That is, different pigments had been used to produce the same colour in contiguous patches. Piero had been successful, but what he did flies in the face of the usual advice. In fresco the exact colour depends on both the pigment and the (damp) plaster to which it is applied, making it doubly unpredictable. So one is generally advised to prepare exactly as much of both as one will need for a particular colour for a specified item, say the red for a cloak.

Perhaps, however, we should not be surprised that as a painter Piero had a stubborn streak: he had somehow persuaded his strong-minded father, Benedetto di Piero della Francesca (1375/6–1464), who was intent on improving the family's social standing, that he, the eldest son, would become an artisan.[18] If there is life after death, Benedetto probably forgave Piero some time in the 1920s.

Details of Piero's pictures and details of those of Netherlandish masters such a Jan van Eyck (fl. 1422, d. 1441) and Hugo van der Goes (d. 1482), show many parallels. Parallels

17 Bert W. Meijer, 'Piero and the North', in *Piero della Francesca and his Legacy*, Studies in the History of Art, 48, Center for Advanced Study in the Visual Arts, Symposium Papers XXVIII, ed. by M. A. Lavin (Washington, DC: National Gallery of Art, 1995), pp. 143–59.

18 See Banker, *Piero della Francesca*.

become much less apparent when one looks at the overall composition. Piero is a master not only of composition in the picture plane, which was an age-old skill passed more or less successfully from master to apprentice over the millennia, but also of the new skill of composition in three dimensions (though the third one is illusory). As anyone who has tried to take photographs knows, elements included as 'background' or 'visual context' can make a nonsense of a picture. Architecture is a particular problem, because much of it is very large compared with human figures. In the late thirteenth and early fourteenth century, Giotto di Bondone (*c.* 1266–1337), whose human figures look thoroughly solid, regularly made the architectural or landscape setting small and fitted it round the figures. Details are naturalistic, but relative sizes are not. His art was hugely admired by his contemporaries and successors. And even in an age accustomed to photography, Giotto still comes across as a vivid storyteller. Piero, however, is working within more naturalistic conventions with regard to relative sizes. But architecture is necessarily important because it offers straight lines that can be used to indicate depth, characteristically by means of a regular ground plan or square-tiled floors. Such elements can be seen in Piero's *Flagellation* (Figure 2), where he has been careful to break then up a little at floor level and, on the coffered ceiling, disrupt the pattern with strong cast shadows.

Apart from a couple of indeterminate houses on the right and a slight narrowing of the wide white orthogonal strip of marble near the centre line of the picture, the perspective of the *Flagellation* seems to be absolutely correct; it has even come through the indisputably dispassionate analysis preformed by a sophisticated artificial vision program (designed to read photographs).[19] However, Piero has adopted the old convention of making the architecture very small. One of his main sources for classical architecture seems to have been the Baptistery of Florence, so there is no doubt he knew how large this kind of architecture should really be.[20] To see how more realistically sized quasi-Roman architecture dwarfs human figures, one can turn to the more archaeologically-minded painters of the next generation, such as Andrea Mantegna (1431–1506). There is, also, a problem with the viewing distance of the *Flagellation*. The correct viewing distance is rather large: two and a half times the picture width. This is convenient for looking at a small reproduction in a book, but I have never seen a viewer of the original panel withstand the temptation to go much closer, to look at details and admire the exquisite finish. Perhaps the picture was intended for a site where the viewer could not get close to it. In any case, although experience shows that in general the correct viewing distance acts as a minimum at which the perspective construction works and the eye feels comfortable, for the *Flagellation* the sense of depth is surprisingly robust when the picture is viewed from far too close. I think this is probably because the lighting gives additional definition to the forms within the picture, producing cast shadows, highlights and strong diffuse reflections. This suggests that the detailed rendering of the effects of light that we have noted in other pictures, and which are indeed characteristic of Piero's art, should really be regarded as deliberately introduced as visual clues to the overall illumination of the scene.

19 Antonio Criminisi, *Accurate Visual Metrology from Single and Multiple Uncalibrated Images*, Distinguished Dissertation Series (London: Springer, 2001).

20 Christine Smith, 'Piero's Painted Architecture: Analysis of his Vocabulary', in *Piero della Francesca and his Legacy*, ed. by M. A. Lavin, pp. 223–53.

Of the pictures we have looked at here, this effect is particularly significant in the painting showing Federigo da Monefeltro kneeling before the Virgin and Child (Figure 7). This picture is usually called the *Monefeltro Altarpiece* because it has been suggested that it was intended as an altarpiece, perhaps for the high altar of the church of San Bernardino, just outside Urbino. The painted architecture does indeed resemble that of the church, which was designed by Francesco di Giorgio Martini (1439–1501/2), who had done a great deal of work in Urbino and surely knew Piero well enough to discuss the architecture with him before the church was built (drawings recording Francesco's designs for the building seem to date from the early 1480s). However, that does not prove Piero's picture was painted as an altarpiece. We have already noted that the barrel vault is a tribute to Masaccio's *Trinity* fresco. In the fresco the barrel vault is the roof over a sarcophagus on which God the Father is standing — the tomb is that of Adam and the composition associating it with the Trinity is a standard one known as 'The Throne of Grace'. Thus, in Masaccio's fresco, we are looking into a funerary chapel. Barrel vaults are found in a number of Roman structures that in the fifteenth century were identified as funeral monuments, so the architectural form confirms that this is a mortuary chapel. At its base, Masaccio's fresco has a skeleton with a *memento mori* inscription. Masaccio's fresco was an altarpiece, but its symbolism has funerary connections of which Piero must have been aware.

Masaccio's barrel vault covers a rectangular area whose sides are in the ratio 4:3. Piero made the area under his vault square, which makes it very easy to calculate the correct viewing distance for the picture, which turns out to be twice the picture width. So the reproduction in this book gives one an idea of what the picture looks like when viewed from far too great a distance. The effect of depth is much more striking when one stands somewhere near the correct station point, that is about three and a half metres from the picture. Also, the height of the centric point is close to the eye height of the standing figures. This is unusual for Piero and implies the picture was placed rather low down (as it is at present in the gallery in Milan). There would be room for a low tomb below it, but not room for an altar. Even the officiant's viewpoint would be too low (other paintings by Piero known to have been parts of altarpieces seem to have been designed for the eye height of the officiant). So the mathematical perspective is against this painting being an altarpiece.

So too is the lighting in the picture. The prominent cast shadow, and the reflections in the armour, imply the light is coming from above left. This was indeed the case for the position in which the picture hung for many years, on the left wall as one enters the church through the main (liturgical West) door. In this position light falls on the picture from the round window above the door. Moreover, one would then see the picture from an appropriately short distance, across the width of the church rather than down its length. Alternatively, the picture may not have been painted for this church at all. it could have been intended for a small mortuary chapel inside Federigo's palace that, in the event, was never built.[21]

21 See Ugolini, *La pala dei Montefeltro*.

This argument assumes that Piero would have wanted the lighting in his picture to correspond to the light falling on it. There is plenty of evidence that this was indeed his normal practice. All but one of the pictures in *The Story of the True Cross* show light whose direction follows that of the natural light. Figures standing near the picture plane have highlights that correspond to light coming from the window. Even in the picture of Constantine's dream, where the main source of light in the picture is the golden cross held in the hand of the descending angel, the direction has been chosen to agree with that of the natural light from the window to the left of the picture. The use of the natural lighting, and its correspondence with lighting shown in the frescos, is obvious (once one has noticed it) when one sees the originals, and it is sufficiently pronounced to be visible in photographs also.

A similar care for the natural lighting is shown by other painters, notably by Mantegna in his frescos portraying members of the Gonzaga family in the *Camera degli Sposi* in their palace in Mantua (painted between 1465 and 1474). Not only does the lighting in the pictures follow the natural light, but Mantegna seems to have arranged for new windows to be put in, presumably to improve the lighting conditions, and the room itself may have been chosen because it looks out over the moat, so the light is partly reflected (giving it something of the quality of the light for which Venice is famous).

Piero took care to coordinate the perspective schemes and compositions of the scenes that make up *The Story of the True Cross*.[22] Following the natural lighting allows him to impose a further degree of visual unity. Frescos still *in situ* allow one to relate the light in the picture to the light falling on it. For movable pictures one can look only at the former. In all his pictures Piero shows self-consistent lighting and the light helps to define the spatial relationships between the objects shown in the picture. This holds throughout his career, from early pictures such as the *Baptism of Christ* and *St Jerome as a Penitent* (Figure 1) to late ones such as the *Madonna di Senigallia* (Urbino, Galleria Nazionale delle Marche) and the *Montefeltro Altarpiece* (Figure 7). It is no accident that in putting together an exhibition to honour the five-hundredth anniversary of Piero's death, the Uffizi Gallery (Florence) gave the exhibition a title that explicitly refers to the treatment of light.[23]

Bibliography

Primary Sources

Dürer, Albrecht, *Underweysung der Messung mit dem Zirkel und Richtscheyt* (Nuremberg 1525, Latin version tr. Camerarius, 1536); French translation Albrecht Dürer, *La géométrie*, translated from the German and with an introduction by J. Peiffer (Paris: Éditions du Seuil, 1995).

Vasari, Giorgio, *Le Vite dei piu eccellenti architetti, pittori e scultori italiani da Cimabue a tempi nostril*, 2nd edn. (Florence, 1568; 1st edn. 1550).

22 See Field, 'Theory and Practice of Perspective' and Field, *Piero della Francesca*.

23 *Una scuola per Piero: Luce, colore e prospettiva nella formazione fiorentina di Piero della Francesca*, ed. by Luciano Bellosi, exh. cat., Uffizi, Florence (Venice: Marsilio, 1992). For a wider view see Paul Hills, *The Light of Early Italian Painting* (London: Yale University Press, 1987).

Secondary Works

Banker, James R., 'A Manuscript of the Works of Archimedes in the Hand of Piero della Francesca', *The Burlington Magazine*, 147 (2005), 165–69.

Banker, James R., *Documenti fondamentali per la conoscenza dellla vita e dell'arte di Piero della Francesca* (Selci-Larma: Editrice Pliniana, 2013).

Banker, James R., *Piero della Francesca: Artist and Man* (Oxford: Oxford University Press, 2014).

Bellosi, Luciano (ed.), *Una scuola per Piero: Luce, colore e prospettiva nella formazione fiorentina di Piero della Francesca*, exh. cat., Uffizi, Florence (Venice: Marsilio, 1992).

Criminisi, Antonio, *Accurate Visual Metrology from Single and Multiple Uncalibrated Images*, Distinguished Dissertation Series (London: Springer, 2001).

Federici Vescovini, Graziella, Graziella, 'Contributo per la storia della fortuna di Alhazen in Italia: il volgarizzamento del MS Vat. 4595 e il commentario terzo del Ghiberti', *Rinascimento*, 2nd Series, 5 (1965), 17–49.

Field, J. V., 'Piero della Francesca's Treatment of Edge Distortion', *Journal of the Warburg and Courtauld Institutes*, 49 (1986), 66–99.

Field, J. V., R. Lunardi and T. B. Settle, 'The Perspective Scheme of Masaccio's "Trinity" Fresco', *Nuncius*, 4 (2) (1988), 31–118.

Field, J. V., *Piero della Francesca: A Mathematician's Art* (New Haven: Yale University Press, 2005).

Field, J. V., 'Theory and Practice of Perspective in the Works of Piero della Francesca', *1492, Rivista della Fondazione Piero della Francesca, Anno 1* (1–2) (2008), 35–54.

Field, J. V., 'Lo sguardo matematico nell' arte rinascimentale Italiana', in *La Matematica, Vol. 3, Suoni, forme, parole*, ed. by Claudio Bartocci and Piergiorgio Odifreddi (Turin: Einaudi, 2011), pp. 319–43.

Hills, Paul, *The Light of Early Italian Painting* (London: Yale University Press, 1987).

Lavin, M. A., *Piero della Francesca: The Flagellation of Christ* (New York: Viking Press, 1972; repr. (with additional bibliography) Chicago: University of Chicago Press, 1990).

Meijer, Bert W., 'Piero and the North', in *Piero della Francesca and his Legacy*, Studies in the History of Art, 48, Center for Advanced Study in the Visual Arts, Symposium Papers XXVIII, ed. by M. A. Lavin (Washington, DC: National Gallery of Art, 1995), pp. 143–59.

Rommevaux, Sabine, Philippe Vendrix and Vasco Zara, *Proportions. Science, musique, peinture et architecture: actes du LIe colloque international d'études humanistes, CESR, Tours, June–July 2008* (Turnhout: Brepols, 2011).

Smith, Christine, 'Piero's Painted Architecture: Analysis of his Vocabulary', in *Piero della Francesca and his Legacy*, Studies in the History of Art, 48, Center for Advanced Study in the Visual Arts, Symposium Papers XXVIII, ed. by M. A. Lavin (Washington, DC: National Gallery of Art, 1995), pp. 223–53.

Taylor, Brook, *Linear Perspective: Or, a New Method of Representing Justly All Manner of Objects as They Appear to the Eye in All Situations* (London: R. Knaplock, 1715).

Ugolini, Guido, *La pala dei Montefeltro: una porta per il mausoleo dinastico di Federico* (Pesaro: Nobili, 1985).

PAUL HILLS

The Venetian Optics of Light and Geometry of Proportion

Introduction: Pacioli (*c.* 1447–1517), Carpaccio (*c.* 1465–1526), and Euclidian Culture

The practice of perspective in Venetian painting around 1500 is marked by a distinctive geometry and a special regard for light. Geometry and lighting are conceived in close relation and together they constitute a new kind of pictorial structure. This relation between geometry and light is distinct from what is commonly found in Central Italian art of the same period.

In a remarkable essay, published thirty years ago, Margaret Daly Davis drew attention to the depiction of regular bodies in the work of the Venetian painter Vittore Carpaccio.[1] Davis pointed to an eye-catching example displayed on the threshold of the large canvas of the *Reception of the English Ambassadors* (Figure 1 and 2, detail), in the cycle from the *Scuola di Sant'Orsola*. This belongs to a type of perspective exercise performed by Paolo Uccello, namely a *mazzocchio* ('hat-ring'). The version circling the base of Carpaccio's column is stellated in a manner comparable to one in Daniele Barbaro's manuscript, *La pratica della prospettiva*, written more than half a century later. Since Barbaro recorded that both Giovanni Bellini and Carpaccio were taught by one Hieronimo Malatini, Davis posits a plausible genealogy of geometric expertise that runs from Piero della Francesca to Luca Pacioli, through the mysterious Malatini to Carpaccio, as well as from Piero and Pacioli to Barbaro himself.[2]

Pacioli's importance for Venetian pictorial culture has been increasingly recognised.[3] A native of Borgo Sansepolcro, Pacioli first stayed in Venice in the 1460s, where he was employed as a tutor in a mercantile family. Here he soon became familiar with the accounting practices of the Venetian world of trade and banking. He returned to the lagoon repeatedly

1 Margaret Daly Davis, 'Carpaccio and the Perspective of Regular Bodies', in *La prospettiva rinascimentale*, ed. by Marisa Dalai Emiliani (Florence: Centro Di, 1980), I, pp. 183–200.

2 Davis, 'Carpaccio', I, pp. 183–84. Barbaro's reference to Malatini, Bellini, and Carpaccio is in his *Pratica della prospettiva*, in Marciana Library, Venice, MS 5446, fol. 2[r]. For Malatini, see Sara Menato, *Per la giovinezza di Carpaccio* (Padua: Padova University Press, 2016), pp. 47–60.

3 For example by Oskar Bätschmann, *Giovanni Bellini* (London: Reaktion, 2008), pp. 36–38, 187.

Paul Hills The Courtauld Institute of Art, London, Paul.Hills@courtauld.ac.uk

Perspective as Practice. Renaissance Cultures of Optics, ed. by Sven Dupré, Turnhout, 2019 (*Techne. Knowledge, Technique, and Material Culture*, 1), p. 409-427

DOI 10.1484/M.*Techne*-EB.5.117735

Fig. 1 Carpaccio, *The Reception of the English Ambassadors* (*c.* 1500), canvas, 278 x 589 cm. Venice, Gallerie dell'Accademia. Photograph by Didier Descouens. [See colour plate 37]

over the next fifty years.[4] He was there in 1470, and again in 1494 to oversee publication of his *Summa de arithmetica*; then briefly in 1499–1500, and again in 1508 when he edited the fifth book of Euclid's *Elements* for the Venetian press. His lecture on Euclid, delivered in the church of San Bartolomeo at the Rialto on the eighth of August 1508 was attended by almost all the learned world of Venice, including the sculptor Pietro Lombardo.[5] That lecture was the prelude to the publication of his edition of Euclid and his *Divina proportione* in Venice the following year.

Pacioli's visits span the period of Giovanni Bellini's rise to maturity and the years of his greatest achievement, and it is clear that these decades were crucial for the formation of the younger master, Carpaccio. The Paciolian culture that was so congenial to Venetians was less concerned with perspective as a system for delineating recessional space than with the projection of regular bodies, together with questions of proportion and interval. This essentially Euclidian phase in Venetian art and intellectual culture was stimulated by the arrival of Giorgio Valla in 1481. A scholar profoundly versed in mathematical and geometrical learning, Valla stayed on as teacher until his death in Venice in 1500. His encyclopedic work, *De expetendis et fugiendis rebus*, published posthumously in 1501 by Aldus Manutius in a handsome illustrated edition, gives great prominence to Euclidian theorems. Printing the geometrical figures positioned at the appropriate place in the text,

---

4 For brief accounts: S. A. Jayawardene, 'Towards a Biography of Luca Pacioli', in *Luca Pacioli et la matematica del rinascimento*, ed. by E. Giusti (Città di Castello: Petruzzi, 1998), pp. 19–28, and Augusto Marinoni's introduction to the facsimile of the copy in the Ambrosiana Library of *De divina proportione* (Milan: Silvana Editore, 1982), pp. 5–6.

5 For Pacioli at the Rialto, see: Fernando Lepori, 'La scuola di Rialto dalla fondazione alla metà del cinquecento', in *Storia della cultura veneta*, ed. by G. Arnaldi and M. P. Stocchi (Vicenza: Neri Pozza, 1980), Vol. II, Part 2, pp. 597–605; Carlo Maccagni, 'Le scienze nello studio di Padova e nel Veneto', in *Storia della cultura veneta*, Vol. III, Part 3, pp. 161–63; Martin Lowry, *Nicholas Jenson and the Rise of Venetian Publishing in Renaissance Europe* (Oxford: Blackwell, 1991), pp. 215–18. Also see *Aldo Manuzio in Renaissance Venice*, Exhibition catalogue, Gallerie dell'Accademia, Venice (Venice: Marsilio, 2016), Section VIII, pp. 311–44: 'Divine Proportions'.

Fig. 2 Detail of Figure 1, showing column with stellated ring around the base.

Aldus brought word and image together with elegance and clarity.[6] A precedent for this had been set the year after Giorgio Valla's arrival in Venice, when Erhard Ratdolt published the very first edition of Euclid's *Elements* in a thirteenth-century Latin translation. In a

6 Furio Rinaldi, *'De ludo geometrico': La matematica e la geometria di Leonardo* (Milan: De Agostini, 2013), cat. no. A5, p. 50.

technological innovation, the German émigré blazed a trail by using moveable type to print the 420 marginal figures at the same time as the text.[7]

Luca Pacioli played a key part in the dissemination of this Euclidian culture. Nurtured in a world of business and trade, his mathematics often served practical ends in art as well as business and accounting. The full title of his first book was *Summa de arithmetica, geometria, proportioni, & proportionalita,* and his writings dwell time and again on proportion and *proportionalita* ('proportioning') as central to artistic procedure. In his *Summa* Pacioli declares that 'the painter will never use his colour well if he does not attend to the strength of this one and that'.[8] The *potentia* of colour, he implies, is governed by the *proportionando* ('proportioning') of colour and tone.

Proportion embraces musical qualities of harmony, tone, colour, and balance.[9] In *Divina proportione* Pacioli developed a suggestive analogy:

> [...] if they say that music satisfies the hearing which is a natural sense, so does perspective the sense of sight which is more noble being the first gate of the intellect. If they say that music employs heard number and measure as expressed in the time of its continuances, so too does perspective use natural number according to any definition and measure as represented by the line of sight. If music recreates the soul by its harmony so too does perspective delight by the measurement of distance and the variation of colours. If the one art exploits harmonic proportions, the other exploits arithmetical and geometrical ones.[10]

Like Francesco Colonna in his contemporary dream narrative, *Hypnerotomachia poliphili,* Pacioli proposes that harmony, whether of proportion or colour, could be embodied and made visible. Although of very different stamp, both writers emphasised spacing rather than the more abstract, difficult to grasp, concept of space.[11] Pacioli dedicated the second part of *Divina proportione* to proportion in architecture, the human body, and the letters of the alphabet. In the expanding culture of Venetian printing the proportion of Roman letter-forms was a hot topic. Alberti had recommended that the ascenders of each letter should be twelve

---

7 *Preclarissimus liber elementorum Euclidis prespicacissimi in artem Geometrie incipit qua[m] foelicissime,* trans. by Campano de Novara (Venice: Erhard Ratdolt, 1482). For discussion of this edition as well as the relations between Pacioli and Leonardo, see Rinaldi, '*De ludo geometrico*', esp. cat. no. A2, pp. 49–50.

8 'el pictore mai ben dispone suoi colori, se non atende a la potentia de luno, e de laltro'. Translation from Michael Baxandall, *Patterns of Intention* (New Haven: Yale University Press, 1985), p. 113; original text in Luca Pacioli, *Summa de arithmetica geometria proportioni et proportionalita,* (Venice: Paganino de' Paganini, 1494), p. 68b.

9 For a summary, see Argante Ciocci, *Luca Pacioli tra Piero della Francesca e Leonardo* (Sansepolcro: Aboca 2009), pp. 223–24: subheading 'Le proporzione; il linguaggio universale delle arti e delle scienze'.

10 'Se questi dicano la Musica contentare l'udito uno di sensi naturali, e quella el vedere quale tanto è più degno quanto egli è prima porta a l'intellecto; se dicano quella s'atende al numero sonoro e a la mesura importata nel tempo de sue prolationi, e quella al numero naturale secondo ogni sua diffinitione e la misura de la linea visuale. Se quella recrea l'animo per l'armonia, e questa per debita distantia e varietà di colori molto delecta; se quella sue armoniche proportioni considera, e questa le arithmetici e geometrici'. Pacioli, *De divina proportione,* Chapter 3, p. 16. Translation from John Onians, 'On How to Listen to High Renaissance Art', *Art History,* 7 (1984), 411–37 (pp. 413–14). See also the discussion of 'harmony' in Bätschmann, *Giovanni Bellini,* 187. For the later *fortuna* of the analogy between music and pictorial harmony, see Michael Cole, 'Harmonic Force in Cinquecento Painting', in *Animationen/Transgressionen. Das Kunstwerk als Lebewesen,* ed. by Ulrich Pfisterer and Anja Zimmermann (Berlin: Akademie, 2005), pp. 73–94.

11 For the distinction between space and spacing, see Andrew Benjamin, *Art, Mimesis and the Avant-Garde* (London: Routledge, 1991), pp. 43–59.

times as high as they are wide, whereas the Paduan scribe Felice Feliciano had proposed a ratio of 1:10. Pacioli in turn recommended a slightly broader ratio for letters of 1:9.[12]

Essentially, then, proportion and perspective for Pacioli is founded upon measure, yet it also embraces more elusive pictorial qualities of tone and strength of colour. In this respect Pacioli's understanding of proportion accords with the distinctive geometry and the modulation of colour found in the best Venetian paintings around 1500, such as Giovanni Bellini's *Sacred Allegory* (Figure 7), a work we will examine presently. Influence undoubtedly flowed in two directions — from the mathematician to the artists and from the artists to the mathematician — for throughout his career Pacioli fraternised with artists and was keen to learn from them. His admiration for Piero della Francesca and his friendship with Leonardo is well documented. In the dedicatory epistle to his *Summa de arithmetica* he lists artists he knew, including 'in Venice the blood brothers Gentile and Giovanni Bellini, and in perspective drawing Hieronimo Malatini', praising their works for being *proportionando a perfection* ('proportioned to perfection').[13]

Proportion, Perspective and Shadows in Carpaccio

In their virtuoso projections of complex regular bodies in perspective, fifteenth-century artists normally employ a conceptual system of shading when representing pure geometric forms. A case in point is the segmented *mazzocchio*, depicted on the marble pavement just inside the north door of San Marco (Figure 3), which was probably designed by Paolo Uccello. Here the darkest face of each of the hexagonal segments is always the one facing towards the centre. As this system precludes consistent directional light from left or right it presents problems for an artist who attempts to incorporate this conceptual projection into a narrative scene. Carpaccio confronted the problem with remarkable daring when he included the stellated *mazzocchio* around the base of the column in *The Reception of the English Ambassadors* for the Saint Ursula cycle. Abandoning conceptual shading, he rendered the shadows cast by the geometric shapes in conformity with the directional lighting of the rest of the scene. This light falls from in front and from the left.[14] The railings at the threshold of the space throw shadows according to the same lighting.

Carpaccio's cast shadows are simplified in shape and rather attenuated in reach. Their direction is not always plotted with complete accuracy. Their primary purpose is to clarify the relative placement of bodies or objects in space. In *The Reception of the English Ambassadors,* for example, the position of the shutter on the right, which projects forward in very steep

12 Lowry, *Nicholas Jenson*, 217.

13 The dedicatory epistle is first in Italian, then Latin. The Italian reads: 'Come qui in Vinegia Gentile e Giovan Bellini carnal fratelli, e in prospectivo desegno Hieronimo Malatini, [...] quali sempre con libella e circino lor opere proportionando a perfection mirabile conducano, in modo che non humane, ma divine negli occhi nostri s'apprestano, et a tutte lor figure solo el spirito par che manchi'; the Latin opens: 'Quorum e numero Venetiis fuere Gentilitas et Ioannes Bellini fraters, et in perspectiva praxi Hieronomus Malateni [...]'. Pacioli, 'Epistola al duca Guidobaldo', in *Summa di arithmetica*, n.p. It is noteworthy that 'prospectivo desegno' is rendered in Latin as 'perspectiva praxi' — the practice of perspective.

14 For illustration and discussion of stellated bodies, see Kim H. Veltman, *Studies on Leonardo da Vinci, Vol. I, Linear Perspective and the Visual Dimensions of Science and Art* (Munich: Deutscher Kunstverlag, 1986), pp. 181 ff.

Fig. 3 Marble floor with segmented *mazzocchio* (design attributed to Paolo Uccello) (*c.* 1450), inside the Porta di San Pietro in the left aisle of San Marco, Venice. Photo by Paul Hills.

foreshortening in order to open up a view into Ursula's private bedchamber, is only elucidated by the shadow it casts across the adjacent steps. Without that shadow its radically oblique position would be barely intelligible. Similarly, the structure of the open gate near the column with the stellated ring is only legible thanks to the shadows of each upright that track up and across the steps to its right. Carpaccio was evidently so proud of this piece of virtuoso shadow-play that he affixed the *cartellino* bearing his signature to the front of this same gate.

Although one might well relate this fascination with the fall of shadow to the science of sundials, of gnomics, and with the projection of shadows which are present in Pacioli's writings and — fifty years later — much more prominently in Daniele Barbaro's texts, particularly in his manuscript *Pratica della prospettiva*, it is more important to acknowledge the creative contribution of Carpaccio as a painter self-consciously playing with different modes of representation.[15] Carpaccio's shadows subtly modify and transform the regular geometry of solid bodies: like the shadows cast by a sundial they intimate the passage of time within the extended narrative of arrivals and departures that constitute the story of Saint

15 Daniele Barbaro discusses clocks and sundials in Part IX of his *Pratica della prospettiva*.

Ursula, as well characterizing the pattern of daily life in the maritime republic of Venice. As Stefan Neuner has demonstrated, Carpaccio incorporated tell-tale signs of how the geometry of sight-lines — and by implication cast shadows — guided navigators entering the lagoon.[16] Like a navigator, Carpaccio is alert to the interplay of horizontal water and vertical towers, and the perspective of distance and interval in the lagoon environment. That keen navigational sense informs the Venetian optic at this period.

A monumental clock is visible on the tower in the background at the right-hand edge of *The Reception of the Ambassadors*, and the golden sunburst on its face is reiterated in steep foreshortening on the darkened ceiling of the foreground loggia. As our eyes travel across this broad canvas, the daily passage of time from dawn to sunset intimated by the slant of shadow is accompanied by signs of far longer passages of time such as the span of human life. On the steps that thrust forward across the threshold on the right, the shadow cast by the foreshortened shutter points towards an old woman, probably Ursula's nurse, who gazes into space. In an eloquent *contrapposto*, her aged head finds its antithesis in the young head of Ursula above it (Figure 4). Art historians writing of this scene, in which the pious princess enumerates to her father the conditions on which she will marry, have noted that the panel of the Madonna and Child fixed high on the wall is indicative of the position of devotional images in the bedrooms of Venetian patricians.[17] Less attention has been paid to how cunningly Carpaccio has contrasted this devotional painting with two other types of image, one ancient and one modern.

To the right, below the gold-ground Madonna, a dark rectangular object is suspended from the wall. Pictorially its tone provides a dark foil to the Princess's pale complexion. The rectangle is square and Ursula's head and shoulders are artfully inscribed within its elementary geometry. But what is this object? Although one must acknowledge that the canvases in the Saint Ursula cycle are worn and have suffered over the centuries, it is clear that this square object — unlike the red book nearby — does not rest on the ledge or cornice but hangs just above it. Mottled almost like marble, it might conceivably be a mirror of marble, since discs of highly polished marble set into buildings in fifteenth-century Venice were known as *specchi*, or mirrors; but it is more likely to be a mirror made of metal — of the type that soon would be eclipsed by the modern flat-glass mirrors that were manufactured in Murano from the mid-1490s onwards.[18] In Carpaccio's pictorial conceit the placement of the head and bust of the Princess within the square of metal transforms the mirror into a simile of a portrait — indeed the kind of portrait that might be sent to a prospective royal husband. A few years later, when Marcantonio Michiel started to write his descriptions of collections in Venice and the Veneto, he would list mirrors alongside portraits.[19]

Carpaccio underscores the analogy and points to an antique precedent by inserting roundels with *all'antica* low relief heads of various Caesars nearby on the pier and on its

---

16 Stefan Neuner, 'Malerei und Navigation. Kleines Logbuch zu Carpaccio's "Ursula-Zyklus"', *Wallraf-Richartz-Jahrbuch*, 72 (2011), 137–92. I am indebted to Stefan Neuner for sharing his innovative research on Carpaccio with me.

17 Ronda Kasl, 'Holy Households: Art and Devotion in Renaissance Venice', in *Giovanni Bellini and the Art of Devotion*, ed. by R. Kasl (Indianapolis: Indianapolis Museum of Art, 2004), pp. 58–89.

18 For mirrors in Venetian pictorial culture, and further bibliography, see Paul Hills, *Venetian Colour: Marble, Mosaic, Painting and Glass* (New Haven: Yale University Press, 1999), pp. 130–31, 186.

19 Marcantonio Michiel, *Der Anonimo Morelliano*, ed. and trans. by Theodor Frimmel (Vienna: Graeser, 1888), p. 96.

Fig. 4 Detail of right-hand side of *The Reception of the English Ambassadors* (Figure 1) showing the bedroom of St Ursula.

base to the left of Ursula's bedchamber. Beside these frontal images, he depicted heads adjacent to them in sharp foreshortening on the receding face of pier and base. Viewed as an ensemble, the roundels, the Madonna panel and Ursula's head create an exemplary triad of images: the antique relief, the devotional icon, and the modern picture. Whereas the icon is dignified by gold, the modern picture is distinguished by divine proportion. And just as the *all'antica* heads are inscribed within the circle, so the Princess's head and shoulders are inscribed within the square — a highly unusual format for painting at this date and therefore a deliberate choice on Carpaccio's part.

We may note here that the Euclidian problem of squaring the circle was prominently laid out in Pacioli's *Summa* of 1494 and taken up tirelessly by Leonardo in Milan. Carpaccio painted the canvas of *The Reception of the Ambassadors*, including the scene of Saint Ursula's bedchamber, after the completion of the Clock Tower in the Piazza San Marco in 1499, and possibly as late as 1501, a date that coincides with the publication of Giorgio Valla's *De expetendis et fugiendis rebus*. So it is conceivable that the chapter of that celebrated work devoted to the squaring of circles, *De quadrato circuli,* which proved of such great interest to Leonardo also caught the eye of Carpaccio as well. Or it could it be that the mysterious Hieronimo Malatini was the intermediary through whom the Euclidian learning of both Pacioli and Valla came to inform Carpaccio's art in this period? The Venetian painter's command of geometry is noticeably more virtuoso in this last phase in the execution of the Saint Ursula cycle than at any other time in his career, therefore we might speculate that in his later years he no longer had Malatini to guide him.

High above the old woman and the open shutter in Carpaccio's picture a grid of bars protects the bedchamber of the princess. They furnish a metaphor for her unassailable virginity. Together, shutter and grid heighten the sense of inside and outside, of what is perspicuous and what is closed, of solid and void. There is no need here to suppose that Carpaccio was mindful of Alberti's recommendation that a veil divided into a squared grid might be used by painters to realise the plane of intersection of the visual pyramid, for these bars replicate the iron bars in regular grids that commonly protected the windows of Venetian palaces, such as can be seen in Gentile Bellini's *Miracle at the Bridge of San Lorenzo* (Figure 5).[20] And yet the grid, like Alberti's veil, realises a plane of intersection and sets off the dramatic recession of the beams of the ceiling that is depicted behind it. In Carpaccio's painting geometry typically becomes embodied in material form.

Transparency, Reflections, and Shadows

Carpaccio's control of lighting also relates his painting to the culture of transparency characteristic of Venice in the last decades of the fifteenth century. He distinguishes the direction of lighting inside the bedroom from the exterior light of the rest of the painting. A window at the rear, cut off at the right margin and glazed with Venetian *vetro a rui* (bottle glass) is the source of the glow that spreads across the ceiling. In addition a hidden source, most likely a doorway, off-camera to the right would account for the raking shadows cast by the frame of the gold Madonna and the foot-lighting of the cornice above it. A series of rectangles all frontally disposed — the perspicuous grid of bars, the mirror with the head of Ursula, the gold-backed Madonna, and even the window — together demonstrate the pictorial value of incident light, reflection and transparency.

To comprehend what is most distinctive about the Venetian optic of light around 1500, we need to examine the relation between dark reflections and shadows. Carpaccio once more offers a useful starting point as another of his large canvases from the Saint Ursula cycle presents in schematic form what Giovanni Bellini and Giorgione will render with more subtlety. In the *Arrival of St Ursula in Cologne* (Figure 6) Carpaccio envisaged the walls of Cologne flanked by the river Rhine in terms of the topography of the Venetian

20 Leon Battista Alberti, *On Painting and on Sculpture, the Latin Texts of 'De pictura' and 'De statua'*, ed. by Cecil Grayson (London: Phaidon, 1972), 'De pictura', II, 31 (pp. 67–69).

Fig. 5 Gentile Bellini, *The Miracle of the True Cross at the Bridge San Lorenzo* (1500), canvas, 323 x 420 cm. Venice, Gallerie dell'Accademia. Photograph by Didier Descouens.

lagoon. The walls themselves appear to be based on views of the Venetian Arsenal along the Rio della Tana, and yet the visual experience of the lagoon has been radically re-ordered. Like most of the canvases in the Saint Ursula cycle, the horizon is high — well above the heads of the figures standing in the foreground, and much higher than what Alberti would have recommended. Near the centre an arched bridge with flanking towers is reflected in the calm water to form an oval. The optics of what we are shown here are explained in a passage in Leonardo's *Codex Urbinas*:

> The shadows of bridges are never seen [lying directly] on top of their areas of water if the water has not first lost its property of reflectiveness as a result of turbulence, and this is proved because clear water has a lustrous and polished surface and reflects the bridge in all places at which the eye and the bridge subtend equal angles, and it reflects the air under the bridge at the place where it is shaded by this bridge. This cannot occur with water that is stirred up because it is not reflective but is very receptive to the shadow, as would occur with a dusty road.[21]

21 Leonardo da Vinci, *Codex Urbinas Latinas 1270*, Vatican Library, fol. 72^{v}, trans. by Martin Kemp and Margaret Walker in: Leonardo da Vinci, *Leonardo on Painting: An Anthology of Writings by Leonardo da Vinci, with a Selection of Documents Relating to his Career as an Artist*, ed. by Martin Kemp (New Haven: Yale University Press, 1989), p. 170.

Fig. 6 Carpaccio, *The Arrival of St Ursula at Cologne*, (1490), canvas, 279 x 254 cm. Venice, Gallerie dell'Accademia. Photograph by Didier Descouens.

Given the abundance of bridges in Venice, Carpaccio probably noticed this effect, but his inclusion of the reflection on water has more to do with pictorial construction than empirical observation. He reiterated the oval curve of the bridge in the billowing sails above, and below it he enlarged the same shape in a series of arcs down the left side of the picture. These expansive reflections of solid objects play an essential role in his picture-making. They are quite different from the concentrated reflections of light sources known as lustre that were painted with such virtuosity by Netherlandish artists in the fifteenth century. Instead of receding, as the converging orthogonals of linear perspective typically recede, Carpaccio's dark reflections extend downwards — leading *towards* rather than away from the viewer.

In the Albertian system, as well as in common Florentine pictorial practice, bodies are modeled by light falling from one side.[22] Carpaccio and his workshop assistants also use lateral lighting, but the attenuated cast shadows falling across the painting from right to left are inconspicuous: instead the dark reflections descending downwards dominate. Of course, the pale glassy surface of the water reflects the light of the sky, and this is what Leonardo means when he writes that 'the clear water […] reflects the air under the bridge' rather than the shadow of the bridge.[23] In this situation, the darker reflections — of the towers and of the boats — that interrupt the pale reflection of the bright sky, can easily be mistaken for shadows. Looking out onto the Venetian lagoon, especially on slightly overcast days, it is often difficult to distinguish between dark reflections and shadows. The difference is that reflections on water descend towards the viewer rather than depend upon the direction of light. Shifting with our vantage point, they appear to address the eye.[24] In the *Arrival of St Ursula in Cologne* dark reflections sustain the narrative dynamic of the composition.

What is schematic in the large canvases of Carpaccio is finessed with great subtlety in the paintings of Giovanni Bellini. Already in his early paintings, such as the *Crucifixion* in the Correr Museum, which is just fifty-four centimetres high, Bellini engaged with the culture of transparency and reflection, water and glass. New manufactures, such as mirrors in flat glass and the translucent *cristallo* glass, circulating in the social world refined, and redirected perception. For the painters as well as their clients, the novel products in glass heightened discrimination of translucency and opacity, as well as lustre and reflections.[25] It was this technical knowledge and aesthetic discrimination that had already prompted Donatello, during his time in Padua from 1443–53, to experiment with inlaying glass into his sculpture.[26] Over the following half-century the glassmakers of Murano were refining their techniques to perfect *cristallo* glass and to diversify the range of translucent and opaque colours. An early sixteenth-century collection of recipes peppered its chemistry with emotive value judgments. It tells how to make glass 'clear and transparent like water […] just as limpid'[27] — and it describes how to achieve *rosechiero*, a rare translucent red.

Glassware was not the only Venetian luxury that played a part in refining the cognitive skills of Venetian artists and their elite clients. Many different manufactures and proto-industrial processes that were highly developed on the lagoon involved varieties of surface finish from matt to highly polished. These processes included soap making, the damascening of metals and the finishing of textiles, especially in the fast expanding silk

22 Alberti, 'De pictura', ed. by Grayson, II, 47.

23 Da Vinci, *Leonardo on Painting*, 170.

24 For the discussion of the difference between *lume* (illumination) and *lustro* (lustre or highlight), see E. H. Gombrich, 'Light, Form and Texture in Fifteenth-Century Painting North and South of the Alps', in *The Heritage of Apelles*, ed. by E. H. Gombrich (Oxford: Phaidon, 1976), pp. 19–38, which is invaluable but neglects the phenomenon of *dark* reflections.

25 For a fuller discussion see Hills, *Venetian Colour*, pp. 109–31: 'Transparency, Lucidezza and the Colours of Glass'.

26 Amy R. Bloch, 'Donatello's "Chellini Madonna", Light and Vision', in *Renaissance Theories of Vision*, ed. by J. Shannon Hendrix and C. H. Carman (Farnham: Ashgate, 2010), pp. 63–88.

27 'cristallo 'chiaro e trasparente come l'acqua […] così limpido'. *Ricette vetrarie del Rinascimento: Trascrizione di un manoscritto anonimo veneziano*, ed. by Cesare Moretti and Tullio Toninato (Venice: Marsilio, 2001), pp. 69, 79.

industry.[28] Stone-masons and plasterers too were masters of texture. The masons were skilled at tempering the luminosity of Istrian limestone, the standard building stone of Venice; they controlled its response to light by rubbing it smooth or by using a hammer-punch to achieve varieties of surface grain from rough to smooth. The hammering of the surface created a micro-structure of shadow that darkened the tone, and this lightening or darkening was used in a systematic manner to accent edges and changes in the orientation of each moulding. Although little studied today, this differential texturing of the surfaces of stone is one of the keys to the lucidity of Venetian architecture.[29]

Apart from the refinement of glass and esteem for lucidity, the other broad cultural influence that shaped the Venetian optic in the mature years of Giovanni Bellini was the rapid establishment of the Serenissima as the leading centre of printing in Europe following the introduction of certain new technology in 1469. In the 1470s the French printer Nicolas Jenson pioneered the printing of classical texts with wide margins and a more orderly layout than the typical pages of the manuscript era. In his early printed editions, and later in those of Aldus Manutius, divisions between chapters are signaled by spacing rather than by changes of colour. The proportion of the letters and the spacing of the layout on the page becomes key to the elegance and intelligibility of the text. Just as the manuals of classical rhetoric, that were being closely studied by fifteenth-century humanists, stressed the importance of *dispositio* ('disposition') in the effective ordering of a speech, so Jenson and Aldus — at least in their productions for the educated elite — emphasised spacing, proportion and interval.[30]

Bellini and Giorgione: The Perspective of Light

By the late 1480s — the decade when the *Sacred Allegory*, presently in the Uffizi Gallery, was probably painted — Giovanni Bellini was matching lucidity with care for proportion and interval (Figure 7).[31] He displayed mastery of linear perspective in the paving of the marble terrace, but this perspective exists unmoored from the eye of any viewer. The centric point, where the orthogonals meet, falls well to the right of centre, just above the line of hills and much higher than the horizon implied by the water and the landscape. Indeed, the pattern of the pavement functions primarily as a symbolic form: at its centre a square of pale marble is divided into four by a cross shape of coloured marbles, and at the centre of the cross Bellini positioned a circular urn from which grows the Tree of Life. Placing the infant Christ on a cushion to the left of this geometric centre, he introduces

28 Luca Molà, *The Silk Industry of Renaissance Venice* (Baltimore: Johns Hopkins, 2000).

29 For hammer work (*lavoro a martellina*), see Andrea Benedetti, 'La pietra d'Istria e l'intonaco nei paramenti esterni dell'architettura veneziana', in *Restauro e tecniche: Saggi e ricerche sulla costruzione dell'architetura a Venezia*, ed. by Giuseppe Cristinelli (Venice: Arsenale, 1992), pp. 30–45 (pp. 40–41). For surface decoration of Venetian architecture, see Wolfgang Wolters, *Architettura e ornamento: la decorazione nel Rinascimento veneziano* (Sommacampagna: Cierre, 2007).

30 For a fuller account see Hills, *Venetian Colour*, pp. 161–62; see also Lowry, *Nicholas Jenson*; Martin Lowry, *The World of Aldus Manutius* (Oxford: Blackwell, 1979); *Aldo Manuzio in Renaissance Venice* (see note 5).

31 For the dating and the probable trimming of the panel, see Peter Humfrey's catalogue entry in *Giovanni Bellini*, ed. by Mauro Lucco and Giovanni Carlo Federico Villa, Exhibition catalogue, Scuderie del Quirinale, Rome (Milan: Silvana, 2008), cat. no. 30, pp. 236–38.

Fig. 7 Giovanni Bellini, *Sacred Allegory*, (*c.* 1494), panel, 73 x 119 cm. Florence, Uffizi. [See colour plate 38]

a narrative turn or subtle change of register from the abstract truth of mathematics to the figurative allegory of Salvation. In the broader scheme of the painting, the Venetian master sets up a series of shifts that neutralise or soften the insistence of linear perspective and invite the viewer's eye to roam over the scene. Despite the high viewpoint, the top of the balustrade at the rear is all but invisible; indeed the balustrade reads as the simplest vertical plane, devoid of all mouldings. Behind the balustrade an expanse of calm water, faintly modulated by the blurred reflections of the cliffs and buildings beyond, foils any attempt to measure distance with linear precision. As in Carpaccio, though less insistently, the reflections on water descend downwards, moving forward towards the eye. Near and far are related through proportion and modulation rather than the convergent orthogonals of linear perspective. Tellingly, the proportions of the dark intervals in the balustrade are repeated in the dark rectangle of the doorway of the building on the far shore. Close to the central axis of the painting, this single doorway is strangely compelling, unobtrusively holding near and far in balance.

Giovanni Bellini's *Sacred Allegory* has often been seen as a forerunner of Giorgione's *The Tempest* (Figure 8) on account of its enigmatic subject and the enlarged role given to a setting in which landscape, water, sky, and distant buildings all work together to create a singular mood.[32] In the present context, we need not join the unending quest for *The Tempest*'s subject, but instead should note how Giorgione deploys reflections and shadows, and how he regulates the proportion of the intervals between light and dark

32 For a recent catalogue entry on *The Tempest*, with bibliography, see the entry by Rosella Lauber in *Giorgione*, ed. by Enrico Maria Dal Pozzolo and Lionello Puppi (Milan: Skira, 2009), cat. no. 46, pp. 427–30.

Fig. 8 Giorgione, *Tempest*, (c. 1505–06), 82 x 73 cm. Venice, Accademia. [See colour plate 39]

forms to fashion a painterly perspective of light. The composition is strikingly similar to Carpaccio's *Arrival of St Ursula in Cologne* — with the curve of the bank on the right leading to the towers of a city, an expanse of water running as a division down the centre of the picture, and a bridge in the distance. The scale of the two pictures is vastly different and the similarity probably fortuitous, but the comparison underlines how purposefully Giorgione blurs the distinction between shadows and dark reflections. He weaves them together in the fabric and texture of his painting.

Vasari, in a famous anecdote included in his life of Giorgione, told how the master from Castelfranco

> painted a man in the nude and with his back turned and, at his feet, a limpid stream bearing his reflection. To one side was a burnished cuirass [...] and this reflected his left profile [...]. On the other side was a mirror reflecting the other profile of the nude figure.[33]

If the story is not apocryphal it indicates Giorgione was well aware of the value of reflections on water, as well as the *paragone* between painting and sculpture that from the time of Leonardo had become a favourite topos in writing about art. Even if it is a fiction it suggests that the Tuscan Vasari noticed the ubiquity of watery mirrors in Venetian painting.

In *The Tempest* the light-catching towers of the city and the darker trees to their right are reflected in the stream, and the vertical supports of the bridge continue downwards a little way as tapering reflections. But Giorgione's stream is not as calm and mirror-like as Carpaccio's expanse of water, for the reflections are ruffled and broken. Under the bridge an area of dark shadow is difficult to differentiate from the dark reflections. The situation here does not correspond to the conditions under which, as Leonardo noted, 'the shadows of bridges are never seen on top of their areas of water' since the rougher surface of the water permits the shadow to be registered. As Leonardo further observed: in flowing water whose course is rapid reflections will be 'more extended and with more muddled contours'.[34]

I quote Leonardo here to direct attention to what is described in the painting rather than to posit Leonardo's influence. Instead the relation to a Venetian sensibility, and especially to Bellini, is clear. In *The Tempest* Giorgione builds proportional relations between shapes lying at different depths of space, notably the truncated columns, the dark supports of the bridge and the pale towers beyond, just as Bellini had done in the *Sacred Allegory*. Where the older master set up a spatial relation between dark rectangles within a pale surround, Giorgione introduced a zigzag rhythm in which light and dark alternate: for instance he leads the viewer in jumps forward and back from the pale marble columns on their platform to the dark supports of the bridge and on again to the paler faces of the distant towers. The distinctive geometry of proportion is now softened by the blending of dark reflections and shadows. This is perspective as painterly practice, a perspective of light.

Coda: The Optics of Light in Titian

After Giorgione's death in 1510, Titian introduced fresh dynamism into this perspective of colour and light. In his great *Assumption of the Virgin*, unveiled in the Frari in 1518, perspective in its linear mode as fixed pictorial scaffold or framework is swept away in the unfolding drama of light and of figures in vigorous movement. Titian's friend and advocate, Pietro Aretino, would perfectly comprehend how this painterly perspective served to heighten the affective power of Christian mysteries. In his account of the Annunciation, first published in 1535, he described the narrative in Titianesque terms. He envisages the angel Gabriel's

---

33 Giorgio Vasari, *Le opere di Giorgio Vasari*, ed. by G. Milanesi (Florence: Sansoni, 1906), IV, p. 98; Giorgio Vasari, *Lives of the Artists*, trans. by George Bull (Harmondsworth: Penguin, 1965), p. 276.

34 Da Vinci, *Leonardo on Painting*, 170.

arrival as a cosmic drama in which the spatial relations of nearness and distance are not fixed but changing before the beholder's eyes:

> a knot of seraphim spread out through the air and, beating their wings, lifted the veils that lay across the path that the angel had to follow. And a good space of pure blue was revealed [...] The Messenger of the Father of the Gods appeared as a small point of light, which hurt the eyes [with its brightness] when he left the empyrean realm. [He appeared small] because of the great distance and was already beginning to grow larger to those watching.[35]

A decade later, Aretino penned a letter in which he describes a sunset seen from the balcony of his lodgings on the Grand Canal as though he were viewing a painting by Titian. Despite its rhetorical patterning, the letter displays an acute understanding of the Venetian spatial mobility induced by colour and light, and again the interplay between nearness and distance is crucial. Moving his gaze from the canal to the rooftops and up to the sky, Aretino writes,

> Imagine also the wonder I felt at the sight of those great clouds, masses of condensed humidity: partly low and heavy on the roofs of the houses and partly in middle distance, so that the right of the scene was all greys and blacks [...] Oh, with what beautiful strokes did Nature's brush sweep back the atmosphere, clearing it away from the palaces in the way Titian distances it in his landscapes! In certain areas there appeared a bluish green, and in others a greenish blue, truly mixed by the fancy of Nature, the mistress of the masters. With lights and darks she created the effects of distance and relief so convincingly that I, who know how your brush is the very soul of her spirits, cried out three or four times: "Oh, Titian, where are you now?"[36]

The geometry of proportion that was given such critical value by the teaching of Luca Pacioli and Giorgio Valla is less salient here, but the Venetian optics of light, with its subtle blending of shadows and dark reflections, which was characteristic of Giovanni Bellini and Giorgione, still underlies the more dynamic optic of the mature Titian that Aretino

35 '[...] un groppo fiammegiante di Seraphini si sparse per l'aere; e scotendo l'ali disgrombravano con esse i veli, che si attraversavano al sentiero, che dovea far l'Angelo, et discoperto buono spatio di azuro puro [...] Il Nuntio del Padre de gli Dei, la cui luce offendeva le luci; di un picciol punto, che egli parve per la gran distantia ne lo uscire de le magioni empiree, già cominciava a crescere a gli occhi de i riguardanti'. Translation from Una Roman d'Elia, *The Poetics of Titian's Religious Paintings* (Cambridge: Cambridge University Press, 2005), p. 108; for the Italian text, see p. 220 n. 3. For the full text see Pietro Aretino, *I Quattro libri de la humanità di Christo* (Venice: Francesco Marcolini, 1539), fols 5^{v}–10^{r}.

36 Translation from *Titian: His World and His Legacy*, ed. and trans. by David Rosand (New York: Columbia University Press, 1982), p. 21; Italian text in Pietro Aretino, *Lettere sull'arte*, ed. by F. Pertile and E. Camesasca (Milan: Edizioni del Milione, 1957–60), II, pp. 16–18: 'Considerate anco la meraviglia, ch'io ebbi de' nuvole composti d'umidità condensa. I quali in la principal veduta, mezzi si stavono vicini a'tetti, degli edifizj, e mezzi nella penultima, perocché la diritta era tutto d'uno sfumato pendent in bigio nero [...] Oh con che belle tratteggiature i pennelli naturali spingevano l'aria in là, discostandola dai palazzi con il modo che la discosta il Vecellio nel far de' paesi. Appariva in certi lati un verde azzurro, ed in alcuni altri un azzurro verde veramente composto dale bizzarrie della natura maestra dei maestri. Ella con i chiari e con i scuri sfondava e rilevava in maniera ciò che le pareva di rilevare e di sfondare, chi io, che so come il vostro pennello è spirito de' suo spiriti, e tre o quattro volte esclamai, oh Tiziano dove siete mo'? .

described so well. The perspective of tone and colour would be the lasting legacy of the Venetians to the European tradition of painting in oil on canvas.

Bibliography

Manuscript and Archival Sources

Marciana Library, Venice, Barbaro, *Pratica della prospettiva*, MS 5446, fol. 2ʳ.

Primary Sources

Alberti, Leon Battista, *On Painting and on Sculpture, the Latin Texts of 'De pictura' and 'De statua'*, ed. by Cecil Grayson (London: Phaidon, 1972).

Aretino, Pietro, *I Quattro libri de la humanità di Christo* (Venice: Francesco Marcolini, 1539).

Aretino, Pietro, *Lettere sull'arte*, ed. by F. Pertile and E. Camesasca (Milan: Edizioni del Milione, 1957–60).

Michiel, Marcantonio, *Der Anonimo Morelliano*, ed. and trans. by Theodor Frimmel (Vienna: Graeser, 1888).

Novara, Campano de (trans.), *Preclarissimus liber elementorum Euclidis prespicacissimi in artem Geometrie incipit qua[m] foelicissime* (Venice: Erhard Ratdolt, 1482).

Pacioli, Luca, *Summa de arithmetica geometria proportioni et proportionalita,* (Venice: Paganino de' Paganini, 1494).

Pacioli, Luca, *De divina proportione* (Milan: Silvana Editore, 1982).

Rinaldi, Furio, *'De ludo geometrico': La matematica e la geometria di Leonardo* (Milan: De Agostini, 2013).

Vasari, Giorgio, *Le opere di Giorgio Vasari*, ed. by G. Milanesi (Florence: Sansoni, 1906).

Vasari, Giorgio, *Lives of the Artists*, trans. by George Bull (Harmondsworth: Penguin, 1965).

Vinci, Leonardo da, *Codex Urbinas Latinas 1270*, Vatican Library, fol. 72ᵛ, trans. by Martin Kemp and Margaret Walker in: Leonardo da Vinci, *Leonardo on Painting: An Anthology of Writings by Leonardo da Vinci, with a Selection of Documents Relating to his Career as an Artist*, ed. by Martin Kemp (New Haven: Yale University Press, 1989).

Secondary Works

Aldo Manuzio in Renaissance Venice, Exhibition catalogue, Gallerie dell'Accademia, Venice (Venice: Marsilio, 2016).

Bätschmann, Oskar, *Giovanni Bellini* (London: Reaktion, 2008).

Baxandall, Michael, *Patterns of Intention* (New Haven: Yale University Press, 1985).

Benedetti, Andrea, 'La pietra d'Istria e l'intonaco nei paramenti esterni dell'architettura veneziana', in *Restauro e tecniche: Saggi e ricerche sulla costruzione dell'architetura a Venezia*, ed. by Giuseppe Cristinelli (Venice: Arsenale, 1992), pp. 30–45.

Benjamin, Andrew, *Art, Mimesis and the Avant-Garde* (London: Routledge, 1991).

Bloch, Amy R., 'Donatello's "Chellini Madonna", Light and Vision', in *Renaissance Theories of Vision*, ed. by J. Shannon Hendrix and C. H. Carman (Farnham: Ashgate, 2010), pp. 63–88.

Ciocci, Argante, *Luca Pacioli tra Piero della Francesca e Leonardo* (Sansepolcro: Aboca 2009).

Cole, Michael, 'Harmonic Force in Cinquecento Painting', in *Animationen/Transgressionen. Das Kunstwerk als Lebewesen*, ed. by Ulrich Pfisterer and Anja Zimmermann (Berlin: Akademie, 2005), pp. 73–94.

D'Elia, Una Roman, *The Poetics of Titian's Religious Paintings* (Cambridge: Cambridge University Press, 2005).

Davis, Margaret Daly, 'Carpaccio and the Perspective of Regular Bodies', in *La prospettiva rinascimentale*, ed. by Marisa Dalai Emiliani (Florence: Centro Di, 1980), I, pp. 183–200.

Gombrich, E. H., 'Light, Form and Texture in Fifteenth-Century Painting North and South of the Alps', in *The Heritage of Apelles*, ed. by E. H. Gombrich (Oxford: Phaidon, 1976), pp. 19–38.

Hills, Paul, *Venetian Colour: Marble, Mosaic, Painting and Glass* (New Haven: Yale University Press, 1999).

Jayawardene, S. A., 'Towards a Biography of Luca Pacioli', in *Luca Pacioli et la matematica del rinascimento*, ed. by E. Giusti (Città di Castello: Petruzzi, 1998), pp. 19–28.

Kasl, Ronda, 'Holy Households: Art and Devotion in Renaissance Venice', in *Giovanni Bellini and the Art of Devotion*, ed. by R. Kasl (Indianapolis: Indianapolis Museum of Art, 2004), pp. 58–89.

Lepori, Fernando, 'La scuola di Rialto dalla fondazione alla metà del cinquecento', in *Storia della cultura veneta*, ed. by G. Arnaldi and M. P. Stocchi (Vicenza: Neri Pozza, 1980), Vol. II, Part 2, pp. 597–605.

Lowry, Martin, *The World of Aldus Manutius* (Oxford: Blackwell, 1979).

Lowry, Martin, *Nicholas Jenson and the Rise of Venetian Publishing in Renaissance Europe* (Oxford: Blackwell, 1991).

Lucco, Mauro, and Giovanni Carlo Federico Villa (eds.), *Giovanni Bellini*, Exhibition catalogue, Scuderie del Quirinale, Rome (Milan: Silvana, 2008).

Maccagni, Carlo, 'Le scienze nello studio di Padova e nel Veneto', in *Storia della cultura veneta*, ed. by G. Arnaldi and M. P. Stocchi (Vicenza: Neri Pozza, 1980), Vol. III, Part 3, pp. 161–63.

Maria Dal Pozzolo, Enrico, and Lionello Puppi (eds.), *Giorgione* (Milan: Skira, 2009).

Menato, Sara, *Per la giovinezza di Carpaccio* (Padua: Padova University Press, 2016).

Molà, Luca, *The Silk Industry of Renaissance Venice* (Baltimore: Johns Hopkins, 2000).

Moretti, Cesare, and Tullio Toninato (eds.), *Ricette vetrarie del Rinascimento: Trascrizione di un manoscritto anonimo veneziano* (Venice: Marsilio, 2001).

Neuner, Stefan, 'Malerei und Navigation. Kleines Logbuch zu Carpaccio's "Ursula-Zyklus"', *Wallraf-Richartz-Jahrbuch*, 72 (2011), 137–92.

Onians, John, 'On How to Listen to High Renaissance Art', *Art History*, 7 (1984), 411–37.

Rosand, David (ed.), *Titian: His World and His Legacy* (New York: Columbia University Press, 1982).

Veltman, Kim H., *Studies on Leonardo da Vinci, Vol. I, Linear Perspective and the Visual Dimensions of Science and Art* (Munich: Deutscher Kunstverlag, 1986).

Wolters, Wolfgang, *Architettura e ornamento: la decorazione nel Rinascimento veneziano* (Sommacampagna: Cierre, 2007).

GEORGES FARHAT

Topographic Perspective as Constructed Optics

Landscape Design and the Grand Canal at Versailles

Introduction

While acknowledging the foundational role played by perspective in shaping the history of landscape and garden design in the West, landscape studies have paradoxically endorsed a late modern disjuncture of perspective from optics and sensory experience. As a result, not only is perspective reduced to its representational functions, its examination is also severed from the material processes of design practice. A case in point is the Grand Canal of Versailles designed by André Le Nôtre (1613–1700). Close inspection of its construction suggests an alternative framework for the historiography of perspective (Figure 1).

The starting point for such a revision is to unravel the confusing relationship woven, over the emblematic site, between optics and levelling operations. To that end, I first reveal the fictitious nature of the narrative of the Grand Canal's construction that was fabricated by capitalizing on the authority of the telescope and the *Académie des Sciences*. I then show why the actual performance of surveying, along with its instruments, material procedures, and conceptual framework, must be brought back into the history of perspective. The latter will thus appear related to the worksite and design practices, of which I reconstruct some processes by performance using topographic data, drawings, and calculation. Ultimately, my analysis points to a specific and inventive appropriation of optical knowledge by surveyors, gardeners, and designers. I term this practice 'topographic perspective'. It consists of inventing and constructing built-in optical devices, which include visual alignment chains (collimation) and metric distension schemes (anamorphosis).

Such spatial devices did not merely result from applying a set of formulas or conventional graphic construction methods. They rather constituted site-specific responses to topographic conditions, structural changes in manorial economy, and subsequent shifts in scale. In fact, this constructive modality of geometrical optics is ingrained in a long professional history and informed by cultural transfers, operated from the decorative arts to garden design, by craftsmen, royal officers, and artists. Interestingly, between 1650 and 1700, this history intersected with many scholarly debates over optical matters that unfolded at the three French Academies of Painting and Sculpture, Science, and Architecture. Landscape design practice was able to concretely address and solve the very issues of perspective that lay at the heart of these theoretical disputes, and this has major historiographical implications. Indeed, counter to what a late modern historiography would reflect, in the

Georges Farhat University of Toronto, georges.farhat@daniels.utoronto.ca

Perspective as Practice. Renaissance Cultures of Optics, ed. by Sven DUPRÉ, Turnhout, 2019 (*Techne. Knowledge, Technique, and Material Culture*, 1), p. 429-465

DOI 10.1484/M.*Techne*-EB.5.117736

Fig. 1 View of Grand Canal vista at Versailles, seen from Grande Terrasse eastern end, with Tapis Vert and Apollo Basin in the middle ground. Photograph by Christian Milet, 2012. [See colour plate 40]

field of landscape design, perspective and optics could only overlap, while projective functions, spatial deployment, and bodily experience conflated.

Perspective, Garden History, and Landscape Studies

In a 1955 article, the art and garden historian Marguerite Charageat — who was cognizant of Baltrusaitis's *Anamorphoses* — observed that geometric figures of parterres, pools, and canals in the gardens designed by André Le Nôtre had been subject to optical adjustments.[1] She rightly pointed out that, when lined up axially, at Versailles (along the Grand Canal) or in other gardens (Tuileries, Palais-Royal), such figures had 'increasing dimensions'.[2] Charageat identified a 'decelerated perspective' which she paralleled with Pierre Le Muet's canal design at Tanlay (in the 1640s); she loosely connected this kind of perspective to 'gardenists', such as Olivier de Serres and Salomon de Caus; she linked it also to contemporary 'doctrines' and 'mindsets' assumed to be related to René Descartes's *La dioptrique* (1637) and Jean Francois Niceron's *La perspective curieuse* (1638).[3] In other essays, Charageat insightfully acknowledged that Le Nôtre's work was primarily 'about optics depending on

1 Jurgis Baltrusaitis, *Anamorphoses*, 3rd edn. (Paris: Flammarion, 1984; 1st edn. 1955); Marguerite Charageat, 'André Le Nôtre et l'optique de son temps. Le Grand Canal de Tanlay par Pierre Le Muet', *Bulletin de la Société de l'histoire de l'art français* (1955–56), 66–78.

2 Charageat, 'André Le Nôtre et l'optique de son temps'.

3 Charageat, 'André Le Nôtre et l'optique de son temps'.

topography'[4]. However, she would never attempt to check design against terrain, that is, interrogate the material condition of the practice of optics as perspective in the garden. The same lack of interest in how topographic and perspectival technicalities were articulated in design practices would subsequently mark three major historiographical trends in garden design history. These can be respectively associated with semantics, aesthetics, and philosophy. In the first stream, scholars scrutinise the historical iconography (drawings, engravings, paintings) and privilege the visual rhetoric (viewpoint, discrepancies, sequence) and social functions (display, encomiastic, legitimisation) of perspectival images, including their role in the reception of gardens (circulation, experience).[5] In the second trend, cross-examination of site and archival evidence asks how perspective helped express categories such as order (Humanism) or illusionism (Baroque).[6] The third stream reaches out to politics, epistemology, and metaphysics.[7]

Though dissimilar in aim, such approaches overlap in a few aspects. Their understanding of perspective is shaped by the latter's late nineteenth-century evolution towards a unified, and invariant means of graphic figuration. Perspective is thus exclusively related to representational and symbolic functions. As such, it is even ascribed the capacity to inform the landscape as a 'way of seeing' and thinking (a worldview).[8] But its complex agency within the process of design is overlooked.[9] Interpretation and theory are prioritised over socio-technical procedures and material contingency in a wide range of praxes, mediums, scales, and experiences.[10] Accordingly, landscape historians endorse and consolidate the

---

4 Marguerite Charageat, 'André Le Nôtre et ses dessins', *La gazette illustrée des amateurs de jardins* (1953–54), 21–27. See also Marguerite Charageat, 'Pourquoi Le Nôtre a-t-il créé, à Versailles, le grand canal?', *Jardin des arts*, 42 (1958), 371–78.

5 See among others: Erik de Jong and Marleen Dominicus-van Soest, 'Tuijngesigten en perspective', in *Aardse paradijzen. De tuin in de Nederlandse kunst, 15de tot 18de eeuw* (Ghent: Snoeck-Ducaju & Zoon, 1996), pp. 11–123; Hervé Brunon, 'De l'image à l'imaginaire: notes sur la figuration du jardin sous le règne de Louis XIV', *XVII*^e^ *siècle*, 209 (4) (2000), 671–90; Dianne Harris, *The Nature of Authority: Villa Culture, Landscape, and Representation in Eighteenth-Century Lombardy* (University Park: Pennsylvania State University Press, 2003), pp. 19–39; Denis Ribouillault, 'Towards an Archeology of the Gaze: The Perception and Function of Garden Views in Italian Renaissance Villas', in *Clio in the Italian Garden*, ed. by Mirka Beneš and Michael G. Lee (Washington, DC: Dumbarton Oaks, 2011), pp. 203–32.

6 Franklin H. Hazlehurst, *Gardens of Illusion: The Genius of André Le Nostre* (Nashville: Vanderbilt University Press, 1980); Jean Guillaume, 'Le jardin mis en ordre: jardin et château en France du XVe au XVIIe siècle', in *Architecture, jardin, paysage. L'environnement du château et de la villa aux XVe et XVIe siècles*, ed. by Jean Guillaume (Paris: Picard, 1999), pp. 103–36.

7 See among others: Leonardo Benevolo, *La cattura del'infinito* (Roma: Laterza, 1991); Allen Weiss, *Mirrors of Infinity: The French Formal Garden and 17th-Century Metaphysics* (New York: Princeton Architectural Press, 1995); Horst Bredekamp, *Leibniz und die Revolution der Gartenkunst: Herrenhausen, Versailles und die Philosophie der Blätter* (Berlin: Verlag Klaus Wagenbach, 2012).

8 See, Denis Cosgrove, *Social Formation and Symbolic Landscape* (Madison: The University of Wisconsin Press, 1984), and contibutions in *Sites Unseen: Landscape and Vision*, ed. by Dianne Harris and D. Fairchild Ruggles (Pittsburgh: University of Pittsburgh Press, 2007).

9 Among a few exceptions for which the process of design is central, see: Robin Evans, *The Projective Cast: Architecture and its Three Geometries* (Cambridge, MA: MIT Press, 1995); Wouter Reh and Clemens Steenbergen, *Architecture and Landscape. The Design Experiment of the Great European Gardens and Landscapes* (Munich: Prestel, 1996), pp. 136–235; and Marvin Trachtenberg, *Dominion of the Eye: Urbanism, Art, and Power in Early Modern Florence* (New York: Cambridge University Press, 1997).

10 For a display of this variety of praxes, see Masaccio, *Nel segno di Masaccio. L'invenzione della prospettiva*, ed. by Filippo Camerota (Florence: Giunti, 2001); *Quadratura: Geschichte, Theorie, Techniken*, ed. by Matthias Bleyl and Pascal Dubourg Glatigny (Berlin: Deutscher Kunstverlag, 2011).

disjuncture between optics (science) and perspective (technique) that Erwin Panofsky posited long ago, but which does not stand up against the proof of site-specific cases.[11]

Optics and Perspective in the Historiography of Versailles

Ever since the architectural historian Sigfried Giedion identified the construction of Versailles, which unfolded from 1661 to 1715, as one of the main origins of modern urbanism, the Grand Canal has been associated with the role that the telescope played in astronomy, geodesy, cartography, and surveying during the so-called Scientific Revolution.[12] An interpretive tradition thus flourished through a variety of disciplines ranging from hydraulic engineering to urban history, philosophy, and science studies.[13] Expanding on Charageat's insights, Michel Baridon has offered this modernist narrative an elaborate expression by hinging optics and perspective while contradictorily building on their disjuncture.[14] Extended axial prospects in the gardens ('long perspectives'), he claims, became possible due to progress in levelling instruments developed at the *Académie des Sciences* (founded in 1666) and equipped with telescopes.[15] This would be best illustrated at Versailles with the construction of the Grand Canal (between 1668 and 1672), whose entire length and three basins' 'gradual lengthening' were supposedly 'calculated as accurately as possible' by Le Nôtre 'with the help of "wonderful telescopes" [...]' (Descartes's phrase).[16] Thus would be determined the 'proper ratio between the length and breadth of the allées' as well as 'the angle of their slope'.[17] Yet, Baridon does not recognise the anamorphic scheme he is precisely describing. And instead of grappling with design techniques, he looks at contemporaneous theories of physics, light, and the cosmos. In doing so, like his predecessors, he blindly builds on the

11 Erwin Panofsky, 'Die Perspektive als "Symbolische Form"', in *Vorträge der Bibliothek Warburg, 1924–25* (1927), 258–330. English translation: Erwin Panofsky, *Perspective as Symbolic Form* (New York: Zone Books, 1991). The disjuncture hypothesis is variously questioned in: Sven Dupré, 'The Historiography of Perspective and "Reflexy-Const" in Netherlandish Art', *Nederlands Kunsthistorisch Jaarboek*, 61 (2011), 35–60; and Dominique Raynaud, *Optics and the Rise of Perspective: A Study in Network Knowledge Diffusion* (Oxford: The Bardwell Press, 2014).

12 Siegfried Giedion, *Space, Time and Architecture: The Growth of a New Tradition*, 5th edn. (Cambridge, MA: Harvard University Press, 1967; 1st edn. 1941), pp. 107–62. For a discussion of the epistemic changes implied by the telescope, see Philippe Hamou, *La mutation du visible: essai sur la portée épistémologique des instruments d'optique au XVIIe siècle* (Villeneuve-d'Ascq: Presses universitaires du Septentrion, 1999).

13 Hubert Loriferne, 'L'influence de Picard dans les travaux d'alimentation en eau du château de Versailles sous Louis XIV', in *Jean Picard et les débuts de l'astronomie de précision au XVIIe siècle*, ed. by Guy Picolet (Paris: CNRS, 1987), pp. 274–311; Benevolo, *La cattura del'infinito*; Thierry Mariage, *The World of André Le Nôtre*, (Philadelphia: University of Pennsylvania Press, 1999); Hélène Vérin, 'Technology in the Park: Engineers and Gardeners in Seventeenth-Century France', in *The Architecture of Western Gardens*, ed. by Monique Mosser and Georges Teyssot (Cambridge, MA: MIT, 1991), pp. 135–46; Chandra Mukerji, *Territorial Ambitions and the Gardens of Versailles* (Cambridge: Cambridge University Press, 1997), pp. 181–93; Anthony Gerbino, 'Introduction', in *Geometrical Objects: Architecture and the Mathematical Sciences 1400–1800*, ed. by Anthony Gerbino (New York: Springer, 2014), pp. 1–41 (pp. 20–29).

14 Michel Baridon, 'The Scientific Imagination and the Baroque Garden', *Studies in the History of Gardens & Designed Landscapes*, 18 (1) (1998), 5–19. Expanded in Michel Baridon, *A History of the Gardens of Versailles* (Philadelphia: University of Pennsylvania Press, 2008; French edn. 2003), pp. 61–117 and explicitly inspired by Erwin Panofsky, 'Galileo as a Critic of the Arts: Aesthetic Attitude and Scientific Thought', *Isis*, 47 (1956), 3–15.

15 Baridon, *A History of the Gardens of Versailles*, 88–89.

16 Baridon, *A History of the Gardens of Versailles*, 88–89.

17 Baridon, *A History of the Gardens of Versailles*, 88–89.

account that Charles Perrault delivered, towards the end of his life, in the *Parallèle des Anciens et des Modernes*, though without questioning the reliability and motivations of its author.[18] Indeed Perrault, embittered as he was, had good reasons to tell stories. He was prompted to resign, from 1682 onward, from key positions he had held for twenty years in the administration of Jean-Baptiste Colbert (1619–83). As such, Perrault had not only been instrumental in organizing the royal academies and directing cultural and scientific production. He had also been involved, as *contrôleur général* of the *Bâtiments du Roi*, in supervising works in the gardens at Versailles.[19]

A Self-Instituted Historiographer's Account

The *Parallèle* stages a series of dialogues, between an abbot, a knight, and a president, in the gardens of Versailles which are deemed, by one of the three protagonists, an appropriate setting for demonstrating the superiority of the Moderns over the Ancients in the arts, technology, science, rhetoric, and poetry during the reign of Louis XIV.[20] The fourth volume begins with a presentation of the merits of the telescope and ensuing progress in astronomy and geodesy.[21] Then the Grand Canal provides a pretext for describing one of the two instruments for measuring angles and elevations fitted with lenses that were developed from 1667 onward by members of the *Académie des Sciences* and under the supervision of the astronomer Jean Picard (1620–82).

The first of these instruments was developed to carry out geodetic works, trace the Paris meridian, and create a new map of France.[22] It consisted of an astronomical quadrant (with two telescopes) mainly used for measuring horizontal angles in triangulation operations. The second instrument was a levelling device that, Perrault claims, was used to survey the Grand Canal's site before excavation (Figures 2 and 3). But, it should be noted, contrary to what he asserts, in 1671 this level was still a theoretical proposal that Picard discussed conditionally at the end of a lavishly illustrated thirty-page pamphlet on measuring the Earth published the same year.[23] As Philippe de La Hire would make clear, it was not effectively put into practice until 1674, well after the Grand Canal had been dug; it was actually used for measuring elevations and slopes along the Seine and Loire rivers to convey water to the gardens of Versailles and their ever-multiplying fountains.[24] Moreover, measurements

18 Charles Perrault, *Parallèle des Anciens et des Modernes [...]*, 4 vols (Paris: J.-B. Coignard, 1688–97), IV, pp. 81–85.

19 Charles Perrault, *Mémoires de ma vie* (Paris: Macula, 1993), pp. 229–38, (with an essay by Antoine Picon, 'Un moderne paradoxal', pp. 1–107).

20 Perrault, *Parallèle*, I, p. 10. On the *Parallèle*'s context, see Picon, 'Un moderne paradoxal' and Marc Fumaroli, *Le sablier renversé: des Modernes aux Anciens* (Paris: Gallimard, 2013), pp. 257–467.

21 Perrault, *Parallèle*, IV, pp. 26–46, 66–80.

22 Monique Pelletier, *Les cartes de Cassini. La science au service de l'État et des régions* (Paris: Éditions du CTHS, 2002).

23 Jean Picard, *Mesure de la terre* (Paris: L'imprimerie royale, 1671), p. 27. On Picard's life and work, see: Guy Picolet, *Jean Picard et les débuts de l'astronomie de précision au XVIIe siècle* (Paris: CNRS, 1987).

24 Philippe de La Hire, 'Preface', in Jean Picard and Philippe de La Hire, *Traité du nivellement par M. Picard ... avec une relation de quelques nivellements faits par ordre du Roy... Mis en lumière par les soins de M. de La Hire...* (Paris: Michallet, 1684). On the son of the painter Laurent La Hyre, an astronomer and cartographer who worked with Picard at the Académie des Sciences and taught structure at the Académie d'Architecture from 1687, see: *Philippe de La Hire (1640–1718), entre architecture et sciences*, ed. by Antonio Becchi, Hélène Rousteau-Chambon and Joël Sakarovitch (Paris: Picard, 2013).

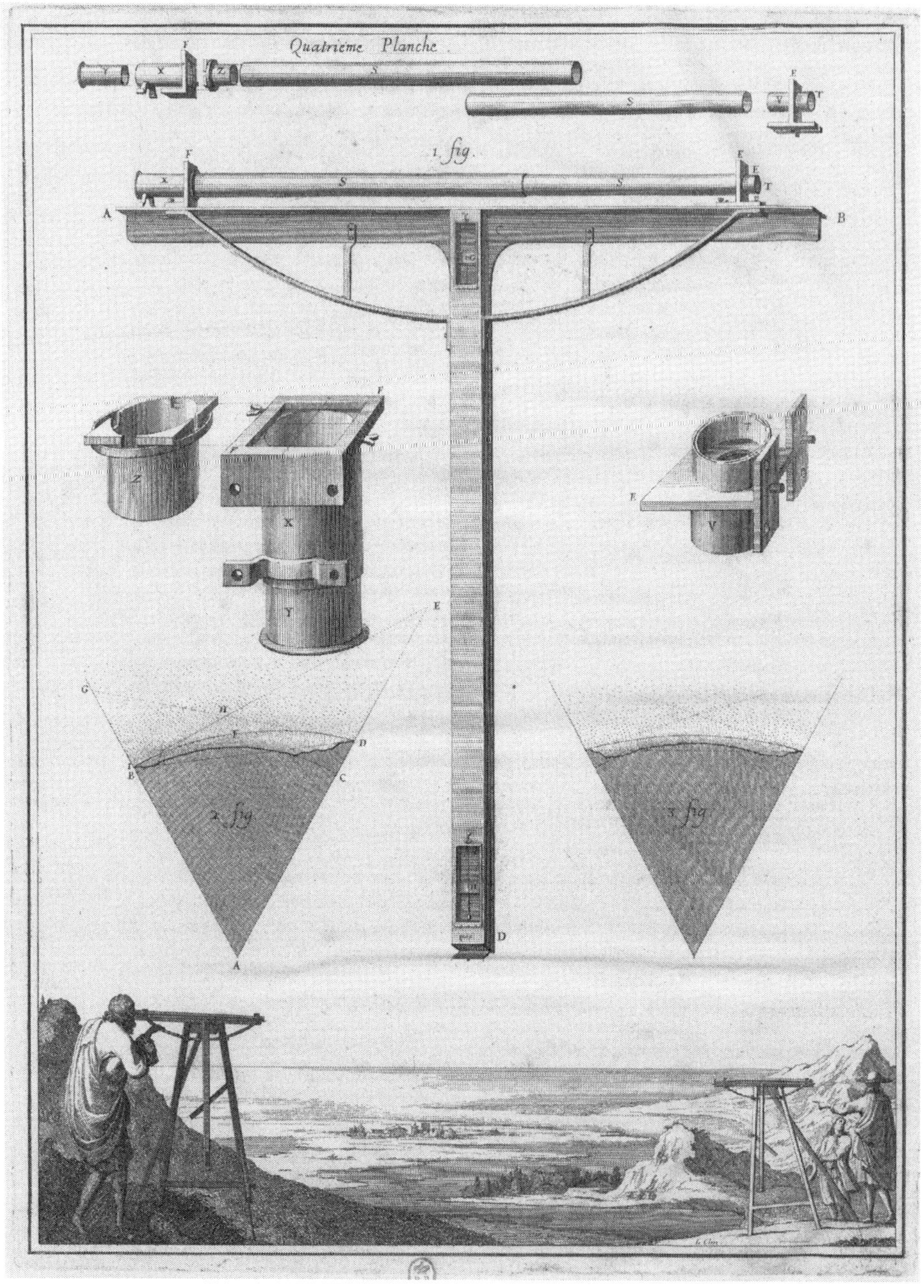

Fig. 2 Survey level fitted with a telescope. Drawing and engraving by Sébastien Leclerc, in Jean Picard, *Mesure de la Terre* (Paris, 1671) Plate 4, Bibliothèque nationale de France, Paris.

were corrected in later, more exhaustive survey campaigns while the instrument was virtually improved, as La Hire stated in a 1689 treatise on surveying in which precedents and competing proposals were discussed.[25] The technical innovation in Picard's instrument was reportedly twofold. It consisted in combining a telescope with a reticule (cross-wires) and a more stable perpendicular comprised of a wind-protected, lead-weighted woman's hair sealed in a square tube.[26] According to Perrault, only the accuracy resulting from such an innovation could allow for correctly surveying the site and digging such a long canal of 1675 metres, whereas the 'ordinary levels in the old fashion' led to erroneous measurements by an unspecified company of *Maçons & Fontainiers* ('masons and fountaineers').[27]

Historical Falsification and its Motivations

However, data on measurement in Perrault's *Parallèle* prove to be at total variance with those in his unpublished *Mémoires*.[28] Moreover, hitherto understudied quotes, payments, and memoranda relative to earthworks fully belie Perrault's figures, arguments, and collapsed chronology.[29] In addition to anonymous masons and navvies, these documents highlight an unacknowledged cohort of professional protagonists and a series of stages in the Grand Canal's construction. One of them, hydraulic engineer Denis Jolly, had overtly been in conflict with Perrault.[30] Featuring prominently as well were two famous gardeners to the King, Marin Trumel and Henry Dupuis, the latter of which was recognised by the *Bâtiments* as an 'expert in alignment'.[31] Finally, contractor Jean Legendre appears to have taken charge of the second phase of the Grand Canal's construction (1671–72).[32] When cross-examined with historical maps and the current state of the terrain, archival documents also reveal a major discrepancy between Perrault's assertions (about the

25 See Philippe de La Hire, *L'école des arpenteurs* (Paris: Thomas Moette, 1689), pp. 139–55, where La Hire recalled and reproduced precedents and improvements to Picard's lens-fitted instrument. Such were the 'liquor levels' of Melchisedech Thevenot, *Machine nouvelle pour la conduite des eaux, pour les bâtimens, pour la navigation, et pour la pluspart des autres arts* (Paris: Mabre-Cramoisy, 1666); Giovanni Battista Riccioli, *Geographiae et hydrographiae reformatae libri duodecim*, 2nd edn. (Venice: J. La Noū, 1672); and Edme Mariotte, *Traité du nivellement* (Paris: J. Cusson, 1672). They were completed with La Hire's and Roemer's own inventions as well as Christiaan Huygens's, which featured a box with floating telescope and was first publicised in the *Journal des savants* (1680), 21–24. The period saw a proliferation of other hypothetical prototypes published by royal engineers and instrument makers like Butterfield and Chapotot, who worked for the Academy: see *Journal des savants* (1678), 440–43 and (1680), 174–76.

26 Picard, *Mesure de la terre*, 27.

27 'niveaux ordinaires, & à l'ancienne mode'. Perrault, *Parallèle*, IV, pp. 81–85.

28 Charles Perrault, *Mémoires: Memoirs of My Life*, ed. and trans. by J. Morgan Zarucchi (Columbia, MO: University of Missouri Press, 1989).

29 See some of these documents listed in *Comptes des Bâtiments du Roi sous le règne de Louis XIV*, ed. by Jules Guiffrey, 5 vols (Paris: Imprimerie nationale, 1881–1901), I, 195, 290, 255–56, 336, 339–40, 424, 518, 633.

30 Suffice it here to cite one of the many memoranda written by Charles Perrault, 'Evidence of Sir Jolly's misconduct in the delivery of plumbs and solder to Vincennes and to Versailles from 1664 and through April 6th, 1667', in Archives nationales, Paris, O1, 1887.

31 *Comptes*, ed. by Guiffrey, I, 290. Payment for 'officers' pledge and appointment: 'To Henry Dupuis who is in charge of conducting works at the grand canal of Versailles, for five months of his appointment [i.e., from May to October 1668] befell on the 15th of October last year: 500 livres', 16 August–20 October 1668.

32 *Comptes*, ed. by Guiffrey, I, 518.

craftsmen's and the academicians's respective measurements) and the terrain's topography before and after the Grand Canal's excavations (Figures 3 and 10). In the end, Perrault's accounts turn out to be but falsifications by a former state officer with intricate, personal motivations.

The polished and heroic narrative in the *Parallèle* was meant to promote the academicians's and sovereign's joint credit. In an uninterrupted chain of legitimization, running from science (astronomy) to technique (levelling), the telescope and the Academy lent authority.[33] Also, Versailles provided the most prestigious site of display.[34] But, in praising the ventures in whose supervision he had been involved, Perrault as well aimed to regain social status and reputation. He had fallen into disgrace under Colbert, in 1682.[35] Shortly after, he was ousted from the *Petite Académie* by Louvois, Colbert's successor as superintendent.[36] These setbacks urged him to first write his famous long poem celebrating the century of Louis the Great (1687) and then expand it into the four-volume prose of the *Parallèle*.[37] But there was more to this publication; it was actually part of a larger counterattack against the recent empowering of Racine and Boileau (the Ancients party) at Court.[38] Under such conditions, a famous storyteller like Perrault, who argued for the moral superiority of modern tales, waged war at the expense of historical truth.[39] He gave preference to more efficacious weapons: virtual innovation and theoretical precision over expertise, knowhow, and ordinary practice in both surveying and design at the Grand Canal worksite.[40]

33 On the telescope's role in social strategies, see Mario Biagioli, *Galileo's Instruments of Credit: Telescopes, Images, Secrecy* (Chicago: University of Chicago Press, 2007). On its multiple origins, material fabrication, circulation, and impacts on optical theories, see *The Origins of the Telescope*, ed. by Albert van Helden, Sven Dupré, Rob van Gent, and Huib Zuidervaart (Amsterdam: Amsterdam University Press, 2010) as well as Hamou, *La mutation du visible*.

34 *Sciences et curiosités à la Cour de Versailles*, ed. by Béatrix Saule and Catherine Arminjon (Paris: Réunion des musées nationaux, 2010).

35 In an attempt to install his own son Jules Armand as superintendent of the *Bâtiments*, Colbert ended up calling into question Perraults's management methods. Perrault, *Mémoires*, 229–33; Thierry Sarmant, *Les demeures du soleil. Louis XIV, Louvois et la surintendance des Bâtiments du Roi* (Seyssel: Champ Vallon, 2003), pp. 44–46, 51–54.

36 Perrault, *Mémoires*, 233–35. On the *Petite Académie*, a powerful organ which oversaw the inscription of Louis XIV's official history in buildings and the decorative arts, see Marc Fumaroli, 'L'Académie des Inscriptions et Belles-Lettres dans la République des Lettres', *Comptes rendus des séances de l'Académie des Inscriptions et Belles-Lettres*, 150 (4) (2006), 2073–81.

37 Charles Perrault, *Le siècle de Louis le Grand* (Paris: Jean-Baptiste Coignard, 1687); Perrault, *Mémoires*, 235–38.

38 The French quarrel of the Ancients and the Moderns stemmed from lingering, deep tensions in social, intellectual space between classical authority (Court, Academy) and individual freedom (salons). See Fumaroli, *Le sablier renversé*, 417–41 and Larry F. Norman, *The Shock of the Ancient* (Chicago: University of Chicago Press, 2011).

39 Ironically enough, the author of *Tales of my Mother Goose* (1697) had also curated the material representation of the fables of Aesop in the fountains of the Bosquet du Labyrinthe at Versailles built in 1669: Charles Perrault, *Recueil de divers ouvrages en prose et en vers* (Paris: J.-B. Coignard, 1675), pp. 225–76: 'Le labyrinthe de Versailles'.

40 For a more comprehensive study of this intricate case, I refer the reader to Georges Farhat, 'Optical Instrumenta[liza]tion and Modernity at Versailles: From Measuring the Earth to Leveling in French Seventeenth-Century Gardens', in *Technology and the Garden*, ed. by Kenneth Helphand and Michael Lee (Washington, DC: Dumbarton Oaks, 2014), pp. 23–50.

Fig. 3 Topographic plan of the Grand Canal's future site at Versailles, with gardens and Grand Parc in their 1667 state prior to canal's digging; reconstruction based on historic maps and a 2010 GPS survey campaign. Reconstruction and drawing by Georges Farhat, DG by P. Robert, 2011–13.

Measurement, Vision, and Perspective

Although mentioned in a number of surveying manuals from the second half of the seventeenth century, lens-fitted levels would not be of common use in professional practice for some time.[41] Nor were the physical and physiological optics they implied taken for granted immediately and without controversy. Indeed, geometrical optics and monocular vision, which also guided perspective methods and their Euclidian grounds, were to remain effective for measuring elevations and distances by sight well into the eighteenth century. This was strongly argued for in the technical literature by authors such as Sébastien Leclerc.

'The line, the semicircle, the proportional compass and the plane table' were the main surveying instruments employed for direct and indirect measuring that Leclerc, a military engineer and geographer, expounded on in his 1690 manual of practical geometry.[42] Known among gardeners as the *boussole* ('compass'), the semicircle or 'graphometer' consisted of a semicircular protractor fitted with a sea-compass and two alidades, one fixed and one hinged.[43] Handy and precise, it was apt at measuring angles, distances, and elevations. It was particularly favoured for long alignments (and line networks), as the royal officer Claude Mollet had earlier specified in his gardening treatise; it supplemented the mason level fit only for smaller-scale operations such as grading a place and plotting a parterre or a tree grid onto the ground.[44] Equally suited for long distances and large layouts, the square plane-table was fitted with a free and removable alidade as well as a sea-compass. It was used to determine elevations or slope angles, with its upper edge acting as a level line. Occasionally, from the 1660s onward, the plane-table's alidade would be replaced by a telescope. Strange as it may seem, Leclerc did not mention this improvement or Picard's apparatuses of which he was nevertheless cognizant. Indeed, not only did he, as an attaché of the *Académie des Sciences*, execute all the engraved plates in both of Picard's 1671 and 1684 treatises; a prolific artist appointed to the *Cabinet du Roi*, he also reproduced two lens-fitted instruments in the middle and background of the *Académie des Sciences et des Beaux Arts* he engraved in 1698 (Figure 4). These were complemented with more traditional instruments which, made for measuring by sight, all equally apply the principles of Euclidian optics and geometry: using the proportionality of similar triangles determined by the plane section of visual pyramids in aiming operations.[45] At the forefront of his technological tableau, Leclerc more conspicuously exhibited perspectival devices including demonstration tools as well as direct, catoptric, and dioptric anamorphoses.

---

41 Among other famous manuals featuring lens-fitted instruments, worth mentioning is Allain Manesson-Mallet, *La géométrie pratique* (Paris: Anisson, 1702), whose publication was commissioned by Jean-Paul Bignon, the then *Académie*'s president.

42 'Le Cordeau, le Demicercle, le Compas de proportion, & la Planchette'. Sébastien Leclerc, *Traité de geometrie* (Paris: Jombert, 1690), pp. 218–49, which expands on Leclerc's *Pratique de la géométrie [...]* (Paris: Jolly, 1669).

43 Philippe Danfrie, *Declaration de l'usage du graphomètre [...]* (Paris: Danfrie, 1597).

44 Claude Mollet, *Theatre des plans et jardinages* [MS 1620s–30s] (Paris: Charles de Sercy, 1652), Chapter LIII, pp. 320–28.

45 Oronce Finé, *Protomathesis* (Paris: Gérard Morrhi and Jean Pierre, 1532), 'De geometria libri'; Reiner Gemma Frisius, *Cosmograficus liber Petri Appiani* (Antwerp: Arnold Birckman, 1533), 'De locorum describendorum ratione'; *The Worlds of Oronce Fine: Mathematics, Instruments, and Print in Renaissance France*, ed. by Alexander Marr (Donington: Shaun Tyas, 2009).

Fig. 4 Sébastien Leclerc, *L'Académie des Sciences et des Beaux Arts*, 1698, print, Bibliothèque nationale de France, Paris, Estampes et Photographie.

Such gathering and display could not be neutral. It echoed the academic and professional quarrel raging between two camps, respectively headed by the painter Charles Le Brun and the engraver Abraham Bosse, over the role played by perspective in the visual arts. The quarrel was triggered after Leonardo da Vinci's treatise on painting was published by Roland Fréart in 1651.[46] Leclerc would later engage in these matters as a member of the *Académie Royale de Peinture et Sculpture* and further develop Bosse's stance.[47] Like his contemporaries, he consistently conceived of measuring by sight and of representing in perspective within one and the same optical realm. Such consistency is reflected for instance in another of Fréart's publications: the 1663 French version of Euclid's *Optics*. Indeed, in this composite work, in which Fréart referenced the treatises of Witelo (1270–78), Porta (1593), and Aguilon (1613) in optics, mathematic demonstrations were complemented with topics ranging from metric distension in the helical frieze of Trajan's column to practice in painting including linear perspective diagrams.[48]

46 *Traité de la peinture, de Léonard de Vinci, donné au public et traduit d'Iitalien en Ffrançois par R. F. S. D. C.*, trans. and ed. by Roland Fréart de Chambray (Paris: J. Langlois, 1651).

47 On the socio-professional stakes of perspective during this particular period, see Marianne Leblanc, *D'acide et d'encre: Abraham Bosse (1604?*Paris: check if page number indication is correct here *–1676) et son siècle en perspectives* (Paris: CNRS, 2004).

48 Roland Fréart de Chambray, *La perspective d'Euclide* (Le Mans: Jacques Isambart, 1663), pp. 11–14, 26, 43–44. This work was actually based on Jean Pena's 1557 Greek edition of Theon's *Recension*.

Made for the unaided eye, the traditional levelling instruments that Leclerc promoted supposed a linear trajectory of light and the visual alignment of points (collimation). When using alidades and levelling over short distances, there was no need to take into consideration any discrepancy, between apparent and real levels, resulting from the Earth's curvature and atmospheric refraction. Indeed, both these factors, as Picard himself demonstrated, turn out to be negligible. Over distances shorter than two thousand metres, the curvature remains insignificant; moreover, in practice, distances were commonly divided into small sections of less than two hundred metres, which curtails the effect of refraction.[49]

Sticking to traditional procedures, as Leclerc did, actually fit into a broader optical framework. Levelling operations were performed in monocular vision which, as Leclerc stated in a 1679 pamphlet, is the only valid way of seeing. Furthermore, it translates well into the double projection method of perspectival drawing.[50] On such premises, Leclerc would push Bosse's strict conformity to geometric rules in visual matters to the limits. He countered the pragmatic approach to painting expressed by Gregoire Huret who denounced the use of central perspective as corrupt because of the lateral distortions (anamorphic effects) it involves.[51] This was a time when binocularity remained an open-ended question. Therefore, Leclerc went as far as to question the binocular nature of vision on the grounds that (quasi-)image fusion amounts to monocular vision. He thus renewed Egnatio Danti's argumentation, disputing Johannes Kepler's *pinceaux optiques* ('optical paintbrush') and Descartes's physiological optics as well as blending perspectival principles with Epicurean 'species'.[52]

While Leclerc resisted the use of lens-fitted instruments by omission, others adduced them to arouse doubt or express distrust in recent inventions. Pierre Bullet, architect to the city of Paris, did so as a preamble to promoting his high-precision version of the mason level with which, he claimed, he had carried out the capital's 1673–76 survey plan prior to laying out its first boulevards. He recommended the levelling of sections shorter than one hundred toises (ca. 195 metres) to avoid the need for corrections brought about by the use of lenses.[53] More explicitly, the physicist and garden theorist Dezallier d'Argenville would favour water levels with vials in levelling operations over the use of telescopes and on the grounds that there was not yet, as late as 1747, a way of producing lenses of reliable quality.[54]

---

49 Picard, *Mesure de la terre*, 26, 28.

50 Sébastien Leclerc, *Discours touchant le point de vue, dans lequel il est prouvé que les choses qu'on voit distinctement ne sont veuës que d'un œil* (Paris: T. Jolly, 1679), later expanded on in Sébastien Leclerc, *Système de la vision fondée sur de nouveaux principes* (Paris: Delaulne, 1712).

51 Abraham Bosse, *Traité des pratiques géométrales et perspectives enseignées dans l'Académie Royale de Peinture et Sculpture* (Paris: The author, 1665); Grégoire Huret, *Optique de portraiture et peinture* (Paris: The author, 1670).

52 Leclerc, *Discours touchant le point de vue*, 42–46, 51, 57–86. Jacoppo Barozzi da Vignola and Egnatio Danti, *Le due regole della prospettiva pratica* (Rome: F. Zannetti, 1583), pp. 54–54. On the revival of Skepticism and Epicurean optics during the seventeenth century, in particular in Gassendi's thought, see Saul Fisher, *Pierre Gassendi's Philosophy and Science* (Leiden: Brill, 2005).

53 Bullet discarded all levelling instruments of various manners lately set on the King's worksites, considering them 'pas si universels que l'on ne pût mieux faire' ('not so universal that one could not do better'). Pierre Bullet, *Traité du nivellement* (Paris: Langlois, 1688), pp. 22–36, 9–10.

54 Antoine-Joseph Dezallier d'Argenville, *La théorie et la pratique du jardinage*, 4th edn. (Paris: Jean Mariette, 1747; 1st edn. 1709), Book 4, Chapter IV.

In sum, none of the recently developed instruments were necessary at the Grand Canal of Versailles other than for a thought experiment whose sociopolitical and technical procedures were borrowed from different circumstances to forge one single, fictitious event.[55] These procedures involved: performance — of triangulation (for the Paris Meridian) and level measurements (for water conveyance); authority — of performers (academicians) and witnesses (La Hire and Perrault); reproduction — of levelling campaigns (over a decade) and instruments (theoretical emulation); representation — in images and text; and dissemination — in engravings, books, and journals. By contrast, as archival evidence, topography, and technical literature show, traditional instruments and craftsmanship, that were employed since the sixteenth century for aqueducts and canals, proved useful enough at Versailles. Moreover, the same Euclidean principles of optics that were applied in measuring by sight were appropriated by designers and gardeners alike for constructing architecture and landscape.

Topography, Visual Alignment, and Motion

Though little acknowledged in architectural and landscape historiographies, the use of visual alignment of points (or collimation) that was necessary in surveying and leveling operations, seems to have been common practice in both phases of designing and laying out buildings as well as their grounds during the seventeenth century. Illustrative of this is a cross-section drawn along the courtyard of Colbert's chateau at Sceaux (1670–73), whose gardens were redesigned by Le Nôtre (1670–early 1690s). The drawing shows the use of visual alignment in composing an outdoor sequence (Figure 5).[56] Gateway and chateau facade were optically proportioned to each other by two imaginary viewers. These are represented as standing at opposite ends and different elevations: along the approach avenue and on the central balcony. They define a field of vision with their sightlines, which among other factors results from grade (earthwork profile). Such visual control was undoubtedly understood by both designer and patron as attested by line drawings for other contexts.

In a project dating from 1673, for restructuring the Palais Royal Theatre in Paris the stage designer Carlo Vigarani makes use of a *ligne de veüe* ('sightline') similar to those in the drawings for Sceaux[57] (Figure 6). He seeks to demonstrate to Colbert that the proposed elevated building's attic would remain invisible in two instances: from within the courtyard and from a sixty-metre distance on a fronting square.

Likewise, earlier designs by the architect François Mansart for the Val-de-Grâce complex (1645–46) display sightlines. Here, ornament in the building's volumes — 'top of plinths of

---

55 On experiment: Steven Shapin and Simon Schaffer, *Leviathan and the Air-Pump: Hobbes, Boyle, and the Experimental Life* (Princeton, N.J.: Princeton University Press, 1985); Lorraine Daston, 'The Empire of Observation', in *Histories of Scientific Observation*, ed. by Lorraine Daston and Elizabeth Lunbeck (Chicago: University of Chicago Press, 2011), pp. 81–113.

56 Nationalmuseum, Stockholm, Anonymous, CC 139 (and variant, CC 2202); reproduced with interpretations differing from mine in Hazlehurst, *Gardens of Illusion*, 235–37 and Gérard Rousset-Charny, 'Colbert, ses architectes et praticiens', in *Sceaux, architectures pour un domaine*, ed. by Cécile Dupont-Logié (Sceaux: Domaine de Sceaux, 2006), I, pp. 19–52 (pp. 34–35 and 40–41).

57 Bibliothèque nationale de France, Paris, VA-419 J, 6-FT.

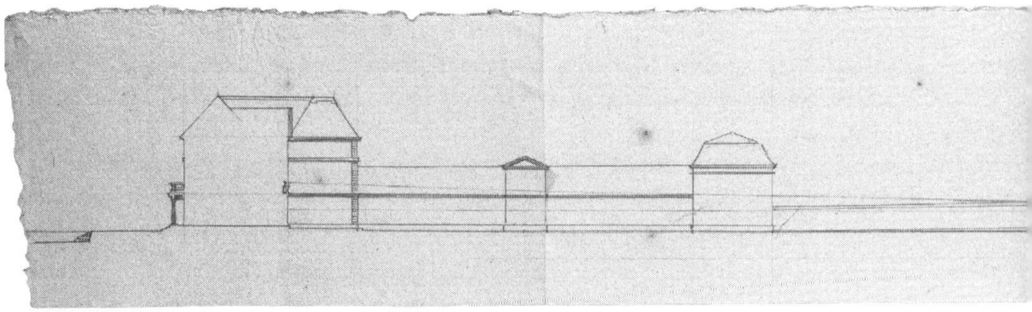

Fig. 5 Anonymous, Cross section along gateway, courtyard, and house, château de Sceaux, c. 1670–73. Drawing, 13 x 93 cm, Nationalmuseum, Stockholm, CC 2202. Photo Hans Thorwid/Nationalmuseum. [See colour plate 41]

attic's pilasters' and 'last cornice under great dome' — is related to two viewpoints. The latter are duly marked on the design plan: respectively, B on the street and A in the courtyard.[58]

In all three instances (Sceaux, Palais-Royal, Val-de-Grâce), the aim was not only to determine the 'distance point' and necessary step back to grasp the ideal image of a building.[59] More interestingly, whether in plan or cross-section, represented sightlines suppose similar triangles and control of a desired effect using both drawing and simple calculation (rule of three). As a result, areas of interest are bracketed between strategic standpoints within a spatial sequence; building and surrounding grounds are dynamically interrelated for and by a moving viewer.

While cross-section diagrams with sightlines are missing from surviving drawings attributed to Le Nôtre and his collaborators, a number of their design plans bear aiming lines and viewpoints similar to those in Mansart's.[60] However, here the main focus shifts entirely from building masses to graded and terraced terrain, earthworks and ground figures. It is therefore not so adventurous to conjecture that visual alignments actually observed in many sites are not random but were obtained from forming similar triangles based on levelling data, which had to be well covered if only for hydraulic networks and their fountains.[61] In other words, just as in levelling operations, vision was geometrically controlled by double projection. Distance and elevation were thus coordinated. Less useful would be in that case a perspectival rendering. This is precisely what visual alignments (collimations) at Versailles demonstrate.

58 'le dessus des socqes des pilastres de l'atique'; 'la dernierre corniche sous le grand dosme'. François Mansart, Design Plan, for the Val-de-Grâce, Bibliothèque nationale de France, Paris, VA-443-FT 6, repr. in: *François Mansart, Le génie de l'architecture*, ed. by Jean-Pierre Babelon and Claude Mignot (Paris: Gallimard, 1998), pp. 183–87.

59 This static reading of 'distance point' can be found in Claude Mignot, 'Orthographie, scénographie et jardinages', in *François Mansart*, ed. by Babelon and Mignot, pp. 75–81.

60 See drawings reproduced and introduced by Georges Farhat and Patricia Bouchenot-Déchin in *André Le Nôtre in Perspective*, ed. by Patricia Bouchenot-Déchin and Georges Farhat (Paris: Hazan; Versailles: Château de Versailles, 2013), 'Designs and Drawings. Notebook One. Sketches, Roughs, Preliminary Drawings' (n.p.).

61 Georges Farhat, 'Great Vistas in the Work of Le Nôtre', in *André Le Nôtre in Perspective*, ed. by Bouchenot-Déchin and Farhat, pp. 170–87.

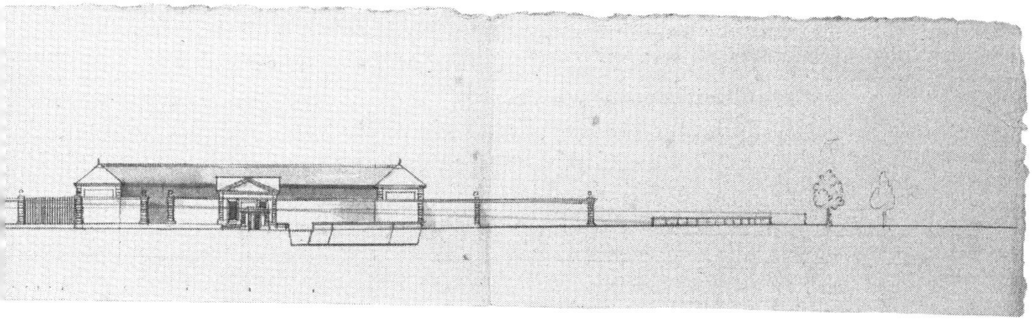

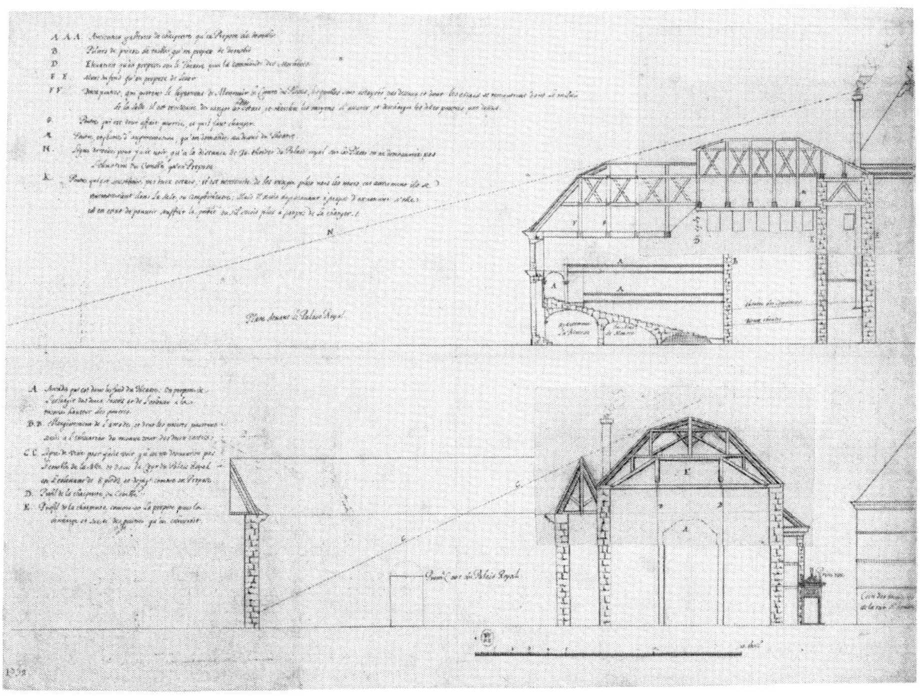

Fig. 6 Carlo Vigarani (and collaborators), Cross sections along fronting plaza and courtyard of Palais Royal theatre, with overlays, Paris, 1673. Drawing with annotations, pen and brown ink, 42 x 58.4 cm, Bibliothèque nationale de France, Paris, VA-419 J, 6-FT.

Fig. 7 Stations 4 to 7 along Grand Canal Axis at Versailles, standpoints taken between Chateau stairway and Parterre d'Eau allée. Cross sections showing visual alignments (in red) of Grande Terrasse end with different Grand Canal segments. Optical analysis and drawings by Georges Farhat, DG by N. Bimbra, A. Bragagnini, and A. Home-Douglas, 2012–14. [See colour plate 42]

Fig. 8 Stations 8 to 10 along Grand Canal Axis at Versailles, standpoints taken between midway and end of Parterre d'Eau allée. Detail. Cross sections showing visual alignments (in red) of Grande Terrasse end with Grand Canal front basin (station 8), Apollo Basin (station 9), head of Latona statue (station 10). Optical analysis and drawings by Georges Farhat, DG by N. Bimbra, A. Bragagnini, and A. Home-Douglas, 2012–14. [See colour plate 43]

station 7 $\frac{H_2H_1}{H_2K_2} = \frac{K_1K_2}{K_1P_1}$

station 6 $\frac{G_2G_1}{G_2K_2} = \frac{K_1K_2}{K_1Q_1}$

station 5 $\frac{F_2F_1}{F_2K_2} = \frac{K_1K_2}{K_1U_1}$

station 4 $\frac{E_2E_1}{E_2K_2} = \frac{K_1K_2}{K_1T_1}$

station 10 $\frac{J'_2J'_3}{J'_2K_2} = \frac{K_1K_2}{K_1V_1}$

station 9 $\frac{J_2J_3}{J_2K_2} = \frac{K_1K_2}{K_1M_1}$

station 8 $\frac{F_2F_3}{F_2K_2} = \frac{K_1K_2}{K_1U_1}$

Visual Alignments at Versailles

Along the Grand Canal axis at Versailles, between the château and the Latona stairway, starting at the western end of the Grand Terrace, walkers can experience ten remarkable collimations from as many stations. Though not detailed in visit accounts, these stations pertain to the garden itinerary that the Swedish royal architect Nicodemus Tessin the Younger (1654–1728) followed in 1687, shortly after the final versions of the Parterre d'Eau (on the Grand Terrace) and Latona Fountain (right below) were completed.[62] Tessin was then walked by Le Nôtre who had taught him garden design some years prior, between 1678 and 1680. They actually anticipated the tours that the king would soon prescribe for official visitors, from 1691 to 1695, with designated 'viewpoints' and places from where to 'consider' scenes.[63]

Moving forward along the main east-west axis on the Terrace, one gradually uncovers the great vista's foreground. The top step of the Latona stairway (1665–66) visually, and successively, comes into line with ever closer parts of the Grand Canal, the (multilobed) Apollo Basin, and the Tapis Vert (Figures 7 and 8, refer also to Figures 9 and 10). Nearest to the Terrace, the Latona Fountain only moves into the field of view once the end of the Parterre d'Eau *allée* is reached. However, from there onward to the terrace edge, the Latona statue never comes to obstruct the viewer's vista onto the Grand Canal (Figure 9). Instead, the fountain's water-shoots end up filling the visual gap up to the Apollo Basin image. As can be verified by calculation, such precision in the whole sequence suggests a better coordinated project than usually assumed.

Certainly, by perching the Latona figure on a four-pool pyramid in 1687, the architect Jules Hardouin-Mansart disrupted the low horizontal balance of the sculpted group in the original Oval Basin designed by Charles Le Brun and constructed from 1665 to 1667.[64] But the need for dramatic change must have been obvious to everyone once the château's new garden façade was completed. By then, the Grand Canal had been through various phases: initially built as one straight stretch of water (1668), it was soon doubled in width and extended with end basins and transverse arms (1671–72).[65] Also, the first Parterre d'Eau — whose elaborate pattern conceived by Le Brun (and set in place between 1672 and 1683) had succeeded an embroidery parterre (only existent between 1665 and 1669) — was in turn replaced by two simpler oblong basins (1683–84) '*du dessein de M Le Nôtre*' ('designed by Mr Le Nôtre').[66] The final stage reached by the Grand Canal in 1672 must have called for

62 Nicodemus Tessin the Younger, *Nicodemus Tessin the Younger: Sources, Works, Collections: Travel Notes 1673–77 and 1687–88*, ed. by Merit Laine and Borje Magnusson (Stockholm: Nationalmuseum, 2002).

63 Louis XIV, *Manière de montrer les jardins de Versailles* (Paris: RMN, 1992); Robert Berger and Thomas Hedin, *Diplomatic Tours in the Gardens of Versailles under Louis XIV* (Philadelphia: University of Pennsylvania Press, 2008).

64 Payments of marble work for Latona Fountain, 21 March–5 December 1688: *Comptes*, ed. by Guiffrey, III, p. 106. See latest discussion of this change in Thomas Hedin, 'From Basin to Fountain: Le Nôtre before Le Brun in the Gardens of Versailles' in *André Le Nôtre in Perspective*, ed. by Bouchenot-Déchin and Farhat, pp. 213–25.

65 Bibliothèque nationale de France, Paris, Manuscrits, *Mélanges Colbert*, 148bis, fol. 718. See also 'Quote of the Excavation Needed to Ensure the Increase in Length and Width of the Canal of Versailles [18 March 1671]', in Archives nationales, Paris, O[1], 1793[1].

66 Payments of construction works for the short-lived embroidery parterre: *Comptes*, ed. by Guiffrey, I, pp. 82, 337, 338; first Parterre d'Eau: *Comptes*, ed. by Guiffrey, I, pp. 588, 615; second and final Parterre d'Eau: *Comptes*, ed. by Guiffrey, II, pp. 339, 458, 617–28. Le Nôtre is acknowledged as the 1683–84 oblong basins' designer on a late seventeenth-century drawing: Institut de France, Paris, MS 1307, fol. 51.

the reconfiguration of the foregrounds in accordance with the scale change it brought into the vista. However, although this hypothesis would now seem quite obvious, the question remains as to how so many massive alterations could cohere into one unified grand vista.

As the analytical cross-sections show, the composition along the Grand Canal axis extends over a three thousand metre distance with a relatively small thirty metre elevation drop concentrated in the garden area (between the Grand Terrace and Apollo Basin). By far the largest section, the Grand Canal was dug across the shallow Gally Creek valley. Given the sheer scale of the composition and its resulting acute optical angles, the perceptive condition is one of flat vision from château and garden onto the Grand Canal. It translates into extreme foreshortening of ground figures laid out on different levels and meant to be experienced by a moving viewer. If coordinating elevation and distance through collimation was one possible response to such a design challenge, metric distension (anamorphosis) was another. It was meant to counter visual collapse and flatness while achieving an effect of grandeur. Such distension could be modelled after a variety of perspectival devices.

Anamorphosis at Versailles

It now appears necessary to revisit and specify Charageat's (as well as Hazlehurst's and Baridon's) vague observations that, along the major axis at Versailles, the farther away from the château the bigger the geometric figures.[67] To identify such a metrical distension, one can refer to a variety of procedures. These can be mechanical (using thread or light-aided projection) or geometric (applying distance point or double projection methods). They cover a wide range of productions, from flat hidden or three-dimensional images to cabinet curios, gallery frescoes, and stage sets.[68] The corpus of anamorphosis abounds with such optical devices that deploy, spatially and physically, the principles and procedures of perspective. But, in landscape architecture, because of topography, extent, and materiality, solutions turn out to be even more site-specific than in other mediums.

To conduct thorough geometric analysis of the layout along the Grand Canal axis at Versailles, one needs to re-enact part of the design process through the performance of drawing (both by hand and digitally). Reconstruction here includes tracing the different stages of the Grand Canal's development along with attendant terrain manipulation and terracing. Focusing on the geometrical structure of the axial composition reveals that it is organised by an unprecedented, rudimentary version of the common divergent anamorphic grid scheme (Figure 10). In this case, two simple angular sectors are fit into one another,

67 Charageat, 'André Le Nôtre et l'optique de son temps'; Hazlehurst, *Gardens of Illusion*, 126–47; Baridon, *A History of the Gardens of Versailles*, 88–89.

68 Among other sources available to seventeenth-century designers, some of which included illustrations with garden settings, were the following: Jacoppo Barozzi da Vignola and Egnatio Danti, *Le due regole della prospettiva pratica* (Rome: F. Zannetti, 1583), pp. 94–96; Salomon de Caus, *La perspective avec la raison des ombres et des miroirs* (London: R. Barker, 1611), Chapters 27–31; Pietro Accolti, *Lo inganno degl'occhi* (Florence: Pietro Cecconcelli, 1625), pp. 48–50; J.-L. Sr de Vaulezard, *La perspective cylindrique et conique* (Paris: J. Jacquin, 1630); Jean-François Niceron, *La perspective curieuse* (Paris: P. Billaine, 1638); Mario Bettini, *Apiaria [...]* (Bologna: J. B. Ferroni, 1641–42), II, pp. 2–32, Apiarium V: 'Paradoxa et arcana opticae scenographicae'; Jean Du Breuil, *Perspective practique [...]* (1649), III, fols 92–107.

Fig. 9 View of Grand Canal vista at Versailles with Latona Fountain in the foreground, seen from Grande Terrasse western end. Photograph by Georges Farhat, 2011.

with their respective apexes located in the main courtyard of the château and at the Latona Basin. The outer sector successively brackets the transverse widths of the *Parterre d'Eau allée*, Latona stairway, Latona Basin, and Apollo Basin diameters; included as well are three virtual circles that respectively circumscribe the Grand Canal's front, crossing, and end basins. In the last two features (crossing and end basin), the angular sector determines longitudinal distension which unlike width is not screened off by tree lines edging the Canal. Within the inner angular sector the following elements are inscribed: Apollo Basin, width of Grand Canal's straight body as well as rectangle at crossing and side of square in end basin. The ideal visual effect of inscribing the aforementioned features within the outer sector is rendered in a conjectural view (Figure 11). It shows how, while aimed at offsetting the foreshortening, the distension scheme would make key elements appear of the same width if theoretically seen from above the sector's apex. Similar control is achieved with the inner angular sector (Figure 12).[69]

The angular sector device conforms well to a Euclidian principle according to which things 'seen within equal angles appear to be of the same size'.[70] However, it can also be translated into perspectival terms. Indeed, while all orthogonals to a picture plane will have

69 Reconstructed anamorphic scheme and collimation chains are visualized in a didactic animation which was produced to accompany a fifteen-metre long all-glass model of the Grand Canal Axis layout at the 2013–14 Le Nôtre exhibition that I co-curated at the Château de Versailles: *Construction du grand axe de Versailles*, online video, YouTube, 14 January 2014: https://www.youtube.com/watch?v=-RnjYn-pWWU [accessed 22-03-2019].

70 Euclid, *The Optics of Euclid*, trans. by Harry Edwin Burton, *Journal of the Optical Society of America*, 33 (5) (1943), Definition 4, 357.

their image converge to a single point on the horizon, the reverse is also true: straight lines that radiate from the point of the eye in the ground plane (our sector's apex) should be imaged as parallel verticals in the picture plane. Despite irreducible differences, optics and perspective are brought into collaboration here in a way that was familiar to contemporary practitioners and theorists, as we saw above with Fréart's *perspective d'Euclide* and Leclerc's multifaceted work. Contrary to what Panofsky would claim, and using his own words, one realises that 'angle axiom' and 'distance axiom' complement each other and coexist in a comprehensive field of practice.[71] Did not de Caus and Niceron, to name but two, equally understand perspective as a projective modality of optics? Indeed, in the preambles to their respective treatises, they each supplemented Euclidean theorems with new ones relating to the principle of visual cone section. Niceron even more explicitly equated projective lines with 'visual rays', and the picture plane with the retina.[72]

As well as such theoretical considerations, the methodological benefits of anamorphosis for design practice were manifold. By pairing an angular sector and its straightened-up image, for instance, one could shape both the appearance and physical size of spaced out geometrical features along a compositional axis. This could actually be done without using conventional, tedious perspective drawing (in any event, the latter would have been impossible to perform given the layout's scale). Had such an inventive appropriation of anamorphosis as a design tool been disclosed in its time, it might have featured among the many shortening procedures which were promoted and debated in France from the 1630s onward.[73] However, the aim here was not to produce a fixed picture but to form a prototype grid or template (vertical lines). The latter would allow for the dimensioning and proportioning of figures and the construction of corresponding components (*allées*, basins, greensward, and canal) based on their perspectival appearance. This inverted use of projective perspective had two major implications that may seem counterintuitive to someone with a strictly representational understanding of perspective. 1). The position of each sector's apex was by necessity defined to not be occupied by any viewer's eye, as was usually the case in other large-scale anamorphic devices.[74] 2). The spatial distension scheme made it possible to accommodate various temporalities such as viewer motion (position change) and an open-ended design process (multiple, unforeseen, and contradictory phases).

A number of other layouts by Le Nôtre exhibit different types of distension methods such as a divergent modular grid (Grand Canal at Sceaux) and cylindrical curving grids (*Tapis Vert* at Sceaux) or a straight non-divergent and irregular polyhedral scheme

71 Panofsky, *Perspective as Symbolic Form*, 35–36.

72 De Caus, *La perspective*, Chapters 'Theoresme nœufuiesme' and 'Theoresme dixiesme'; Jean-François Niceron, *La Perspective curieuse*, augmented edn. by Marin Mersenne, ed. by G. P. de Roberval (Paris: F. Langlois, 1652), pp. 27, 29.

73 Girard Desarges, *Exemple de l'une des manieres universelles du S.G.D.L. touchant la pratique de la perspective sans emploier aucun tiers point, de distance ny d'autre nature, qui soit hors du champ de l'ouvrage* (Paris: Jacques Dugast, 1636); Jacques Aleaume and Étienne Migon, *La perspective spéculative et pratique* (Paris: Tavernier et Langlois, 1643). On the controversies over perspective methods see, among others: Leblanc, *D'Acide et d'encre*.

74 See mechanical methods of anamorphic construction for gallery frescoes with differed projection apex in: Jean-François Niceron, *Thaumaturgus opticus* (Paris: F. Langlois, 1646), pp. 123–26, Plate 33, and Emmanuel Maignan, *Perspectiva horaria* (Rome: P. Rubei, 1648), pp. 438–49.

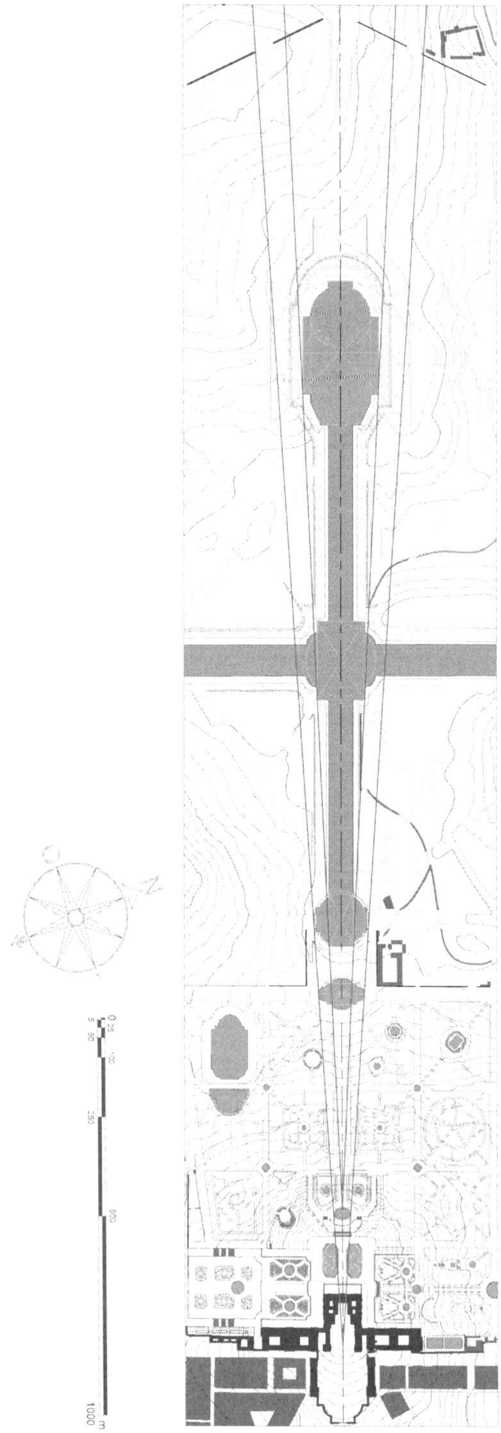

Fig. 10 Angular sectors fit one within another, with respective apexes at chateau courtyard and Latona Basin, traced on topographic plan of 1688 state Grand Canal axis at Versailles. Optical analysis and drawing by Georges Farhat, DG by P. Robert and U. Tanebe, 2012–13. [See colour plate 44]

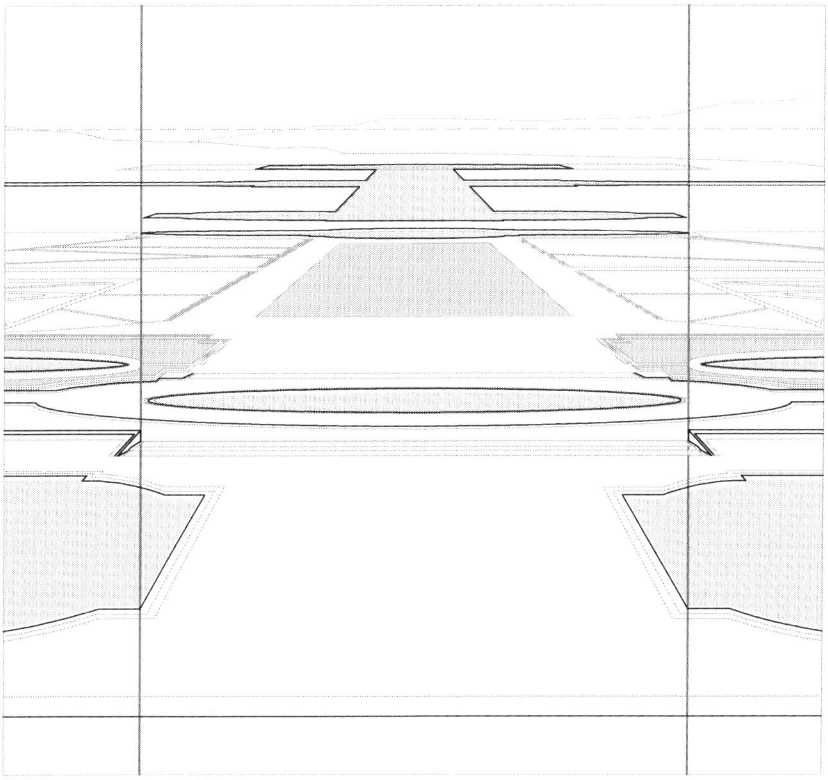

Fig. 11 Conjectural view along Grand Canal axis at Versailles, taken above apex of external angular sector. Optical analysis and drawing by Georges Farhat, DG by M. Faure, 2013. [See colour plate 45]

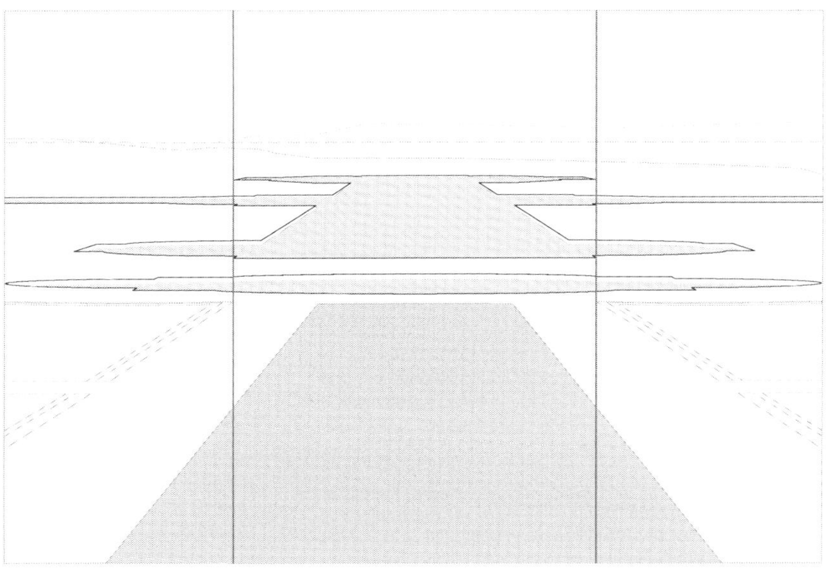

Fig. 12 Conjectural view along Grand Canal axis at Versailles, taken above apex of internal angular sector. Optical analysis and drawing by Georges Farhat, DG by M. Faure, 2013. [See colour plate 46]

(Saint-Germain-en-Laye).[75] But only a few historical design drawings have been found with anamorphic schemes traced on.[76] One such document is for a grand parterre and canal layout at Saverne (near Strasbourg) commissioned by Cardinal de Rohan; it was designed by Robert de Cotte (1656–1735) and carried out by Charles Le Bouteux and Charles Carbonnet, all three of whom had formerly worked with Le Nôtre (Figure 13).[77] Dating from around 1712, this early-stage drawing features three interlocked angular sectors which, unlike Versailles', share the same apex.[78] Obviously, the Saverne scheme owed much to Le Nôtre's project at Versailles. Yet, achievement at this royal worksite was the most spectacular outcome of a long-term pursuit in design practice.

Topographic Perspective: A Genealogy in Design Practice and Literature

The construction of optical schemes in the landscape — through earthworks and ground figures informed by perspectival principles — is what I call topographic perspective. This design practice included collimation chains and anamorphic schemes. It aimed to identify correct proportions of parts in both the materialization and the physical experience of a landscape layout. It resulted in a joint construction of gaze and motion. As such, topographic perspective ranged beyond the mere application of a graphic representation technique. It rather formed a fully-fledged field of invention with the elaboration of unprecedented and site-specific optical-spatial devices. Nevertheless, such devices involved a number of cultural transfers from architecture and the decorative arts to landscape design. Among others, topographic perspective expanded Vitruvian *scaenographia* and optical adjustments, or *temperaturae*, into garden design.[79] It thus addressed an old but unresolved conflict between physical reality and its appearance.[80]

---

75 Georges Farhat, 'Pratiques perspectives et histoire de l'art des jardins. L'exemple du Grand Canal de Sceaux', *Revue de l'art*, 129 (2000), 28–40; Georges Farhat, 'Optique topographique: la Grande Terrasse de Saint-Germain-en-Laye', in *Le Nôtre, un inconnu illustre?* (Paris: Éd. du Patrimoine, 2003), pp. 122–35.

76 See Farhat, 'Great Vistas', in *André Le Nôtre in Perspective*, ed. by Bouchenot-Déchin and Farhat.

77 Institut de France, Paris, MS 1604, fol. 7. For more detail, see Georges Farhat, 'La perspective à l'épreuve du jardin. Le canal de Saverne au XVIIIe siècle', in *L'artiste et l'œuvre à l'épreuve de la perspective*, ed. by Marianne Cojannot-Le Blanc, Marisa Dalai Emiliani and Pascal Dubourg Glatigny (Rome: École française de Rome; Paris: Blanchard, 2006), pp. 179–98. As *Premier architecte du Roi*, Robert de Cotte directed the Drawing Office at the *Bâtiments*, from 1688 to 1708, while Le Nôtre still held his position as *contrôleur général* in the same administration. De Cotte and his collaborators subsequently surveyed or redesigned many sites Le Nôtre had worked on: Saint-Maur, Maintenon, Anet, Saint-Cloud, and the Tuileries.

78 Construction at Saverne developed over a forty-year period: after 1730, the axial layout reached a 4,500-metre stretch for only a thirteen-metre elevation drop. My analysis was based on the current, deeply altered topography of the site and a host of additional drawings from Institut de France, Paris, among which: *Plan de la ville, du château, des jardins, du canal et des parcs de Saverne*, c. 1725 (MS 1042, fol. 22) and a measured section of the Grand Canal project (MS 1039, fol. 18).

79 Vitruvius, *De architectura libri decem*. On *entasis* and *scamilli*, see Vitruvius, *De architectura*, III, 2, 3, 5; on elongation, see Vitruvius, *De architectura*, VI, 2; on the notion of *temperatura*, see Vitruvius, *De architectura*, III, 3, 13. See also *Vitruve, De l'architecture, livre III*, ed. by Pierre Gros (Paris: Les Belles Lettres, 1990), pp. 122–23.

80 As Gros recalls, this conflict's optical grounds was defined by Plato and theorised by Euclid (*Optics*, in particular Propositions-Theorems 4, 5, 8, 10–12, 22); it opposed the Epicureans to all other Hellenistic doctrines in Rome during Vitruvius's time. See Pierre Gros, 'The Theory and Practice of Perspective in Vitruvius's "De architectura"',

As a practice and loose set of rules, topographic perspective can be traced back to research developed by garden professionals who, from 1550 onward, enlarged their design purview: from isolated parterres to the layout of an entire garden and park in a manorial demesne taken as a whole. Accordingly, topographic perspective appears to have initially stemmed from an ornament problem, consisting of the transfer of patterns (knots, arabesques, foliage, tessellation) from a range of decorative arts (tapestry, lace, marquetry, goldsmithery) and inert materials (wood, stone, metal) to the display of plants on horizontal ground surfaces. Translation here implied the necessity to deal with biological dynamics (both spatial and temporal) as well as scale and experiential shifts.

In its time, Jacques du Cerceau's engraved oeuvre (from the 1540s to 1580s) covered this pattern migration from a variety of ornaments to garden parterres, though without theorizing any of its related technical issues.[81] Such specifications were first to be found in the comments to parterre models published in *La maison rustique* (in 1583) by the physiologist Jean Liébault. He discussed at length how the distribution, height, and colour of vegetation should be coordinated relative to spectator vision, life cycles of plants, and the latter's maintenance so that 'the beauty of the Parterre can be seen and appear more easily'.[82]

Soon after, the agronomist Olivier de Serres, seigneur du Pradel, expanded the topic of garden design and maintenance into a useful speculation on 'Perspective'. When it came to 'looking at compartments from afar' he recommended 'that their [planting] beds be arranged farther away from each other [...] as what is seen diminishes in size relative to distance, by reason of Perspective'.[83] The practical purpose of this was to prevent a pattern from being confused. Accordingly, drawing on technical literacy, de Serres extensively discussed perspective as a geometrical modality of optics. As we know from his registers, he owned Jean Cousin's 1560 *Livre de perspective* from which he seems to have appropriated a few notions.[84] Indeed, the French painter, who conceived of visual foreshortening within a modular box showing a diminishing scale, debated lateral distortion and related visual corrections in the wake of Piero della Francesca and Leonardo da Vinci.[85] Most interestingly, like many of his contemporaries, Cousin insisted on the hybrid nature of this

in *Perspective, Projections & Design: Technologies of Architectural Representation*, ed. by Mario Carpo and Frédérique Lemerle (London: Routledge, 2008), pp. 5–17.

81 Jacques Androuet du Cerceau, *Parquets ou mosaïques*, 26 engraved plates of knot models, in Bibliothèque de l'Inha, Paris, 4° Res 84 (2); Jacques Androuet du Cerceau, (3e) *Livre d'architecture [...]* (Paris: The author, 1582); Jacques Androuet du Cerceau, *Le premier (-second) volume des plus excellents bastiments de France* (Paris: The author, 1576–79).

82 '... à fin que la beauté du Parterre se puisse veoir & paroisse plus aysement'. Charles Estienne and Jean Liébault, *L'agriculture et maison rustique* [1564], augmented edn. by J. Liébault (Paris: Du Puys, 1583), fols 143^{r}–55^{r}.

83 'Ceci est notable, qu'estant question de regarder de loin les compartimens, pour l'assiete des Jardins, convient faire les rangees d'iceux, plus loin l'une de l'autre [...] s'appetissant la chose regardee, à mesure de l'esloignement, par la raison de Perspective'. Olivier de Serres, *Le théâtre d'agriculture et mesnage des champs* (Paris: Jamet-Métayer, 1600), pp. 581–82.

84 *Le livre de raison d'Olivier de Serres*, ed. by Dominique Margnat (Grenoble: Presses universitaires de Grenoble, 2004), p. 175.

85 Jean Cousin, *Livre de perspective* (Paris: Jehan le Royer, 1560), see in particular fols cij^{r}–ciij^{v} and Dj^{r}–Dij^{v}. For a discussion of perspective as a graphic means of representation and invention in sixteenth-century France, see: Valérie Auclair, *Dessiner à la Renaissance: la copie et la perspective comme instrument de l'invention* (Rennes: Presses universitaires de Rennes, 2010). On the circulation of technical knowledge from Piero's and Leonardo's

Fig. 13 Robert de Cotte (and collaborators), Château, Grand Parterre and Grand Canal of Saverne, with three angular sectors inserted one within another, *c.* 1712. Drawing, pen and black ink, watercolour, and graphite, 35,9 x 45,7 cm, Paris, Bibliothèque de l'Institut de France, Ms 1604, fol. 7. Photo copyright: RMN-Grand Palais (Institut de France) / Gérard Blot.

'art of Perspective [being] derived from things Natural and things artificial', that is, from vision (light) and geometry (line).[86] But de Serres is also likely to have read Sebastiano Serlio who — following Albrecht Dürer, and later emulated by de Caus — reversed incidental distortion into intended distension, applied either vertically or horizontally, to make different heights or distances all appear equal from a given viewpoint.[87] Overall, de Serres's discussion included the impact of open air, distance, and elevation on judgement. Closely following the terminology of the French Vitruvius (1547), he thus betrays some debt to its illustrator and commentator, the sculptor Jean Goujon. The latter envisaged perspective (*scenographia*) in a spatialised mode. He presents it as a method for optically adjusting the proportions of a building's ornament to fix its physical dimensions, relative to some viewpoint, in the process of material construction.[88]

De Serres's book was illustrated with the layout of parterres planted in royal gardens by Claude Mollet whose later practice was to be informed in return by the agronomist's thoughts. Mollet claimed to have established, with the help of the architect Étienne Dupérac, a new fashion of 'large volume' designs in intricate boxwood *broderie* in order to achieve unity and permanence of pattern in lieu of a hitherto favoured variety of small compartments.[89] But, more importantly, in so doing, the discrepancy between ground figures and their appearance must have become obvious to him. This called for optical adjustment which Mollet addressed empirically. To present a square-shaped appearance to a viewer looking at a parterre, he proposed designing the latter as an oblong figure 'three or four toises longer than wide, because Perspective foreshortens'.[90]

Such observations were equally extended to *allées* and pathways by Mollet and by Jacques Boyceau, superintendent of the royal gardens and visual art expert.[91] In a much more elaborate fashion than his colleague, Boyceau reflected on the interrelatedness of height, width, and length in the spatial characteristics of tree-planted walks for different contexts: covered or uncovered; with or without bordering hedges; in gardens, wildernesses, and fields.[92] This typology, which he introduced with Vitruvian considerations of the need for optical adjustment, abounds with measurements, some generic and others drawn from concrete examples. But the method remains empirical, as well reflected in Boyceau's

manuscripts, see: J. V. Field, *Piero della Francesca: A Mathematician's Art* (New Haven: Yale University Press, 2005), pp. 353–72; Robert Kretzchmar and Lorenz Sönke, *Leonardo da Vinci und Heinrich Schickhardt: zum Transfer technischen Wissens im vormodernen Europa* (Stuttgart: W. Kohlhammer, 2010).

86 '[…] cest Art de Perspective est extraict des choses Naturelles & artificielles'. Cousin, *Livre de perspective*, fol. 14v.

87 Albrecht Dürer, *Géométrie. Underweysung der messung […]* [1st edn. 1525; 2nd edn. 1538], ed. and trans. by Jeanne Peiffer (Paris: Seuil, 1995), III, [28], p. 278; Sebastiano Serlio, *Le premier livre d'architecture, le second livre de perspective* (Paris: Jean Barbé, 1545), fols 9v–10r; De Caus, *La perspective*, fols 37, 38.

88 Jean Goujon, 'Sur Vitruve', in *Architecture, ou art de bien bastir de Marc Vitruve Pollion*, trans. by Jean Martin (Paris: J. Gazeau, 1547), fols Diii–Dv.

89 Mollet, *Theatre des plans et jardinages […]*, 199–203.

90 '[…] trois ou quatre toises plus long que large, à cause que la Perspective fait accourcir'. Mollet, *Theatre des plans et jardinages […]*, 324.

91 Mollet, *Theatre des plans et jardinages […]*, 113–14. Boyceau is mentioned in 1621 and 1622 as a member of the expert committee asked to judge Rubens's submissions to different French royal commissions: Alexis Merle du Bourg, *Peter Paul Rubens et la France, 1600–1640* (Villeneuve d'Ascq: Presses universitaires du Septentrion, 2004), pp. 36, 96–99.

92 Jacques Boyceau de La Barauderie, *Traité du jardinage selon les raisons de la nature et de l'art* (Paris: Vanlochom, 1638), pp. 72–73.

discussion of 'perspective' in the central elm-planted *allée* of the Tuileries garden. This reveals how optical phenomena were understood in design practice.

Finally, André Mollet, Claude's son, would tackle the garden as a whole rather than a collection of separate components, offering the first two comprehensive model plans ever published in a French gardening treatise.[93] He defined a multi-scalar distension principle of optical consequence that was to shape a major characteristic of French gardens. Firstly, he assigned the garden's overall layout and all resulting subdivisions a 3:4 ratio rectangular figure. Though directional, such elongation was not set relative to a specific viewpoint. Nor was it gradual. It was rather fixed and non-divergent, echoing a scheme analogous to Samuel Marolois's anamorphic devices.[94] Additionally, Mollet directed, 'the parterres furthest from sight must be given greater volume than those which are closer'.[95] He paired this with a then canonical, gradual scaling down of visual density of pattern, dependent on how far away the viewer stood: from foreground intricate embroidery, to middle ground widely spaced cutwork, to remote plain grass plats.

Towards Landscape Design's Contribution to the Historiography of Perspective

Albeit approximate, the discussion of a Vitruvian theme, such as optical adjustments, was highly strategic for the technical literature on the garden published between 1580 and 1655. It supported growing claims by master gardeners for recognition of their professional expertise and status. Competition with architects was at stake, in particular, when it came to large-scale layouts. Indeed, contemporary landscape compositions designed by architects, such as the grounds at Richelieu, also featured steady enlargement of axial figures.[96] But the long sought after rule for optic-metrical distension would only come to fruition with the work of a royal garden designer, Le Nôtre, who ironically never published a word on his métier.

The son and grandson of gardeners to the king, Le Nôtre was to take advantage of his family's networks and further achieve social mobility. His well-planned career included training in draftsmanship, between 1628 and 1635, in the workshop of the *Premier peintre du Roi,* Simon Vouet (1590–1649). Apprenticeship during those years allowed Le Nôtre to mine an even more valuable resource for design than mere graphic representation. Effectively,

93 André Mollet, *Le jardin de plaisir* (Stockholm: H. Kayser, 1651), Chapter XI, Plates 1 and 2.

94 Samuel Marolois, *Opera mathematica* (The Hague: Hondius, 1614), Plate 80, Fig. CC.LXXV.

95 '[...] les parterres les plus esloignez de la veüë doivent estre mis en plus grand volume, que ceux qui en sont plus proches [...]'. André Mollet, *Le jardin de plaisir,* Chapter XI.

96 Designed during the 1630s by the architect Jacques Lemercier, the grounds of the Chateau de Richelieu are best approached through two graphic documents: 'Design Plan of Château, Gardens and Petit Parc of Richelieu (*c.* [1638–40(?)]), Bibliothèque nationale de France, Paris, Ge BB 246, XI, fol. 68^{v}–69, and engraved plates by Jean Marot, *Le magnifique chasteau de Richelieu* ([Richelieu(?)]: 1657–59). On the complex history of this project, see *Richelieu à Richelieu: architecture et décors d'un château disparu,* exhibition catalogue, Musée des Beaux-Arts d'Orléans, Musée municipal de Richelieu and Musée des Beaux-Arts de Tours, 12 March–13 June 2011 (Milan: Silvana, 2011).

Vouet happened to be tightly connected to research on perspective and anamorphosis.[97] In turn, Le Nôtre himself would contribute to the corpus of perspective by addressing challenges of unprecedented scale and physical experience in landscape layouts.

Indeed, whether for royal or private commission, Le Nôtre's designs primarily responded to the dynamics of manorial economy and its interaction with geomorphology within the *ancien regime* setting of land institutions (in an order-based society).[98] In this context, gardens and parks formed distinct and complementary entities in any manorial demesne. Following a steady trend since the early 1600s, demesnes underwent increases in both size and economic values relative to production, taxation, and social credit. Given the low sedimentary reliefs of the Paris Basin, such a complex change resulted in flat vision conditions and the need to offset excessive foreshortening. This is the framework that all seigneurs, including the king who privately owned Versailles, would conform to. It prompted the invention of topographic perspective schemes by Le Nôtre and his predecessors in landscape design.

We saw here how topical this practice remained as it intersected with at least two major mid-seventeenth-century academic controversies around optical matters: surveying instruments and procedures as well as visual distortions in painting for large decors. But, at a time when modern buildings were reaching unprecedented scales, equally relevant was the quarrel over proportions in architectural ornament. The well-known conflict chiefly opposed two members of both academies of Science and Architecture: Claude Perrault (Charles's brother) and François Blondel.[99] It peaked with the last section of Blondel's *Cours d'architecture* (1683). There, arguing in favour of metric distension, Blondel first recalled perspectival schemes that had been formerly proposed by Serlio and Bosse for the right proportioning of an entire building's elevation (Figure 14).[100] Then, after countering Claude Perrault's rejection of Vitruvius's opinion on 'perspective' and visual adjustments, he strived to demonstrate their necessity using examples of ancient and modern buildings (including Mansart's Val-de-Grâce).[101] However, his conclusion is at once disconcerting and consistent. He had to admit he did not know how to precisely find the middle way between 'the effect of optical rules and that of natural proportions.'[102] But why not adduce here Le Nôtre's designs, which he could not but know, as successful experiments in metric

97 On Le Nôtre's training, see Jacques Thuillier, *Vouet* (Paris: RMN, 1990), pp. 39–40. On Vouet's connections with anamorphosis and the Minime Order in Rome and Paris, see Baltrusaitis, *Anamorphoses*, 37–58, 131–77; Thuillier, *Vouet*, 26–31; *La Trinité-des-Monts redécouverte. Arts, foi et culture*, ed. by Yves Bruley (Rome: De Luca, 2002).

98 Georges Farhat, 'Manorial Economy and French Seventeenth-Century-designed Landscapes: The Formal Type by Savot (1624) and at Sceaux (1670–1690)', *Landscape Research*, 40 (5) (2015), 566–85.

99 Claude Perrault, *Les dix livres d'architecture de Vitruve* (Paris: Coignard, 1673), VI, 2, pp. 194–95 n. 1; Claude Perrault, *Abrégé des dix livres d'architecture de Vitruve* (Paris: J.-B. Coignard, 1674); Claude Perrault, *Ordonnance des cinq espèces de colonnes selon la méthode des Anciens* (Paris: J.-B. Coignard, 1683), II, p. 7; François Blondel, *Cours d'architecture* (Paris: Auboin & Clouzier, 1675–83), Part 5, Book 4, pp. 703–26. On these two protagonists and their dispute, see, respectively, Antoine Picon, *Claude Perrault 1613–1688 ou la curiosité d'un classique* (Paris: Picard, 1988), pp. 138–56, and Anthony Gerbino, *François Blondel: Architecture, Erudition, and the Scientific Revolution* (London: Routledge, 2010), pp. 148–65.

100 Blondel, *Cours*, 711–13.

101 Blondel, *Cours*, 714–22.

102 '[…] l'effet des regles de l'Optique & celuy des proportions naturelles'. Blondel, *Cours*, 722–26.

distension? The answer lies in an earlier publication in which Blondel made it clear he was against recent trends in garden refashioning.[103]

However different their timeframes and stakes, the three abovementioned academic debates overlapped, not least because of ubiquitous players such as Charles Perrault. Now, in the light of Blondel's opposition to his brother, questions resurface as to Charles's multiple motivations to silence the use of anamorphic devices in landscape layouts. Did not his interest lie also in upholding his brother's condemnation of optical adjustments? Indeed, in the second dialogue of the *Parallèle*'s first volume, as they walked in the gardens of Versailles, the Abbot and President characters overtly re-enacted the dispute between Claude Perrault and François Blondel over *temperaturae*. Two years after Blondel's death, this artifice allowed Charles to sum up Claude's arguments, add to them his own observations, and definitively signify the rejection of anamorphosis in architecture and sculpture as a distinctive trait of the Moderns.[104] He therefore could have even less accepted it in landscape design. Thenceforth, with emphasis on the role played by the telescope in hypothetical surveying instruments, Perrault obscured a major episode in the development of topographic perspective. In so doing, he paved the way to a blinded history that still cannot recognise the garden as a field of invention for perspective. Yet this state of affairs can be reversed. To that end, as suggested here, one needs to use specific methods of inquiry such as acquiring and processing terrain data as well as reconstructing historical surveying and design procedures within specific environmental, technical, and institutional frameworks, including land economy. Only at this cost could one understand the relation of perspective to optics in landscape design. As a result, landscape studies should be able to more broadly contribute to the ongoing revision of the historiography of perspective.

Bibliography

Manuscript and Archival Sources

Archives nationales, Paris, 'Quote of the Excavation Needed to Ensure the Increase in Length and Width of the Canal of Versailles [18 March 1671]', O¹, 1793¹.

Archives nationales, Paris, Charles Perrault, 'Evidence of Sir Jolly's Misconduct in the Delivery of Plumbs and Solder to Vincennes and to Versailles from 1664 and through April 6th, 1667', O¹ 1887.

Bibliothèque de l'Inha, Paris, Jacques Androuet du Cerceau, *Parquets ou mosaïques*, 26 engraved plates of knot models, 4° Res 84 (2).

Bibliothèque nationale de France, Paris, 'Design Plan of Château, Gardens and Petit Parc of Richelieu (*c.* [1638–40(?)]), Ge BB 246, XI, fol. 68v–69.

103 In the reedition of Louis Savot, *L'architecture françoise des bastimens particuliers* (Paris: F. Clouzier & P. Auboüin, 1673, notes a, p. 175 and a, p. 178; 1st edn. 1624), which he personally curated in his capacity as director of the *Académie d'Architecture*, Blondel commented negatively on recent changes brought about in existing gardens (like the Tuileries in Paris) where wooded areas were cleared by Le Nôtre (whom he tellingly does not name) in order to open stately vistas.

104 Perrault, *Parallèle*, I, pp. 132–61.

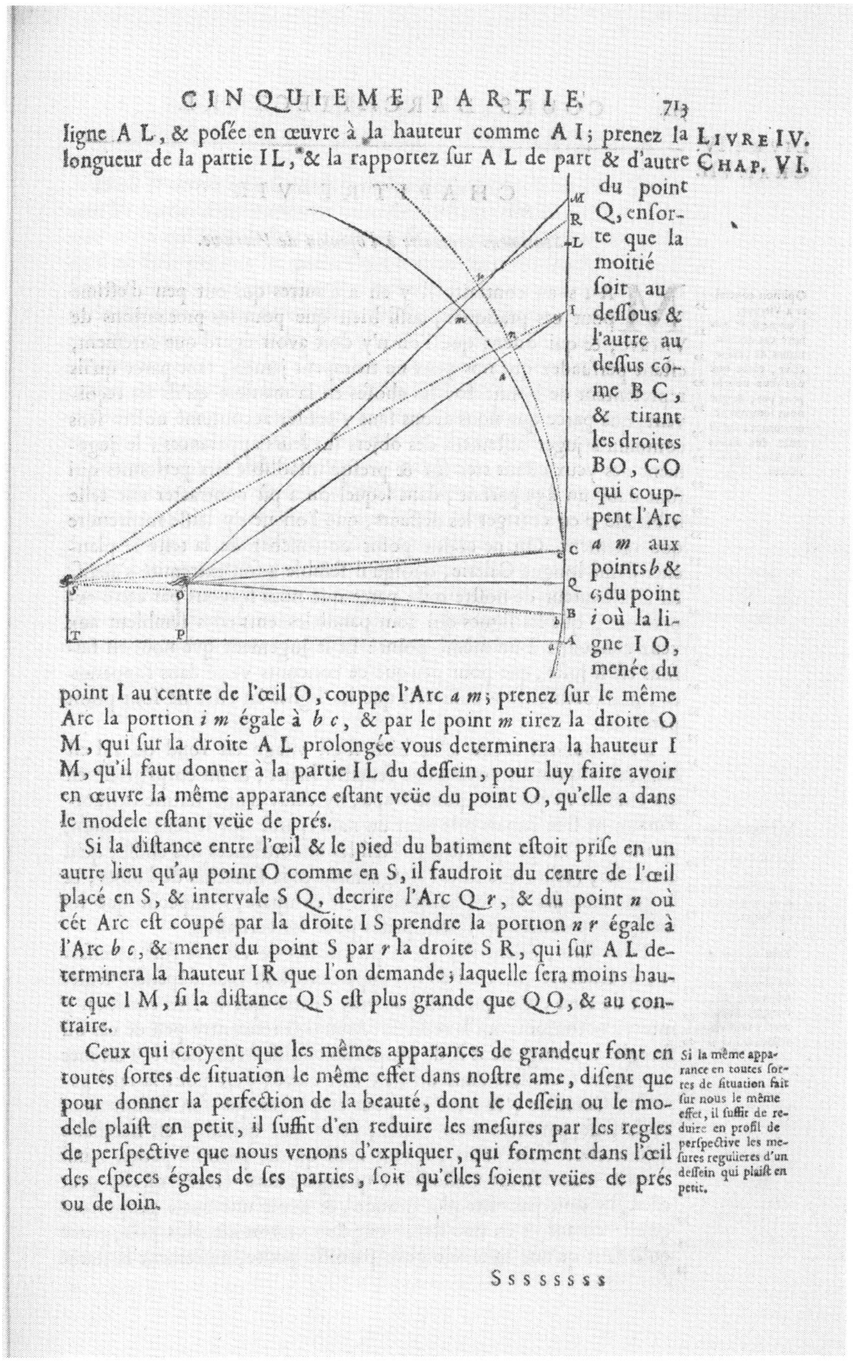

CINQUIEME PARTIE. 713

ligne A L, & poſée en œuvre à la hauteur comme A I; prenez la longueur de la partie I L, & la rapportez ſur A L de part & d'autre du point Q, enſorte que la moitié ſoit au deſſous & l'autre au deſſus cõme B C, & tirant les droites B O, C O qui coupent l'Arc *a m* aux points *b* & *c*; du point *i* où la ligne I O, menée du point I au centre de l'œil O, couppe l'Arc *a m*; prenez ſur le même Arc la portion *i m* égale à *b c*, & par le point *m* tirez la droite O M, qui ſur la droite A L prolongée vous determinera la hauteur I M, qu'il faut donner à la partie I L du deſſein, pour luy faire avoir en œuvre la même apparance eſtant veüe du point O, qu'elle a dans le modele eſtant veüe de prés.

LIVRE IV. CHAP. VI.

Si la diſtance entre l'œil & le pied du batiment eſtoit priſe en un autre lieu qu'au point O comme en S, il faudroit du centre de l'œil placé en S, & intervale S Q, decrire l'Arc Q *r*, & du point *n* où cét Arc eſt coupé par la droite I S prendre la portion *n r* égale à l'Arc *b c*, & mener du point S par *r* la droite S R, qui ſur A L determinera la hauteur I R que l'on demande; laquelle ſera moins haute que I M, ſi la diſtance Q S eſt plus grande que Q O, & au contraire.

Ceux qui croyent que les mêmes apparances de grandeur font en toutes ſortes de ſituation le même effet dans noſtre ame, diſent que pour donner la perfection de la beauté, dont le deſſein ou le modele plaiſt en petit, il ſuffit d'en reduire les meſures par les regles de perſpective que nous venons d'expliquer, qui forment dans l'œil des eſpeces égales de ſes parties, ſoit qu'elles ſoient veües de prés ou de loin.

Si la même apparance en toutes ſortes de ſituation fait ſur nous le même effet, il ſuffit de reduire en profil de perſpective les meſures regulieres d'un deſſein qui plaiſt en petit.

Sssssss

Fig. 14 François Blondel, Presentation of Abraham Bosse's method of metric distension in architecture, 'Regles de Perspective de Serlio & de Bosse', Chap. VI of Book IV, 'Des changements que la hauteur ou l'éloignement peuvent apporter aux parties de l'Architecture', in *Cours d'architecture enseigné dans l'Academie royale d'architecture...*, (Paris, 1675–83), Part V (1683), p. 713. Thomas Fisher Rare Book Library, University of Toronto.

Bibliothèque nationale de France, Paris, François Mansart, Design Plan , cation is correct herefor the Val-de-Grâce, cation is correct here, VA-443-FT 6.
Bibliothèque nationale de France, Paris, Manuscrits, *Mélanges Colbert*, 148bis, fol. 718.
Bibliothèque nationale de France, Paris, VA-419 J, 6-FT.
Institut de France, Paris, MS 1604, fol. 7.
Institut de France, Paris, MS 1042, fol. 22; MS 1039, fol. 18.
Institut de France, Paris, MS 1307, fol. 51.
Nationalmuseum, Stockholm, Anonymous, CC 139, CC 2202.

Primary Sources

Accolti, Pietro, *Lo inganno degl'occhi* (Florence: Pietro Cecconcelli, 1625).
Aleaume, Jacques, and Étienne Migon, *La perspective spéculative et pratique* (Paris. Tavernier et Langlois, 1643).
Barauderie, Jacques Boyceau de La, *Traité du jardinage selon les raisons de la nature et de l'art* (Paris: Vanlochom, 1638).
Bettini, Mario, *Apiaria [...]* (Bologna: J. B. Ferroni, 1641–42).
Blondel, François, *Cours d'architecture* (Paris: Auboin & Clouzier, 1675–83).
Bosse, Abraham, *Traité des pratiques géométrales et perspectives enseignées dans l'Académie Royale de Peinture et Sculpture* (Paris: The author, 1665).
Breuil, Jean Du, *Perspective practique [...]* (1649).
Bullet, Pierre, *Traité du nivellement* (Paris: Langlois, 1688).
Bullet, Pierre, *Traité du nivellement* (Paris: Langlois, 1688).
Caus, Salomon de, *La perspective avec la raison des ombres et des miroirs* (London: R. Barker, 1611).
Cerceau, Jacques Androuet du, *Le premier (-second) volume des plus excellents bastiments de France* (Paris: The author, 1576–79).
Cerceau, Jacques Androuet du, (3e) *Livre d'architecture [...]* (Paris: The author, 1582).
Chambray, Roland Fréart de (ed.), *Traité de la peinture, de Léonard de Vinci, donné au public et traduit d'Iitalien en Ffrançois par R. F. S. D. C.* (Paris: J. Langlois, 1651).
Chambray, Roland Fréart de, *La perspective d'Euclide* (Le Mans: Jacques Isambart, 1663).
Cousin, Jean, *Livre de perspective* (Paris: Jehan le Royer, 1560).
Danfrie, Philippe *Declaration de l'usage du graphomètre [...]* (Paris: Danfrie, 1597).
Desarges, Girard, *Exemple de l'une des manieres universelles du S.G.D.L. touchant la pratique de la perspective sans emploier aucun tiers point, de distance ny d'autre nature, qui soit hors du champ de l'ouvrage* (Paris: Jacques Dugast, 1636).
Dezallier d'Argenville, Antoine-Joseph, *La théorie et la pratique du jardinage*, 4th edn. (Paris: Jean Mariette, 1747; 1st edn. 1709).
Dürer, Albrecht, *Géométrie. Underweysung der messung [...]* [1st edn. 1525; 2nd edn. 1538], ed. and trans. by Jeanne Peiffer (Paris: Seuil, 1995).
Estienne, Charles, and Jean Liébault, *L'agriculture et maison rustique* [1564], augmented edn. by J. Liébault (Paris: Du Puys, 1583).
Euclid, *The Optics of Euclid*, trans. by Harry Edwin Burton, *Journal of the Optical Society of America*, 33 (5) (1943).
Finé, Oronce, *Protomathesis* (Paris: Gérard Morrhi and Jean Pierre, 1532).

Frisius, Reiner Gemma, *Cosmograficus liber Petri Appiani* (Antwerp: Arnold Birckman , 1533).
Goujon, Jean, 'Sur Vitruve', in *Architecture, ou art de bien bastir de Marc Vitruve Pollion*, trans. by Jean Martin (Paris: J. Gazeau, 1547).
Guiffrey, Jules (ed.), *Comptes des Bâtiments du Roi sous le règne de Louis XIV*, 5 vols (Paris: Imprimerie nationale, 1881–1901).
Hire, Philippe de La, 'Preface', in Jean Picard and Philippe de La Hire, *Traité du nivellement par M. Picard … avec une relation de quelques nivellements faits par ordre du Roy… Mis en lumière par les soins de M. de La Hire…* (Paris: Michallet, 1684).
Hire, Philippe de La, *L'école des arpenteurs* (Paris: Thomas Moette, 1689).
Huret, Grégoire, *Optique de portraiture et peinture* (Paris: The author, 1670).
Leclerc, Sébastien, *Pratique de la géométrie […]* (Paris: Jolly, 1669).
Leclerc, Sébastien, *Discours touchant le point de vue, dans lequel il est prouvé que les choses qu'on voit distinctement ne sont veuës que d'un œil* (Paris: T. Jolly, 1679).
Leclerc, Sébastien, *Traité de geometrie* (Paris: Jombert, 1690).
Leclerc, Sébastien, *Système de la vision fondée sur de nouveaux principes* (Paris: Delaulne, 1712).
Louis XIV, *Manière de montrer les jardins de Versailles* (Paris: RMN, 1992).
Maignan, Emmanuel, *Perspectiva horaria* (Rome: P. Rubei, 1648).
Manesson-Mallet, Allain, *La géométrie pratique* (Paris: Anisson, 1702).
Mariotte, Edme, *Traité du nivellement* (Paris: J. Cusson, 1672).
Marolois, Samuel, *Opera mathematica* (The Hague: Hondius, 1614).
Marot, Jean, *Le magnifique chasteau de Richelieu* ([Richelieu(?)]:1657–59).
Masaccio, *Nel segno di Masaccio. L'invenzione della prospettiva*, ed. by Filippo Camerota (Florence: Giunti, 2001).
Mollet, André, *Le jardin de plaisir* (Stockholm: H. Kayser, 1651).
Mollet, Claude, *Theatre des plans et jardinages* [MS 1620s–30s] (Paris: Charles de Sercy, 1652).
Niceron, Jean-François, *La perspective curieuse* (Paris: P. Billaine, 1638).
Niceron, Jean-François, *Thaumaturgus opticus* (Paris: F. Langlois, 1646).
Niceron, Jean-François, *La Perspective curieuse*, augmented edn. by Marin Mersenne, ed. by G. P. de Roberval (Paris: F. Langlois, 1652).
Perrault, Claude, *Les dix livres d'architecture de Vitruve* (Paris: Coignard, 1673).
Perrault, Claude, *Abrégé des dix livres d'architecture de Vitruve* (Paris: J.-B. Coignard, 1674).
Perrault, Charles, *Recueil de divers ouvrages en prose et en vers* (Paris: J.-B. Coignard, 1675).
Perrault, Claude, *Ordonnance des cinq espèces de colonnes selon la méthode des Anciens* (Paris: J.-B. Coignard, 1683).
Perrault, Charles, *Le siècle de Louis le Grand* (Paris: Jean-Baptiste Coignard, 1687).
Perrault, Charles, *Parallèle des Anciens et des Modernes […]*, 4 vols (Paris: J.-B. Coignard, 1688–97).
Perrault, Charles, *Mémoires: Memoirs of My Life*, ed. and trans. by J. Morgan Zarucchi (Columbia, MO: University of Missouri Press, 1989).
Perrault, Charles, *Mémoires de ma vie* (Paris: Macula, 1993).
Picard, Jean, *Mesure de la terre* (Paris: L'imprimerie royale, 1671).
Riccioli, Giovanni Battista, *Geographiae et hydrographiae reformatae libri duodecim*, 2nd edn. (Venice: J. La Noũ, 1672).
Savot, Louis, *L'architecture françoise des bastimens particuliers* [1624], annotated edn. by François Blondel (Paris: F. Clouzier and P. Auboüin, 1673).

Serlio, Sebastiano, *Le premier livre d'architecture, le second livre de perspective* (Paris: Jean Barbé, 1545).
Serres, Olivier de, *Le théâtre d'agriculture et mesnage des champs* (Paris: Jamet-Métayer, 1600).
Tessin the Younger, Nicodemus, *Nicodemus Tessin the Younger: Sources, Works, Collections: Travel Notes 1673–77 and 1687–88*, ed. by Merit Laine and Borje Magnusson (Stockholm: Nationalmuseum, 2002).
Thevenot, Melchisedech, *Machine nouvelle pour la conduite des eaux, pour les bâtimens, pour la navigation, et pour la pluspart des autres arts* (Paris: Mabre-Cramoisy, 1666).
Vaulezard, J.-L. Sr de, *La perspective cylindrique et conique* (Paris: J. Jacquin, 1630).
Vignola, Jacoppo Barozzi da, and Egnatio Danti, *Le due regole della prospettiva pratica* (Rome: F. Zannetti, 1583).

Secondary Works

Auclair, Valérie, *Dessiner à la Renaissance: la copie et la perspective comme instrument de l'invention* (Rennes: Presses universitaires de Rennes, 2010).
Babelon Jean-Pierre, and Claude Mignot (eds.), *François Mansart, Le génie de l'architecture* (Paris: Gallimard, 1998).
Baltrusaitis, Jurgis, *Anamorphoses*, 3rd edn. (Paris: Flammarion, 1984; 1st edn. 1955).
Baridon, Michel, 'The Scientific Imagination and the Baroque Garden', *Studies in the History of Gardens & Designed Landscapes*, 18 (1) (1998), 5–19.
Baridon, Michel, *A History of the Gardens of Versailles* (Philadelphia: University of Pennsylvania Press, 2008; French edn. 2003).
Becchi, Antonio, Hélène Rousteau-Chambon and Joël Sakarovitch (eds.), *Philippe de La Hire (1640-1718), entre architecture et sciences* (Paris: Picard, 2013).
Benevolo, Leonardo, *La cattura del'infinito* (Roma: Laterza, 1991).
Berger, Robert, and Thomas Hedin, *Diplomatic Tours in the Gardens of Versailles under Louis XIV* (Philadelphia: University of Pennsylvania Press, 2008).
Biagioli, Mario, *Galileo's Instruments of Credit: Telescopes, Images, Secrecy* (Chicago: University of Chicago Press, 2007).
Bleyl, Matthias, and Pascal Dubourg Glatigny (eds.), *Quadratura: Geschichte, Theorie, Techniken* (Berlin: Deutscher Kunstverlag, 2011).
Bourg, Alexis Merle du, *Peter Paul Rubens et la France, 1600–1640* (Villeneuve d'Ascq: Presses universitaires du Septentrion, 2004).
Bredekamp, Horst, *Leibniz und die Revolution der Gartenkunst: Herrenhausen, Versailles und die Philosophie der Blätter* (Berlin: Verlag Klaus Wagenbach, 2012).
Bruley, Yves, *La Trinité-des-Monts redécouverte. Arts, foi et culture* (Rome: De Luca, 2002).
Brunon, Hervé, 'De l'image à l'imaginaire: notes sur la figuration du jardin sous le règne de Louis XIV', *XVIIe siècle*, 209 (4) (2000), 671–90.
Charageat, Marguerite, 'André Le Nôtre et ses dessins', *La gazette illustrée des amateurs de jardins* (1953–54), 21–27.
Charageat, Marguerite, 'André Le Nôtre et l'optique de son temps. Le Grand Canal de Tanlay par Pierre Le Muet', *Bulletin de la Société de l'histoire de l'art français* (1955–56), 66–78.
Charageat, Marguerite, 'Pourquoi Le Nôtre a-t-il créé, à Versailles, le grand canal?', *Jardin des arts*, 42 (1958), 371–78.

Cosgrove, Denis, *Social Formation and Symbolic Landscape* (Madison: The University of Wisconsin Press, 1984).

Daston, Lorraine, 'The Empire of Observation', in *Histories of Scientific Observation*, ed. by Lorraine Daston and Elizabeth Lunbeck (Chicago: University of Chicago Press, 2011), pp. 81–113.

De Jong, Erik, and Marleen Dominicus-van Soest, '"Tuijngesigten en perspective"', in *Aardse paradijzen. De tuin in de Nederlandse kunst, 15de tot 18de eeuw* (Ghent: Snoeck-Ducaju & Zoon, 1996), pp. 11–123.

Dupré, Sven, 'The Historiography of Perspective and "Reflexy-Const" in Netherlandish Art', *Nederlands Kunsthistorisch Jaarboek*, 61 (2011), 35–60.

Evans, Robin, *The Projective Cast: Architecture and its Three Geometries* (Cambridge, MA: MIT Press, 1995).

Farhat, Georges, 'Pratiques perspectives et histoire de l'art des jardins. L'exemple du Grand Canal de Sceaux', *Revue de l'art*, 129 (2000), 28–40.

Farhat, Georges, 'Optique topographique: la Grande Terrasse de Saint-Germain-en-Laye', in *Le Nôtre, un inconnu illustre?* (Paris: Éd. du Patrimoine, 2003), pp. 122–35.

Farhat, Georges, 'La perspective à l'épreuve du jardin. Le canal de Saverne au XVIIIe siècle', in *L'artiste et l'œuvre à l'épreuve de la perspective*, ed. by Marianne Cojannot-Le Blanc, Marisa Dalai Emiliani and Pascal Dubourg Glatigny (Rome: École française de Rome; Paris: Blanchard, 2006), pp. 179–98.

Farhat, Georges, and Patricia Bouchenot-Déchin, in *André Le Nôtre in Perspective*, ed. by Patricia Bouchenot-Déchin and Georges Farhat (Paris: Hazan; Versailles: Château de Versailles, 2013).

Farhat, Georges, 'Optical Instrumenta[liza]tion and Modernity at Versailles: From Measuring the Earth to Leveling in French Seventeenth-Century Gardens', in *Technology and the Garden*, ed. by Kenneth Helphand and Michael Lee (Washington, DC: Dumbarton Oaks, 2014), pp. 23–50.

Farhat, Georges, 'Manorial Economy and French Seventeenth-Century-designed Landscapes: The Formal Type by Savot (1624) and at Sceaux (1670–1690)', *Landscape Research*, 40 (5) (2015), 566–85.

Field, J. V., *Piero della Francesca: A Mathematician's Art* (New Haven: Yale University Press, 2005).

Fisher, Saul, *Pierre Gassendi's Philosophy and Science* (Leiden: Brill, 2005).

Fumaroli, Marc, 'L'Académie des Inscriptions et Belles-Lettres dans la République des Lettres', *Comptes rendus des séances de l'Académie des Inscriptions et Belles-Lettres*, 150 (4) (2006), 2073–81.

Fumaroli, Marc, *Le sablier renversé: des Modernes aux Anciens* (Paris: Gallimard, 2013).

Gerbino, Anthony, *François Blondel: Architecture, Erudition, and the Scientific Revolution* (London: Routledge, 2010).

Gerbino, Anthony, 'Introduction', in *Geometrical Objects: Architecture and the Mathematical Sciences 1400-1800*, ed. by Anthony Gerbino (New York: Springer, 2014), pp. 1–41.

Giedion, Siegfried, *Space, Time and Architecture: The Growth of a New Tradition*, 5th edn. (Cambridge, MA: Harvard University Press, 1967; 1st edn. 1941).

Gros, Pierre, 'The Theory and Practice of Perspective in Vitruvius's "De architectura"', in *Perspective, Projections & Design: Technologies of Architectural Representation*, ed. by Mario Carpo and Frédérique Lemerle (London: Routledge, 2008), pp. 5–17.

Gros, Pierre, *Vitruve, De l'architecture, livre III* (Paris: Les Belles Lettres, 1990).

Guillaume, Jean, 'Le jardin mis en ordre: jardin et château en France du XVe au XVIIe siècle', in *Architecture, jardin, paysage. L'environnement du château et de la villa aux XVe et XVIe siècles*, ed. by Jean Guillaume (Paris: Picard, 1999), pp. 103–36.
Hamou, Philippe, *La mutation du visible: essai sur la portée épistémologique des instruments d'optique au XVIIe siècle* (Villeneuve-d'Ascq: Presses universitaires du Septentrion, 1999).
Harris, Dianne, *The Nature of Authority: Villa Culture, Landscape, and Representation in Eighteenth-Century Lombardy* (University Park: Pennsylvania State University Press, 2003).
Harris, Dianne, and D. Fairchild (eds.), *Sites Unseen: Landscape and Vision* (Pittsburgh: University of Pittsburgh Press, 2007).
Hazlehurst, Franklin H., *Gardens of Illusion: The Genius of André Le Nostre* (Nashville: Vanderbilt University Press, 1980).
Helden, Albert van, Sven Dupré, Rob van Gent, and Huib Zuidervaart (eds.), *The Origins of the Telescope* (Amsterdam: Amsterdam University Press, 2010).
Kretzchmar, Robert, and Lorenz Sönke, *Leonardo da Vinci und Heinrich Schickhardt: zum Transfer technischen Wissens im vormodernen Europa* (Stuttgart: W. Kohlhammer, 2010).
Leblanc, Marianne, *D'acide et d'encre: Abraham Bosse (1604?*Paris: *–1676) et son siècle en perspectives* (Paris: CNRS, 2004).
Loriferne, Hubert, 'L'influence de Picard dans les travaux d'alimentation en eau du château de Versailles sous Louis XIV', in *Jean Picard et les débuts de l'astronomie de précision au XVIIe siècle*, ed. by Guy Picolet (Paris: CNRS, 1987), pp. 274–311.
Margnat, Dominique (ed.), *Le livre de raison d'Olivier de Serres* (Grenoble: Presses universitaires de Grenoble, 2004).
Mariage, Thierry, *The World of André Le Nôtre*, (Philadelphia: University of Pennsylvania Press, 1999).
Marr, Alexander, *The Worlds of Oronce Fine: Mathematics, Instruments, and Print in Renaissance France* (Donington: Shaun Tyas, 2009).
Mukerji, Chandra, *Territorial Ambitions and the Gardens of Versailles* (Cambridge: Cambridge University Press, 1997).
Norman, Larry F., *The Shock of the Ancient* (Chicago: University of Chicago Press, 2011).
Panofsky, Erwin, 'Die Perspektive als "Symbolische Form"', in *Vorträge der Bibliothek Warburg, 1924–25* (1927), 258–330.
Panofsky, Erwin, 'Galileo as a Critic of the Arts: Aesthetic Attitude and Scientific Thought', *Isis*, 47 (1956), 3–15.
Panofsky, Erwin, *Perspective as Symbolic Form* (New York: Zone Books, 1991).
Pelletier, Monique, *Les cartes de Cassini. La science au service de l'État et des régions* (Paris: Éditions du CTHS, 2002).
Picolet, Guy, *Jean Picard et les débuts de l'astronomie de précision au XVIIe siècle* (Paris: CNRS, 1987).
Picon, Antoine, *Claude Perrault 1613–1688 ou la curiosité d'un classique* (Paris: Picard, 1988).
Picon, Antoine, 'Un moderne paradoxal', in Charles Perrault, *Mémoires de ma vie* (Paris: Macula, 1993), pp. 1–107.
Raynaud, Dominique, *Optics and the Rise of Perspective: A Study in Network Knowledge Diffusion* (Oxford: The Bardwell Press, 2014).
Reh, Wouter, and Clemens Steenbergen, *Architecture and Landscape. The Design Experiment of the Great European Gardens and Landscapes* (Munich: Prestel, 1996).

Ribouillault, Denis, 'Towards an Archeology of the Gaze: The Perception and Function of Garden Views in Italian Renaissance Villas', in *Clio in the Italian Garden*, ed. by Mirka Beneš and Michael G. Lee (Washington, DC: Dumbarton Oaks, 2011), pp. 203–32.

Richelieu à Richelieu: architecture et décors d'un château disparu, exhibition catalogue, Beaux-Arts d'Orléans, Musée municipal de Richelieu and Musée des Beaux-Arts de Tours, 12 March–13 June 2011 (Milan: Silvana, 2011).

Rousset-Charny, Gérard, 'Colbert, ses architectes et praticiens', in *Sceaux, architectures pour un domaine*, ed. by Cécile Dupont-Logié (Sceaux: Domaine de Sceaux, 2006).

Sarmant, Thierry, *Les demeures du soleil. Louis XIV, Louvois et la surintendance des Bâtiments du Roi* (Seyssel: Champ Vallon, 2003).

Saule, Béatrix, and Catherine Arminjon (eds.), *Sciences et curiosités à la Cour de Versailles* (Paris: Réunion des musées nationaux, 2010).

Shapin, Steven, and Simon Schaffer, *Leviathan and the Air-Pump: Hobbes, Boyle, and the Experimental Life* (Princeton, N.J.: Princeton University Press, 1985).

Thuillier, Jacques, *Vouet* (Paris: RMN, 1990).

Trachtenberg, Marvin, *Dominion of the Eye: Urbanism, Art, and Power in Early Modern Florence* (New York: Cambridge University Press, 1997).

Vérin, Hélène, 'Technology in the Park: Engineers and Gardeners in Seventeenth-Century France', in *The Architecture of Western Gardens*, ed. by Monique Mosser and Georges Teyssot (Cambridge, MA: MIT, 1991), pp. 135–46.

Weiss, Allen, *Mirrors of Infinity: The French Formal Garden and 17th-Century Metaphysics* (New York: Princeton Architectural Press, 1995).

Plates

Plate 1

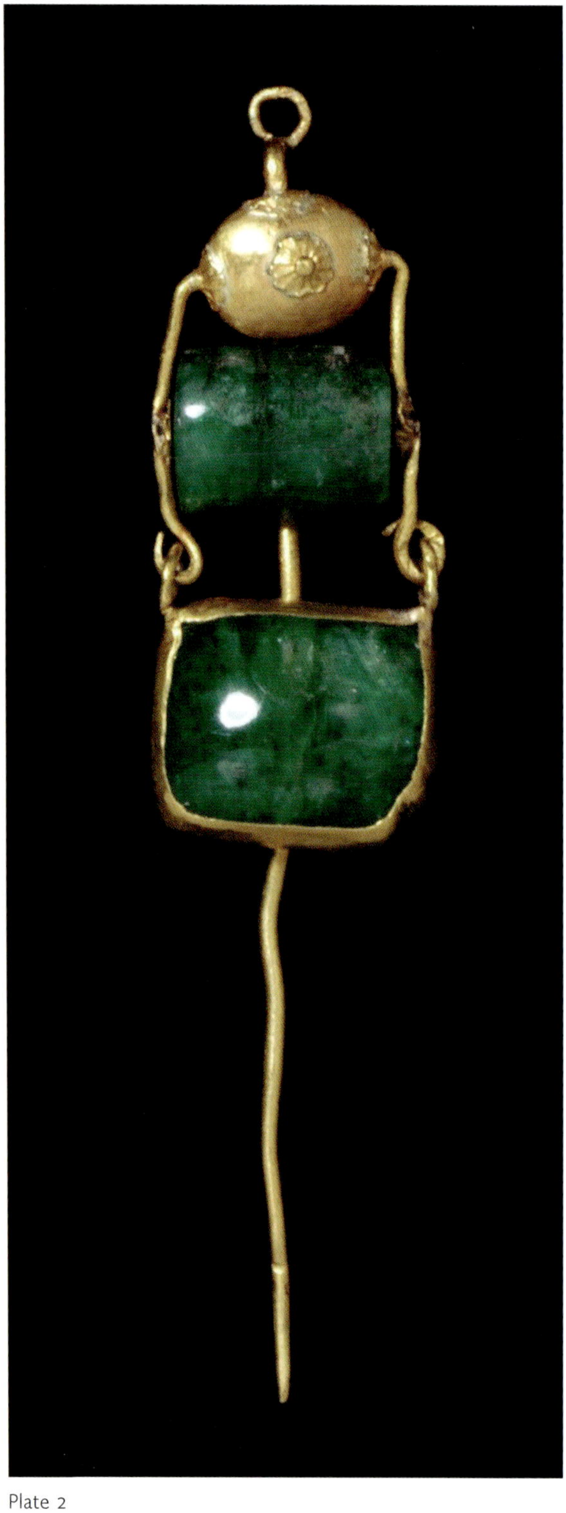

Plate 2

Plate 3

Plate 4

Plate 5

Plate 6

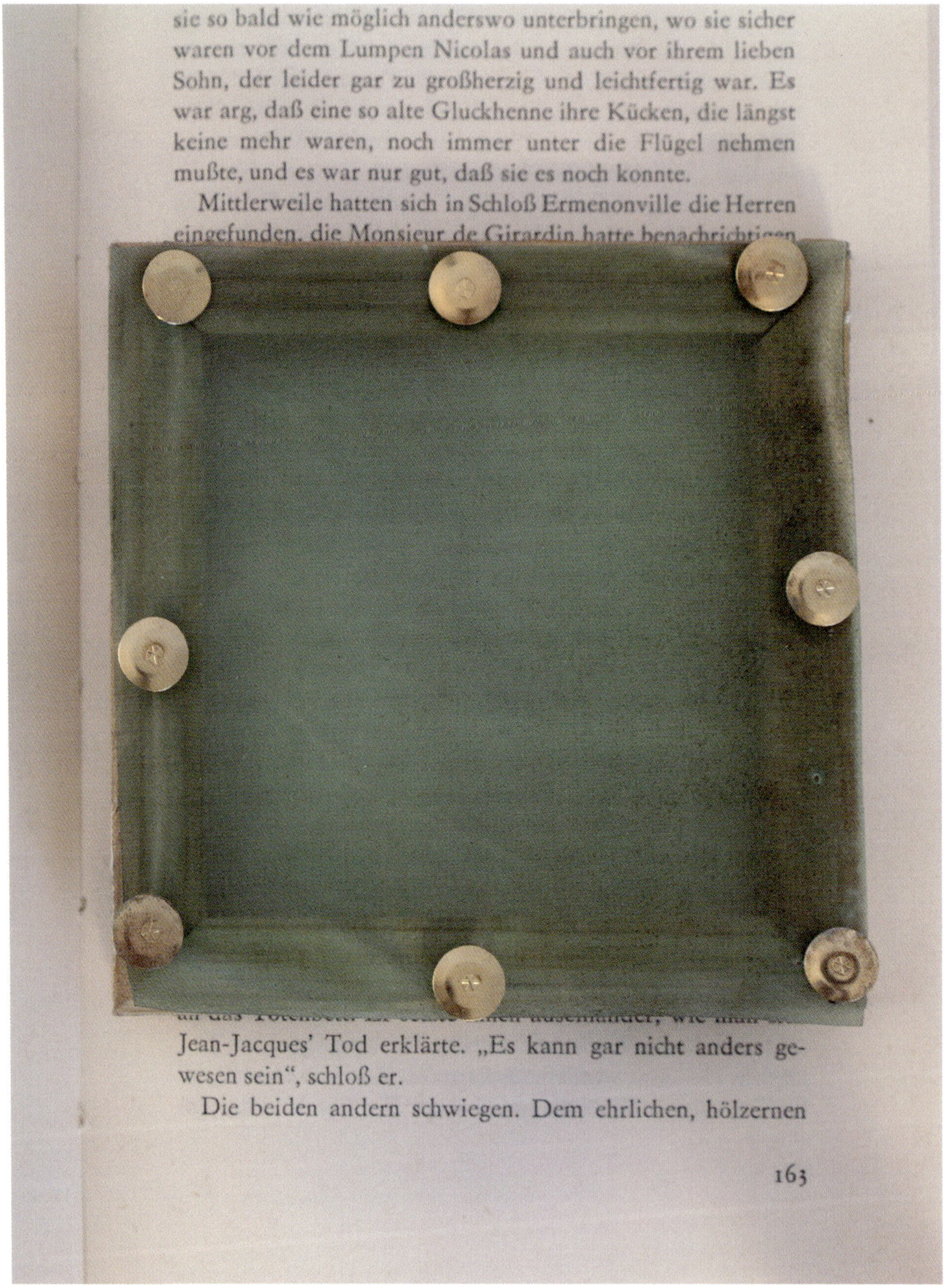
sie so bald wie möglich anderswo unterbringen, wo sie sicher
waren vor dem Lumpen Nicolas und auch vor ihrem lieben
Sohn, der leider gar zu großherzig und leichtfertig war. Es
war arg, daß eine so alte Gluckhenne ihre Kücken, die längst
keine mehr waren, noch immer unter die Flügel nehmen
mußte, und es war nur gut, daß sie es noch konnte.
Mittlerweile hatten sich in Schloß Ermenonville die Herren
Jean-Jacques' Tod erklärte. „Es kann gar nicht anders ge-
wesen sein", schloß er.
Die beiden andern schwiegen. Dem ehrlichen, hölzernen
163

Plate 7

sie so bald wie möglich anderswo unterbringen, wo sie sicher waren vor dem Lumpen Nicolas und auch vor ihrem lieben Sohn, der leider gar zu großherzig und leichtfertig war. Es war arg, daß eine so alte Gluckhenne ihre Kücken, die längst keine mehr waren, noch immer unter die Flügel nehmen mußte, und es war nur gut, daß sie es noch konnte.

beiden nächsten Freunde Jean-Jacques', Lebègue und Ducis, an das Totenbett. Er setzte ihnen auseinander, wie man sich Jean-Jacques' Tod erklärte. „Es kann gar nicht anders gewesen sein", schloß er.

Die beiden andern schwiegen. Dem ehrlichen, hölzernen

163

Plate 8

Plate 9

Plate 10

Plate 11

Plate 12

HORIZON ERECTVS
Aries
Taurus
Gemini
Cancer
Leo
Libra
Scorpius
Sagittarius
Capricornus
Tropicus
Canis maior
Orionis sinister pes
Cingulum Orionis

Plate 13

Imagines coeli Septentrionales cum duodecim imaginibus zodiaci.
Aratus Cilix
Ptolemeus Aegyptius
M. Manili-
us Romanus
Azophi
Arabus
Ursa maior
Cancer
Leo
Virgo
Libra
Scorpio
Sagittarius
Aquila
Bootes

Plate 14

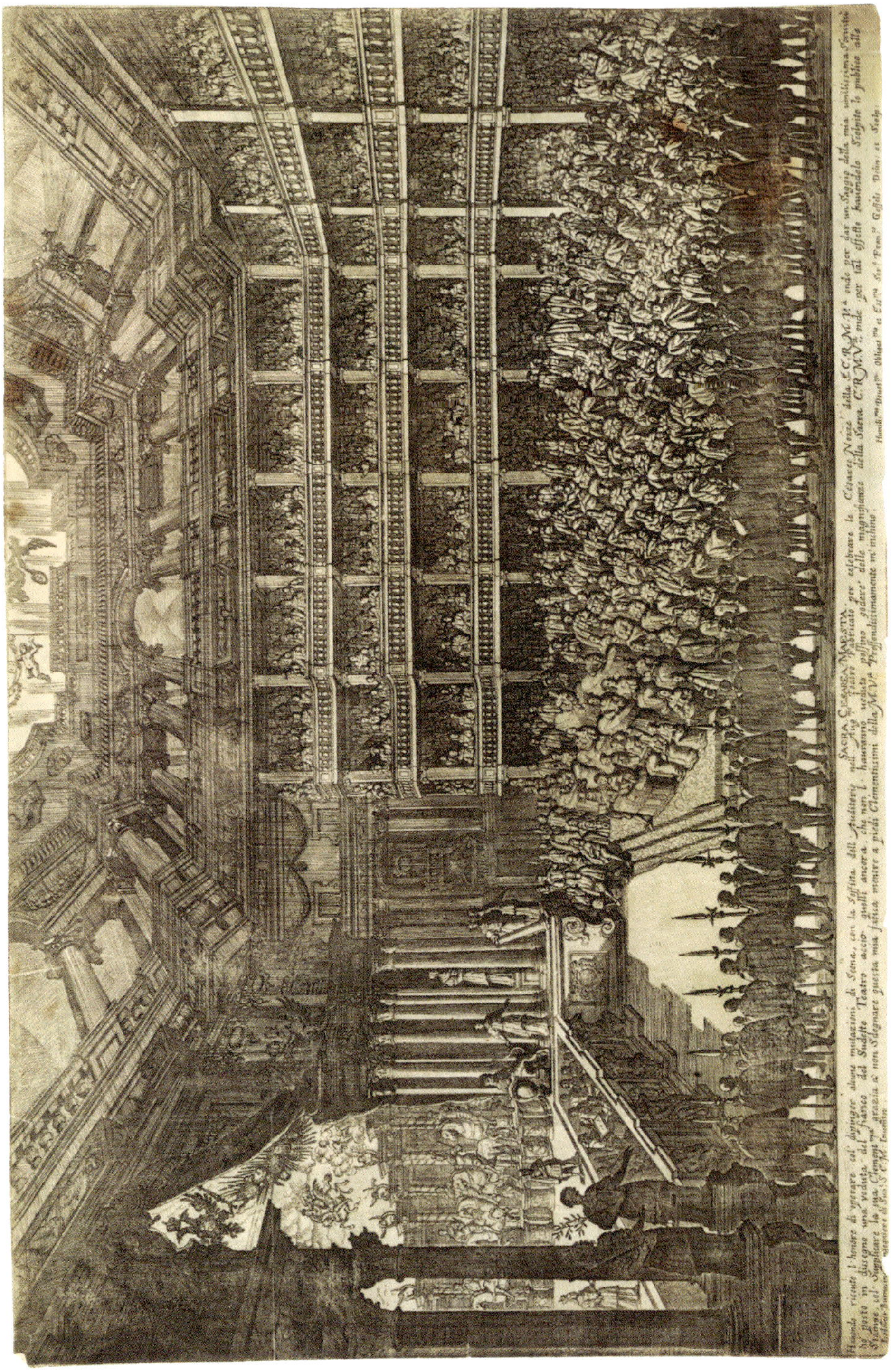

Plate 15

Plate 16

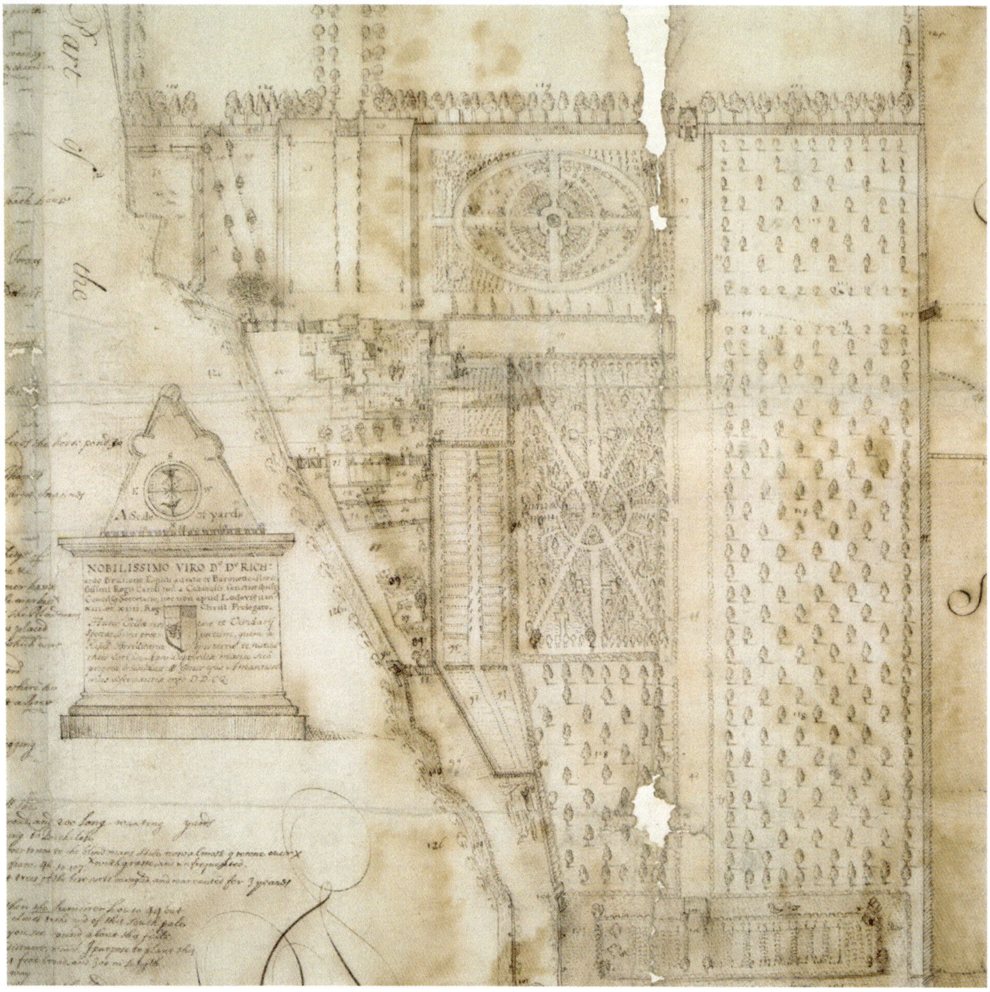

Plate 17

Plate 18

Plate 19

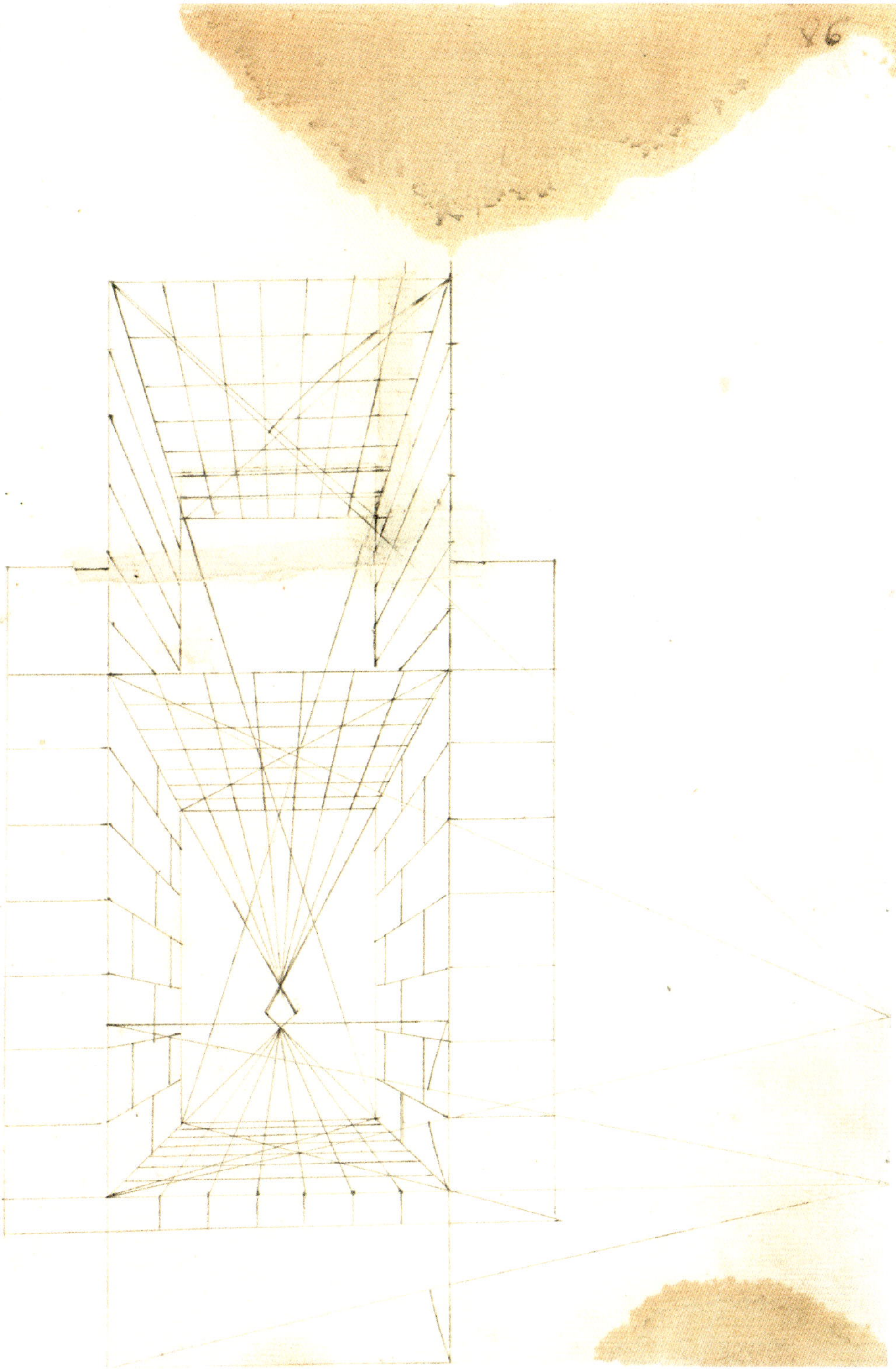

Plate 20

Plate 21

a

b

Plate 22

Plate 23

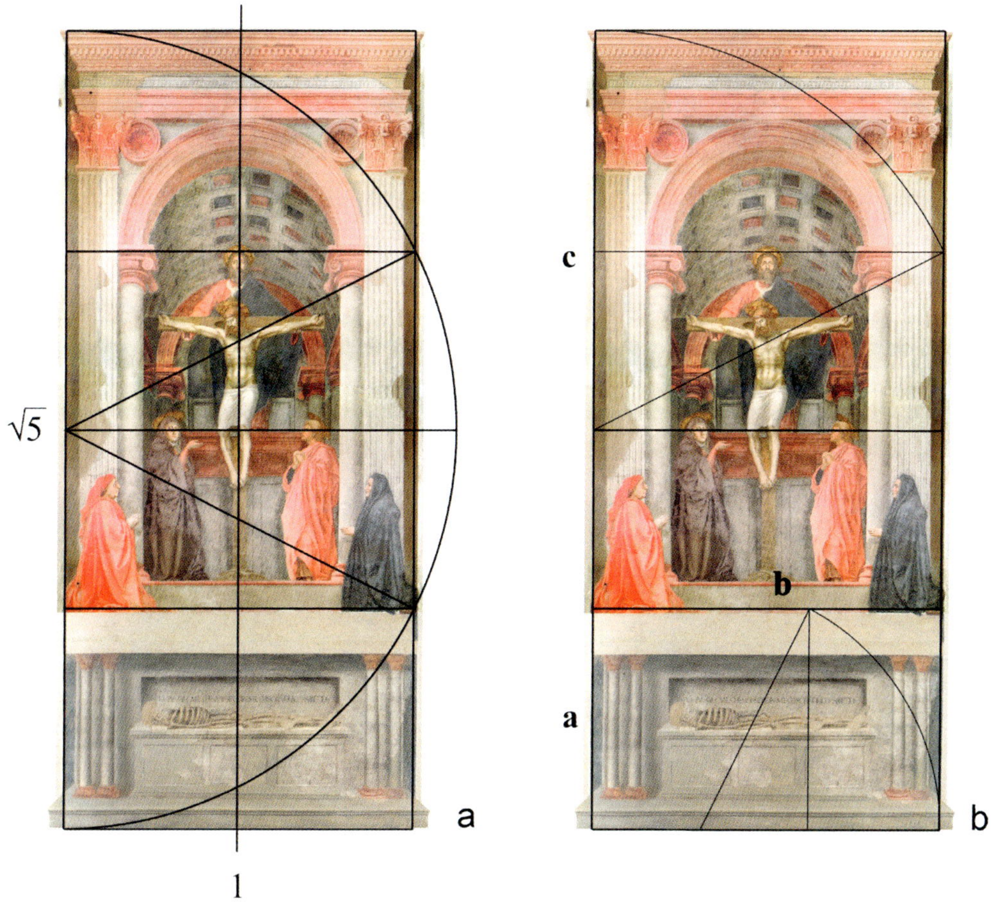

Plate 24

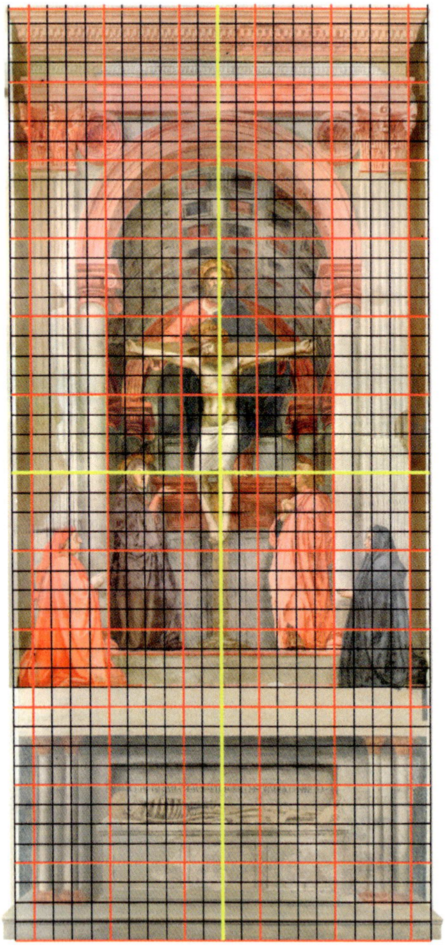

a

b

Plate 25

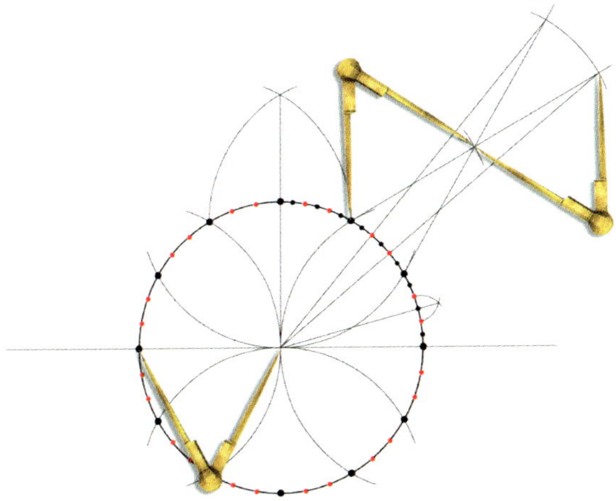

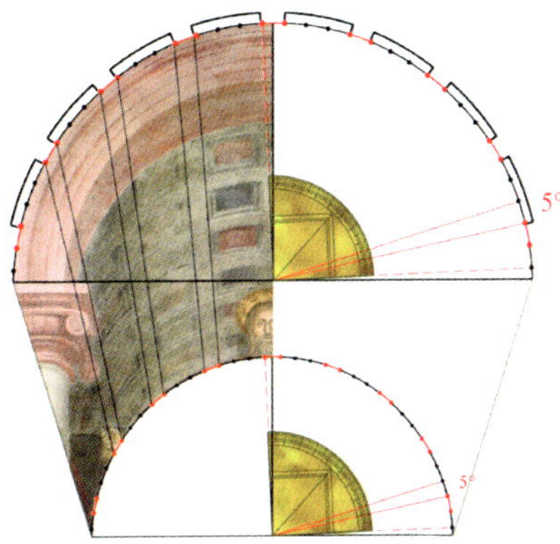

Plate 26

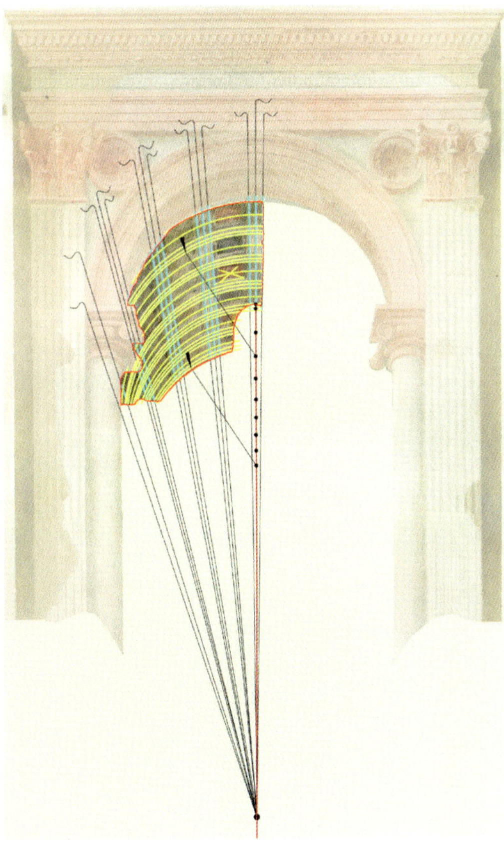

Plate 27

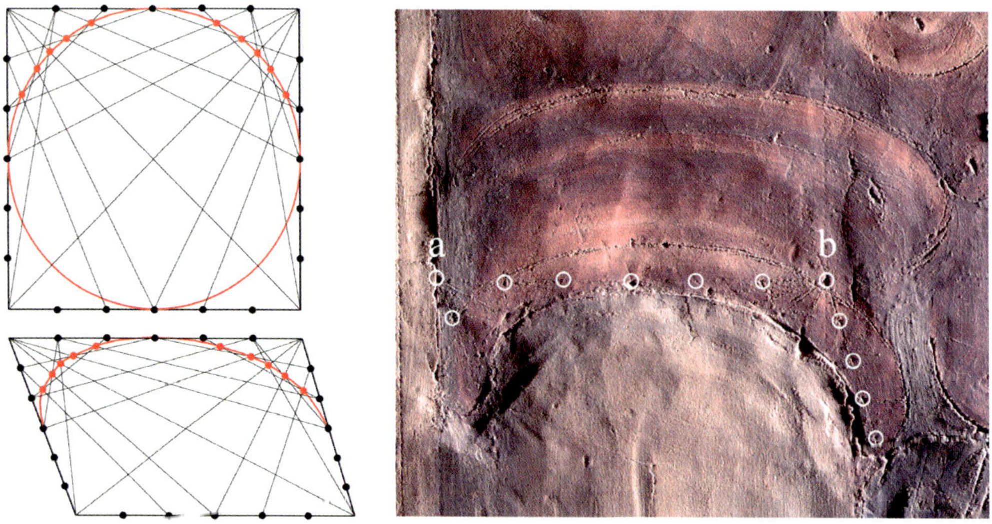

Plate 28

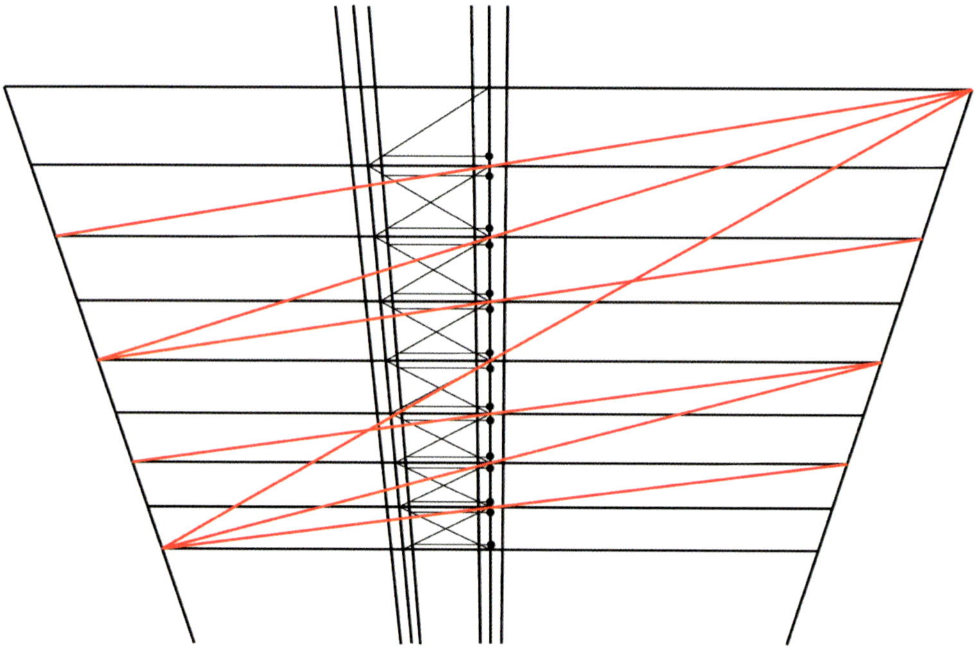

Plate 29

Plate 30

Plate 31

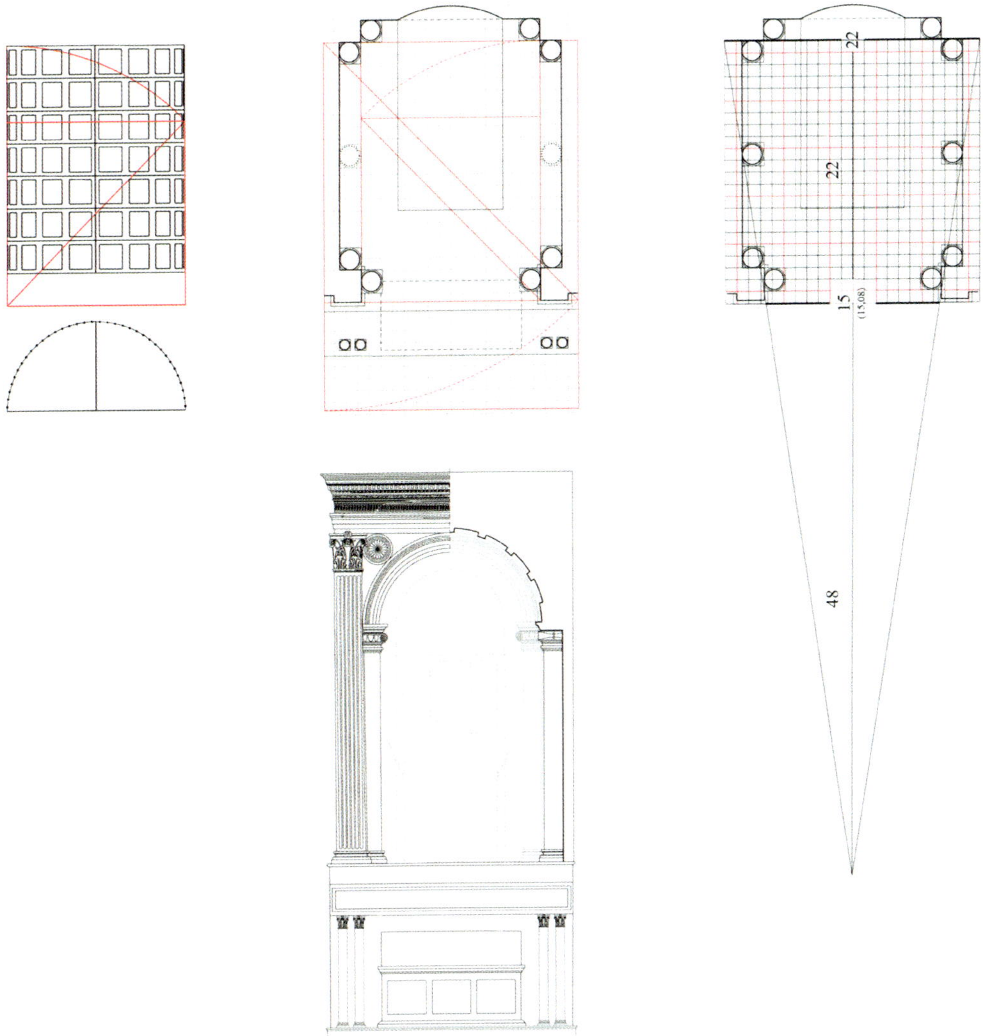

Plate 32

Plate 33

Plate 34

Plate 35

Plate 36

Plate 37

Plate 38

Plate 39

Plate 40

Plate 41

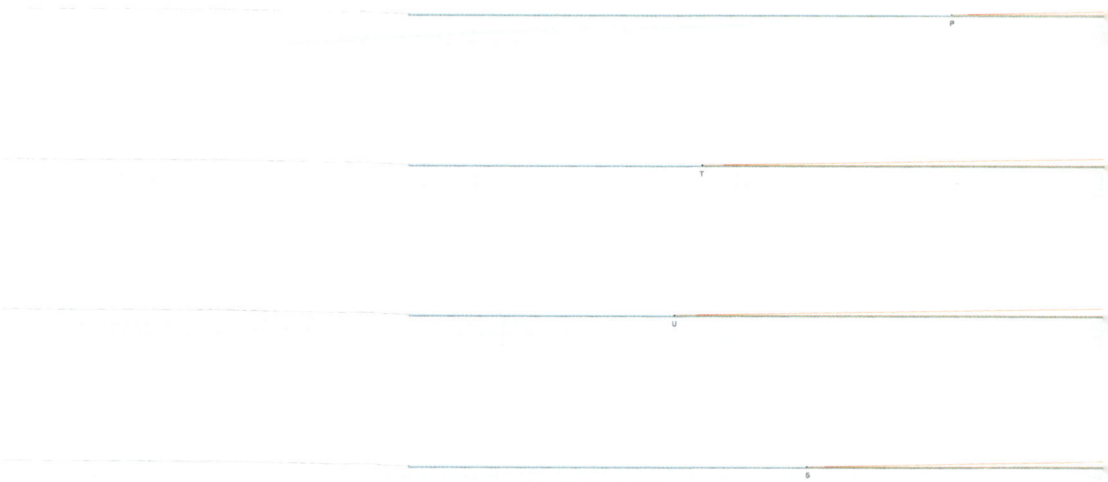

Plate 42

Plate 43

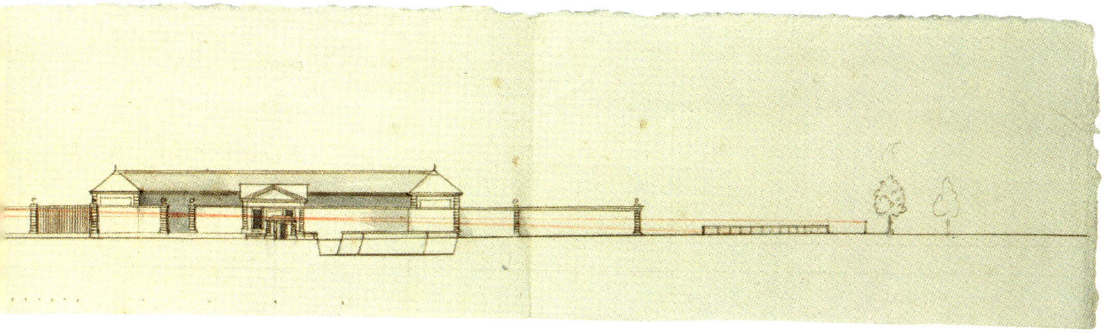

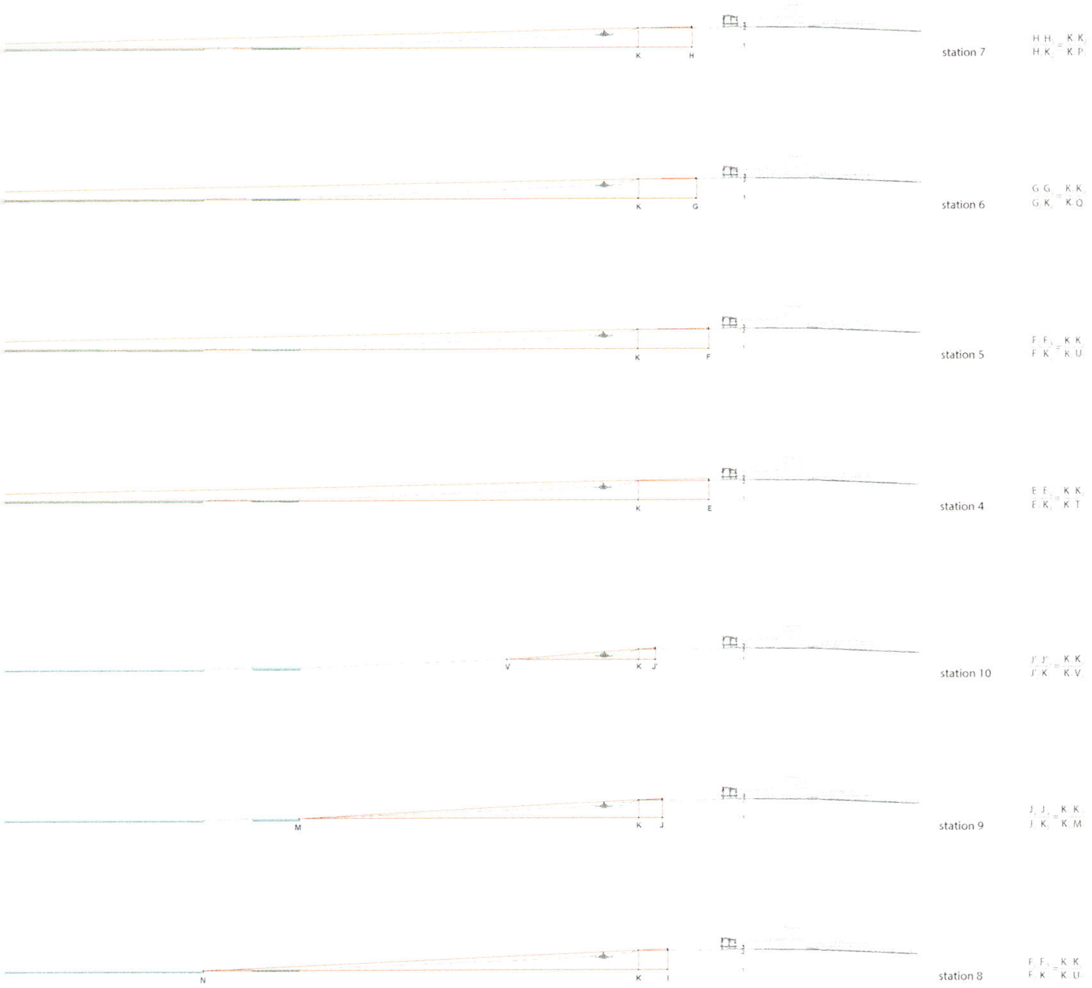
station 7
station 6
station 5
station 4
station 10
station 9
station 8

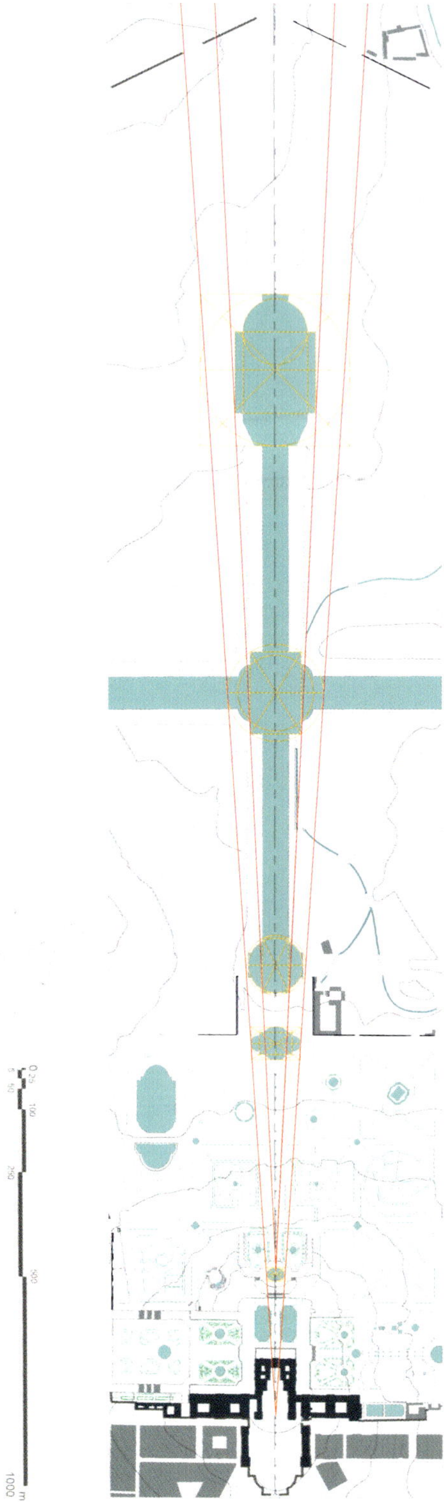

Plate 44

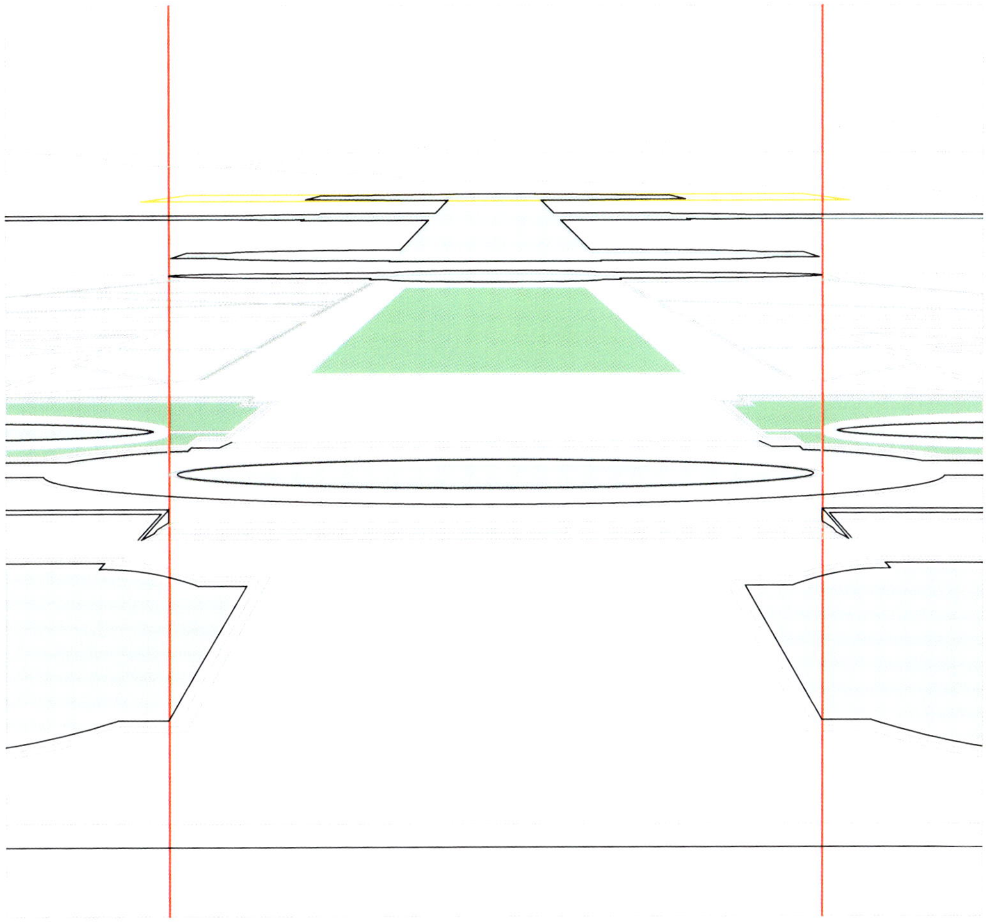

Plate 45

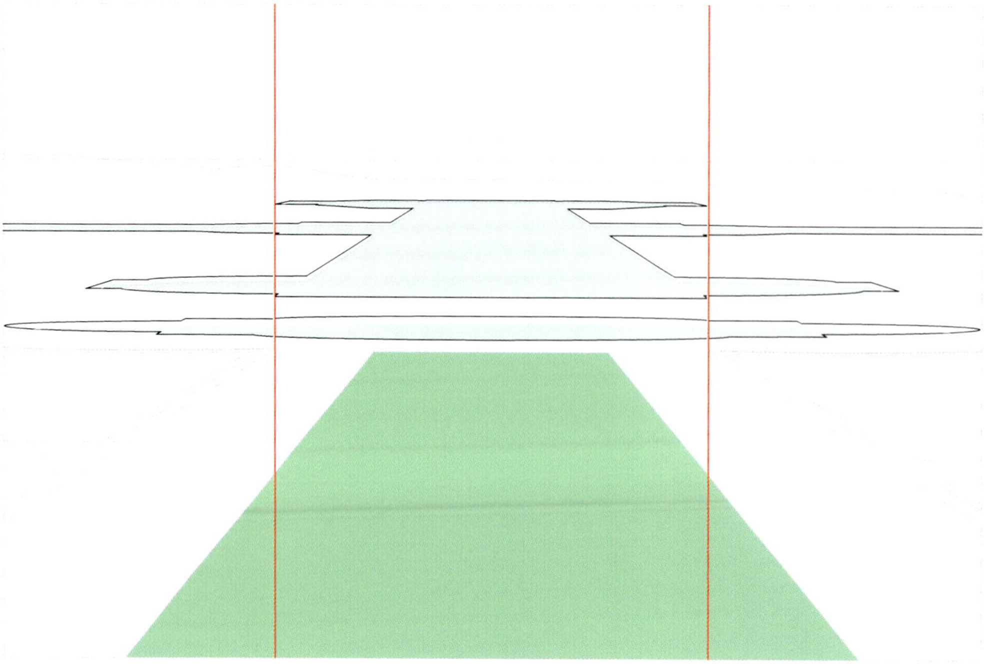

Plate 46

Techne

Knowledge, Technique, and Material Culture

The Techne series publishes books in all fields of history that examine the skills, techniques, technologies or material cultures of knowledge, including media. With an emphasis on the notion of "technique", the series includes topics and approaches that go beyond the traditional history of technology.

Techniques and skills have been developed and employed by artisans and inventors, and are present in a wide variety of settings. They play a role in grinding natural materials to create dyes and paints, in the playing of musical instruments, in the intricate gestures of a magical performance, in the tacit knowledge involved in a scientific experiment, as well as in the spiritual techniques of many religions. The editors invite contributions that study techniques, skills and training regimes in a broad variety of fields and cultural domains.

Knowledge is often produced at the intersection of a diversity of material cultures that have developed around artefacts, natural objects, or even the human body. The editors encourage the combination of practice-based and material culture approaches and invite submissions with new analytical and historiographical approaches, including reflexive contributions to the historiography of techniques and studies of technique-related concepts.

The series welcomes book proposals that focus on the time period from the Renaissance up to the present. Studies of any geographical region are welcome. The editors are open to a variety of historiographic traditions, as well as to sociological, anthropological, ethnographic, and especially transdisciplinary approaches. The series publishes volumes in English, French, or German. We publish single- or multi-authored volumes, edited collections, as well as source editions.

Titles in Series

Arnaud Dubois, *La vie chromatique des objets. Une anthropologie de la couleur de l'art contemporain* (2019)

Céline Le Gall, *Giovanni Poleni (1683-1761) et l'essor de la technologie maritime au siècle des Lumières* (2019)